AF764546

GEOLOGY

GEOLOGY

Editor

Lennart Martz

scitus
academics

Geology

Edited by **Lennart Martz**

This Book is Printed on Acid-Free Paper.

This book contains information obtained from highly regarded resources. Copyright for individual articles remains with the authors as indicated. All chapters are distributed under the terms of the Creative Commons Attribution License, which permits unrestricted use, distribution, and reproduction in any medium, provided the original author and source are credited.

ISBN: 9781681178974

Library of Congress Control Number: 2017948926

© 2018 by

SCITUS Academics LLC,

616, Corporate Way, Suite 2, 4766,

Valley Cottage, NY 10989

www.scitusacademics.com

Notice

Reasonable efforts have been made to publish reliable data and views articulated in the chapters are those of the individual contributors, and not necessarily those of the editors or publishers. Editors or publishers are not responsible for the accuracy of the information in the published chapters or consequences of their use. The publisher believes no responsibility for any damage or grievance to the persons or property arising out of the use of any materials, instructions, methods or thoughts in the book. The editors and the publisher have attempted to trace the copyright holders of all material reproduced in this publication and apologize to copyright holders if permission has not been obtained. If any copyright holder has not been acknowledged, please write to us so we may rectify.

Printed in United States of America.

Preface

Although the study of geology has been around for centuries, at least in its modern form, the science and practice of engineering geology only commenced as a recognized discipline until the late 19th and early 20th centuries. The field of geology involves the study of earth materials and processes. Discover the importance of studying geology and why is geology important in life and to civilization. Geology is important for our global food supply through identifying ideal growing areas and conditions as well as monitoring soil quality; geology's impact is commonly forgotten in the areas of construction and infrastructure whether in consideration of concrete and masonry for roads or the gypsum that drywall is made of. And geology is important in securing our worlds demand for energy, building infrastructure, advancing science and technology, sustaining the global food supply, predicting and reacting to natural disasters and everything in between. Geology plays an essential role in many areas of the economy. Economic growth and sustainability, as well as societal well-being, will require reliable supplies of energy and mineral resources, a dependable supply of clean water and the secure and sustainable production of food. Modern industry, technology and consumer products require a vast array of minerals, both abundant and rare. Their extraction and trade forms a major part of our national and global economy. As the population and demand for resources grow, innovative technologies are required to locate and extract minerals.

This advanced guide 'Geology', dedicated to the latest advancements in geology, is ideal for advanced graduate students, researchers and practitioners, or anyone wishing to explore this fascinating subject. There are many reasons why geology is important in life and to civilization. Think about earthquakes, landslides, floods, droughts, volcanic activity, ocean water currents (el Niño), types of soil, fuel oils, minerals (gold, silver, uranium), and so on. Geologists study all of these concepts. Thus, the study of geology plays an important role in modern life and civilization.

Table of Contents

CHAPTER 1

The Research and Application of Bright Spot Quantitative Interpretation in Deepwater Exploration

Yi He, Jitian Zhu, Yingzhao Zhang, Aiqun Liu, Guangchao Pan

Zhanjiang Branch of China National Offshore Oil Corporation Ltd., Zhanjiang, China

ABSTRACT

Prospectors usually recognize the event with strong amplitude in seismic sections as bright spots. However, such a simple and rough method can't distinguish whether these bright spots are related to favorable gas layer or water layer directly. In this paper, for the high correlation between reservoir gascontent and amplitude anomaly in research area, based on rock physical analysis of the wells drilled, using forward modeling technique respectively simulates and analyzes the seismic amplitude of gas layer and water layer. Then, combining the simulation result with corresponding statistics amplitude obtains the numerical relationship between each layer amplitude. At last, using the display technique directly recognizes the bright spots of gas layer in seismic profile and gets rid of those false bright spots caused by water layer, which improved the robustness in the bright spots interpretation and provided reliable basis for reducing exploration risks. Moreover, applying the method to the target zone, we obtain huge success.

Keywords: Bright Spot, Wedge Model, Tuning Thickness, Rock Physical Analysis, Amplitude Quantitative Interpretation

1. INTRODUCTION

Bright Spot Technology [1] [2] is a method used to detect oil and gas accumulated in traps based on real amplitude, which became an important or indispensable means in oil-gas exploration with good results in practical application [3] [4] . This technology is based on the change of seismic wave amplitude versus hydrocarbon in reservoir. Usually when searching the favorable target with bright spot, we mostly used the energy strength of seismic amplitude to distinguish the bright spot, and eventually achieved the target evaluation. This method to determine whether the stratigraphic is a favorable target is simple and rough, which can't actually judge whether the stratigraphic really contains gas on the basis of the strong amplitude from seismic profile. According to bright spot identification, the upper and lower groups of strata are characterized by strong amplitude and continuous reflection in the central canyon target area of the deep water area of the South China Sea. Based on that, the target in two strata has been evaluated and drilled. The drilling result indicated that the lower bright spot is thick sandstone that was full of high-yield gas. It was found that both gas layer and water layer drilled the well L1 and well L2 show bright spot features in seismic profile with horizontal contrast (Figure 1).

The practice has proved that not all the bright spots are the reflection of oil and gas reservoirs and all the oil and gas reservoirs do not necessarily show the bright spot characteristic. This bright spot feature in seismic profile is the expression of not only oil and gas, but also some special geological bodies like water layer. In 1975 [5] , according to the success or failure experience in Bright Spot technology, Royal Dutch Shell considered most bright spots have been not resulted in by the hydrocarbon reservoirs which have commercial value, and warned the industry that there have been some traps in bright spots in time. Bright Spot technology has become one of the key technologies in oil and gas exploration, which has been used to discover lots of large and medium oil-gas fields successfully. With the deepening of exploration more and more cases of failure in Bright Spot technology have come, whose drawback of multi-solution has been exposed gradually [6] [7] [8] [9] . By analyzing the reason of the failure, it is attributed to the multi-solutions of the geophysics that is the equivalent of petrophysical properties. Therefore, in order to better serve the exploration, Bright Spot technology also needs to keep pace with the times. According to the change of oil and gas exploration environment and in order to meet the new demand, it is necessary to further study and improve the applicability of Bright Spot technology.

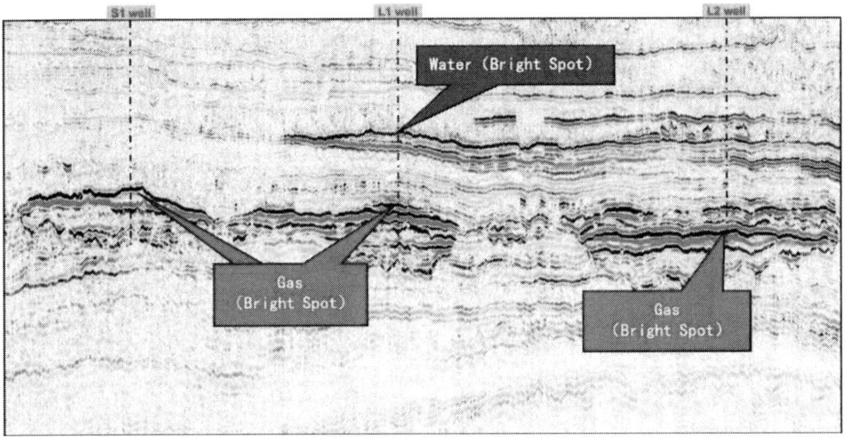

Figure 1. The true bright spots in conventional seismic section of three wells drilled.

By analyzing the petrophysical data of the wells drilled in the study area, it is found that there is a great difference between the elastic parameters of the reservoir contained gas and water. The low impedance and low density characteristics have been showed in both of gas layer and water layer, but the acoustic impedance and density of gas layer is lower than water layer, which caused that the impedance difference between gas layer and the overlying mudstone is far greater than the difference between aquifer and mudstone. There is a close relationship between the difference of P-wave impedance and the reflection coefficient, which is that the impedance difference is big, the reflection coefficient is large, and the impedance difference is small, the reflection coefficient is small. The value of the reflection coefficient determines the amplitude value of the seismic profile. Therefore, gas layer and water layer will show different amplitude values in the seismic profile. In this case, whether gas layer and water layer can be identified directly on the seismic profile by analyzing the amplitude of bright spot, so as to reduce the multi-solution of bright spot identification [10] [11] [12] [13] ? Based on rock physics analysis and using the current amplitude fidelity of seismic data, the quantitative interpretation research of bright spot has been carried out from the perspective of amplitude quantification for gas layer and water layer, which directed the identification of hydrocarbon reservoir in study area and laid the foundation for the success of subsequent wells.

2. PRINCIPLE OF METHOD

In order to realize the purpose of explaining the bright spot directly into gas layer or water layer in seismic profile, first of all, based on the petrophysical analysis for the wells drilled and according to the relationship between the impedance of the reservoir and the surrounding rock, determine the relative relationship of amplitude between the reservoir and the surrounding rock by using simulation method, combine the relationship with the statistical results of the actual seismic amplitude value corresponding to the drilled reservoir, distinguish bright spots shown by gas layer and water layer with the quantitative display means, eliminate the false bright spots caused by water layer and highlight the true bright spots caused by gas layer [14] [15] [16] .

Then, based on the above understanding and aimed at the different situation of reservoir contained gas and water, carry out the petrophysical analysis deeply, find the sensitive elastic parameters of gas layer and water layer, conduct the forward modeling of reservoir fluid, physical properties and thickness tuning by using wedge model, further simulate and analyze the relationship of amplitude value caused by the impedance difference between gas layer, water layer and surrounding rock, determine the numerical value of gas layer and water layer amplitude under the same surrounding rock.

Ultimately, combined with the actual seismic amplitude statistics of gas layer and water layer in wells drilled, establish the quantitative relationship between reservoir and amplitude [17] , form a quantitative interpretation and evaluation technique for Huangliu Formation reservoirs in the central canyon.

2.1. Analysis of Seismic Rock Physics

Based on the petrophysical analysis of the well drilled in the study area (Figure 2), we know that the gas group in the well L1 shows the significant characteristics of low impedance and low density compared with the overlaying rock (mudstone), meanwhile, which still presents the feature of low impedance and low density compared with the lower water layer (sandstone). However, the P-wave impedance difference between gas layer and water layer is smaller than that between gas layer and mudstone.

The analysis results of Huangliu Formation in other two wells indicate that gas layer still displays the characteristics of low impedance and low density, and the P-wave impedance difference between water layer and mudstone is much smaller than that of gas layer and mudstone. Therefore, the petrophysical analysis results of three wells give that once the reservoir is filled with gas, the impedance of it will be far less than that of water layer, and there is obvious difference between these layers of Huangliu Formation with the density of mudstone, water layer and gas layer decreasing.

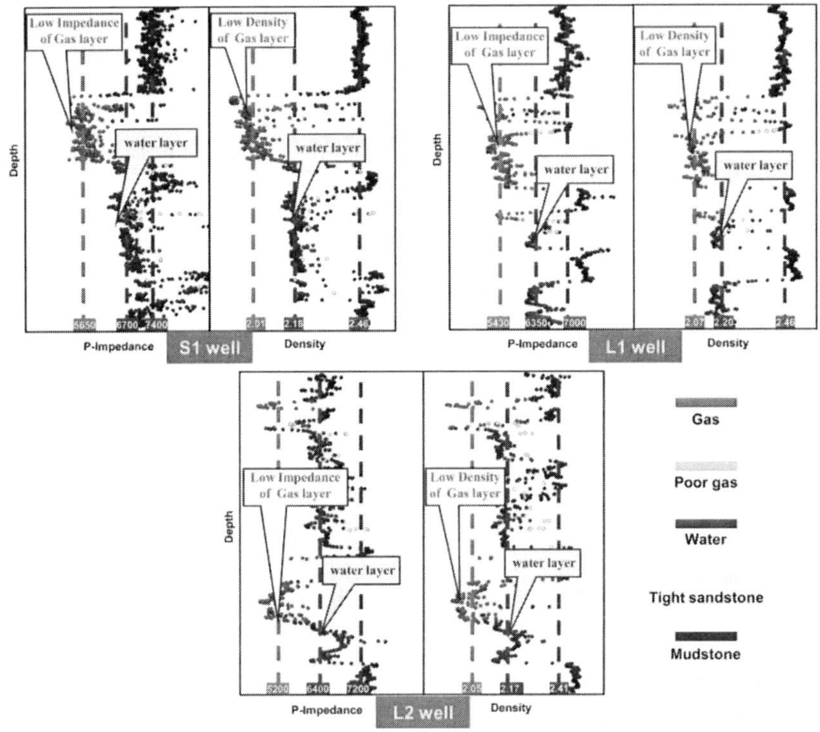

Figure 2. The petrophysical analysis of the well drilled in the Central Canyon.

2.2. Forward Modeling of Amplitude of Reservoir

The principle of seismic exploration tells that the amplitude value of the interface between the underground geological bodies is proportional to the reflection coefficient size of it and there is the following relationship between the reflection coefficient size and the impedance value of upper and lower media [18] .

$$R_p = \frac{\rho_2 V_2 - \rho_1 V_1}{\rho_2 V_2 + \rho_1 V_1} \tag{1}$$

Here, R_P is reflection coefficient, $\rho_1 V_1$ is P-wave impedance of overlaying rock, $\rho_2 V_2$ is P-wave impedance of underlying rock.

Known from the above describing, there is a numerical relationship between the seismic amplitude and the reservoir. However, what is this relationship? It needs to be ascertained by forward modeling. Previous studies [19] [20] indicated that the seismic amplitude is mainly affected by the following three factors: acquisition, processing and geological reasons which include the reflection coefficient, the thickness tuning, multi-layer interference, etc. It is considered that the influence factors of the seismic amplitude is mainly due to the geological factors such as the difference of the reflection coefficient caused by the difference of the fluid and the physical property, the thickness tuning caused by the thickness variation, etc. Because multilayer interference as the influence factors of the thickness can increase the complexity of seismic amplitude, only the physical property difference, the change of the fluid and the thickness tuning will be considered as the main influence factors of seismic amplitude in this study, but not the multilayer interference. Based on the petrophysical characteristics of the reservoir and the influence factors of the amplitude, the forward modeling is studied from three aspects.

1) Physical property

According to the physical property data of the reservoir of the wells drilled, designed porosity within the range from 23% to 30%, conducted forward modeling of gas layer and water layer for each value of the porosity, got the statistics for each simulation results (Figure 3), obtained the relationship with porosity changes between the reservoir (including gas layer and water layer) and amplitude (Figure 4). In Figure 4, the Red color indicates the maximum tuning amplitude of gas layer, the Magenta color indicates the non-tuning amplitude of gas layer, the Blue color presents the maximum tuning amplitude of water layer and the Purple color presents the non-tuning amplitude of water layer. As can be seen from Figure 4, i) with the increase of porosity, the amplitude of gas layer and water layer increases gradually; ii) under the same porosity, the maximum tuning amplitude of gas layer is greater than that of other three layers, the maximum tuning amplitude of water layer is equivalent to that of gas layer, and it is less than the maximum tuning amplitude of gas layer, the non-tuning amplitude of water layer is smallest; iii) When the difference between two different porosities is in the range of 3% to 4%, the non-tuning amplitude of gas layer with small porosity is still greater than the non-tuning amplitude of water layer with large porosity.

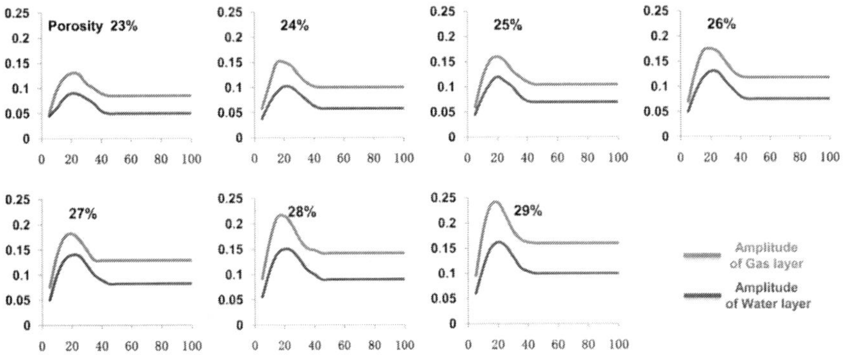

Figure 3. The physical property, the thickness and the amplitude relationship analysis.

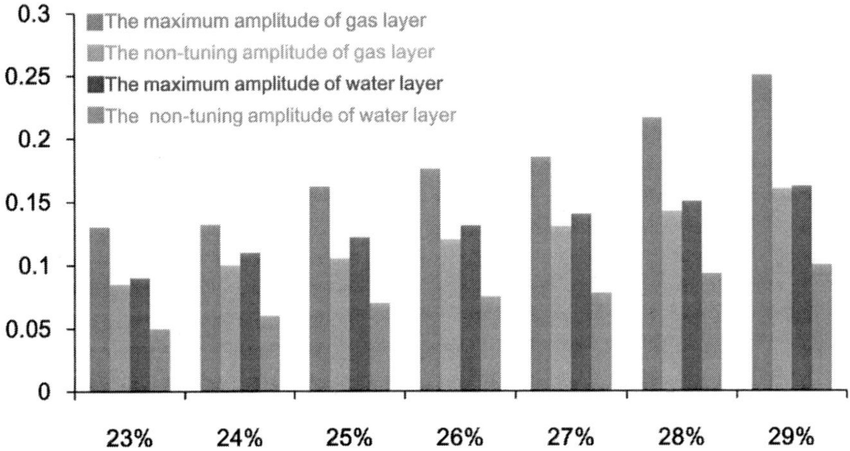

Figure 4. The amplitude relationship analysis of gas layer and water layer in different porosity.

2) Fluid

The reservoir drilled in Huangliu Formation are mainly explained as gas layer and water layer, so gas layer and water layer with different depths were chosen in three wells drilled respectively, and the amplitude of these layers was simulated. It is known from the simulation results (Figure 5) that the amplitude of gas layer is much larger than that of water layer, which is consistent with the amplitude value of gas layer and water layer extracted from the seismic profile. These results show that the top surface of gas layer can lead to stronger seismic amplitude, and there is a great difference of amplitude caused by the top surface of gas layer and that of water layer in the seismic profile, which can be distinguished by naked eyes.

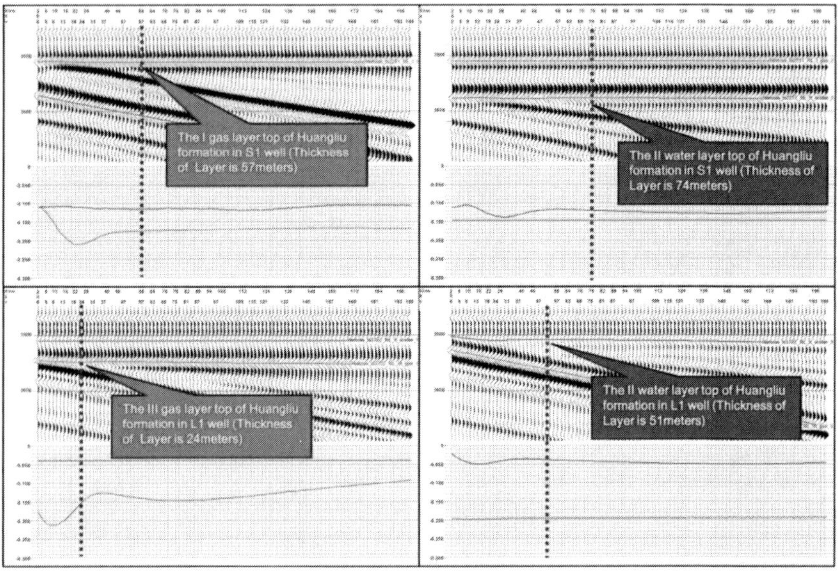

Figure 5. The forward modeling of gas layer and water layer in the central canyon.

3) Thickness tuning

The corresponding wedge models have been designed respectively for gas layer and water layer chosen above, meanwhile, the forward modeling of the effect of thickness tuning on amplitude has been carried out for these layers. The steps of simulation are as follows: firstly, according to the actual logging data, the corresponding velocity and density curves of gas layer and water layer have been turned into square wave; secondly, referring to the thickness of the sand which has been drilled, the thickness of the wedge model is changed from 0 m to 2 times the maximum thickness of reservoir; finally, the wedge model has been simulated by using theoretical wavelet(Ricker) and the wavelet extracted from Well-to-Seismic Calibration referring to convolution theory. The frequency of Ricker wavelet is determined in the light of the main frequency of the objective layer obtained from frequency analysis of the seismic data.

The simulation result shows that the amplitude values of gas layer and water layer have increased with the increasing of thickness gradually, when the thickness increases to a certain value, the amplitudes of gas layer and water layer reach the maximum value (the maximum tuning amplitude), then the amplitude with thickness decreases, at last, the amplitude tends to be stable after the thickness reaches a certain greater value. As a whole, the amplitude of gas layer is stronger than that of water layer, even if water layer is in the maximum tuning amplitude, the value of which is still smaller than that of gas layer (Figure 6). Further analyzing the relationship between the amplitude values of gas layer and water layer deeply, pick up the maximum tuning amplitude value of gas layer, the non-tuning amplitude value of gas layer , the maximum tuning amplitude value of water layer and the non-tuning amplitude value of water layer, think that there are multiple relationship between the amplitude values of gas layer and water layer like this: 1.1 ~ 2.6 after comparing and calculating the four kinds of data (Figure 7). In addition, when the gas layer and water layer in the study area get the maximum tuning amplitude, the corresponding thickness is 16 m and 19 m respectively; when the amplitude of gas layer and water layer is transferred from tuning to non-tuning, the thickness of them is 32 m and 38 m respectively; both of the amplitudes tend to be stable and remain unchanged when the thickness of them is more than 32 m and 38 m respectively.

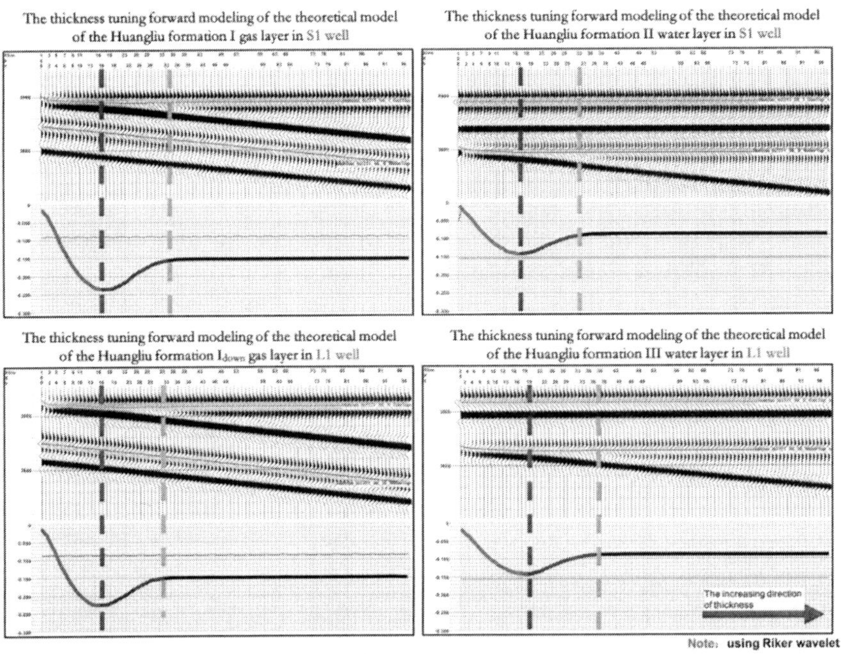

Figure 6. The influence analysis of the reservoir amplitudes during the thickness tuning.

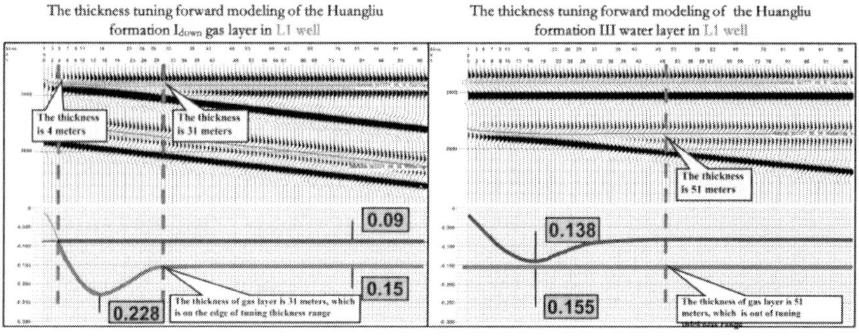

Figure 7. The numerical relationship analysis of gas layer and water layer.

2.3. Bright Spot Quantitative Interpretation and Evaluation

After determining the numerical relationship between the amplitudes of different reservoirs, according to the amplitude threshold values of gas layer and water layer, eliminate the false bright spots caused by water layer on the seismic profile, retain the true bright spots of gas layer, finally, identify the real highlights of gas layer quickly and accurately.

1) The amplitude statistics of actual reservoir drilled

It is necessary to achieve the amplitude statistics of seismic trace near wells drilled corresponding to the reservoir drilled before determining the gain parameters. After the well-seismic calibration done, each reservoir is corresponded to the event in seismic profile one by one, and then the seismic amplitude value corresponding to each reservoir is picked up. The amplitude statistics of gas layer and water layer of three wells drilled in this study are as follows (Table 1), which has been obtained by the method above.

Combined with the multiple relations between the maximum tuning amplitude or the non-tuning amplitude of gas layer and that of water layer on the basis of forward modeling analysis as mentioned in the article, the range of the amplitude of water layer is calculated, then after comprehensive considered, the threshold used to remove the bright spots caused by water layer will be confirmed.

2) Confirming the thresholds of gas layer and water layer

The multiple relation between the amplitude value of gas layer and that of water layer is as follows: 1.1 ~ 2.6 from the forward modeling, which shows that the absolute value of the amplitude of gas layer is still greater than that of water layer even if the amplitude of water layer is the maximum tuning amplitude and that of gas layer is the non-tuning amplitude. It is known that the average amplitude of water layer is about −0.35e + 08 from the statistical data of the actual amplitude of the wells drilled, therefore, combined with the results of forward modeling and the amplitude statistics of the reservoir, the threshold value of water layer can be set to −0.35e + 08. Next, the amplitude values of the event are weakened by reducing the gain of seismic profile, which is greater than the threshold of water layer, and the event that still shows very strong amplitude in the profile is the top interface of gas layer (Figure 8).

Table 1. The amplitudes and the porosity statistics of the well drilled in the central canyon.

Well name	Sand	Amplitude in well	Amplitude average around well	Thickness in well (m)	Porosity in well (%)	Note
S1	I	-5.02×10^8	-5.3×10^8	58.4	27.49	gas
	II	-4.05×10^8	-3×10^8	68.2	30.18	water
L1	I_{up}	-7.2×10^8	-8.1×10^8	12	28.8	gas
	I_{down}	-5.2×10^8	-7.2×10^8	30.9	27.49	gas
	II	-3.85×10^8	-5.1×10^8	4.3	27.85	gas
	III	-3.27×10^8	-3.0×10^8	62	25.02	water
	IV	-6.75×10^8	-4.15×10^8	6.2	23.98	gas
L2	I	-6.0×10^8	-5.9×10^8	7.4	27.1	gas
	II	-2.0×10^8	-2.4×10^8	26.3	25.72	water
	III	-10.2×10^8	-8.5×10^8	22.1	27.73	gas
	IV	-4.5×10^8	-4.0×10^8	12.2	28.62	Water (gas contained in top of layer)

3. APPLICATION EFFECT ANALYSIS IN PRACTICE

According to the petrophysical analysis and study on regional geological conditions, studied the quantitative relationship between the reservoir and the amplitude by the forward modeling; formed the quantitative interpretation evaluation technology of Bright Spot; identified the real bright spot; then, by using the technique combined with pre-stack joint inversion, judged the gas bearing of bright spot in the undrilled structure of Huangliu formation of the Central Canyon; finally, proved the gas-bearing prediction accuracy by practical drilling.

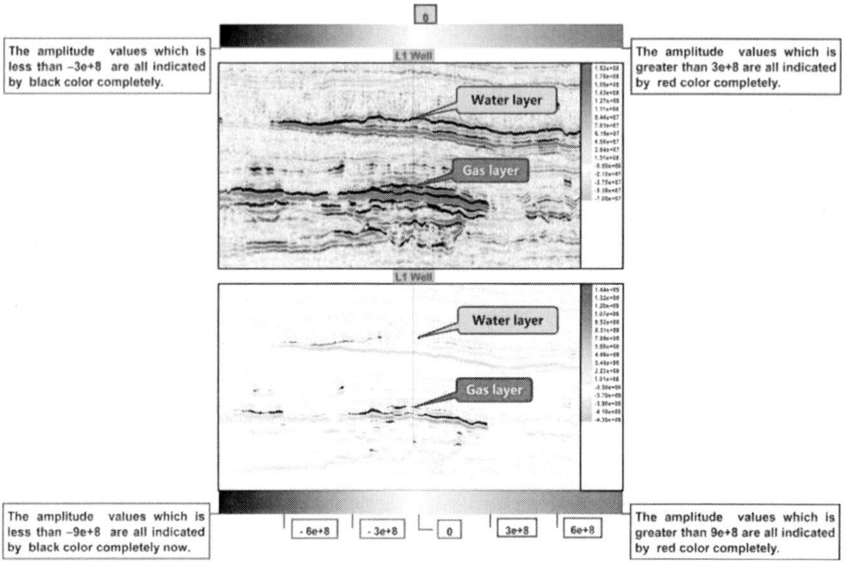

Figure 8. The quantitative visualization of the amplitudes for seismic section.

3.1. Application of Gas Bearing Prediction

In this paper, the quantitative analysis of bright spot and the gas bearing prediction of it are carried out by using the technique of bright spot and pre-stack joint inversion before drilling well L3/L4 (Figure 9). It can be seen from the cross section of two wells in Figure 9 that the seismic amplitude of each gas group at the well L4 point has been preserved after the treatment of amplitude quantitative weakening, which is still showing strong energy. At the same time, the I and IV gas groups in the Huangliu formation of the well L3 point show strong amplitude energy. However, the II and III gas groups present weak amplitude. According to the principle of quantitative interpretation of the bright spot, it is considered that each of these gas groups with strong amplitude feature in the two wells is the true bright spot.

On the basis of fine description of sand body, pre-stack joint inversion is carried out by pre-stack seismic data with interpretative processing. It can be shown (Figure 10) that each of these gas groups distinguished corresponds to remarkable low P-wave impedance and low density anomaly. Combined with the results of rock physics analysis, it is thought that there is a great possibility of gas bearing in these gas groups mentioned above. After drilling, logging interpretation shows that the I, II, III and IV gas groups of Huangliu formation in the well L4 are all high productive gas layers, the I and IV gas groups of Huangliu formation in the well L3 are also high productive gas layers, however, the II and III gas groups are all water layers. The result of log interpretation is completely consistent to the prediction before drilling. Furthermore, high industrial gas-flow is obtained from these gas groups of two wells after gas testing.

Similarly, the above techniques are used in the section through well L8. It is predicted that the I gas group of Huangliu formation in well L8 have been filled with gas, and the II gas group of Huangliu formation in well L8 has been filled without gas (Figure 11). After drilling, the log interpretation results are consistent with the prediction. High industrial gas-flow is also obtained from the well.

3.2. Application Prospection

In addition to being able to distinguish the true and false bright spots in the seismic profile, the quantitative interpretation of bright spot can be used to guide the coring test, well completion and drilling target selection in the drilling process. Practice has proved that the application of this technology in the recognition of gas layer and water layer and in drilling operations has achieved good results. It is not only presented that the method is simple and practical and it plays an important role in the objective assessment, but also that its application brings direct economic benefits for production operations, especially the increase of reserves, saving the drilling cost.

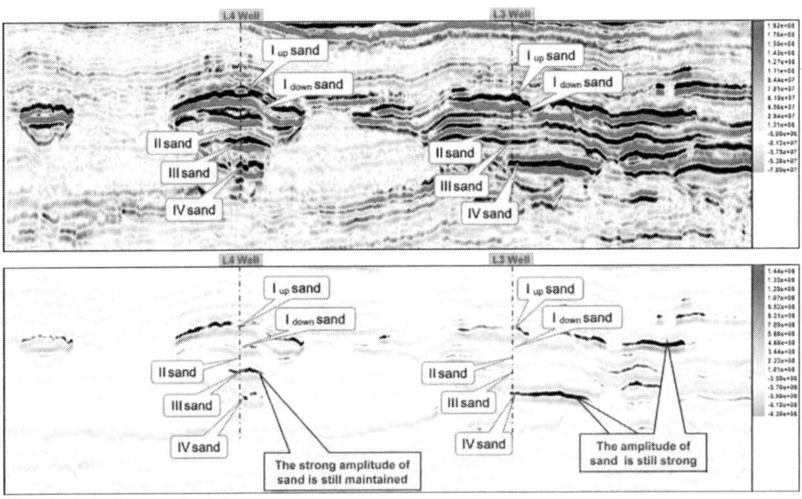

Figure 9. The amplitude quantitative interpretation of well L3 and L4 before drilling.

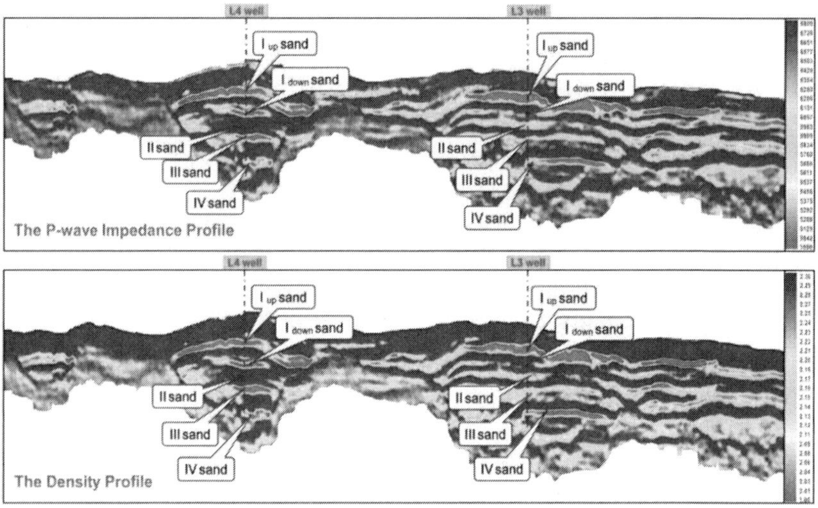

Figure 10. The pre-stack inversion of the section through well L3 and L4 before drilling.

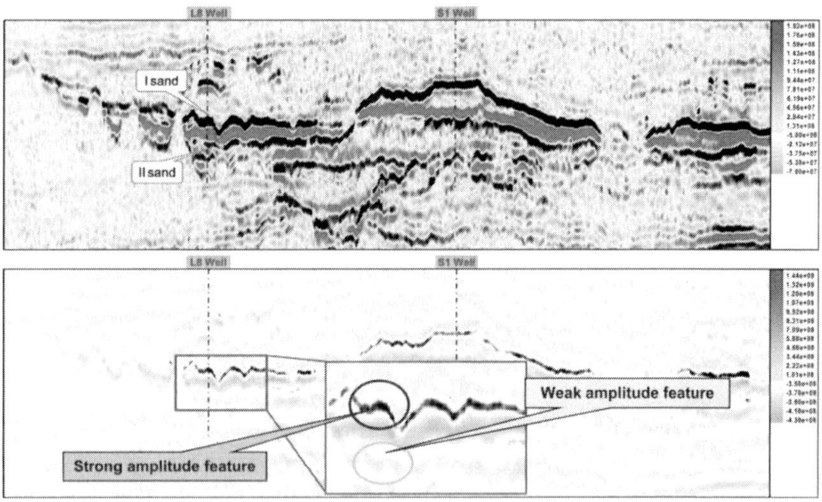

Figure 11. The amplitude quantitative interpretation of well L8 before drilling.

4. CONCLUSIONS AND SUGGESTION

Through the quantitative analysis and case study, we can draw the following conclusions.

For a certain depth and porosity, whether the reservoir thickness is tuned or not, seismic amplitude of gas layer showed stronger than that of water layer in seismic profile, which established theory foundation of identifying "true" and "false" bright spots in the seismic section.

Application effect of exploration example demonstrates that this method of amplitude quantitative identification has strong practicability and robustness, which has provided a new idea for distinguishing true and false bright spots and reduced the risk of oil and gas exploration to some extent.

However, there are still some limitations in the quantitative identification of bright spot, which does not take into account the influence of reservoir buried depth on seismic amplitude. In the future, we will study the reservoir amplitude combined with the diagenesis.

CITE THIS PAPER

He, Y., Zhu, J.T., Zhang, Y.Z., Liu, A.Q. and Pan, G.C. (2017) The Research and Application of Bright Spot Quantitative Interpretation in Deepwater Exploration. Open Journal of Geology, 7, 588-601. https://doi.org/ 10. 4236/ojg.2017.74040

REFERENCES

1. Fu, C.F. (1975) "Bright Spot" Technology—The Method of Directly Searching Oil and Gas by Using Seismic Profile. Oil Geophysical Prospecting, 10, 9-34.
2. Calculating Station of Chengdu Institute of Geology, et al. (1977) "Bright Spot" Model Technique. Geophysical Prospecting for Petroleum, 4, 46-53, 55-62.
3. Wang, L.Q. (1984) The Bright Spot in Traditional Seismic Profile. Oil Geophysical Prospecting, 19, 80-86.
4. Wang, D.C. (1982) The Detection Effect of Bright Spot in Continental Facies Strata. Oil Geophysical Prospecting, 17, 30-35.
5. The Intelligence Department of Geophysical Prospecting Bureau of China Petroleum (1975) The Warning of Bright Spot—There Is Not Really without Mistake. Oil Geophysical Prospecting, 2, 40-41.
6. Almoghrabi, H., et al. (1986) Layers & Bright Spots. Geophysics, 51, 699-709.
7. Pan, Y.J. (1988) The Relationship between Bright Spot, Dark Spot and Oil-Gas. Oil Geophysical Prospecting, 6, 737-744.
8. He, Z.H. and Ma, C.D. (1989) The Feasibility of Finding Gas by "Bright Spot" in Junggar Basin, Xinjiang Petroleum. Geology, 10, 14-23.

9. 9. Yang, Y.L., Sun, H.F. and Wang, H.Z. (1991) The Research on Bright Spot and Its Application of Shallow Gas Exploration in the Dongying Sag. Geophysical Prospecting for Petroleum, 1, 39-49.
10. Wang, Y.J. and Wang, X.M. (1995) The Feature and the Recognition of Bright Spot Gas Reservoir in the 3-D Seismic Data of Jiyang Depression. Geophysical Prospecting for Petroleum, 4, 66-72.
11. Wei, W., Li, H.M., Yang, Y.L., et al. (2011) The Recognition Technique of True and False Bright Spot in the Shallow of Chenjiazhuang Area. Reservoir Evaluation and Development, 1, 12-15.
12. Zhao, S. and Gao, Q. (2013) Using Adjacent Seismic Reflection Characteristics of the Energy Mix to Exclude "Bright Spot" Trap—Taking Deyang Syncline in Western Sichuan Depression as an Example. Natural Gas Geoscience, 24, 1253-1259.
13. Liu, W. and Cao, S.Y. (2015) Identification of Gas Reservoir Related with Bright Spots and Thin-Bed Tuning. Oil Geophysical Prospecting, 50, 678-683, 705.
14. Zhu, H.T., Hu, X.Q. and Chen, X.J. (2002) Preliminary Study on the Application of Bright Spots Oil and Gas Testing Technique. Offshore Oil, 3, 27-32.
15. Li, S.E., Zhang, Y.Z. and Wen, Y.J. (2003) Shallow Formation Bright Spot and Gas Exploration. Natural Gas Geoscience, 14, 291-294.
16. Cai, X.B. (2012) The Effectiveness Analysis of Using "Bright Spot", "Dark Spot" Technique to Test Oil and Gas in Putaohua Reservoir of Taipingchuan Region. Science Technology and Engineering, 12, 2132-2136.
17. Xiao, Z.B., Zhang, J.M. and Cao, X.Y. (2013) The Application Analysis of Bright Spot Technology Based on the Statistics of the Physical Rock. Computerized Tomography Theory and Applications, 22, 447-454.
18. Lu, J.M. (1993) Principle of Seismic Exploration. Petroleum Industry Press, Beijing, 19-20.
19. Jiang, B.H., Cai, Y.Q. and Wang, Y.Y. (2012) Analysis of Influence Factors on Seismic Amplitude in Bohai Sea Area. Petroleum Geology and Engineering, 26, 47-51.
20. Zhou, J.X., Liu, W.W. and Ma, G.K. (2013) Fine Description and Prediction of Seismic Attributes of HPHT Gas Reservoirs in the Yinggehai Basin. Natural Gas Industry, 33, 7-11.

CHAPTER 2

The Effect of Alluvial Foundation on the Earth Dams Settlement

Masoud Ghaemi[1], Kaveh Ahangari[1*], Ali Noorzad[2], Kamran Goshtasbi[3]

[1]*Department of Mining Engineering, Science and Research Branch, Islamic Azad University, Tehran, Iran*

[2]*Faculty of Water and Environmental Engineering, Shahid Beheshti University, Tehran, Iran*

[3]*Faculty of Mining Engineering, Tarbiat Modares University, Tehran, Iran*

ABSTRACT

Careful monitoring in the earth dams, to measure deformation caused by settlement and movement has always been a concern for engineers in the field. In order to measure settlement and deformation of earth dams, usually the precision instruments of settlement set and combined Inclinometer that is commonly referred to IS instrument, will be used. In some dams, because the thickness of alluvium is high and there is no possibility of alluvium removal (technically and economically and in terms of performance), there is no possibility to place the end of IS instrument (precision instruments of Inclinometer-settlement set) in the rock foundation. Inevitably, have to accept installing pipes in the weak and the deformable alluvial foundation that this leads to errors in the calculation of the actual settlement (absolute settlement) in different parts of the dam body. The purpose of this paper is to present new and refine criteria for predicting settlement and deformation in earth dams. The study is based on conditions in three dams with a deformation quite alluvial (Agh Chai, Narmashir and Gilan-e Gharb) to provide settlement criteria affected by alluvial foundation. To achieve this goal, the settlement of dams was simulated by using finite difference method with FLAC3D software and then the modeling results were compared with reading IS instrument. In the end, the caliber of the model and validate the results, by using regression analysis techniques and scrutinized modeling parameters with real situations and then by using MATLAB software and Curve Fitting Toolbox, a new criteria for the settlement based on elasticity modulus, cohesion, friction angle, density of earth dam and alluvial foundation was obtained. The results of these studies show that, by using the new criteria measures, the amount of settlement and deformation for the dams with alluvial foundation can be corrected after instrument readings and the error rate in reading IS instrument can be greatly reduced.

Keywords: Earth Dam, Alluvial Foundation, Settlement, Finite Difference, FLAC3D, MATLAB, Curve Fitting, Refine Criteria, IS Instrument

1. INTRODUCTION

There are always parameters in the design of earth dams that are considered the leading cause of dam design, including leakage, settlement, and permeability, type of spillway and dynamic resistance of the dam are considered the important parameters of the dam [1] .

Perhaps it can be said that among aforementioned parameters, the importance of settlement and deformation is an important factor in the behavior of the dam during construction, impounding dam and in operational stage. In order to measure settlement and horizontal deformation of dams usually the precision instruments of settlement set and combined Inclinometer that is commonly referred to IS instrument, will be used [2] . Installing IS instrument (Inclinometer- settlement set) in the rock foundation of earth dams with alluvial layers in fixed point or in other words, to determine the point at which the movement or deformation of the pipes will be zero in it, is very difficult. In terms of another, because of the correct placement of IS instrument in the bedrock, the results are always relative; however, it is necessary to compare this reading with the base magnet to analyze settlement set reading. This question has also been raised that what is the appropriate criteria for the results of IS pipe, according to the different physical characteristics in the alluvial layers of dam foundation.

Many researchers and engineers have suggested different methods to analyze the settlement and horizontal deformation of the dam and they have divided them into five general groups that include: 1) numerically; 2) the experimental method; 3) instrumentation; 4) micro geodesy; 5) smart methods (Fuzzy Network, Neural Network, Genetic Algoritm [3] . Chrzanowski carried out the development of the first research on the analysis of the earth dam deformation. In mentioning research, the transformations were more considered caused by the stresses imposed on the dam body [4] . Kelaf, et al. were the first people that modeled an earthen dam in 1997 by using finite element method. He calculated the stress and strain in a gravel dam by using linear elastic behavior model [5] .

Then Duncan, et al. in 1997 presented a few papers and introduced nonlinear hyperbolic model, a new behavioral model for settlement and deformation analysis of earth dams. [6] In their results, they indicated that the behavioral model provides results that are more realistic [7] .

Marandi, M., et al., in 2012, tried to estimate the extent of settlement in dam crest by using GEP. They have studied on 30 dams in seven countries. The results showed that the method of GEP is able to estimate the dam crest settlement based on four characteristics: the porosity of the dam e, height H, vertical deformation modulus E_v and shape factor Sc [8] .

In this regards, in this paper, first the evaluation and case studies of Geo- me- chanical parameters will be discussed. After explaining the modeling and its steps in FLAC3D software, the settlement plots, and initial analysis of dams will be shown based on three intended history points, which respectively, include M1 (dam bottom), M2 (one third of the dam body height) and M3 (two thirds of the dam body height). In the following, preliminary modeling results will be compared with the instrumentation results of IS instrument and in the next step, after determining the deviation parameters for numerical modeling, modeling parameters will be reviewed by using back analysis technique. Finally, by using the MATLAB and Curve Fitting Toolbox [9] , a new equation will be presented for correction values settlement of earth dam, with alluvial foundation, based on the basic parameters of the earth dams [10] .

2. EVALUATION OF CASE STUDIES

As it was mentioned earlier, in this article, three case studies have been used to determine the criteria measures, that by modeling, it is included Agh Chai Dam, [11] Narmashir dam and Gilan-e Gharb dam. In addition, the main specifications of all three dams have been briefly brought in Table 1.

In addition, plan and longitudinal sections of all three dams have been respectively brought [12] .

Table 1. Main Specifications of Agh Chai dam, Narmashir dam and Gilan-e Gharb [10] [11] [12] .

Gilan-e Gharb dam	Narmashir dam	Agh Chai dam	Project name / Specifications
Earth dam	Rockfill dam	Arch dam embankment	Type of dam
60	108.5	111.5	Dam height (m)
610	720	1240	*X* length of dam (m)
9	16	19	The dam crest width (m)
5	7.9	9.3	Embankment volume (million cubic meters)
21.5	32	25	Maximum thickness of alluvium (m)
2002	2004	2005	Construction time

3. MODELING EARTH DAM AND FOUNDATION

3.1. Modeling Process

In this part, modeling and the process are generally examined. According to the principles of numerical modeling as well as the fact dam, the dam modeling of Agh Chai, Narmashir and Gilan-e Gharb with Flac 3D program is as follows:

1) Determining the size, scope and number of meshes in dam geometry modeling;
2) Assign materials to different parts of the model (alluvium, foundation and the dam body);
3) Reticulation of dam body, alluvial foundation and bedrock by 15 node triangular elements (plane strain condition);
4) Apply static initial and boundary conditions;
5) Create the initial stresses in foundation and alluvium dam;
6) Allocation and defining gravity;
7) Solving the model to achieve a basic balance;

8) Fix the displacement and speed in knots;

9) Construction of the dam body and overall solution to the stable conditions.

3.2. Geomechanical Properties of Earth Dam

In this part, the geomechanical properties of the earth dam, alluvial foundation and bedrock components have been surveyed, given the prevailing sandy texture and low clay content and rock foundation in the main specifications are in Tables 2-6 below.

Table 2. Geotechnical properties of components AghChay dam, Narmashir and Gilan-e Gharb [10] [11] [12] .

	Specifications			Components of dam	
Dam name	Parameter		unit	Core	Crust
	Particularly dry weight	dry	kN/m^3	17	21
	Special saturated weight	sat	kN/m^3	18	22
AghChay	Permeability	*k*	cm/s	10 - 6	10 - 3
	Cohesion	*C*	kPa	28	1
	Angle of friction		deg	25	42
	Particularly dry weight	dry	kN/m^3	16	22
	Special saturated weight	sat	kN/m^3	18	24
Narmashir	Permeability	*k*	cm/s	10 - 7	10 - 3
	Cohesion	*C*	kPa	30	1.5
	Angle of friction		deg	28	39
	Particularly dry weight	dry	kN/m^3	15	24
	Special saturated weight	sat	kN/m^3	16	25
Gilan-e Gharb	Permeability	*k*	cm/s	10 - 8	10 - 3
	Cohesion	*C*	kPa	28	1
	Angle of friction	□	deg	25	42

To model the dam body, every three dams, along with alluvial foundation and bedrock have taken place as an effective stress analysis, together with analysis of consolidation. It should be noted that, for three-dimensional modeling of all three dams, Fish functions in software FLAC3D was used and a complete coding has been done, which is very difficult and time-consuming, because it does not have a default like tunnel modeling, so the built models are time-consuming and difficult because of coding. In this model, behavior criteria taken into account for modeling the dams are hardening Soil Model, because the hardening soil model is an advanced Elasto-plastic model to simulate the behavior of different soils contains both soft soils and hard soils, in other words, it is not-united Mohr-Coulomb model. In fact, Mohr-Coulomb major disadvantages will be resolved by adding a cap level to model the pulp flow under the identical stresses, and pulb flow expression before crash with identical hardening law enforcement [13] . Then, for example, the construction of the Narmashir dam has been done as shown in Figures 1-4.

Table 3. The shear and bulk modulus in Agh Chai Dam, Narmashir and Gilan-e Gharb [10] [11] [12] .

	Specifications		Components of dam	
Dam name	Parameter	unit	Core	Crust
AghChay	Gm0	MPa	115	185
	K	MPa	560	440
Narmashir	Gm0	MPa	125	170
	K	MPa	550	410
Gilan-e Gharb	Gm0	MPa	110	190
	K	MPa	620	515

Table 4. The Geotechnical properties of alluvial foundation in the Agh Chai dam [10] .

Specifications		Depth (m)			
Parameter	Unit	0 - 5	5 - 10	10 - 15	15 - 25
Special weight	kN/m^3	17	18	19	19
Relative density	%	62	65	74	81
The initial shear modulus	MPa	65	87	110	140
Shear modulus (after construction of the dam)	MPa	98	110	145	170
Bulk modulus	MPa	115	190	235	410
Permeability	cm/s	10 - 3	10 - 3	10 - 3	10 - 3
Cohesion	kPa	0	0	0	0
Angle of friction	deg.	29	33	33	35

Table 5. The Geotechnical properties of alluvial foundation in the Narmashir dam [11] .

Specifications		Depth (m)			
Parameter	Unit	0 - 10	10 - 20	20 - 30	30 - 32
Special weight	kN/m³	14	16	17	18
Relative density	%	56	60	70	77
The initial shear modulus	MPa	92	98	124	155
Shear modulus (after construction of the dam)	MPa	151	151	182	195
Bulk modulus	MPa	190	230	440	520
Permeability	cm/s	10 - 3	10 - 3	10 - 3	10 - 3
Cohesion	kPa	0	0	0	0
Angle of friction	deg.	24	25	27	31

Table 6. The Geotechnical properties of alluvial foundation in the Gilan-e Gharb dam [12] .

Specifications		Depth (m)			
Parameter	Unit	0 - 5	5 - 10	10 - 15	15 - 22
Special weight	KN/m³	13	14	14	15
Relative density	%	48	52	55	58
The initial shear modulus	MPa	80	87	110	142
Shear modulus (after construction of the dam)	MPa	145	160	178	190
Bulk modulus	MPa	160	210	300	350
Permeability	cm/s	10 - 3	10 - 3	10 - 3	10 - 3
Cohesion	kPa	0	0	0	0
Angle of friction	deg.	30	30	33	33

3.3. Back Analysis

After the initial modeling, the results are compared with installed instrumentation. It should be noted that for this purpose, the results of the three settlement sets at the bottom of the dam body, one-third and two-thirds of the dam height are taken into account for all three dams (can be seen in Table 7). The same corresponding points are installed based on the points X, Y, Z, the points are defined in the modeling and you can see the exact coordinates in Table 8 and the results of modeling are compared with instrumentation which the results can be seen in Table 9.

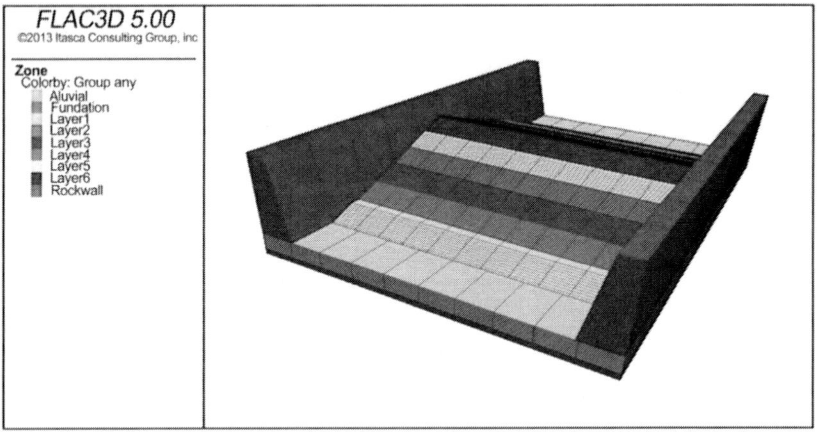

Figure 1. The alluvial and layer foundation modeling of the Narmashir dam.

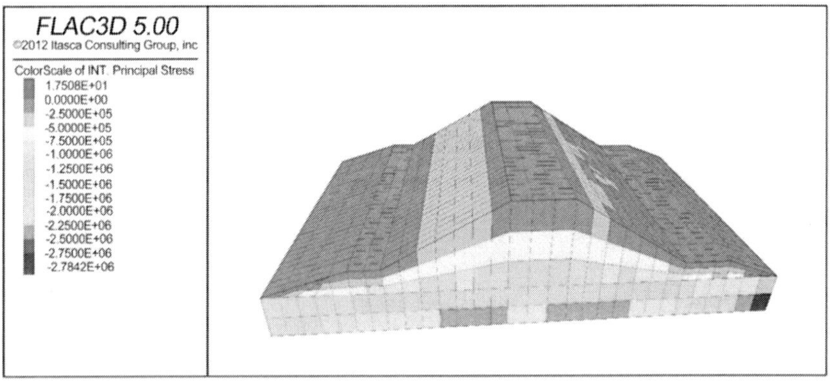

Figure 2. Contour of Z-Displacement in the Narmashir dam.

In the following, in the Tables 10-14, by changing the parameters, the modeling results are presented in tables after the back analysis.

Finally, modeling results after back analysis are reviewed and they are presented at the Table 15 below.

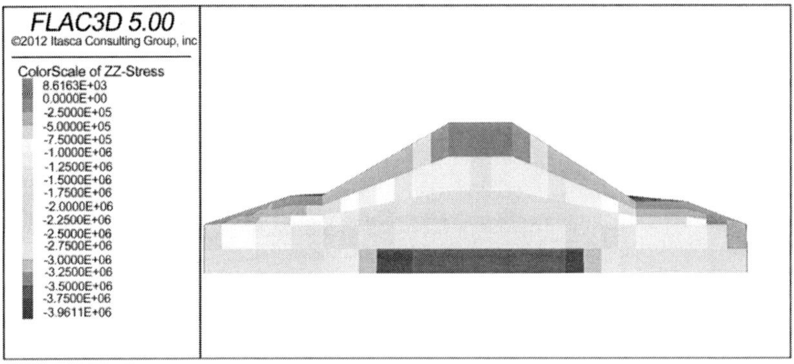

Figure 3. Initial Balance of Narmashir dam structure.

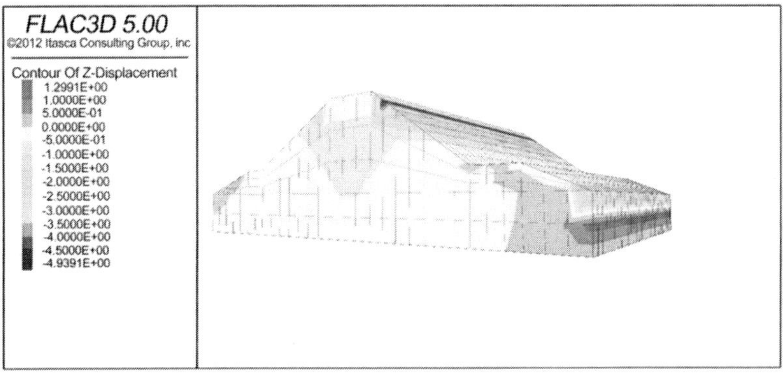

Figure 4. Determination of the general settlement in the Narmashir dam.

Table 7. Position of IS instrument installed in every three dams AghChay, Narmashir and Gilan-e Gharb [10] [11] [12] .

Dam Name	Parameter	Installation position	Installation balance	Section
AghChay	*M1* Bottom	downstream	0	11
	M2 A third of dam body height	downstream	39	11
	M3 Two third of dam body height	downstream	76	11
Narmashir	*M1* Bottom	Upstream	0	13
	M2 A third of dam body height	Upstream	36	13
	*M3*Two third of dam body height	Upstream	74	13
Gilan-e Gharb	M1 Bottom	downstream	1	17
	M2 A third of dam body height	downstream	21	17
	M3 Two third of dam body height	downstream	40	17

Table 8. Position of history points in modeling dams AghChay, Narmashir and Gilan-e Gharb.

Dam Name	Parameter	Coordinate *X*	Coordinate *Y*	Coordinate *Z*
AghChay	*M1* Bottom	122.3	17.34	26
	M2 A third of dam body height	122.3	17.34	63
	M3 Two third of dam body height	122.3	17.34	99
Narmashir	*M1* Bottom	240.2	56	32
	M2 A third of dam body height	240.2	56	68
	*M3*Two third of dam body height	240.2	56	104
Gilan-*e* Gharb	*M1* Bottom	78	44	22
	M2 A third of dam body height	78	44	52
	*M3*Two third of dam body height	78	44	62

Table 9. Compare the settlement of the results between Instrumentation and Modeling.

Dam Name	Parameter	The settlement set of Instrumentation (cm)	The settlement Modeling (cm)
AghChay	*M1* Bottom	40	17
	M2 A third of dam body height	65	34
	M3 Two third of dam body height	91	68
Narmashir	*M1* Bottom	35	11
	M2 A third of dam body height	46	16
	*M3*Two third of dam body height	58	19
Gilan-e Gharb	*M1* Bottom	15	5
	M2 A third of dam body height	28	15
	*M3*Two third of dam body height	33	19

4. PRESENTING SETTLEMENT EQUATION BY USING A CURVE FITTING TOOLBOX

Refine Criteria obtained in this study is designed by using MATLAB software. The database of three dams is considered that is the measure of the same information. It should be noted that in the Curve Fitting Toolbox, only three variables could be used to input data to the Toolbox. That is why by multiplication and division operations on input data, the number of variables became two main parameters, namely A and B. A parameter was inserted as the first input and parameter B was inserted as the second input and the results of the settlement were inserted as the third input. In the same way, it was applied to obtain the settlement of Dam bottom (M1), the settlement in one third of the dam body height (M2), the settlement in two-thirds of the dam body height (M3). Select basis of exposure parameters was the attempt and error, so that the highest correlation coefficient and the most optimized mode will be obtained. The used data are shown with maximum and minimum values in the Table 16.

Table 10. Geotechnical properties of the dam components AghChay, Narmashir and Gilan-e Gharb after back analysis.

Features				Dam Components	
Dam Name	Parameter		Unit	Core	Crust
AghChay	Unit Weight(dry)	□dry	kN/m^3	16.5	20
	Unit Weight(saturated)	□sat	kN/m^3	17.2	21
	Permeability	k	cm/s	10-7	10-3
	Cohesion	C	kPa	19	1
	Friction Angle		deg	19	28
Narmashir	Unit Weight(dry)	dry	kN/m^3	13	18
	Unit Weight(saturated)	sat	kN/m^3	16	21
	Permeability	k	cm/s	10-7	10-3
	Cohesion	C	kPa	24	1
	Friction Angle		deg	22	31
Gilan-e Gharb	Unit Weight(dry)	dry	kN/m^3	9	19
	Unit Weight(saturated)	□sat	kN/m^3	13	18
	Permeability	k	cm/s	10-9	10-3
	Cohesion	C	kPa	16	1
	Friction Angle		deg	17	23

To obtain better results, all data entries were normalized in the range of 0 to 1 by using the following equation. Then they were entered into the software

$$N = \frac{(X - \text{Min})}{(\text{Max} - \text{Min})} \tag{1}$$

In the Curve Fitting Toolbox of X-axis, A parameter values were entered and in the Y-axis, B parameter values and in the Z-axis, dam settlement values were entered. Detailed parameters of A, B and the relation obtained from fitting these settlements of dam Bottom are as follows:

$$A1 = \frac{C_f \gamma_f k_f}{\varnothing_f} + \frac{C_a \gamma_a k_a}{\varnothing_a} \tag{2}$$

Table 11. Shear modulus and the Balkans of Agh Chai, Narmashir and Gilan-e Gharb Dam after back analysis.

	Features		Dam Components	
Dam Name	Parameter	Unit	Core	Crust
AghChay	Gm0	MPa	105	165
	K	MPa	515	410
Narmashir	Gm0	MPa	110	130
	K	MPa	510	390
Gilan-e Gharb	Gm0	MPa	90	170
	K	MPa	585	490

Table 12. Geotechnical properties of AghChay alluvial foundation, after back analysis.

Features		Depth (m)			
Parameter	Unit	0 - 5	5 - 10	10 - 15	15 - 25
Special weight	kN/m³	15	16	17	17
Relative density	%	57	60	69	76
The initial shear modulus	MPa	62	85	107	137
Shear modulus (after construction of the dam)	MPa	95	107	142	167
Bulk modulus	MPa	111	186	231	406
Permeability	cm/s	4 - 10	4 - 10	4 - 10	4 - 10
Cohesion	kPa	0	0	0	0
Angle of friction	deg.	25	29	29	31

Table 13. Geotechnical properties of Narmashir alluvial foundation, after back analysis.

Features		Depth (m)			
Parameter	Unit	0 - 10	10 - 20	20 - 30	30 - 32
Special weight	kN/m³	12	14	15	16
Relative density	%	51	55	66	71
The initial shear modulus	MPa	85	92	129	151
Shear modulus (after construction of the dam)	MPa	145	145	176	189
Bulk modulus	MPa	180	224	431	511
Permeability	cm/s	10-3	10-3	10-3	10-3
Cohesion	kPa	0	0	0	0
Angle of friction	deg.	22	20	21	27

Table 14. Geotechnical properties of Gilan-e Gharb alluvial foundation, after back analysis.

Features		Depth (m)			
Parameter	Unit	0 - 5	5 - 10	10 - 15	15 - 22
Special weight	kN/m³	10	11	11	13
Relative density	%	42	48	48	51
The initial shear modulus	MPa	71	81	105	125
Shear modulus (after construction of the dam)	MPa	140	155	175	185
Bulk modulus	MPa	145	202	196	346
Permeability	cm/s	10 - 3	10 - 3	10 - 3	10 - 3
Cohesion	kPa	0	0	0	0
Angle of friction	deg.	28	28	29	29

$$BI = \frac{H}{B*\nabla} * \frac{\Delta e_f * G_f}{OCR_f} + \frac{\Delta e_a * G_a}{OCR_a} \quad (3)$$

H: Height of the dam W: Width of the crown C: Cohesion Δe: porosity difference γ: density ∇: volume to weight ratio of the Dam (cubic meters per tons) K: bulk modulus

Table 15. Compare the settlement of the results between Instrumentation and Modeling after back analysis.

Dam Name	Parameter	The settlement set of the Instrumentation (cm)	The settlement Modeling (cm)
	M1 Bottom	40	40
AghChay	M2 A third of dam body height	65	62
	M3 Two third of dam body height	91	87
	M1 Bottom	35	35
Narmashir	M2 A third of dam body height	46	42
	M3Two third of dam body height	58	53
	M1 Bottom	15	15
Gilan-e Gharb	M2 A third of dam body height	28	26
	M3Two third of dam body height	33	32

Table 16. The lowest and highest values of used parameters.

Input	Unit	Values range	Input	Unit	Values range
Shear modulus	MPa	160 - 100	Height	m	180 - 90
Angle of friction	deg	50 - 20	Crest width	m	22 - 12
Cohesion	kg/cm^2	18 - 1	Axial length	m	1240 - 610
Bulk modulus	MPa	590 - 400	Porosity difference	%	0.206 - 0.104
Special Weight	kN/m3	21 - 12	Ratio of volume, on the weight of dam	m^3/ton	0-48 - 0.52

Rock foundation parameters of the dam values are shown in f index, core parameter values of the dam are shown in c index and the values of the alluvium parameters are shown in an index

$$M_1 = -116.1 + 31.4 * A1 + 1.149 * B1 \tag{4}$$

M1: the dam Bottom settlement (cm)

To obtain the settlement in one third of the dam's dam body height, which means M2 was acted as M1 equation; input parameters are divided into two parameters of A2, and B2 that the details of which are as follows:

$$A2 = \frac{C_f \gamma_f k_f}{\varnothing_f} + \frac{C_a \gamma_a k_a}{\varnothing_a} + \frac{C_c \gamma_c k_c}{\varnothing_c} \quad (5)$$

$$B2 = \frac{H}{B * \nabla} * \frac{\Delta e_f * G_f}{OCR_f} + \frac{\Delta e_a * G_a}{OCR_a} + \frac{\Delta e_c * G_c}{OCR_c} \quad (6)$$

$$M_2 = 18.33 - 1.135 * A2 + 0.85 * B2 \quad (7)$$

M2: The settlement in one third of the dam's dam body height (cm)

$$A3 = \frac{C_f \gamma_f k_f}{\varnothing_f} + \frac{C_a \gamma_a k_a}{\varnothing_a} + \frac{C_c \gamma_c k_c}{\varnothing_c} + \frac{C_s \gamma_s k_s}{\varnothing_s} \quad (8)$$

$$B3 = \frac{H}{B * \nabla} * \frac{\Delta e_f * G_f}{OCR_f} + \frac{\Delta e_a * G_a}{OCR_a} + \frac{\Delta e_c * G_c}{OCR_c} + \frac{\Delta e_s * G_s}{OCR_s} \quad (9)$$

$$M_3 = 64.01 - 4.38 * A3 + 1.26 * B \quad (10)$$

M3: The settlement in the point near the dam crest (cm)

After calculating each prediction model, it is necessary to examine the ability and the power of different forecasting models. There are diverse criteria for evaluating the performance of different forecasting methods, however, in this study, to compare the prediction power, the average absolute error criteria, standard deviation, coefficient of determination and mean square error root are used in the Table 17. These criteria can be shown as Equations (11) to (14):

$$\sigma = \sqrt{\left(\frac{1}{n}\right) \sum_{i=1}^{n} (e_i - \bar{e})^2} \quad (11)$$

$$MAE = \left(\frac{1}{n}\right) \sum_{i=1}^{n} |m_i - p_i| \quad (12)$$

$$RMSE = \sqrt{\left(\frac{1}{n}\right)\sum_{i=1}^{n}(m_i - p_i)^2} \quad (13)$$

$$R^2 = \left(\frac{\sum_{i=1}^{n}(p_i - \bar{p})(m_i - \bar{m})}{\sum_{i=1}^{n}(p_i - \bar{p})^2 \sum_{i=1}^{n}(m_i - \bar{m})^2}\right)^2 \quad (14)$$

Table 17 is according to the performance of the specified Equations in CURVE FITTING, which represents the accuracy and precision of equations. In addition, Table 17, will be taught based on the instrumentation information and it will calculate the errors in each equation.

Table 17. The statistics range of best fitting.

RMSE	Adjusted R-square	R-square	SSE	
0.0323	0.986	0.987	0.038	M1
0.0963	0.864	0.8718	0.343	M2
0.044	0.961	0.963	0.073	M3

5. DISCUSSION

An equation is presented for dam's settlement in three points of close to the crown, half of the dam body height and the dam's bottom by using the collected data. Based on Equations 4, 7 and 10, dams' settlement of Agh Chai, Narmashir and Gilan-e Gharb was amended and results are shown in Table 18.

In addition, the results of performance obtained from the presented equations are shown in Figure 5 in comparison by using various criteria.

As network analysis results showed, the best correlation coefficient between the predicted settlement and settlement of instrumentation in the Curve Fitting of MATLAB software is related to the M1 equation. It should be noted that the correlation coefficient between the instrumentation and predicted values is obtained by all three equations as according to all the data. The proposed equation can be used easily to estimate the settlement at minimum time for the dams with alluvium that with the use of it, the IS pipes readings can be corrected. To determine the amount of M1 and other proposed equations (M2 & M3), it is trying to present an equation by trial and error that has the highest correlation and the least amount of errors. Also, the presented equations should be closed to the objective function (reading results of instrumentation) in terms of accuracy.

Table 18. Compare the predicted values of settlement and calculated settlement.

	Equation	Predicted settlement	Calculated settlement
Agh Chai	M1	20	21
	M2	26	27
	M3	31	32
Narmashir	M1	28	29
	M2	41	41
	M3	56	57
Gilan-e Gharb	M1	29	29
	M2	41	43
	M3	54	55

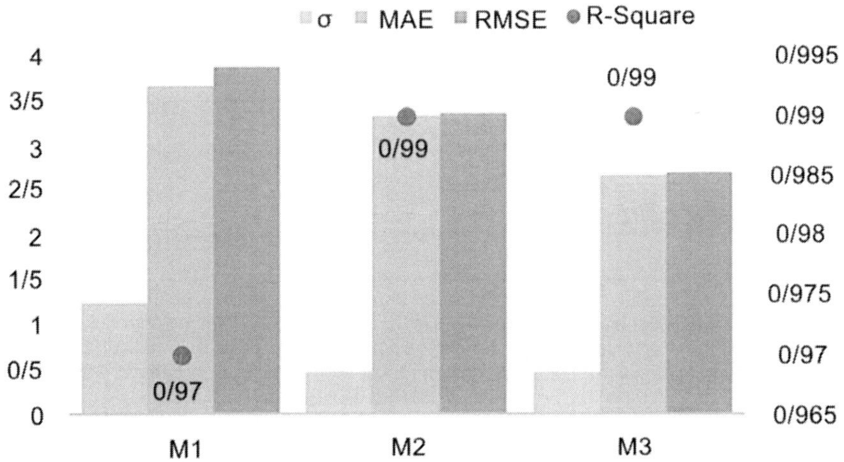

Figure 5. Comparison of different criteria for the presented equations.

6. CONCLUSION

In this study, it has been trying to compare the results of IS instrumentation readings, with the results of numerical modeling, based on behavioral models of hardening Mohr-Coulomb. The original modeling results indicated significant differences between numerical modeling and the IS instrumentation readings, which confirmed the reason of this research. The analysis indicated that alluvial is a major cause of error in modeling and causes the settlement to the dam body. By back analysis, more realistic parameters will be achieved for the behavior of dam materials as well as determining the behavior of parts that there is no information on their behavior, such as alluvium. With a good adaptation of instrumentation and numerical model, during construction and operation, we can predict actual behavior of the dam, in the future, with reasonable accuracy. Also, based on analysis on all three equations, it was found that the M1 equation (as the bottom of the dam) had a better correlation coefficient than the equations of M2 (as a third of the dam height) and M3 (as two thirds of the dam height). The lowest average absolute error, mean square error root and sum of squares residual error were related to equation M1. It is noted that this limitations presented equations, valid only until the construction of earth dams and also the earth dams that are only under static stress.

CITE THIS PAPER

Ghaemi, M., Ahangari, K., Noorzad, A. and Goshtasbi, K. (2017) The Effect of Alluvial Foundation on the Earth Dams Settlement. Open Journal of Geology, 7, 360-373. https://doi.org/10.4236/ojg.2017.73025

REFERENCES

1. Bell, F.G. (1995) Engineering Treatment of Dam. F.N. Spon, Durban.
2. Ahangari, K., Moeinossadat, S.R., Behnia, D., Behnia, M. and Moeinossadat, S.H. (2012) Application of New Methods for Intelligent ANN, ANFIS and GEP for Predicting the Static Elasticity Modulus of Limerock. International Conference on Nonlinear Modeling & Optimization, Shomal University, Amol, 28-29 August 2012.
3. Chrzanowski, A., Chen, Y.Q. and Secord, J. (2007) Geometrical Analysis of Deformation Surveys. Proceedings (MIT), Deformation Measurements Workshop, MIT, Boston, 31 October-1 November 2007, 170-206.
4. Chrzanowski, A., Chen, Y.Q. and Secord, J. (2008). On the Strain Analysis of Tectonic Movements Using Fault Crossing Geodetic Surveys. Tectonophysics, 97, 297-315.
5. Chugh, A.K. and Falvey, H.T. (1978) A Computer Program for Planar Seepage Analysis in a Zoned Anisotropic Medium by the Boundary Element Method. Advances in Engineering Software, 5, 196-201.

6. Duncan, J.M. (1997) State of the Art: Static Stability and Deformation Analysis. In: Seed, R.B. and Boulanger, R.W., Eds., Proceedings Specialty Conference on Stability and Performance of Slopes and Embankments II, ASCE, 222-266.
7. Duncan, J.M. (1996) State of the Art: Limit Equilibrium and Finite-Element Analysis of Slopes, ASCE. Journal of Geotechnical Engineering, 122, 577-595.
8. Marandi, M., Vaezinejad, M. and Khavari, E. (2012) Prediction Of Concrete Faced Rock Fill Dams Settlements Using Genetic Programming Algorithm. International Journal of Geoscience, 3, 601-609.
9. Curve Fitting Toolbox—MATLAB. The Math Works Inc., Natick, Massachusetts.
10. Final Technical Report, Second Phase Studies of AQCHAY Earth Dam (2003) Tehran, Consulting Engineering MAHAB QODS, Repot No. 19, 50-442.
11. Final Technical Report, Second Phase Studies of Narmashir Earth Dam (2000) Tehran, Consulting Engineering MAHAB QODS, Repot No. 3, 720-2840.
12. Final Technical Report, Second Phase Studies of GILANQARB Earth Dam (2001) Tehran, Consulting Engineering AB NIROU, Repot No. 3, 230-460.
13. Itasca Consulting Group, Inc. (2004) FLAC3D User Manual: Version 5.0. USA.

CHAPTER 3

Model Design for AMT Data Inversion*

Shancong Yao[1,2], Ruliang Zhang[2], Baoqiang Qiao[2], Hanping Wan[2], Xiang Yu[2]

[1]*School of Geophysics and Information Technology, China University Geosciences (Beijing), Beijing, China*

[2]*CNNC Key Laboratory of Uranium Resource Exploration and Evaluation Technology, Beijing Research Institute of Uranium Geology, Beijing, China*

ABSTRACT

By analyzing the characters of the mainstream commercial magnetotelluric inversion softwares in dealing with audio magnetotelluric data, a dynamic model-making method for inversion has been developed based on the observed AMT data. This method focusing on model domain can adjust mesh's scale and model's dimension depending on the field data just with a few parameters. By this, it is convenient to study the geo-electrical anomalies variations of different scale or dimensional models. Applying such model-making technique into the known hardrock geological setting, it is easy to obtain a new geo-electrical model which agrees with the resistivity curves of core samples better than before. It is demonstrated that this can increase the recognition of the resistivity contrast and deserves studying further.

Keywords: AMT Exploration, Making-Model Based on AMT Data, Geo-Electric Structure

1. INTRODUCTION

Audio magnetotelluric method (AMT), with simple logistics and no transmitter required, became a key geophysical method accompanying with more kinds of commercial AMT instruments produced in the markets since the 1990s (Yao, 2012) [1] . In the past several decades, one of the sparkles of geophysical exploration is the forward and inversion techniques applied into the field. In the induction electromagnetic exploration field, there were several practicably EM 2D inversion methods applied vastly, such as NLCG, RRI, OCCAM and so on [2] [3] [4] . The hardware and software progress made AMT method to be applied to many fields, such as base metals [5] [6] , uranium exploration [7] and so on.

As general geophysical exploration data process, there are mainly three steps to process AMT data. The first step is to convert each station's time serial data to the frequency domain before calculating the cross power spectrum. Then, the estimation of each station's impedance is followed by seeking a corresponding resistivity model which also satisfies a specified regularization function through inversion. However, many softwares to invert AMT data come from those focusing on MT data. Although the principle is identical for processing two types' data, the anomaly scale to be focused is much different because AMT exploration is about 2000 m depth underground and MT about the deeper and big scale anomalies. Moreover, some commercial software is bounded with several fixed models or will meet memory overflowing when inversing fine meshes. When studying the sand body anomaly among the weak resistivity contrast in uranium exploring by AMT data, an inversion process was outlined using different scale models with checking anomalies [8] . To meet such situations, a simple method of making different scale or dimensional model is required. The paper gives a way of making such models conveniently.

Here, a model-making method based on AMT field data is discussed specially and the main steps for programming are explained in detail. An example of AMT exploring in hardrock area using this model-making method showed that this method was conductive to detailed resistivity anomaly mining and these anomalies agreed with core sample resistivity curves. The paper aims at giving a method to make the model for detailed inversion based on field data, which makes it convenient to study resistivity anomalies variation of models.

2. MODEL DESIGN

The paper here only focuses on the 2-D inversion problem of finding a 2-D electrical resistivity model that can fit the observed data in two mode's impedances (TE and TM). There are many papers on the fundamental equations for 2D modeling (e.g. Swift, et al., 1971; Zhdanov, et al., 1982; Wannamaker, et al., 1985) [9] [10] [11] so that they will not be re-counted here. The inversion process minimizes a regularized function as the trade-off between a structure penalty function and the data residual norm weighted with data variances (Rodi & Mackie, 2001) [2] . The inversion codes based on these or similar modeling schemes are also widely used (e.g. Sasaki, 1989; deLugão, et al., 1997; Rodi, W.L., et al., 2001) [2] [3] [12] . The paper employed the NLCG inversion scheme of Rodi & Mackie (2001) [2] , and electric and magnetic fields are computed using a finite-difference scheme by network analogs to Maxwell's equations (Charles M. Swift, 1971; Madden 1972) [9] [13] . The modelled domain (a user-defined 2D mesh of resistivity blocks incorporating topography) is much larger than the actual region of interest to ensure that the model boundaries are sufficiently far away from resistivity anomalies in order to satisfy the boundary conditions.

There is a coherence relationship between resistivity anomaly scale and that of the model's cell. If the cell grid is too coarse, it may integrate the anomaly and host rock into one cell so that the anomaly is not identified or attenuated. On the contrary, if cell grid is smaller than the scale of the anomaly, it may delineate the boundary between the anomaly and the host rock in detail. However, if the cells are too close, it may produce overburden calculating resource and causes memory overflow sometimes. The problem is that the scale of the resistivity anomaly is unknown before inversion procedure without the prior information.

In order to make the inversion model coherent with the AMT data characteristics, this making model method based on the real field data has the following three steps:

1) To calculate detecting depth of each frequency in all station data of the area or one profile.

2) To configure cells' height. In general, the height of the cells above the minimum detecting depth is set as a constant. The heights of the cells between the minimum and maximum detecting depths are set to a geometric progression, whose first term is that constant height and whose common ratio is based on how much the model dimension to be built. Below the maximum detecting depth, the height distribution is also a geometric progression, but its common ration is at least 0.1 more than that common ration above.

3) To configure horizontal column allocation. Setting two columns in the modeled domain, three columns or more are inserted between neighborly stations based on the AMT observing station space and the model's dimension to be designed. In two lateral zones out of the domain, the spacing between columns distributes also as a geometric progression, whose first term is the spacing between the first and the second column in the domain and whose common ratio is always set as 1.5.

This making model method has several advantages. First, the model made by this way can focus on the modeled domain zone much, especially using bigger common ration beyond the maximum detecting depth and in two lateral zones out of the domain. Second, this method is a solution to designing different scale model. There are two directional parameters to adjust model's dimension. In the vertical direction, one can use different common ratio or set different constant height of cells above minimum detecting depth. In the horizontal direction, one can set the number of the columns inserting between neighborly stations to adjust model's dimension.

For detecting depth is related to apparent resistivity of each frequency, this model-making method is different from that of only using hemi-sphere resistivity value as before and is corresponding to field data's detecting depth.

3. HARDROCK AREA EXAMPLE

Due to the high resistivity value in the hardrock area and lack of significant resistivity contrast, it is more difficult to explore resistivity anomaly. However, in this situation, the weak resistivity contrast is related to some factors with many kinds of ore mineralization or safety so that it's significant to detect such anomalies. The following example is about exploring copper with AMT data in hardrock area, but the emphasis here is about how to mine detailed resistivity anomalies varying using the making model method above.

3.1. Geological Setting and AMT Data Acquiring

The AMT exploring profile (D profile) is in the copper ore distric of Chifuma, Kasempa, North-west province in Zambia. The survey area is mainly composed of Katanga group, which is composed of littoral clastic and carbonate rocks and comprises of Lower Roan formation, Upper Roan, Mwashya formation and Kundelungu formation from bottom to top (Figure

1). Lower Roan formation is the oldest stratum which is comprised of felds-quartz sandstone and dolomite rocks of the survey area and emerges at the southern part of the survey area. Upper Roan formation, whose outcrops are in south-west and south-east parts of the studying area, is sandy slate and dolomite rocks. Mwashya formation is the main ore stratum in central and northern parts and is composed of charry stand stone and dolomite rocks which contain much pyrite, chalcopyrite, pyrrhotite and magnetite. Kundelungu formation is mainly metamorphic sandstone and conglomerate in western and north-eastern parts (Gong, Y.S., et al., 2015; Gong, Y.S., et al., 2016) [14] [15].

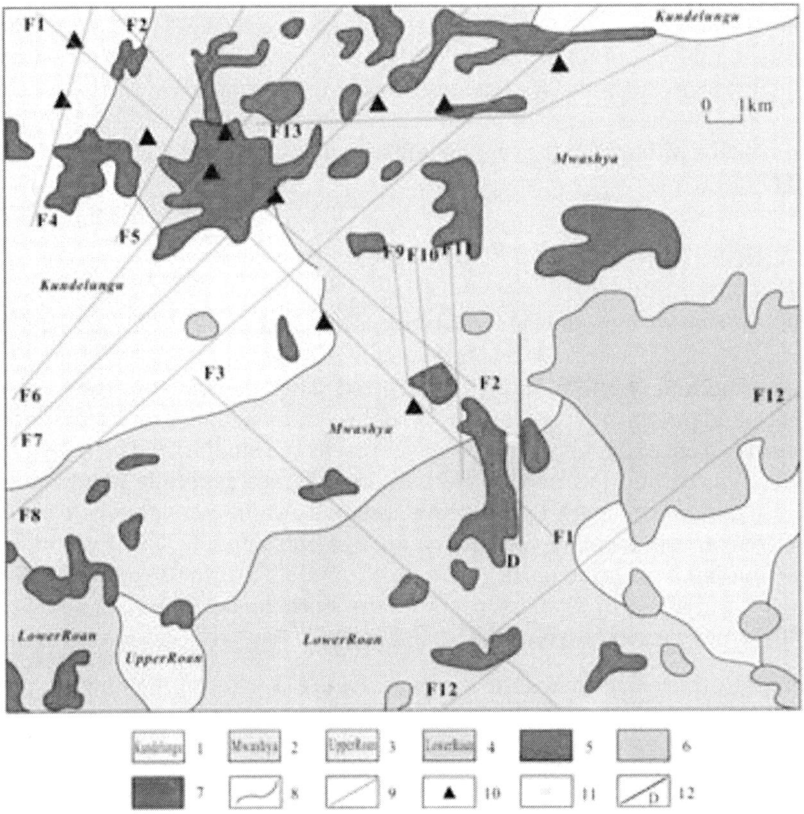

Figure 1. Geology map showing AMT profile and lithology of surveying area. (1, meta- morphic sandstone; 2, charrystandstone and dolomite rocks; 3, sandy slate and dolomite rocks; 4, felds-quartz sandstone and dolomite rocks; 5, granite; 6, monzonite; 7, grabbo; 8, geological boundary; 9, fault; 10, Fe mine; 11, Cu mine; 12, AMT profile.) (after Gong, Y.S, et al., 2016 [15]).

A 6 km long D profile in S-N direction was recorded with V8 multi-function receiver and AMTC-30 induction coils of Phoenix Geophysics Limited. The AMT station space is 50 meters and the electrode spacing is 50 meters with the tensor layout using non polarized electrodes. In order to further improve data quality, each station's data were recorded for over 25 minutes. There is little human interference to AMT data recording. The impedance frequencies range from 7 Hz to 8192 Hz.

In the D profile, there were two boreholes 1301 and 1391 at the distance 3100 m and 3600 m respectively. The depth of borehole 1301 is 847 m deep, and 1391 is 1300 m. Core samples at different depth of the two holes were machined into regular cube shape with 2 cm in length, 2 cm in width and 2 cm in high. For these cubes have three opposite faces, measuring resistivity of different opposite faces can be conducive to understanding whether these samples are electrical anisotropy. The resistivity of different opposite faces with depth of the two boreholes shows as Figure 2. Obvious, there are different degrees of electrical anisotropy with depths. Not only the resistivity, but also the chargeability is measured and will be discussed later.

3.2. Inversion Results Comparing

The geo-electrical models inverted from AMT data of D profile represent in Figure 3 (model A and model B). Model A is inverted from general making model method, while model B is inverted from the model by the above mentioned method based on the profile's data with identical inversion program. These two inversion models present the similarity. The resistivity anomaly area and location in the both models are generally identical, for example, near the mark C and D the resistivity value is relatively low. The high resistivity area and conductive area in both models are identical, but also the left and right parts of the both models are high resistivity. The resistivity values in shallow parts of the both model are relatively low, while in deep are relatively high.

However, there are clear differences in Figure 3 where the model B gives new and more detailed information as following:

The parts, where the two bidirectional arrows (C and D in Figure 3) point to, are the relative low resistivity areas in both models. However, the parts in model B are mixed with relative high resistivity blocks instead of the only relative low resistivity area. Especially for D pointing at part in the model B, there are three relatively low different value anomalies composed while for the corresponding part in model A there is only a block of relative low resistivity.

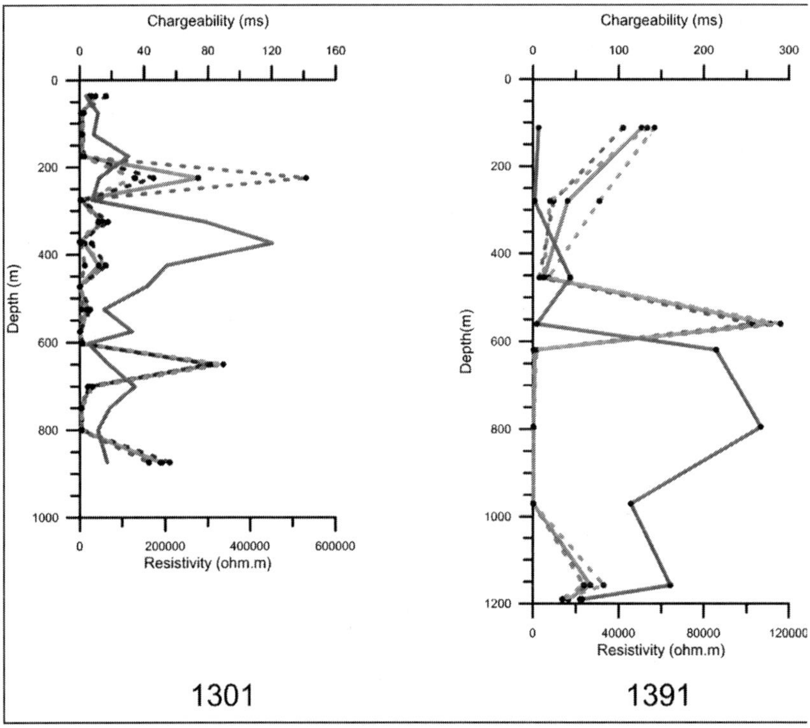

Figure 2. Resistivity and chargeability curves of core samples with depth. (dotted curves, resistivity of different opposite faces; red solid curve, mean of three opposite faces; blue solid curve, chargeability curve) (Data measured indoor).

The inversion geo-electrical model B (in Figure 3) did not change the general electrical anomaly distribution of the model A, but there appeared detailed information which the model A did not show. By applying dynamic model-making method to the inversion procedure, the relatively low or high resistivity blocks are decomposed into several different anomalies and more small scale resistivity variances are probed.

3.3. Reliability Discussion

In the above discussion, the inversion result's reliability was not mentioned. Here, the core sample resistivity curves will be used to discuss about the reliability. For geo-electrical model inverted from AMT data and the resistivity curves of the core samples are reflected the identical geo-electrical structure, both should be met each other. The resistivity curves from core samples are reliable so that it is one of key ways to check inversion reliability by comparing whether both are agreed with each other.

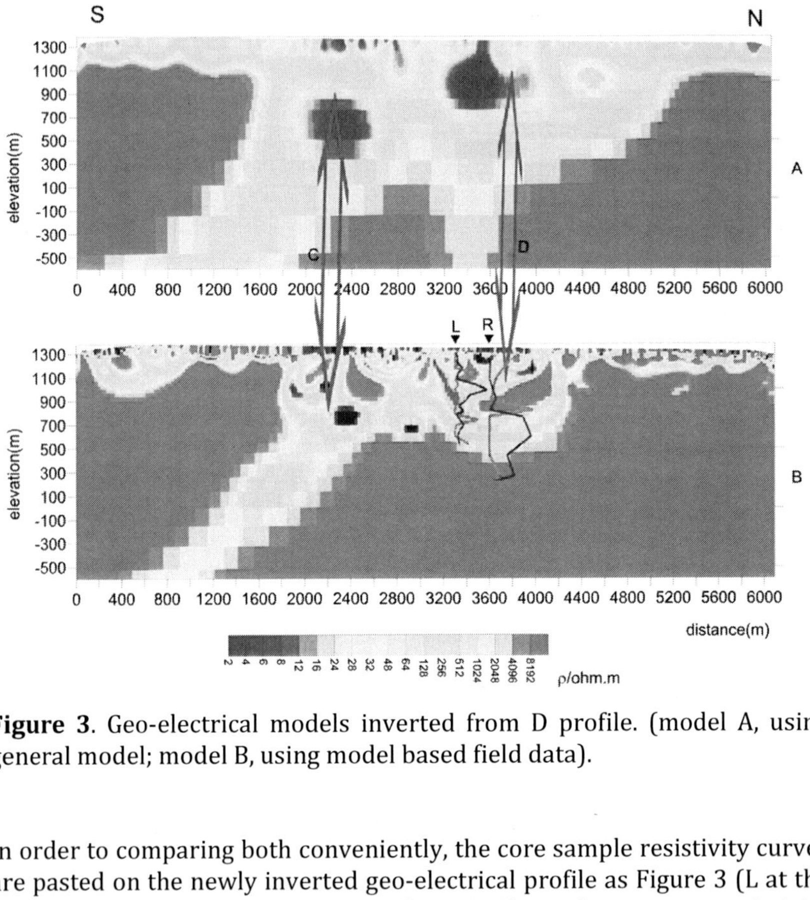

Figure 3. Geo-electrical models inverted from D profile. (model A, using general model; model B, using model based field data).

In order to comparing both conveniently, the core sample resistivity curves are pasted on the newly inverted geo-electrical profile as Figure 3 (L at the location of borehole 1301, R at that of 1391). The pink curves are resistivity varying with the depth and the resistivity value is the mean of three opposite face's resistivity values. The brown curves are chargeability curves.

The newly inverted model (model B) agrees with two borehole core sample resistivity better than model A. The borehole 1391 core sample resistivity curve present decreasing from the depth 550 m downward, then increasing at depth 950m and weakly decreasing at the bottom. At R of the profile in model B, it is clear that nearly at a depth of 500 m there is a high resistivity block whose top and bottom are lower resistivity anomalies. This geo-electrical structure varying agrees with core sample resistivity curve from depth 550 m to depth 1000 m. However, there is a big difference between mode A and core sample resistivity curve from depth 550 m to depth 1000 m. At depths below 600 m, there is a clear chargeability anomaly which is difficult to be explained from the geo-electrical model A; for the high resistivity blocks with high chargeability is unusual. But this is reasonable in the geo-electrical model B because the resistivity is significantly lower than its top and bottom at the corresponding depths.

In Figure 3, the resistivity curve shows higher near depths of 220 m, 650 m and 880 m under the borehole 1301. Near corresponding positions of both models, geo-electrical variation of model B shows such characteristics while that is vague in model A for there is no significantly high resistivity existing.

In this known geo-electrical variation covering the depth of 1200 m, the geo- electrical structure by inverting the model by the dynamic model-making method generally agrees with the core sample resistivity curves. Based on the above analysis and the comparison between geo-electrical models and resistivity cur- ves, the newly inverted model improved the inverting reliability and was close to the real geo-electrical varying underground. There are newly emerging geo- electrical variations in the model B which gives the more detailed geo-electrical varying. Such variation only in some local parts in the model did not change the geo-electrical frame of model A.

4. CONCLUSIONS

In hardrock area, especially in granite rock environment, its resistivity is always much higher and the geo-electrical contrast is not prominent. These facts make it difficult to detect the geo-electrical variations. Mining such electrical resistivity anomalies is significant because this kind of geo-electrical variations is usually related to some key problems such as safety factors, some favorable factors for forming ores, and so on.

During the studying period, the reason why the model A did not agree with the core sample resistivity curves was considered as the decreasing resolution ability with the increase of depth based on the principle of electromagnetic field diffusion. But with further study, it was found that using different scale model to be inverted might mine new information that did not appear before. After checking the new anomaly with the known information such as core sample resistivity curves, the dynamic model-making method based on field data for inversion proved to be a valuable study direction. Besides, the ability of detecting the deep anomaly can be improved if the related methods are used appropriately.

CITE THIS PAPER

Yao, S.C., Zhang, R.L., Qiao, B.Q., Wan, H.P. and Yu, X. (2017) Model Design for AMT Data Inversion. Open Journal of Geology, 7, 420-428.

REFERENCES

1. Yao, S.C. (2012) Analysis of the Impedance Estimatin Process of EH4. Uranium Geology, 28, 42-46.
2. Rodi, W.L. and Mackie, R.L. (2001) Nonlinear Conjugate Gradients Algorithm for 2-D Magnetotelluric Inversion. Geophysics, 66, 174-187
3. de Groot-Hedlin, C. and Constable, S. (1990) Occam's Inversion to Generate Smooth, Two-Dimensional Models from Magnetotelluric Data. Geophysics, 55, 1613-1624. Siripunvaraporn, W. and Egbert, G. (2007) Data Space Conjugate Gradient Inversion for 2-D Magnetotelluric Data. Geophysical Journal International, 170, 986-994.
4. Livelybrooks, D.W., Mareschal, M., Blais, E. and Smith, J.T. (1996) Magnetotelluric Delineation of the Trillabelle Massive Sulfide Body in Sudbury, Ontario. Geophysics, 61, 971-986.
5. Jones, A.G. and Garcia, X. (2003) Okak Bay AMT Data-Set Case Study: Lessons in Dimensionality and Scale. Geophysics, 68, 70-91.
6. Tuncer, V., Martyn, J.U., Weerachai, S. and James, A.C. (2006) Exploration for Unconformity-Type Uranium Deposits with Audiomagnetotellurics Data: A Case Study from the McArthur River Mine, Saakatchewan Canada. Geophysics, 71, 201-209.
7. Yao, S.C, Wang, M., Duan, S.X, Chen, S., Liu, W. and Xu, D.L. (2017) Exploring Sandstone Body in Weak Electrical Resistivity Contrast with AMT Data. International Journal of Geosciences, in Press.
8. Charles, M. and Swift, J.R. (1971) Theoretical Magnetotelluric and Turam Response from Two-Dimensional in Hogeneities. Geophysics, 36, 38-52.
9. Zhdanov, M.S., Golubev, N.G., Spichak, V.V. and Varentsov, I.M. (1982) The Construction of Effective Methods for Electromagnetic Modeling. Geophysical Journal of the Royal Astronomical Society, 68, 589-607.
10. Wannamaker, P.R., Stodt, J.A. and Rijo, L. (1985) PW2D-Finite Element Program for Solution of Magnetotelluric Responses of Two-Dimensional Earth Resistivity Structure: User Documentation. Earth Sci. Lab., Univ. Utah Res. Inst., ESL-158.
11. Sasaki, Y. (1989) Two-Dimensional Joint Inversion of Magnetotelluric and Dipole-Dipole Resistivity Data. Geophysics, 54, 254-262.
12. Madden, T.R. (1972) Transmission Systems and Network Analogies to Geophysical Forward and Inverse Problems, Report No. 72-3. Department of Geology and Geophysics, MIT, Cambridge, MA.
13. Gai, S.S., Biab, Q.T., Liu, Sh.Y., Yi, Z.G., Li, Ch. B., Zhu, X.J., Zhan, Ch.F. and Ran, L. (2015) IOCG Carbonatite Copper Deposits Discovered at North-West Province in Zambia. Geological Review, 61, 644-650.
14. Gong, Y.S., Hou, H.J., Sun, C.J., et al. (2016) Soil Geochemical Survey and Prospecting Effect in Property A in Kasempa of Northwestern Province, Zambia. Geophysical and Geochemical Exploration, 40, 482-487.

CHAPTER 4

M5 Model Tree to Predict Temporal Evolution of Clear-Water Abutment Scour

R. Biabani[1], M. Meftah Halaghi[2], Kh. Ghorbani[2]

[1]*Department of Water Engineering, Faculty of Agriculture, Tabriz University, Tabriz, Iran*

[2]*Department of Water Engineering, Gorgan University of Agricultural Sciences and Natural Resources, Gorgan, Iran*

ABSTRACT

Scour is a natural phenomenon that is created by the rivers streams or the flood which brings about transferring or eroding of bed materials. To have accurate and safe erosion control structures design, maximum scour depth in downstream of the structures gains specific significance. In the current study, M5 model tree as remedy data mining approaches is suggested to estimate the scour depth around the abutments. To do this, Kayaturk laboratory data (2005), with different hydraulic conditions, are used. Then, the results of M5 model were also compared with genetic programming (GP) and pervious empirical results to investigate the applicability, ability, and accuracy of these procedures. To examine the accuracy of the results yielded from the M5 and GP procedures, two performance indicators (determination coefficient (R2) and root mean square error (RMSE)) were used. The comparison test of results clearly shows that the implementation of M5 technique sounds satisfactory regarding the performance indicators (R^2 = 0.944 and RMSE = 0.126) with less deviation from the numerical values. In addition, M5 tree model, by presenting relationships based on liner regression, has good capability to estimate the depth of scour abutment for engineers in practical terms.

Keywords: Abutments, Scour Depth, M5 Model Tree, Genetic Programming Model (GP)

1. INTRODUCTION

Scour is the result of water erosion that causes digging and transferring bed materials and rivers banks. Bridge collapse, due to the total scour in foundation (including abutment and pier), makes the significance of the study about scour prediction and different countermeasures against it completely clear. According to Melville [1] , among 108 bridge collapse that happened between1960-1984 in Newslands, 29 of them were related to scour bridge abutment. Also, 70 percent of bridge collapse expenditure in Newlands was due to this matter. Data show that scour bridge abutment is a serious problem. Studies carried out on 383 bridges in the United State show that 25% of pier destructions and 72% of abutment destructions have been the cause of bridge collapse.

Due to lack of enough information in this issue, studies related to scour pattern began from early 1980 (e.g., at Auckland University). Richardson and Richardson [2] showed that the presented relations estimate the scour amount in abutment more than the real amount. One probable reason is that obtained relations of the extracted results are from rectangular channels; while a natural river mostly consists of compound channels with main channel and flood plains. The other reason for the inaccuracy of the presented data is the lack of attention to scour time development. That is why in recent decades several studies have been conducted in this field and researchers presented some relations for scour time development, such as Gill [3] , Cunha [4] , Cardoso and Bettes [5] , Kothyari and Ranga Raju [6] , Ballio and Orsi [7] , Radice et al. [8] , Oliveto and Hager [9] , Coleman et al. [10] .

The exact estimation of scour depth by the help of laboratory studies is a difficult, costly and time-consuming task. Hence, by developing computer software and using in hydraulic research, the estimation of scour depth has been carried out applying these methods by researchers.

Genetic Programming is one of these techniques that are being used in water engineering in recent decades. Azamathulla et al. [11] [12] have used linear genetic programming for determining pipe line scour. Also, the longitudinal dispersion coefficient in streams was determined by Azmathulla and Ghani [13] .

Salmasi et al. [14] estimated friction factors in pipes (f) according to changes in Reynolds number and relative roughness by the use of genetic programming and artificial neural networks. Guven and Gunal [15] used genetic algorithm to determine the depth of downstream water structure and compared the obtained results with empirical results of researchers.

Recently a new method called M5 decision tree model is presented for solving various problems and predicting output parameters. This model is used to solve engineering problems such as rainfall-runoff modeling [16] , flood forcasting [17] , water level discharge relationship [18] , sedimentation amount measurement in river [19] , sedimentation modeling [20] , suspended sediment load [21] , evaporation aspiration measurement [22] , scour depth measurement around bridge piers [22] , significant wave height prediction [23] and flow discharge prediction in compound channels [24] . In this method, based on the most important input variables, data are divided into different separated groups and for each group a multivariate liner regression equation is presented in order to measure output variable. Simple measurement, accurate results and the generalizability of the results are the great profitable outcomes of this model.

The purpose of this study is to estimate time development of scour around abutments by the use of M5 and GP liner regression technique. The comparison of the obtained results with the empirical results shows the high capability of this software to estimate time development of scour depth.

2. M5 MODEL TREE

The following idea is used by this machine-learning technique: the parameter space is split into areas (subspaces) and in each of them a linear regression model is built. As a matter of fact, the resulting model would be regarded as a modular model, or a committee machine, in which the linear models being specialized on the particular subsets of the input space.

The algorithm called the M5 algorithm is utilized for the sake of inducing a model tree [25] . A collection K of training examples is taken into account. Each example is characterized by the values of a fixed set of (input) attributes and has an associated target (output) value. The main objective is to construct a model that relates a target value of the training cases to the values of their input attributes. The quality of the model will be generally accurately estimated as if it anticipates the target values of the cases which are unseen.

A divide-and-conquer method constructs Tree-based models. The set K is either related to a leaf, or some tests are chosen to split K into subsets corresponding to the test outcomes and a similar procedure is applied recursively to the subsets. The splitting criterion used for M5 model tree algorithm depends on treating the standard deviation of the class values that reach a node as a measure of the error at that node, and calculating the expected reduction in this error as a result of testing each attribute at that node. To compute the standard deviation reduction (SDR), the help of this formula seems necessary:

$$\text{SDR} = sd(K) - \sum \frac{|K_i|}{|K|} sd(K_i) \tag{1}$$

where K indicates a set of examples that reaches the node; K shows the subset of examples that have the ith outcome of the potential set; and sd stands for the standard deviation.

After examining all potential splits, M5 selects the item that enhances the expected error reduction fully. Splitting in M5 ceases when the class values of all the instances that reach a node differ just marginally, or only a few instances are left. The relentless division often creates over-elaborate structures that must be pruned back, namely by substituting a subtree with a leaf. Eventually, a smoothing process is carried out with the aim of compensating for the sharp discontinuities that will unavoidably take place between adjacent linear models at the leaves of the pruned tree, especially for some models constructed from a smaller number of training examples. In this process, the adjacent linear equations are updated in such a way that the projected outputs for the neighboring input vectors corresponding to the different equations are becoming close in terms of value. For more details of this process, Quinlan [25] and Witten & Frank [26] can be referred to.

3. GENETIC PROGRAMMING

Genetic programming, a branch of the genetic algorithm, is a method for acquiring the most "fit" computer programs by taking the full advantage of artificial evolution [27] . The GP optimizes not only the coefficients but also constants in a function and the function type itself. A possible function is determined by given mathematical operators, such as +, −, ×, sin, expand so forth. Each function indirectly includes an assignment to a variable, which paves the way for the use of multiple program outputs in GP; nevertheless, in tree-based GP those side effects need to be integrated explicitly. The GP encodes a function as a tree with nodes and branches, and then optimizes functions according to natural principles. The GP procedure bears some resemblances to a genetic algorithm, which generates solutions as a parent population, and then fortifies solutions by selection, crossover, and mutation processes [28] .

The great merit of GP for the modeling process lies in its ability to produce models that construct an understandable structure, i.e., a formula or equation. Accordingly, for "data rich, theory poor" instances, GP benefits may outweigh other techniques inasmuch as GP can self-modify, via the genetic loop, a population of function trees so as to ultimately produce an "optimal" and physically interpretable model [29] .

The following expression can analyze the fitness of GP algorithm:

$$f = \sum_{j=1}^{j=N} \left| X_j - Y_j \right| \tag{2}$$

where X_j = value returned by a chromosome for the fitness case j and Y_j = expected value for the fitness case j. This configuration has been tested for the proposed GP model and has been found adequate [11] [12] .

4. DIMENSIONAL ANALYSIS

Kayatork [30] laboratory measurements, which investigated the effect of abutment height on scour time development, have been considered as software input data. These tests had been carried out in a rectangular channel with the height of 0.3 meters and the weight of 1.5 meters for investigating the effect of four different heights of abutment on scour time development. For clear-water approach flow conditions, the maximum scour depth at an abutment is a function of the following parameters:

$$(d_s)_{\max} = f\left\{L_a, B_a, U, y, S, g, \rho, \rho_s, d_{50}, \sigma_g, t, B\right\} \tag{3}$$

where, L_a = abutment length, B_a = abutment width, U = mean approach flow velocity, y = flow depth, S = slope of the channel, g = gravitational acceleration, ρ = density of the fluid, ρ_s = density of the sediment, μ = dynamic viscosity of fluid, d_{50} = median particle grain size, σ_g = geometric standard deviation of sediment size distribution, t = time variation of scour when it starts, B = width of channel (Figure 1). In terms of dimensionless parameters Equation (3) can be written as:

$$\frac{d_s}{y} = f\left\{\frac{L_a}{y},\frac{B_a}{y},\frac{\rho_s}{\rho},\frac{U}{\sqrt{gy}},\frac{B}{y},\frac{\mu}{U\rho y},\frac{Ut}{y},S,\sigma_g,\frac{d_{50}}{y}\right\} \tag{4}$$

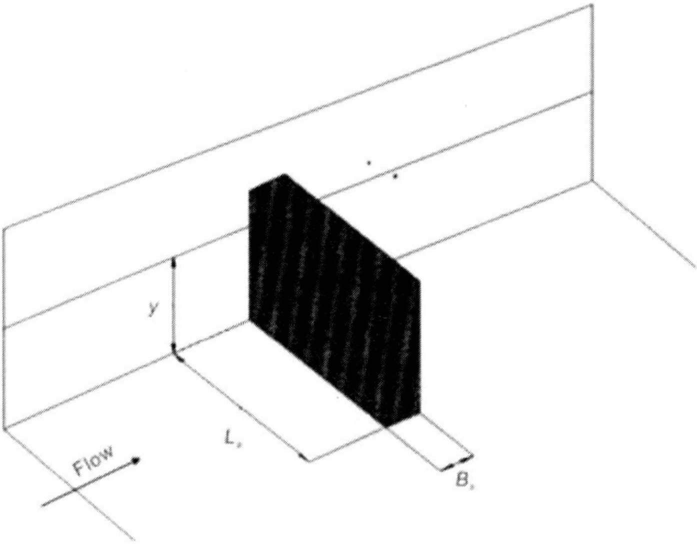

Figure 1. Definition sketch of abutment arrangement.

In this study, channel bed sloop, channel width, sediment particle size, flow depth, and consequently Froude number assumed constant. Finally, the above dimensional analysis is summarized as follows:

$$\frac{d_s}{y} = f\left\{\frac{L_a}{B_a},\frac{Ut}{y}\right\} \tag{5}$$

The first dimensionless relation shows the geometry of the model and represents the fraction of height to weight, and the second dimensionless number is time dimensionless parameter.

Summary

In this study statistical parameters of correlation coefficient (R^2) and root mean square error (RMSE) were used in order to compare the results of two regression methods. The lesser amount of RMSE (0.01) and the larger amount of correlation coefficient (0.0961) introduced GP model better in predicting scour time development in abutments (Table 1). As it was aforementioned about regression model structures, the dependent variable was estimated by breaking computing space into subspaces and presenting liner regression for each subspace and thus computational error increase.

Figure 2 provides the graph plotted between actual and predicted value of scour depth obtained by using M5 model tree. This figure suggests that about 91% of the predicted values lie inside the ±25% lines. Predicted amount versus actual results diagram of time scour depth related to Genetic programming model shows that 93% of data lie within 25% allowed error range (Figure 3).

One major advantage of M5 model tree is the availability of three simple linear relations (Equations (9), (10) and (11)) which can be easily used to predict the scour around the abutments (Figure 4).

The temporal variation of local scour depth can be defined as a function of its independent parameters when dimensionless abutment height (L_a/B_a) is less than 1.12 by the following expressions:

$$\frac{d_s}{y} = 0.0267\left(\frac{Ut}{y}\right) + 12.9903\left(\frac{L_a}{B_a}\right) + 28.345 \qquad (6)$$

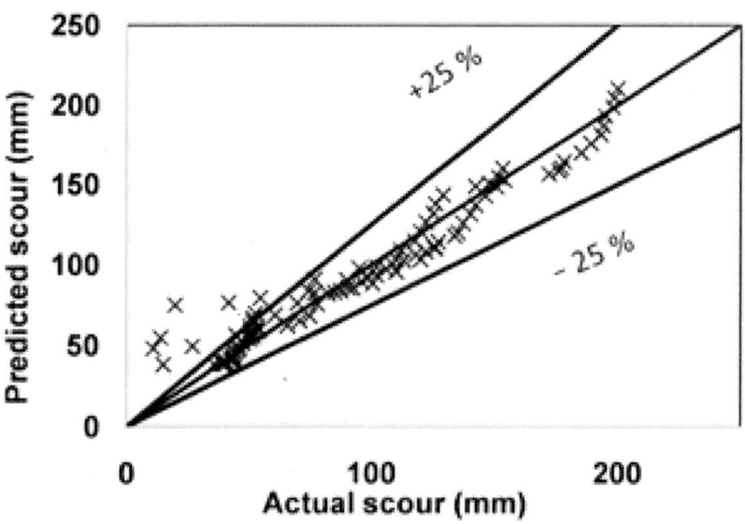

Figure 2. Actual vs. Predicted Scour using M5 Model.

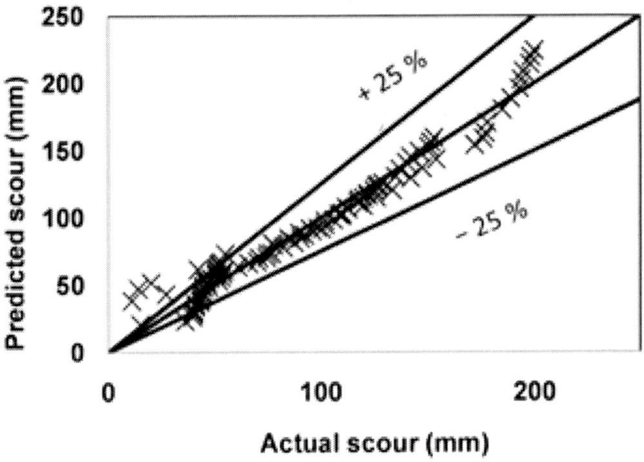

Figure 3. Actual vs. Predicted Scour using GP Model.

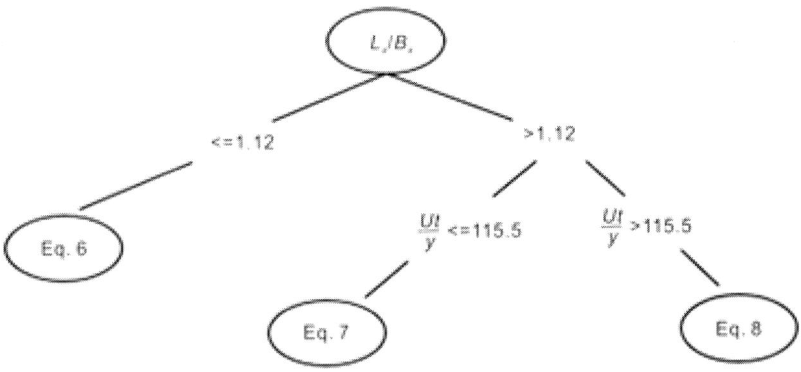

Figure 4. The acquired tree structure from M5 to measurement the momentary scour depth in abutments.

Table 1. Correlation coefficient and root mean square error for M5 Model Tree and Genetic programming.

Modeling Approach	RMSE (m)	Correlation Coefficient
M5 Model Tree	0.126	0.944
Genetic programming	0.1	0.961

On the other hand, if dimensionless abutment height is greater than 1.12, temporal variation of local scour depth is broken into two parts (Equations (7) and (8))

If time dimensionless parameter $\left(\frac{Ut}{y}\right) \leq 115.5$:

$$\frac{d_s}{y} = 0.4472\left(\frac{Ut}{y}\right) + 13.4525\left(\frac{L_a}{B_a}\right) + 26.6258 \quad (7)$$

And for $\left(\frac{Ut}{y}\right) > 115.5$

$$\frac{d_s}{y} = 0.0582\left(\frac{Ut}{y}\right) + 33.419\left(\frac{L_a}{B_a}\right) + 24.1831 \quad (8)$$

The scour development process aroundthe abutment for different heights was shows in Figure 5. For the cases with $L_a/B_a = 0.75$, abutments have no significant effect on flow pattern, therefore equilibrium scour depth and time to reach the equilibrium scour depth decrease and this matter was presented by decision tree model in Equation (6). With increasing abutment height, time development of local scour depth changed rapidly when the time dimensionless parameter is less than 115.5 and after the certain time the depth of scour gradually remains. So, M5 model shows this procedure well by breaking the data into two series.

Non-liner regression relation is represented by Genetic Programming model as follows:

$$\frac{d_s}{y} = \frac{L_a}{B_a}\left[\frac{L_a}{B_a} + \left(8.85 - \left(\frac{L_a}{B_a}\right)^2\right)\right] + \left[\left(\frac{L_a}{B_a} - 5.57\right)^2 \left(\frac{Ut}{y}\right)^{0.5}\right]^{0.5} \left(\frac{L_a}{B_a}\right) + \left[\left(-4.41 - 4.41\right)^2 \left(\frac{Ut}{y}\right)^{0.5}\right]^{0.5} \left(\frac{L_a}{B_a}\right) \quad (9)$$

In contrast with the M5 model results, Genetic Programming model trained by dimensionless data was complicated, thus the regression tree has adaptability and capability to predict the scour depth around the abutments.

Figure 6 represents the comparison between the data reported by Cardoso and Bettes [5] and Coleman et al. [10] and those of the present work for the

M5 model tree. As can be seen, M5 model predict the scour depth more accurately than empirical relations.

5. CONCLUSION

The potential of M5 model tree in predicting the temporal local scour depth around the abutments was investigated in this paper by using Kayaturk laboratory data [30]. A major conclusion from this study is that M5 model tree works equally well to the Genetic Programming model and provides improved results in comparison to all three empirical relations used in this study. Furthermore, M5 decision tree model, besides simple calculation and equations, has good capability in estimating the depth of time scour in abutment.

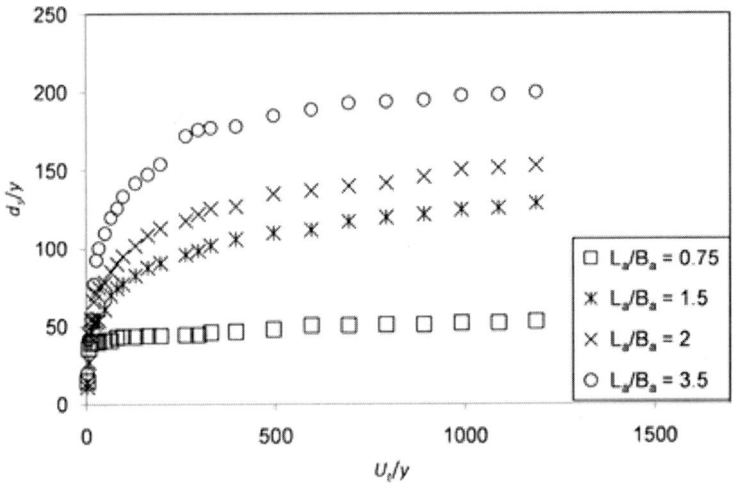

Figure 5. Time development of scour depth for different length of abutment.

CITE THIS PAPER

Biabani, R., Meftah Halaghi, M. and Ghorbani, Kh. (2016) M5 Model Tree to Predict Temporal Evolution of Clear-Water Abutment Scour. Open Journal of Geology, 6, 1045-1054.

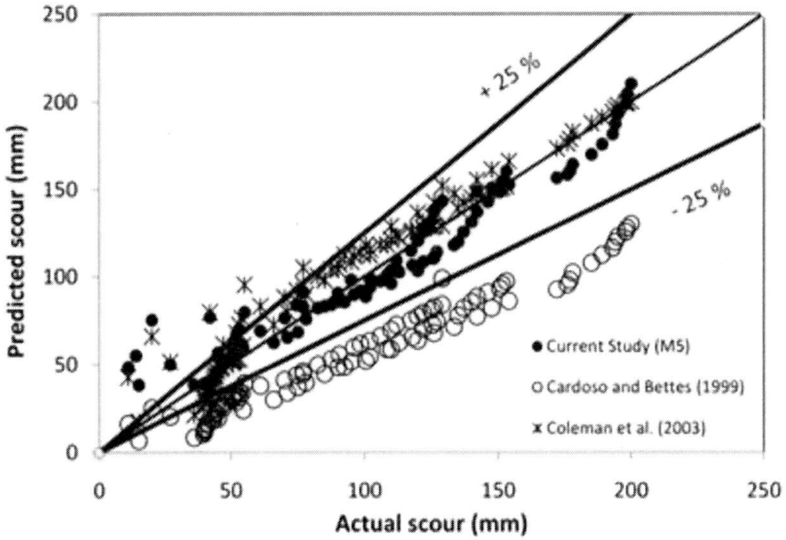

Figure 6. Actual vs. Predicted Scour using Four Empirical Relations.

REFERENCES

1. Melville, B.W. (1992) Local Scour at Bridge Abutments. Journal of Hydraulic Engineering, 118, 615-631.
2. Reddy, M.J. and Ghimire, B.N.S. (2009) Use of Model Tree and Gene Expression Programming to Predict the Suspended Sediment Load in Rivers. Journal of Intelligent Systems, 18, 211-227.
3. Pal, M. and Deswal, S. (2009) M5 Model Tree Based Modelling of Reference Evapotranspiration. Hydrologic Processes, 30, 1437-1443.
4. Etemad Shahidi, A. and Mahjoobi, J. (2009) Comparison between M5 Model Tree and Neural Networks for Prediction of Significant Wave Height in Lake Superior. Ocean Engineering, 36, 1175-1181.
5. Zahiri, A. and Azamathulla, H. (2014) Comparison between Linear Genetic Programming and M5 Tree Models to Predict Flow Discharge in Compound Channels. Journal of Neural Computing and Applications, 24, 413-420.
6. Quinlan, J.R. (1992) Learning with Continuous Classes. Proceedings of Australian Joint Conference on Artificial Intelligence, Hobart 16-18 November 1992, 343-348.
7. Witten, I.H. and Frank, E. (2005) Data Mining: Practical Machine Learning Tools and Techniques with Java Implementations. Morgan Kaufmann, San Francisco.
8. Johari, A., Habibagahi, G. and Ghahramani, A. (2006) Prediction of Soil-Water Characteristic Curve Using Genetic Programming. Journal of Geotechnical and Geoenvironmental Engineering, 132, 661-665.
9. Koza, J.R. (1992) Genetic Programming: On the Programming of Computers by Means of Natural Selection. MIT, Mass.
10. Muttil, N. and Chau, K.W. (2006) Neural Network and Genetic Programming for Modeling Coastal Algal Blooms. International Journal of Environment and Pollution, 28, 223-238.

11. Kayaturk, S.Y. (2005) Scour and Scour Protection at Bridge Abutments. PhD. Thesis on Civil Engineering, Natural and Applied Sciences, Middle East Technical University, Ankara, 213 p.
12. Bhattacharya, B., Price, R.K. and Solomatine, D.P. (2007) Machine Learning Approach to Modeling Sediment Transport. Journal of Hydraulic Engineering, 133, 440-450.
13. Bhattacharya, B. and Solomatine, D.P. (2006) Machine Learning in Sedimentation Modelling. Neural Networks, 19, 208-214.
14. Bhattacharya, B. and Solomatine, D.P. (2005) Neural Networks and M5 Model Trees in Modeling Water Level-Discharge Relationship. Neurocomputing, 63, 381-396.
15. Solomatine, D.P. and Xue, Y. (2004) M5 Model Trees and Neural Networks: Application to Flood Forecasting in the Upper Reach of the Huai River in China. Journal of Hydraulic Engineering, 6, 491-501.
16. Solomatine, D.P. and Dulal, K. (2003) Model Tree as an Alternative to Neural Network in Rainfall-Runoff Modeling. Hydrological Sciences Journal, 48, 399-411.
17. Guven, A. and Gunal, M. (2008) Prediction of Scour Downstream of Grade-Control Structures Using Neural Networks. Journal of Hydraulic Engineering, 134, 1656-1660.
18. Salmasi, F., Khatibi, R. and Ghorbani, M.A. (2011) A Study of Friction Factor Formulation in Pipes Using Artificial Intelligence Techniques and Explicit Equations. Turkish Journal of Engineering and Environmental Sciences, 35, 1-18.
19. Azmathulla, H.M. and Ghani, A. (2009) Genetic Programming for Longitudinal Dispersion Coefficients in Streams. Water Resources Management, 25, 1537-1544.
20. Azamathulla, H.M., Ghani, A.A., Zakaria, N.A. and Guve, A. (2010) Genetic Programming to Predict Bridge Pier Scour. Journal of Hydraulic Engineering, 136, 165-169.
21. Azamathulla, H.M., Guven, A. and Demir, Y.K. (2008) Linear Genetic Programming to Scour below Submerged Pipeline. Ocean Engineering, 38, 995-1000.
22. Coleman, S.E., Lauchlan, C.S. and Melville, B.W. (2003) Clear-Water Scour Development at Bridge Abutments. Journal of Hydraulic Research, 41, 521-531.
23. Oliveto, G. and Hager, W.H. (2002) Temporal Evolution of Clear-Water Pier and Abutments Scour. Journal of Hydraulic Engineering, 128, 811-820.
24. Radice, A., Franzetti, S. and Ballio, F. (2002) Local Scour at Bridge Abutments. International Conference on Fluvial Hydraulics, River Flow, 2, 1059-1068.
25. Ballio, F. and Orsi, E. (2001) Time Evolution of Scour around Bridge Abutments. Water Engineering Research, 2, 243-259.
26. Kothyari, U.C. and Ranga Raju, K.G. (2001) Scour around Spur Dikes and Bridge Abutments. Journal of Hydraulic Engineering, 39, 367-374.
27. Cardoso, A.H. and Bettess, R. (1999) Effects of Time and Channel Geometry on Scour at Bridge Abutments. Journal of Hydraulic Engineering, 125, 388-398.
28. Cunha, L.V. (1975) Time Evolution of Local Scour. The 16th IAHR Congress Proceedings, Sao Paulo.
29. Gill, M.A. (1972) Erosion of Sand Beds around Spur Dikes. Journal of the Hydraulics Division, 98, 1587-1602
30. Richardson, J.R. and Richardson, E.V. (1994) Practical Method for Scour Prediction at Bridge Piers. Proceeding of ASCE National Conference on Hydraulic Engineering, Buffalo, 1-5.

CHAPTER 5

Major and Trace Element Chemical Compositional Signatures of Some Granitic Rocks Related to Metal Mineralization in Japan

Etsuo Uchida*, Kazumasa Osada, Koki Nakao

Department of Resources and Environmental Engineering, Waseda University, Tokyo, Japan

ABSTRACT

We analyzed the major and trace element chemical compositions of 66 granitic rocks from 15 different areas in Japan. The intrusions from which the samples were collected were associated with Pb-Zn, Mo, Cu-Fe, Sn, or W mineralization and, for comparison, samples were also collected from intrusions not associated with any metal mineralization. The analyses indicated that the granitic rocks associated with Pb-Zn, Mo, or Cu-Fe mineralization were granites, granodiorites, or diorites, and that they were all I-type and formed in a volcanic arc tectonic setting. The granitic rocks associated with Sn or W mineralization and barren granitic rocks were classified as granites and as I-type with the exception of a few S-type granitic rocks. Most of the Sn- or W-associated granitic rocks and barren granitic rocks are thought to have formed in a volcanic arc tectonic setting. The Pb-Zn-, Mo-, or Cu-Fe-asso- ciated granitic rocks rarely shows negative Eu anomalies and a few of them are adakitic rocks, whereas all of the Sn- or

W-associated granitic rocks and barren granitic rocks show negative Eu anomalies. For these Japanese granitic rocks, the contents of K_2O, La, Y, Rb, Ta, Pb, Th, U, and REEs other than Eu increase with increasing SiO_2. Conversely, the contents of major components other than Na_2O and K_2O and the trace components V, Zn, Sr, Eu, and Sc decrease with increasing SiO_2. The Zr, Sn, and Hf abundances increase with increasing SiO_2 up to 70 wt%, but their abundances decrease when the SiO_2 exceeds 70 wt%. This suggests that granitic magma is saturated with these elements at 70 wt% of SiO_2, approximately.

Keywords: Granitic Rock, Hydrothermal Mineralization, Chemical Composition, Tectonic Setting, Japan

1. INTRODUCTION

Uchida et al. (2007) investigated the relationship between the chemical composition of biotite and mineralization type associated with representative granitic rocks in Japan. It was found that the average total Al content of biotite, listed by mineralization type, was Pb-Zn = Mo < Cu-Fe < Sn < W < no mineralization. In addition, the study found that the total Al content in biotite increased with the granite's solidification pressure. However, Uchida et al. (2007) only measured the major elements in the granitic rocks; trace elements were not analyzed. In this paper, we determined the whole rock chemical compositions, including trace elements and rare earth elements, for samples of granitic rocks from the same intrusions studied by Uchida et al. (2007). With these analyses, we aim to confirm and clarify the relationships between the chemical composition of the granitic rocks and 1) the type of metal mineralization and 2) the tectonic settings in which the granitic rocks were emplaced (e.g., [1] [2]).

2. GRANITIC INTRUSIONS INVESTIGATED

The granitic rocks analyzed for this study are from the same 15 areas studied by Uchida et al. [3] (Figure 1). Table 1 lists the names of the areas and mining districts from which the samples were collected, rock type, age, associated metal type, and sample number for the samples.

Intrusions in the Taishu [4] , Obira (granite porphyry) [5] and Chichibu mining districts were designated as granitic rocks related to Pb-Zn mineralization. We studied samples from the Ohkawame [6] and Daito-Yamasa [7] mining districts as granitic rocks related to Mo mineralization. As for Cu-Fe mineralization, granitic rocks in the Kamaishi [8] [9] and Yaguki mining districts and granitic rocks in the Tanzawa area [10] were selected for study. The granitic rocks from the Yaguki mining district were the Eastern granodiorite (the Ohisa granodiorite [11]) and the Central granodiorite. For intrusions related to Sn mineralization, samples from biotite granite in the Obira mining district [5] and granitic rocks in the Osuzu [12] and Suzuyama mining districts were selected. As for granitic rocks related to W mineralization, we studied granitic rocks in the Yakushima and the Fujigatani-Kiwida (the Habu granodiorite and the Osogoe complex), Ohtani (the Gyojayama granite [7]) and Yaguki mining districts as well as the Inada coarse-grained granite [13] in the Tsukuba area. In the Yaguki mining district, samples were collected from intrusions of the Western granodiorite (the Yokokawa granodiorite [11]). Barren granitic rocks were collected from the Hidaka metamorphic belt (the Toyonidake cordierite tonalite and hornblende tonalite [14]), the Tsukuba area other than the Inada coarse-grained granite [13] and the Fujigatani-Kiwada mining district (the Nakayamagawa complex and the Shimokuhara granite).

The granitic rocks associated with Pb-Zn, Mo, or Cu-Fe mineralization are almost magnetite-series but some are ilmenite-series whereas the granitic rocks with Sn or W mineralization and those without mineralization are all ilmenite- series (Table 1) [1] .

3. SAMPLE PREPARATION, ANALYSIS, AND RESULTS

Whole-rock chemical analyses were carried out on 66 samples from the granitic rocks mentioned above (Table 1). Broken down by metal type, there were 10 samples related to Pb-Zn mineralization, 12 samples related to Mo mineralization, 11 samples for Cu-Fe mineralization, 9 samples for Sn mineralization, 13 samples for W mineralization, and 11 samples for barren granitic rocks. Samples were pulverized using a tungsten carbide rod mill. About 5 g of the pulverized samples were sent to Activation Laboratories Ltd. (Ancaster, Canada) for analysis by their code "4 Litho" lithogeochemistry package. For those analyses, the granitic rock powders were fused using lithium metaborate/tetraborate and digested in dilute nitric acid. Analyses for a total of 55 elements were then obtained by analyzing the aqueous solutions thus prepared using inductively coupled plasma optical emission spectrometer (ICP-OES) and inductively coupled plasma mass spectrometer (ICP-MS). The analytical results are shown in

Table 2. Because the samples were contaminated with Co and W by the tungsten carbide rod mill during grinding, Co and W values are not listed in Table 2.

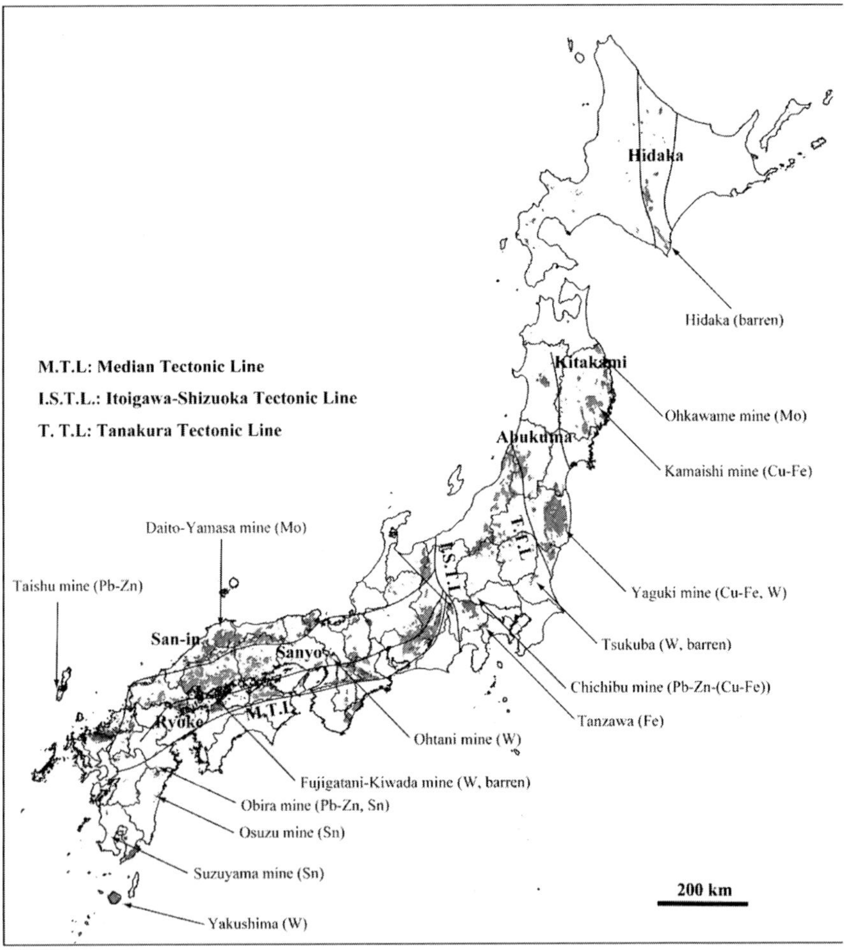

Figure 1. Map showing the localities of studied areas and distribution of Mesozoic and Cenozoic granitic rocks in Japan (Seamless digital geological map of Japan by Geological Survey of Japan).

Table 1. Lithology, age, and associated metal for 66 granitic rocks collected from 15 different localities in Japan.

Area	Granitic rock	Age	Mgt/Ilm series	Metal type	Sample No.
Hidaka	Toyonidake cordierite tonalite	Paleogene to Neogene	Ilm	No mineralization	HD02
	Toyonidake hornblende tonalite	Neogene	Ilm	No mineralization	HD05
Ohkawame mine	Granitic rocks in the zones I and II	Cretaceous	Mgt	Mo	OK01, OK04, OK05, OK07, OK11
Kamaishi mine	Ganidake granodiorite	Cretaceous	Mgt	Cu-Fe	KM03, KM04
	Ganidake diorite/diorite porphyry	Cretaceous	Mgt	Cu-Fe	KM05, KM09
Yaguki mine	Eastern granodiorite	Cretaceous	Mgt	Cu-Fe	YG01, TG04
	Central granodiorite	Cretaceous	Mgt	Cu-Fe	YG08
	Western granodiorite	Cretaceous	Mgt	W	YG16, YG17
Tsukuba	Tsukuba two mica granite	Paleogene	Ilm	No mineralization	TK01
	Inada coarse-grained granite	Paleogene	Ilm	W	TK08, TK09
	Kamishiro fine-grained granodiorite	Paleogene	Ilm	No mineralization	TK11
	Inada medium-grained granodiorite	Paleogene	Ilm	No mineralization	TK12, TK14
	Inada fine-grained granite	Paleogene	Ilm	No mineralization	TK13, TK15
	Tsukuba porphyritic granodiorite	Paleogene	Ilm	No mineralization	TK16
Chichibu mine	Chichibu quartz diorite	Neogene	Mgt	Pb-Zn(-Cu-Fe)	CC03, CC04, CC08
Tanzawa	Yushin tonalite	Neogene	Mgt	Fe	TZ05, TZ08
	Azegamaru tonalite	Neogene	Mgt	Fe	TZ10, TZ11
Ohtani mine	Ohtani or Gyojayama granite	Cretaceous	Ilm	W	OT02, OT03
Daito-Yamasa mine	Yamasa leucocratic granite	Paleogene	Mgt	Mo	DY03, DY04
	Renge granodiorite	Paleogene	Mgt	Mo	DY07, DY08
	Leucocratic granite complex	Paleogene	Mgt	Mo	DY11
	Kawai Hb-Bi hybrid rock	Paleogene	Mgt	Mo	DY10, DY14
Fujigatani-Kiwada mine	Nakayamagawa complex	Cretaceous	Ilm	No mineralization	IW02
	Habu granodiorite	Cretaceous	Ilm	W	IW07, IW08
	Shimokuhara granite	Cretaceous	Ilm	No mineralization	IW13
	Osogoe complex	Cretaceous	Ilm	W	IW18, IW19
Taishu mine	Uchiyama granitic rock and others	Neogene	Mgt-Ilm	Pb-Zn	TS05, TS07, TS08, TS13
Obira mine	Obira granite porphyry	Neogene	Ilm	Pb-Zn	OB02, OB09, OB10
	Obira biotite granite	Neogene	Ilm	Sn	OB01, OB04, OB06
Osuzu mine	Osuzu granodiorite/granite porphyry	Neogene	Ilm	Sn	OS02, OS03, OS04
Suzuyama mine	Suzuyama granite porphyry	Neogene	Ilm	Sn	SZ03, SZ04, SZ05
Yakushima	Yakushima granite	Neogene	Ilm	W	YK01, YK04, YK07

4. DISCUSSION

4.1. Chemical Compositions and Granitic Rock Classification

According to the classification of plutonic rocks based on the SiO_2 vs. (Na_2O + K_2O) diagram (Figure 2) [15] [16] , most of the granitic rocks associated with Sn or W mineralizetion and barren granitic rocks are classified as granites except some of them deviate into granodiorite. The samples associated with Pb-Zn or Mo mineralization fall into granite, diorite and diorite fields, whereas the samples associated with Cu-Fe mineralization are mainly in the diorite to granodiorite fields except one sample fall into granite field.

Table 2. Major, minor, and trace element chemical compositions for 66 granitic rocks collected from 15 different localities in Japan.

Mineralization	Sn									W							
	Obira			Osuzu			Suzayama			Tsukiba		Ohtani		Fijigatani-Kiwada			
Sample No.	OB01	OB04	OB06	OS02	OS03	OS04	SZ03	SZ04	SZ05	TK08	TK09	OT03	OT04	IW07	IW08	IW18	IW19
SiO_2 (wt%)	75.14	76.06	73.15	67.34	67.23	73.17	67.93	70.93	68.63	77.69	74.16	68.75	68.75	66.81	70.98	72.08	74.84
Al_2O_3	12.47	12.28	12.96	14.72	14.83	14.2	14.13	14.16	14.08	12.34	13.44	15.34	15.34	15.7	14.95	13.6	13.56
$Fe_2O_3(T)$*	1.22	1.16	1.63	5.02	5.37	2.6	4.04	2.76	3.4	1.65	1.84	3.15	3.15	4.01	2.96	2.11	2.05
MnO	0.022	0.018	0.027	0.087	0.08	0.055	0.029	0.023	0.038	0.031	0.03	0.067	0.07	0.096	0.082	0.039	0.047
MgO	0.14	0.14	0.2	1.55	1.69	0.75	1.21	0.96	1.06	0.16	0.19	0.93	0.93	1.12	0.74	0.33	0.45
CaO	0.71	0.88	0.87	2.73	2.85	1.09	2.61	2.16	2.49	1.89	1.89	2.04	2.04	3.24	2.2	1.24	1.6
Na_2O	3.25	3.21	3.19	3.02	2.93	2.76	2.75	2.78	2.84	3.06	3.03	2.95	2.95	3.53	3.27	2.9	2.94
K_2O	4.78	4.52	5.02	3.3	3.23	4.24	3.69	3.98	3.56	3.48	4.41	3.92	3.92	3.22	3.98	4.89	4.12
TiO_2	0.149	0.138	0.21	0.803	0.87	0.373	0.524	0.495	0.483	0.143	0.164	0.438	0.44	0.364	0.269	0.186	0.247
P_2O_5	0.03	<0.01	0.02	0.17	0.17	0.17	0.1	0.09	0.1	0.03	0.03	0.23	0.23	0.13	0.1	0.01	0.08
LOI	0.65	0.33	0.84	1.38	1.08	1.46	1.72	2.14	2.11	0.3	0.6	2.07	2.07	1.02	0.92	0.89	0.8
Total	98.57	98.99	98.11	100.4	100.3	100.9	99.23	100.5	98.89	100.8	99.8	100.2	100.2	99.24	100.4	98.27	100.8
Sc (ppm)	6	6	7	13	14	7	15	15	14	5	5	7	7	8	7	5	6
Be	3	2	2	2	3	3	2	3	3	2	2	4	4	3	3	3	3
V	17	17	13	65	77	34	46	46	43	10	12	33	33	50	32	16	22
Ba	216	187	261	494	559	426	515	562	482	958	1314	513	513	513	695	224	351
Sr	43	49	54	158	179	85	180	168	166	159	189	229	229	230	182	84	103
Y	32	24	34	27	30	19	29	28	31	18	21	17	17	15	13	26	19
Zr	105	102	112	232	293	127	188	204	182	120	140	115	115	99	81	63	100
Cr	<20	<20	<20	50	50	40	30	30	30	20	<20	<20	<20	<20	30	<20	20
Ni	<20	<20	<20	<20	<20	<20	<20	<20	<20	<20	<20	<20	<20	<20	<20	<20	<20
Cu	<10	10	<10	<10	10	<10	80	90	40	<10	<10	<10	<10	<10	<10	<10	<10
Zn	40	<30	<30	60	70	40	50	<30	40	40	40	70	70	50	70	<30	30
Ga	18	18	17	19	19	19	20	17	19	16	16	19	19	19	18	18	18
Ge	2	2	2	2	2	2	2	3	2	1	2	1	1	2	2	2	1

Mineralization	W					No mineralization											
	Yakishjoma			Yaguki		Hidaka		Tsukiba								Fijigatani-Kiwada	
Sample No.	YK01	YK04	YK07	YG16	YG17	HD02	HD05	TK01	TK11	TK12	TK13	TK14	TK15	TK16	IW02	IW13	
SiO_2 (wt%)	68.85	68.61	70.33	70.57	73.26	71.54	69.09	68.92	70.04	71.54	70.22	69.84	72.32	69.34	75.53	70.52	
Al_2O_3	14.32	15.15	14.04	14.85	12.44	14.81	15.4	14.9	15.3	14.28	14.3	15.02	14.67	15.18	12.94	14.11	
$Fe_2O_3(T)$*	3.52	3.51	3.58	2.92	3.07	2.36	3.34	2.82	3.32	2.85	2.55	2.45	2.62	3.06	1.17	2.95	
MnO	0.079	0.067	0.067	0.066	0.07	0.046	0.065	0.072	0.063	0.058	0.054	0.053	0.07	0.058	0.063	0.086	
MgO	0.92	0.92	0.97	0.81	0.84	1.04	1.75	0.69	0.6	0.57	0.42	0.45	0.52	0.88	0.13	0.61	
CaO	2.41	2.78	1.79	3.2	2.23	2.79	3.91	2.32	2.8	2.56	2.28	2.33	2.23	2.93	0.81	1.94	
Na_2O	3.33	3.45	2.79	3.42	2.61	3.13	3.36	3.36	3.8	3.36	3.39	3.43	3.22	4.02	3.43	3.07	
K_2O	3.62	3.28	3.82	2.92	3.64	3.06	1.88	3.84	2.75	3.12	3.62	3.93	4.18	1.51	4.34	3.33	
TiO_2	0.525	0.533	0.517	0.309	0.327	0.318	0.483	0.326	0.355	0.32	0.283	0.271	0.302	0.429	0.074	0.289	
P_2O_5	0.17	0.13	0.17	0.06	0.08	0.06	0.08	0.14	0.13	0.13	0.1	0.08	0.12	0.06	0.02	0.06	
LOI	0.92	0.73	1.11	0.7	0.74	0.79	1.23	0.78	0.94	0.46	1.14	0.91	0.67	0.94	0.75	1.23	
Total	98.65	99.14	99.17	99.81	100.4	99.94	100.6	98.18	100.1	99.24	98.35	98.78	100.9	98.42	99.26	98.18	
Sc (ppm)	9	9	8	6	6	8	11	6	8	6	6	5	6	8	4	6	
Be	4	4	7	2	1	1	2	5	2	3	2	3	3	2	4	4	
V	35	38	34	40	36	28	58	30	19	16	15	15	28	44	7	20	
Ba	189	295	248	395	523	811	315	464	867	480	766	760	828	146	235	333	
Sr	133	176	115	213	159	300	262	236	361	290	275	292	247	307	70	116	
Y	28	27	23	13	13	28	13	20	15	22	17	23	22	19	44	31	
Zr	142	183	165	96	103	62	59	122	195	145	155	164	137	205	76	86	
Cr	20	20	30	<20	<20	<20	50	<20	<20	<20	20	<20	<20	<20	<20	<20	
Ni	<20	<20	<20	<20	<20	30	<20	<20	<20	<20	<20	<20	<20	<20	<20	<20	
Cu	<10	30	<10	<10	10	<10	<10	<10	<10	<10	<10	<10	<10	<10	20	<10	
Zn	60	50	50	40	60	40	40	60	50	70	60	60	40	50	50	70	
Ga	20	20	21	15	14	13	13	19	20	20	19	19	16	20	17	18	
Ge	2	2	2	2	1	2	2	2	2	2	1	2	2	2	3	2	

Continued

As	<5	<5	<5	5	16	<5	13	45	137	<5	<5	<5	<5	<5	<5	<5	<5	9	12	<5	<5	<5	<5	<5	<5	<5	<5	<5	<5	<5	<5	<5	<5
Rb	226	206	224	113	116	192	188	181	155	95	113	169	169	145	156	263	192	176	158	201	95	126	65	59	180	97	125	110	138	143	101	176	162
Nb	14	14	11	12	12	11	13	11	13	5	5	10	10	10	9	16	9	12	12	14	5	7	2	1	9	13	9	8	9	8	9	7	9
Mo	<2	<2	<2	<2	<2	<2	<2	<2	<2	<2	<2	<2	<2	<2	<2	<2	2	<2	<2	<2	<2	<2	<2	<2	<2	<2	<2	<2	<2	<2	<2	<2	<2
Ag	<0.5	0.5	<0.5	1	1.6	1	0.9	1.3	0.8	1	0.6	<0.5	<0.5	<0.5	0.6	<0.5	0.7	0.7	0.7	0.7	<0.5	<0.5	0.7	<0.5	0.6	0.9	0.6	1	0.9	1	0.9	<0.5	<0.5
In	<0.2	<0.2	<0.2	<0.2	<0.2	<0.2	<0.2	<0.2	<0.2	<0.2	<0.2	<0.2	<0.2	<0.2	<0.2	<0.2	<0.2	<0.2	<0.2	<0.2	<0.2	<0.2	<0.2	<0.2	<0.2	<0.2	<0.2	<0.2	<0.2	<0.2	<0.2	<0.2	<0.2
Sn	8	2	2	2	4	4	2	7	3	2	1	4	4	3	3	4	4	9	11	5	2	3	2	3	8	2	3	2	3	4	4	8	17
Sb	1.2	0.5	<0.5	<0.5	0.6	1.1	<0.5	<0.5	0.5	0.8	<0.5	<0.5	<0.5	<0.5	0.8	<0.5	1.7	<0.5	<0.5	<0.5	<0.5	<0.5	<0.5	<0.5	<0.5	<0.5	<0.5	0.9	<0.5	<0.5	<0.5	<0.5	<0.5
Cs	11.7	11.1	15.2	5.6	7.9	13.8	24.1	18	19.9	2.9	3.1	8.2	8.2	9.5	8.4	17.2	9	15.7	16.5	16.9	4.1	6.6	3.5	4.6	11.9	5	5.3	2.3	3.5	5.6	7.6	5.8	13.5
La	42.3	40.7	32.9	36.2	38.1	27.2	37.2	37.1	36.4	53.1	40.1	30.6	30.6	29.6	21.8	18.9	36.4	30.4	31.7	33.8	13.7	28.8	25	5	29.9	36.4	37.1	35	37.6	43.5	50.2	15.8	28.5
Ce	88.5	83.7	68.8	73.7	77.9	55.3	76.4	75	74.4	101	77.6	59.4	59.4	54.9	39.3	37.2	70.6	63	64.5	70.7	24.6	50.9	52	10.3	60	69	74.4	67.8	76.5	82.6	96.8	33.4	58.8
Pr	9.97	9.34	8.08	8.23	9.01	6.42	8.45	8.59	8.27	11.3	8.54	6.48	6.48	6.07	4.36	4.5	7.93	6.95	7.36	7.86	2.9	5.18	6.14	1.13	6.77	7.5	8.36	7.75	8.79	8.96	11	4.15	6.44
Nd	37.3	34.6	30.6	31.9	34.2	23.2	31.8	32.1	31.7	39.9	30.2	23.3	23.3	22.3	16.1	17.3	28.9	26.6	27.8	29.8	10.5	17.3	23.7	4.4	25.3	26.8	31.8	28.7	31.7	31.4	38.8	17.4	24.7
Sm	8.1	7	7.1	6.2	6.8	4.9	6.8	6.4	6.7	6.4	5.6	4.7	4.7	4.7	3.4	4.6	5.8	5.7	5.3	6.3	2.4	2.9	5	1.2	5	4.9	6.5	5.4	6.4	5.8	7.2	5	5.8
Eu	0.29	0.29	0.34	1.2	1.28	0.76	1.07	1.06	1.05	0.84	0.93	0.79	0.79	0.8	0.67	0.44	0.63	0.75	0.99	0.75	0.8	0.61	0.92	0.71	0.89	0.99	0.91	0.95	0.97	1.06	1.17	0.31	0.56
Gd	6.2	5.6	5.9	5.3	5.8	4.1	5.7	5.3	5.8	4.8	4.3	3.5	3.5	3.4	3.1	4.5	4.9	5	4.8	5	2.1	2.3	4.3	1.3	3.8	3.9	5.1	4.6	4.9	4.7	5.5	5.7	5.5
Tb	1	0.8	1	0.8	0.9	0.6	0.9	0.8	0.9	0.7	0.6	0.5	0.5	0.5	0.5	0.8	0.7	0.8	0.8	0.8	0.3	0.4	0.7	0.2	0.6	0.5	0.8	0.7	0.7	0.7	0.7	1.1	0.9
Dy	6.2	4.9	5.8	5.2	5.2	3.7	5.4	5.1	5.6	3.7	3.7	3.2	3.2	2.9	2.5	4.3	3.8	5.5	4.8	4.6	2.1	2.3	4.2	1.7	3.7	2.9	4.4	3.6	4.2	3.8	3.7	7.3	5.6
Ho	1.2	1	1.1	1	1	0.7	1.1	1	1.1	0.7	0.8	0.6	0.6	0.5	0.4	0.8	0.7	1.1	0.9	0.9	0.4	0.5	1	0.4	0.7	0.5	0.8	0.7	0.8	0.7	0.7	1.5	1.1
Er	3.7	2.8	3.3	3.1	3.1	2.3	3.4	2.8	3.4	2.1	2.2	1.9	1.9	1.3	1.2	2.2	2	3.2	2.7	2.6	1.3	1.5	2.9	1.2	2	1.5	2.3	1.9	2.2	2.1	1.9	4.1	3
Tm	0.58	0.43	0.5	0.45	0.47	0.37	0.5	0.45	0.52	0.3	0.34	0.33	0.33	0.19	0.17	0.32	0.3	0.48	0.44	0.39	0.22	0.24	0.42	0.21	0.31	0.2	0.33	0.28	0.34	0.33	0.26	0.64	0.44
Yb	3.7	2.7	3.3	2.9	3	2.5	3.1	3	3.2	1.9	2.3	2.1	2.1	1.3	1.1	2	1.8	3.2	2.9	2.4	1.5	1.6	2.5	1.5	2	1.3	2.1	1.7	2	2.2	1.7	4.1	2.7
Lu	0.52	0.4	0.45	0.44	0.45	0.38	0.48	0.45	0.5	0.29	0.33	0.29	0.29	0.19	0.16	0.31	0.26	0.45	0.4	0.36	0.23	0.25	0.37	0.23	0.31	0.18	0.3	0.23	0.3	0.33	0.27	0.6	0.38
Hf	4.5	4.5	3.4	6.5	6.5	4	5.6	4.9	5.8	4	3.9	3.8	3.8	2.8	3	1.9	3.5	4.5	4.4	5.4	2.4	3.2	1.5	1.6	3.2	4.3	4.5	4.6	4.3	3	5	2.9	2.7
Ta	2.3	2.4	2.2	1.2	1.3	2	1.4	1.6	1.6	0.8	1	1.5	1.5	2.2	2	2.8	1.9	1.7	2.1	2	1.3	1.4	0.9	0.7	1.9	2	1.5	1.1	1.6	1.4	1.9	2.8	2.9
Tl	1.2	1	1.2	0.6	0.7	0.6	1.2	1.4	1.1	0.4	0.7	0.8	0.8	0.8	0.7	1.4	0.9	0.9	1.1	1	0.5	0.6	0.3	0.2	1.1	0.7	0.6	0.4	0.7	0.9	0.7	0.9	0.9
Pb	38	28	24	16	16	21	11	10	12	17	17	27	27	20	27	37	31	23	22	21	15	16	20	13	22	20	21	22	23	19	14	34	20
Bi	<0.4	<0.4	<0.4	<0.4	<0.4	<0.4	1.3	0.5	<0.4	<0.4	<0.4	<0.4	<0.4	<0.4	<0.4	<0.4	<0.4	<0.4	<0.4	<0.4	<0.4	<0.4	<0.4	<0.4	<0.4	<0.4	<0.4	<0.4	<0.4	<0.4	<0.4	0.5	1.7
Th	24.2	25.8	17.7	11.9	13.1	10.6	15.5	15.8	15.7	14.8	11.4	12.7	12.7	14.8	11.4	13.8	22.8	12.6	12.5	13.5	10.5	15.4	10.7	0.6	12	8.9	11	8.7	11.1	15.7	21.8	11.1	18.6
U	5.1	4.8	3.7	2.5	2.6	3.9	3.7	4.8	4.1	2	1.8	2.8	2.8	4.3	2.8	3.5	3.3	5.4	5	4	2.2	2.4	1.6	1.7	3.4	1.3	2.2	1.1	1.6	5.5	2.7	2.8	5

*Total Fe as Fe_2O_3

According to the A/NK vs. A/CNK diagram (Figure 3), A/CNK values for the granitic rocks associated with Pb-Zn, Mo or Cu-Fe mineralization range from 0.7 to 1.1; this means that they are metaluminous to slightly peraluminous. They were all classified as I-type granitic rocks [17] . The A/CNK values for the granitic rocks associated with Sn or W mineralization and barren granitic rocks range from 1.0 to 1.3, which corresponds to peraluminous. Most of these rocks are I-type granitic rocks but some are S-type granitic rocks.

4.2. Tectonic Setting

Based on the Rb vs. (Yb + Ta) diagram (Figure 4) [18] , most of the granitic rocks analyzed were classified as volcanic arc granitic rocks. However, some granitic rocks associated with Sn or W mineralization (some Obira and some Fujigatani-Kiwada mining districts samples) were plotted in syn-collision granitic rock field. Because the Rb content of the Tanzawa granitic rocks is extremely low, they were classified as M-type granitic rocks [19] . As the Chichibu granitic rocks are also depleted in Rb, although not as depleted as the Tanzawa granitic rocks, they were also classified as M-type granitic rocks.

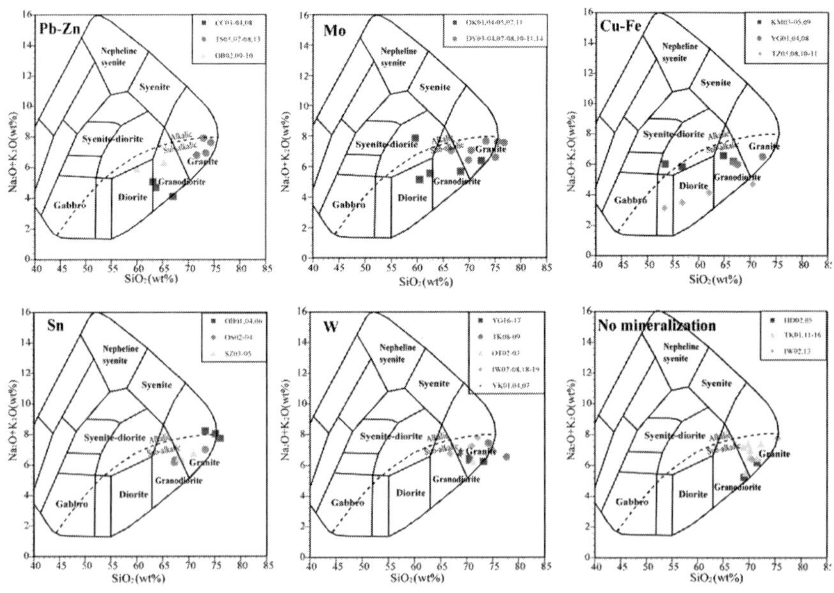

Figure 2. Classification of the analyzed granitic rocks from Japan: Total alkali vs. SiO_2 diagrams (classification boundaries after Cox et al. [15] ; Wilson [16]).

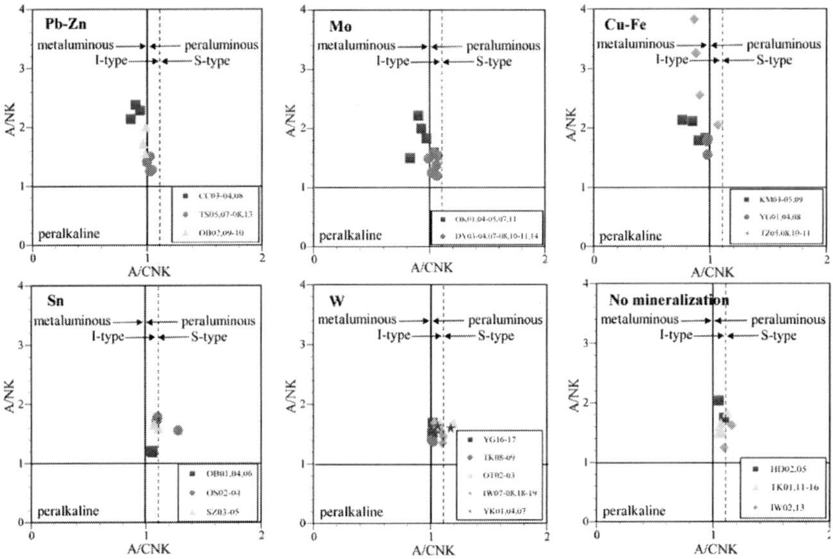

Figure 3. Classification of the analyzed granitic rocks from Japan: $Al_2O_3/(Na_2O + K_2O)$ vs. $Al_2O_3/(CaO + Na_2O + K_2O)$ diagrams.

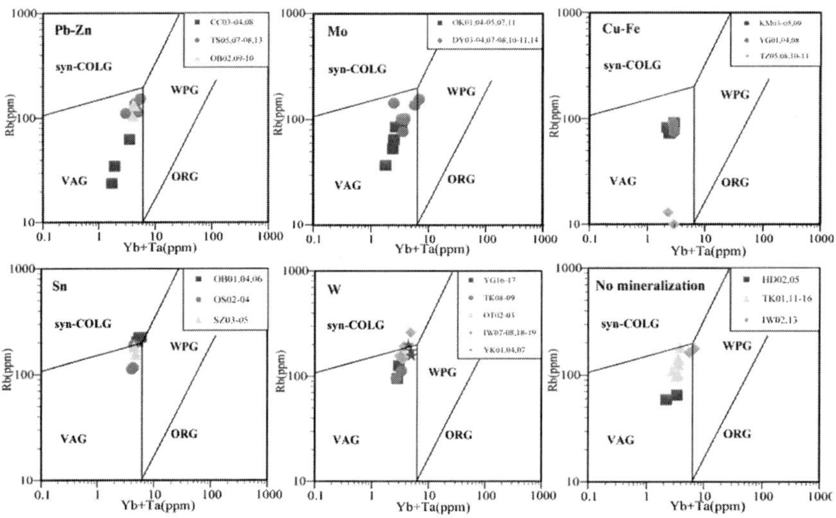

Figure 4. Rb vs. (Yb + Ta) tectonic discrimination diagrams (after Pearce et al. [18]) for the granitic rocks from Japan. ORG: ocean ridge granitic rocks, VAG: volcanic arc granitic rocks, WPG: within plate granitic rocks, and syn-COL: syn-collision granitic rocks.

According to the Sr/Y vs. Y diagram by Defant and Drummond [20] (Figure 5), most of the granitic rocks analyzed in this study are non-adakitic, but a few adakitic rocks are present in the granitic rocks in the Ohkawame and Kamaishi mining districts [21] . For the adakitic rocks analyzed, all the Al_2O_3 content are greater than 15 wt% except for one sample (OK11 at 14.59 wt%). On the SiO_2vs. Zr/TiO_2 diagram (Figure 6) [22] , most of these adakitic rocks were classified as true adakitic rocks which were produced by the partial melting of young subducting oceanic crust.

4.3. REE Patterns

Chondrite-normalized REE patterns for the granitic rocks organized by type of associated metal are shown in Figure 7.

All of the granitic rocks associated with Sn or W mineralization show negative Eu anomalies. The Eu anomalies for the biotite granites associated with Sn mineralization in the Obira mining district are particularly large. Barren granitic rocks also show negative Eu anomalies except for the hornblende granodiorite in the Toyonidake area in the southern part of the Hidaka metamorphic belt. This indicates that the Sn- or W-associated granitic rocks and barren ones were formed by the fractional crystallization of plagioclase from granitic magma under reducing conditions where divalent Eu was stable. The hornblende granodiorite in the Toyonidake area shows a pronounced positive Eu anomaly and it is relatively depleted in REEs. These observations suggest that plagioclase may have accumulated during hornblende granodiorite formation.

Conversely, the majority of the granitic rocks with Pb-Zn, Mo, or Cu-Fe mineralization do not show pronounced negative Eu anomalies. This suggests that these rocks formed under oxidizing conditions where trivalent Eu was stable. The samples associated with Pb-Zn, Mo, or Cu-Fe mineralization that do show significant negative Eu anomalies, the Uchiyama sample TS05 (Pb-Zn) and the Yamasa sample DY04 (Mo), are rich in SiO_2 (>74%) and relatively enriched in REEs. This suggests that they formed from differentiated magmas. The REE patterns for the granitic rocks from the Tanzawa area (TZ samples, Cu-Fe) are almost flat, consistent with the features of M-type granitic rocks [20] [23] . The REE patterns for the granitic rocks from the Chichibu mining district (CC samples, Pb-Zn) are also relatively flat and show the features of M-type granitic rocks.

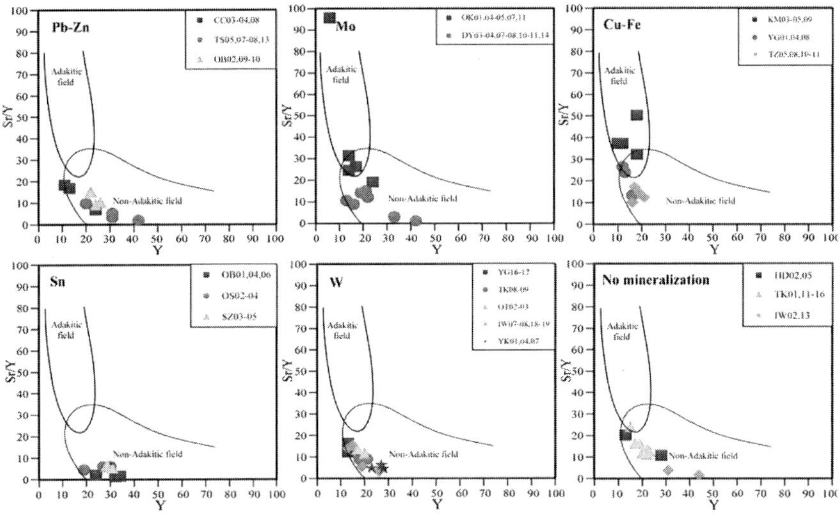

Figure 5. Sr/Y vs. Y diagrams (after Defant and Drummond [20]) for the granitic rocks from Japan.

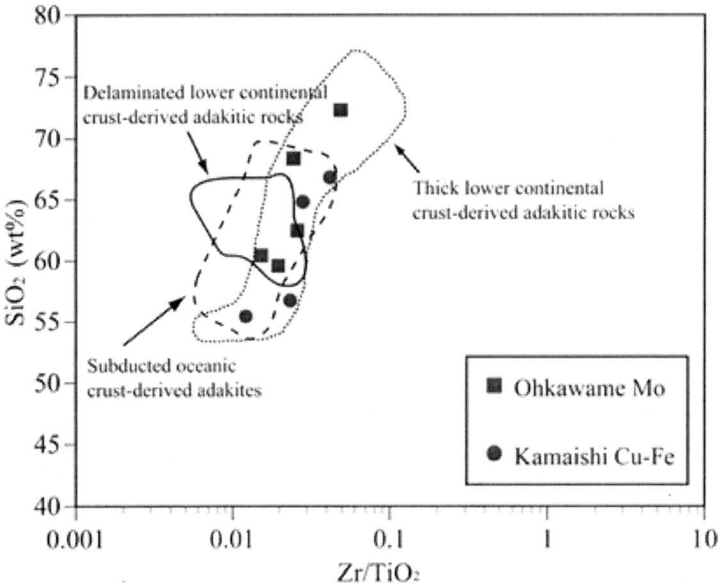

Figure 6. SiO_2 vs. Zr/TiO_2 discrimination diagram for adakitic rocks (after Wang et al. [22]), showing plots for the granitic rocks from the Kamaishi and Ohkawame mining districts, Japan.

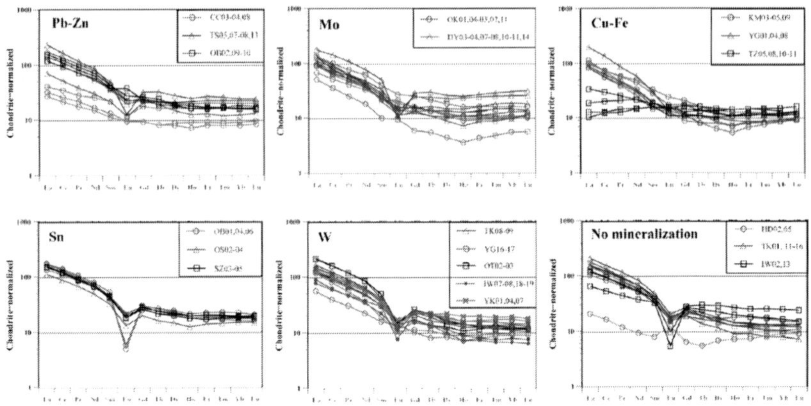

Figure 7. Chondrite-normalized REE patterns for the granitic rocks from Japan.

Figure 8 shows the relationship between the magnitude of the negative Eu anomaly and the differentiation index. The vertical axis in Figure 9 shows the distance (Eu/Eu*) from the line connecting Sm and Gd on the chondrite-norma- lized REE pattern to Eu in Figure 8. The Eu/Eu* is defined by the following equation:

$$\frac{Eu}{Eu^*} = \log\left(\frac{Eu}{Eu_{cn}}\right) - \frac{1}{2}\left(\log\left(\frac{Sm}{Sm_{cn}}\right) + \log\left(\frac{Gd}{Gd_{cn}}\right)\right)$$

where REE with subscript of cn indicates each REE concentration of chondrite.

Figure 8 indicates that the magnitude of the negative Eu anomalies increase as the differentiation index increases.

4.4. Interelement Correlations and SiO_2 Content

Figure 9 shows graphs of SiO_2 vs. the major element oxides and graphs of SiO_2 vs. trace elements. The relationships illustrated in Figure 9 are discussed below.

The Al_2O_3, Fe_2O_3(T), MgO, MnO, CaO, TiO_2, and P_2O_5 show a tendency to decrease with increasing SiO_2, whereas, despite a lot of scatter in the data, K_2O seems to increase. The granitic rocks in the Tanzawa area and the Chichibu mining district have low K_2O and are high in CaO (Table 2), suggesting that they are M-type granitic rocks. It seems that Na_2O content increase slightly or is almost constant with increasing SiO_2.

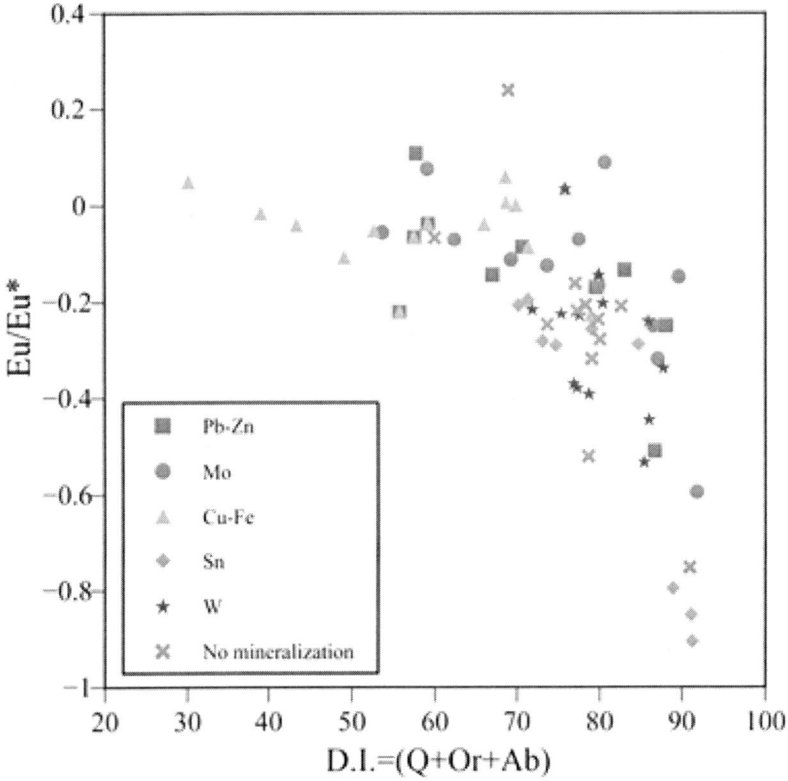

Figure 8. Eu/Eu* vs. differentiation index (D.I.) diagram for the granitic rocks from Japan.

The elements V, Zn, Sr, Eu, and Sc decrease as SiO_2 increases. Vanadium is commonly incorporated in magnetite but is also partitioned into mafic minerals like pyroxenes, amphiboles, and biotite [24] . Because Sc^{3+} is close to Fe^{2+} and Mg^{2+} in ionic radius [25] , it is thought that Sc^{3+} is incorporated in mafic minerals in a similar way to Zn and V. In contrast, Sr is incorporated in plagioclase by substituting Ca.

Yttrium, Nb, Rb, Ta, Pb, Th, U, and REEs other than Eu are elements that increase as SiO_2increases. Among the REEs, Ce, Pr, Nd, and Sm behave like La, and Ga, Tb, Dy, Ho, Er, Tm, Yb, and Lu behave like Y. The granitic rocks associated with Sn or W mineralization and barren granitic rocks have higher SiO_2 contents compared to the granitic rocks associated with Pb-Zn, Mo, or Cu-Fe mineralization, so there is a tendency for La, Y, Nb, Rb, Ta, Pb, Th, U, and REEs other than Eu to be enriched in the Sn- or W-associated granitic rocks and barren granitic rocks (e.g., [1] [26] [27] [28]).

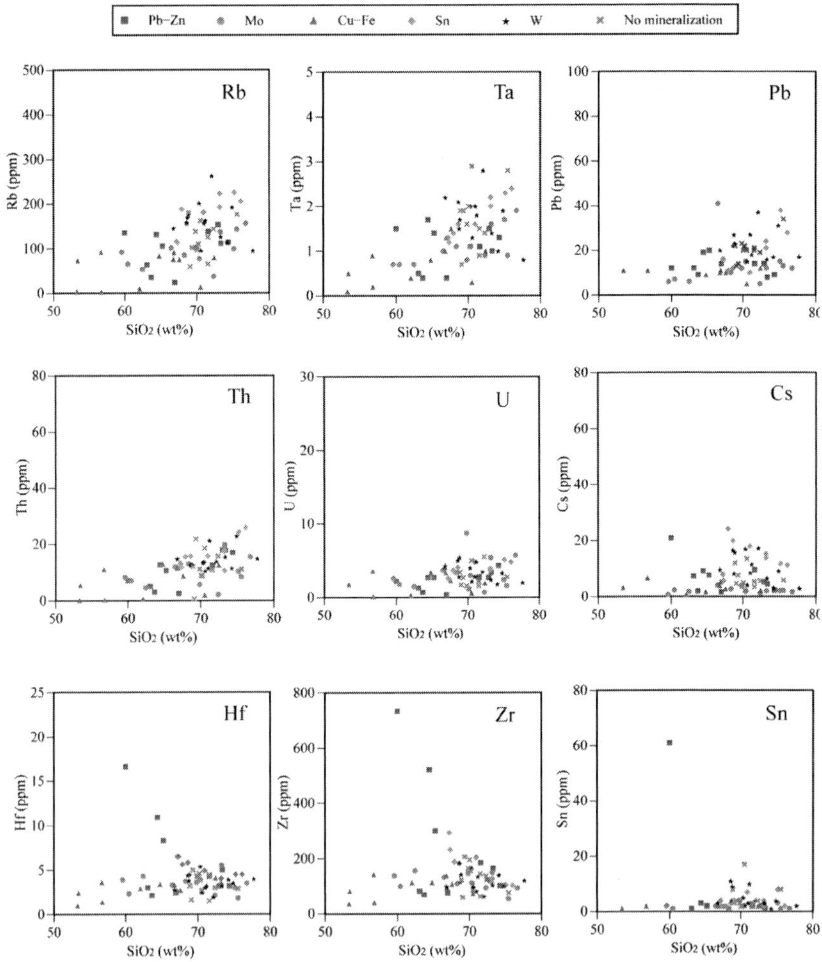

Figure 9. Variation diagrams for the granitic rocks from Japan. Data points are coded for the metal association.

Zirconium, Sn, and Hf increase with increasing SiO_2, but start to decrease when the SiO_2 content exceeds 70 wt%. Zirconium concentrates in the magma during crystal differentiation. However, when the SiO_2 content reaches around 70 wt%, it is likely that magma will be saturated with Zr and zircons will precipitate [29] . Thus above 70 wt% SiO_2, the Zr content in the remaining magma tends to decrease. As shown in Figure 9, the distribution of Hf data points mimics the distribution of the Zr points, indicating that Hf is concentrated in zircon. Tin behaves similarly to Zr and Hf and it is concentrated in the magma as crystal differentiation progresses. Magmas will be saturated with Sn when the SiO_2 content reaches about 70 wt%. A granite porphyry accompanied by Pb-Zn mineralization in the Obira mining district is enriched in Zr, Hf, and Sn compared with other granitic rocks.

The behavior of Ba on Figure 9 is complicated. On the whole, Ba shows a similar behavior to the REEs including La with a larger ionic radius, and Ba content increases as SiO_2 increases. At the same time, however, the Ba content tends to decrease similarly to the alkali earth elements such as Ca and Sr as SiO_2 increases.

5. CONCLUSIONS

Major and trace element contents were determined for 66 samples of granitic rocks from 15 different areas in Japan. The samples were from intrusions that were either associated with Pb-Zn, Mo, Cu-Fe, Sn, or W mineralization or were barren. Examination of the analytical results for the studied granitic rocks allows the following conclusions to be drawn.

The studied granitic rocks associated with Pb-Zn, Mo, or Cu-Fe mineralization were classified as granite to diorite and were magnetite-series and I-type granitic rocks. However, the granitic rocks in the Tanzawa area and the Chichibu mining district were classified as M-type granitic rocks. The granitic rocks in the Ohkawame and Kamaishi mining districts are adakitic rocks and were generated from the partial melting of a subducting oceanic plate. It is thought that all the granitic rocks formed in a volcanic arc tectonic setting. Only a few REE analyses for rocks of this type show negative Eu anomalies.

Most of the studied granitic rocks associated with Sn or W mineralization and the barren granitic rocks were classified as granite and were ilmenite-series and I-type granitic rocks. A few of these rocks are S-type granitic rocks. It is likely that most of the granitic rocks formed in a volcanic arc tectonic setting. None of these granitic rocks is adakitic and most of their chondrite-normalized REE pattern show negative Eu anomalies.

The contents of K_2O, La, Y, Nb, Rb, Ta, Pb, Th, U, and REEs other than Eu in the granitic rocks increase with increasing SiO_2. These elements tend to be enriched in the high SiO_2 granitic rocks associated with Sn or W mineralization and the high SiO_2 barren granitic rocks.

The major components other than Na_2O and K_2O, Sr, Eu, and Sc in the granitic rocks tend to decrease as SiO_2 increases.

Barium has two different trends. For some sets of analyses, Ba tends to increase with increasing SiO_2, for other sets Ba decreases with increasing SiO_2.

The contents of Zr, Sn, and Hf increase with increasing SiO_2 up to approximately 70 wt%, but then decrease when the SiO_2 content exceeds 70 wt%. This phenomenon is thought to be related to the melt becoming saturated with these elements.

ACKNOWLEDGEMENTS

This research was supported in part by a Grant-in-Aid for Scientific Research of the Japan Society for the Promotion of Science (Grant No. 16K06931: E. Uchida). The granitic rocks analyzed in this study were collected together with members of our laboratory: T. Hosono, K. Hirose, Y. Katagiri, S. Katsuno, N. Matsushima, M. Miyoshi, S. Nishi, S. Ozawa, F. Sato, T. Shibayama, and S. Takeda.

CITE THIS PAPER

Uchida, E., Osada, K. and Nakao, K. (2017) Major and Trace Element Chemical Compositional Signatures of Some Granitic Rocks Related to Metal Mineralization in Japan. Open Journal of Geology, 7, 559-576. https://doi.org/10.4236/ojg.2017.74038

REFERENCES

1. Ishihara, S. (1977) The Magnetite-Series and Ilmenite-Series Granitic Rocks. Mining Geology, 27, 293-305.
2. Ishihara, S. (1981) The Granitoid Series and Mineralization. Economic Geology Anniversary, 75, 458-484.
3. Uchida, E., Endo, S. and Makino, M. (2007) Relationship between Solidification Depth of Granitic Rocks and Formation of Hydrothermal Ore Deposits. Resource Geology, 57, 47-56.
4. Shimada, N., Ohyama, Y. and Ikemi, H. (2000) Characterization of Magnetically Zoned Pluton, Tsushima, Japan. Resource Geology, 50, 65-73.
5. Miyahisa, M. (1958) Contact Metasomatic Lead-Zinc Ore Deposits, Obira Mine, Oita Prefecture, Japan (Part III). Journal of Educational Technology & Society, 26, 284-289.
6. Kanisawa, S. and Katada, M. (1988) Characteristics of Early Cretaceous Igneous Activity, Kitakami Mountains, Northeast Japan. Earth Science, 42, 220-236.

7. Ishihara, S. (1971) Major Molybdenum Deposits and Related Granitic Rocks in Japan. Geological Survey of Japan, 239, 183.
8. Hamabe, S. and Yano, T. (1979) Geological Structure of the Kamaishi Minig District, Iwate Prefecture, Japan. Mining Geology, 26, 93-104.
9. Uchida, E. (1986) Relation between Zonal Arrangements of Skarns and Temperatures of Formation at the Kamaishi Mine, Northeastern Japan. Mining Geology, 36, 195-208.
10. Takita, R. (1974) Petrography and the Plutonic History of the Tanzawa Tonalite Complex. Journal of the Geological Society of Japan, 80, 505-523.
11. Ogawa, K. and Shida, A. (1975) Scheelite Mineralization in the Shin-Bu Tungsten Deposit of the Yaguki Mine. Mining Geology, 25, 109-122.
12. Nakada, S. (1978) Geology of the Osuzuyama Acid Rocks, Miyazaki Prefecture, Kyushu, Japan. Journal of the Geological Society of Japan, 84, 243-256.
13. Takahashi, Y. (1982) Geology of the Granitic Rocks in the Tsukuba Area. Journal of the Geological Society of Japan, 88, 77-184.
14. Osanai, Y., Owada, M., Shimura, T., Kawasaki, T. and Hensen, B.J. (1997) Crustal Anatexis and Related Acidic Magma Genesis in the Hidaka Metamorphic Belt. The Journal of the Geological Society of Japan, 47, 29-42.
15. Cox, K.G., Bell, J.D. and Pankhurst, R.J. (1979) The Interpretation of Igneous Rocks. Allen and Unwin, London, 450 p.
16. Wilson, M. (1989) Igneous Petrogenesis. A Global Tectonic Approach. Unwin Hyman, London, 466 p.
17. Chapell, B.W. and White, A.J.R. (1974) Two Contrasting Granite Types. Pacific Geology, 8, 173-174.
18. Pearce, J.A., Harris, N.B.W. and Tindle, A.D. (1984) Trace Element Discrimination Diagrams for the Tectonic Interpretation of Granitic Rocks. Journal of the Petrology, 25, 956-983.
19. Takahashi, M. (1989) Neogene Granitic Magmatism in the South Fossa Magna Collision Zone, Central Japan. Modern Geology, 14, 127-143.
20. Defant, M.J. and Drummond, M.S. (1990) Derivation of Some Modern Arc Magmas by Melting of Young Subducted Lithosphere. Nature, 347, 662-665.
21. Tsuchiya, N. and Kanisawa, S. (1994) Early Cretaceous Sr-Rich Silicic Magmatism by Slab Melting in the Kitakami Mountains, northeast Japan. Journal of Geophysical Research, 99, 22205-22220.
22. Wang, Q., Xu, J.F., Jian, P., Bao, Z.W., Zhao, Z.H., Li, C.F., Xiong, X.L. and Ma, J.L. (2006) Petrogenesis of Adakitic Porphyries in an Extensional Tectonic Setting, Dexing, South China: Implications for the Genesis of Porphyry Copper Mineralization. Journal of Petroleum Science and Engineering, 47, 119-144.
23. Ishihara, S., Kanaya, H. and Terashima, S. (1976) Genesis of the Neogene Granitoids in the Fossa Magna Region in Japan. Marine Sciences Monthly, 8, 523-528.
24. Mason, B. (1966) Principles of Geochemistry. 3rd Edition, Wiley, New York, London, 329 p.
25. Shannon, R.D. and Prewitt, C.T. (1970) Revised Values of Effective Ionic Radii. Acta Crystallogr, 26, 1046-1048.
26. Ishihara, S. (1971) Modal and Chemical Compositions of the Granitic Rocks Related to the Major Molybdenum and Tungsten Deposits in the Inner Zone of South-west Japan. Journal of the Geological Society of Japan, 77, 441-452.
27. Lehmann, B. (1982) Metallogeny of Tin: Magmatic Differentiation versus Geoche-mical Heritage. Economic Geology, 77, 50-59.
28. Sillitoe, R. (2010) Porphyry Copper System. Economic Geology, 105, 3-41.
29. Ishihara, S. (2014) On the Variation Pattern of Zirconium Contents in Some Japanese Granitoids. Journal of the Society of Resource Geology, 64, 127-132.

CHAPTER 6

Arctic Ocean Gas Hydrate Stability in a Changing Climate

ABSTRACT

Recent estimations suggest that vast amounts of methane are locked in the Arctic Ocean bottom sediments in various forms of gas hydrates. A potential feedback from a continued warming of the Arctic region is therefore the release of methane to the atmosphere. This study addresses the relationship between a warming of the Arctic ocean and gas hydrate stability. We apply a theoretical model that estimates the base of the gas hydrate stability zone in the Arctic Ocean considering different bottom water warming and sea level scenarios. We model the present day conditions adopting two different geothermal gradient values: 30 and 40°C/km. For each geothermal gradient value, we simulate a rise and a decrease in seafloor temperature equal to 2°C and in sea level equal to 10 m. The results show that shallow gas hydrates present in water depths less than 500 m would be strongly affected by a future rise in seafloor temperature potentially resulting in large amounts of gas released to the water column due to their dissociation. We estimate that the area, where there could be complete gas hydrate dissociation, is about 4% of the area where there are the conditions for gas hydrates stability.

1. INTRODUCTION

Vast quantities of methane are trapped in oceanic hydrate deposits, and there is a concern that a rise in the ocean temperature will induce dissociation of these hydrate accumulations, potentially releasing large amounts of methane into the atmosphere [1, 2]. In this context, the Arctic should be considered a critical area in a warming climate because massive amounts of methane are currently locked as gas hydrates in ocean sediments and in permafrost that could be released [3–6]. Moreover, in the Arctic ocean, the gas hydrate stability zone is especially sensitive to climate change because the degree of temperature change is greater than at lower latitudes [4]. For example, the gas hydrate stability zone along continental slopes is found to be sensitive to even small changes in ocean bottom temperature [4], and the Arctic Ocean most likely has a larger gas hydrate stability zone compared to other oceans because of its cold water and low geothermal gradients [7]. However, it is still unclear whether the recent discoveries of methane emissions in the Arctic shelf and slope region are a result of better mapping and detection methods or whether they reflect a natural variability or environmental conditions following the present climate change.

Gas hydrates are solid crystalline compounds, in which gas molecules are lodged within the clathrate crystal lattice [8]. They occur when water and adequate quantity of gas combine under conditions of relatively high pressure and low temperature. The gas hydrate stability zone thickness varies widely within Arctic region depending on factors such as ocean bottom water/permafrost temperature, geothermal gradient, salinity of the water, and composition of gas [9].

Both permafrost-associated gas hydrates and the shallowest part of the deepwater marine gas hydrate system are susceptible to dissociation (breakdown to methane and water) under conditions of a warming Arctic climate. The dissociation of these hydrate deposits, which release large quantities of methane, could have dramatic climatic consequences that in turn lead to further atmospheric and oceanic warming through accelerated decomposition of the remaining hydrates [7]. This positive-feedback mechanism has been proposed as a significant contributor to rapid and significant climate changes in the late Quaternary period [10–13]. The "Clathrate Gun Hypothesis" [14] suggests that past increases in water temperatures near the seafloor may have induced such a large-scale dissociation, with the methane spike and isotopic anomalies reflected in polar ice cores and in benthic foraminifera [7]. Methane from dissociating hydrates in ocean sediments has been proposed to never reach the atmosphere, either due to its conversion to carbon dioxide in the water column or from being sequestered in the biosphere [15]. Isotopic studies of air trapped in ice cores question the source of the atmospheric methane during these rapid temperature excursions [16]. Furthermore, Nisbet [17], while suggesting methane as a factor in moderating glaciations, proposed that the methane spikes may follow, not lead, climate change. Simulations of deep ocean hydrates show that deep (>1,000 m) hydrates may be relatively insensitive to ocean temperature changes on short time scales. Dickens [18] calculated that the volume of the GHSZ associated with marine hydrates halved during the Paleocene-Eocene thermal maximum with a temperature increase of 5°C. Here, we define the area where there are the conditions for gas hydrates stability, and we model the effects of warmer ocean bottom temperatures and increased seafloor pressure on the gas hydrate stability zone (GHSZ) in the Arctic Ocean, as shown in Figure 1. Moreover, we focus on the intersection between the base of gas hydrate stability zone and the seafloor, identifying the areas where the complete dissociation of gas hydrates could occur and estimating their extension. The effects on the GHSZ from increased bottom temperatures have previously been modeled in the Antarctic Peninsula [19–21]. The base of GHSZ is calculated as the intersection between the geothermal curve and the gas hydrate stability curve [22, 23] by using the Sloan formula [8]. The following parameters were adopted to estimate the base of gas hydrate stability zone as described in the following sections: geothermal gradient, seafloor temperature, bathymetry, and gas composition.

2. GAS HYDRATES AND THE ARCTIC

Many authors have demonstrated that climate change is occurring and is particularly pronounced in the Arctic [24–26]. Because the Arctic plays a special role in global climate, these changes will also affect the rest of the world [27]. Global climate change is likely amplified in the Arctic by several positive feedbacks, including surface albedo decreased by ice and snow melting, temperature anomalies trapped near the surface by atmospheric stability, and change magnification by cloud dynamics [28, 29]. Consequently, the temperature in the Arctic is increasing at a rate of two to three times that of the global temperature estimated to be 0.4°C over the past 150 years [30].

It is clear that the Arctic can contribute significantly to the amplification of climate changes. So, many efforts should be devoted to improve the knowledge about gas hydrate reservoir located in this area. It is clear that difficulty in carrying out seismic, bathymetric, and geological surveys has yielded little data about gas hydrates in the ice-covered Arctic basin [31]. Investigations are still producing surprising results indicating that our understanding of gas hydrate formation and distribution is incomplete [32].

The available estimates of methane in Arctic hydrates ranging between 1,013 and 1,016 m^3 in onshore regions [33–36] are based mainly on speculations of constant distribution of hydrates within sediments across significant areas in a large depth interval using regional hydrate content percentage. An approach to obtain additional information about gas hydrate deposits, located in this area, could be the theoretical modeling of the GHSZ. In this context, we define the area where there are the conditions for gas hydrate stability zone and we simulate the effects of climate changes on hydrate. The base of the gas hydrate stability zone is calculated as the intersection between the geothermal gradient curve and gas-gas hydrate-pore water equilibrium curve. Moreover, we identify the area where the complete dissociation of gas hydrate could occur, if they were present.

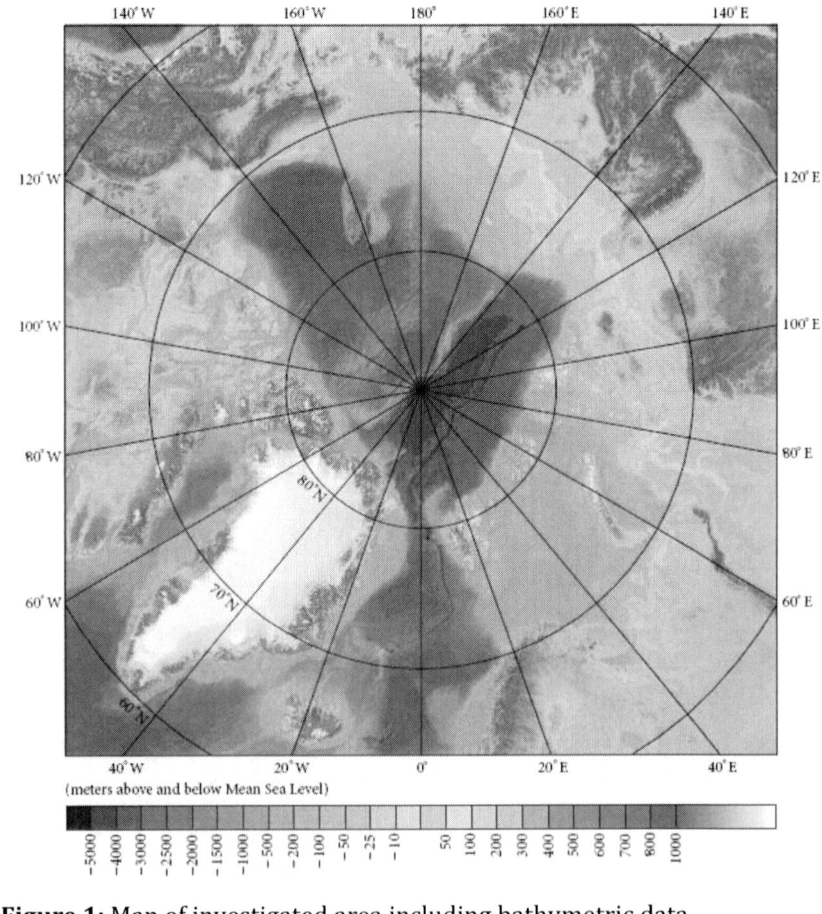

Figure 1: Map of investigated area including bathymetric data.

As already mentioned, gas hydrate stability is mainly affected by ocean bottom water/permafrost temperature, geothermal gradient, salinity of the water, and composition of gas. It is known that the presence of only a small percentage of higher hydrocarbons (such as ethane and propane) shifts the phase boundary to higher temperature (at constant pressure). The effect is that the base of GHSZ at a given temperature is shifted to greater depths [37, 38].

3. GAS HYDRATES MODELING

We consider the main factors affecting the GHSZ and simulate their variation analyzing the effects on gas hydrate stability. By using bathymetric data, sea bottom temperature, and geothermal gradient and considering that the natural gas is methane, we evaluate the theoretical GHSZ, similarly to what was recently evaluated in the Antarctic region by [19–22]. The base of GHSZ was calculated as the intersection between the geothermal curve and the gas hydrate stability curve.

3.1. Seafloor Temperature

The seafloor temperature data are downloaded from the National Oceanographic Data Center website (http://www.nodc.noaa.gov/cgi-bin/OC5/WOA09/woa09.pl). The available water temperatures were sampled at 33 different depths (from 0 to 5,500 m) on one-degree fields. The intersection between the 33 levels and the bathymetric data allow us to extract the seafloor temperature. In this way, we obtained the distribution of seafloor temperature in study area shown in Figure 2.

3.2. Pressure Based on Water Depth

An important factor considered in GHSZ estimation is the water depth because its variation strongly affects the GHSZ thickness especially for deep gas hydrate reservoir (>500 m). The adopted digital bathymetric model was developed by the International Bathymetric Chart of the Arctic Ocean (IBCAO) project (http://www.ngdc.noaa.gov/mgg/bathymetry/ arctic/ downloads.html). The IBCAO map (version 3) is a seamless bathymetric topographic digital terrain model that incorporates three primary data sets: all available bathymetric data; the GMTED 2010 topographic data; and the World Vector Shoreline for coastline representation [39]. The grid cell size of the adopted version is km on a polar stereographic projection. The bathymetric data are translated in terms of pressure considering the average water density equal to 1,046 kg/m^3. We extrapolate the average water density as suggested by Giorgetti et al. [40] for Antarctic Peninsula because of the similarity of water temperature and polar climate conditions.

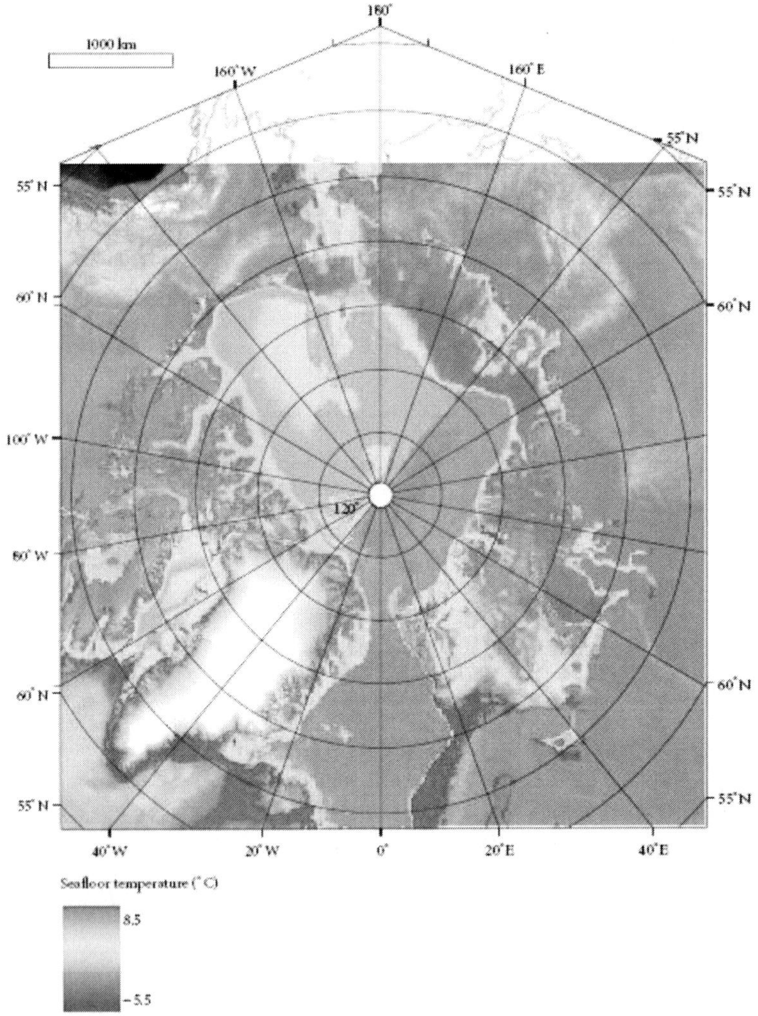

Figure 2: Seafloor temperature (°C) (data processed after downloading from National Oceanographic Data Center website).

3.3. Gas Composition

Knowledge about gas composition of gas hydrates is always poor. In our study, we suppose that the component of gas is pure methane. In this way, we consider the condition for which the base of GHSZ is shallower, so more sensitive to climate changes.

3.4. Geothermal Gradients

The Arctic region lacks geothermal gradient data because of the difficulty to acquire data. So, we consider the following geothermal gradients: 30 and 40°C/km, considering different studies [41, 42]. As suggested by different authors [43, 44], the geothermal gradients can highly range due to different factors such as presence of mud diapirs or salt domes, inversion of the basin resulting in shallow basement rocks, and presence of faults causing focused fluid flow [42]. As previously demonstrated [45] zones characterized by lower geothermal gradients, are more affected by climate change concerning gas hydrate stability.

The base of the GHSZ is estimated as the intersection between geothermal curve and the gas hydrate stability curve calculated using the Sloan formula [8]. First, we simulate the present day condition using the collected parameters to evaluate the zone where there are the conditions for gas hydrate stability. For each geothermal gradient value, we then simulate a rise and a decrease in seafloor temperature equal to 2°C and in sea level equal to 10 m. The rise in seafloor temperature of 2°C is in agreement with temperatures during the latest Paleocene thermal maximum. In fact, the end of the Paleocene epoch was marked by an abrupt episode of global warming coincident with a large perturbation of the global carbon cycle [46]. Within 10 to 30 ky [47, 48], temperatures increased 4 to 5°C in the tropics [49, 50], 6 to 8°C in high latitudes [51], and 4 to 5°C in the deep ocean [52, 53]. The decrease in seafloor temperature is considered to simulate a glacial period. Moreover, results from recent simulations coupling circulation, atmospheric circulation, and atmospheric chemistry (publications in CCSM, 2006) indicate that, under current climate conditions and 1% yr increase in atmospheric CO_2, the temperature at the seafloor would rise by 1°C over the next 100 years and possibly by another 3°C in the following centuries [7]. Regarding the sea-level change, the range of future climate-induced sea-level rise remains uncertain with continued concern that large increases in the twenty-first century cannot be ruled out. The mean rate of sea-level rate measured by high-precision altimeter amounts to mm/year from 1993 to 2009, suggesting that sea-level rate is accelerating compared to the past [54]. Moreover, there is a huge uncertainty due to the potential melting of glaciers, and the possible role of the Greenland and West Antarctic ice sheets, which could amplify the changes in sea level. The Intergovernmental Panel on Climate Change (IPCC) has pointed out that for a warming over Greenland of 5.5°C the Greenland ice sheet contribution is about 3 m in 1,000 years, while for a warming of 8°C the contribution is about 6 m, with the ice sheet being largely eliminated. So, considering 10 m of sea-level variation, it is likely to predict an extreme scenario, but not unrealistic [30]. Anyway, several models have suggested that the sea level dropped by about 100 m during the last glacial maximum [55]. Rohling et al. [56] suggested that the sea level was 120 m below its present value during the last glacial maximum reducing the water-covered shelf and marginal sea areas by approximately 50% [57–59].

4. DISCUSSIONS

We show the modeled base of the GHSZ for all 10 cases analyzed in Figure 3, showing the depth of the base of GHSZ from seafloor. We modeled 5 cases for each geothermal gradient value. Panel (a) shows present day conditions; panel (b) shows temperature decrease of 2 degrees; and panel (c) shows temperature increase, while panels (d) and (e) show decrease and increase of sea level, respectively. As previously mentioned, the base of GHSZ is calculated as the intersection between the gas hydrate stability curve and the geothermal curve. The base of GHSZ is commonly inferred from seismic reflection profiles using a bottom simulating reflector (BSR), which is the boundary between hydrate-bearing sediment above and gassy sediment beneath [60]. As expected, a rise in seafloor temperature causes gas hydrate in the region to become unstable (panel (c) in Figure 3). This could cause the base of the GHSZ to move upward resulting in thinner GHSZ and subsequent methane releases. On the contrary, a decrease in seafloor temperature could cause the base of the GHSZ to move downward. In this condition, the GHSZ could increase in thickness as shown in panel (b). A negative feedback from a warming climate is a rising sea level from melting of glaciers and ice sheets, which would increase the pressure at the seafloor. This could result in the deepening of the base of the GHSZ extending the zone where there are the conditions for gas hydrate stability, as shown in column (e) of Figure 3. On the contrary, the decrease of sea level lead to a shallower base of the GHSZ. However, the effect of 10 m sea-level change on gas stability zone is very low compared to those of a 2-degree temperature change. In Figure 4, we show a profile offshore Svalbard Island, considering the case for which the geothermal gradient is equal to 30°C/km.

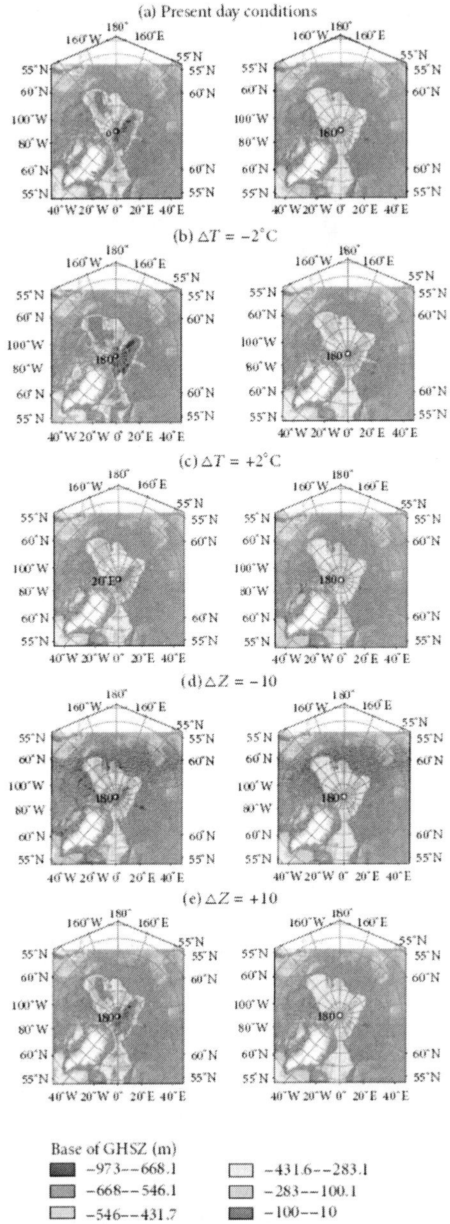

Figure 3: Depth of base of GHSZ from seafloor. In each panel, the left modeling was performed by using a geothermal gradient equal to 30°C/km, while, for the right panel, the geothermal gradient was 40°C/km. For details see the text.

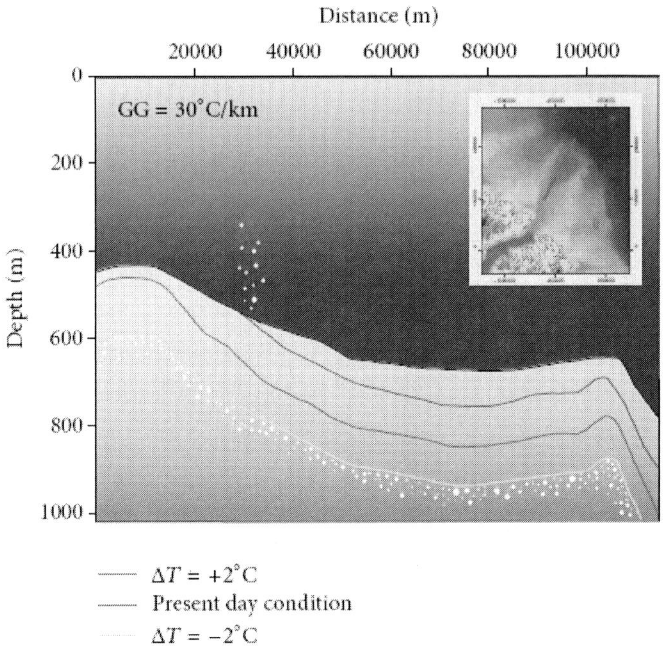

Figure 4: Selected profile offshore Svalbard Island. The black line represents the present day topography along the profile. The shown simulations are performed using a geothermal gradient (GG) equal to 30°C/km.

The black line represents the bathymetry along the profile. Below it, we show three hypothetical future scenarios, which will affect gas hydrate stability: (a) present day condition (purple line), (b) considering a rise of 2°C (green line), and (c) considering a decrease of 2°C (yellow line). The modeling shows clearly that a rise in temperature could affect the gas hydrate reservoir in this location and cause dissociation. In fact, where the depth is deeper than 500 m, as shown in Figure 4, an increase of 2°C in seafloor temperature could cause the complete dissociation of gas hydrate releasing methane in the water column and, maybe, in the atmosphere. Moreover, the free gas trapped below the gas hydrate reservoir could also be released. When there is a partial dissociation of gas hydrates, that is, gas hydrate in water depth shallower than 500 m, because of the increase in temperature or/and the decrease of sea level, there may also be a positive feedback on climate. In fact, the free gas released from gas hydrate dissociation, will be trapped in the sediments because gas hydrate will act as cap. To contribute to the estimation of the effects of gas hydrate on climate change, we identify the area, where there could be a complete gas hydrate dissociation because of a rise of 2°C in sea bottom temperature, and estimate its extension equal to 3.1 × 10^5 km^2 (Figure 5). This area represents about 4% of the area where there are the conditions for gas hydrates stability. To estimate this area, we calculate the present day intersection between the seafloor and the base of gas hydrate zone and that one considering a rise in seafloor temperature equal to 2°C. In the zone included between the two intersections, there could be a complete dissociation of gas hydrate as shown in Figure 5.

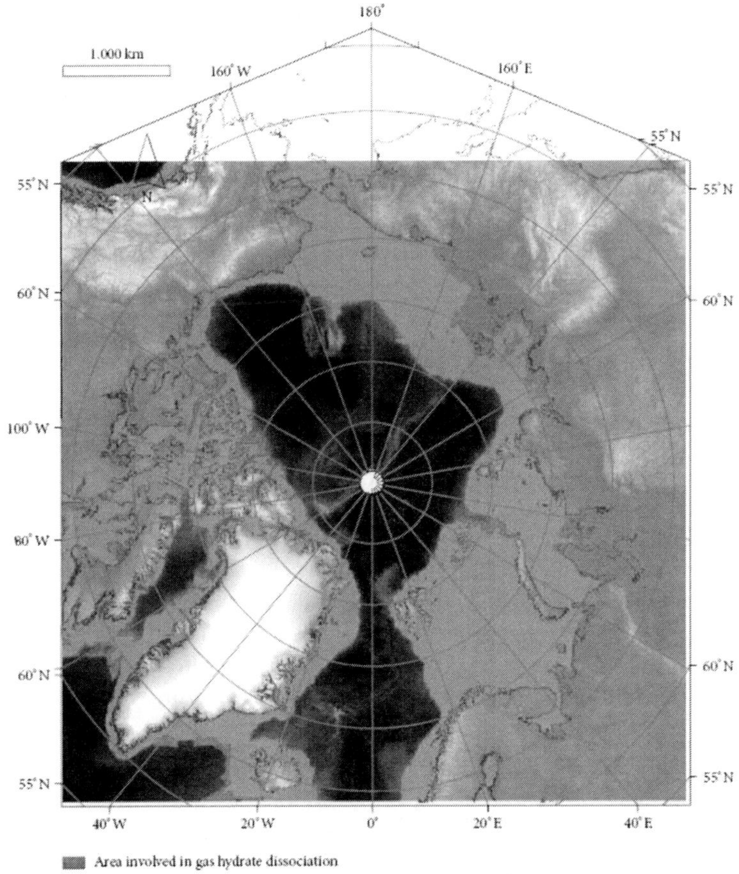

Figure 5: In the red areas, complete dissociation of gas hydrate could talk place with a rise in the seafloor temperature of 2°C. This simulation is performed using a geothermal gradient of 30°C/km.

5. CONCLUSIONS

Our modeling confirms that the shallow gas hydrate deposits (water column < 500 m) could be strongly affected by climate change, mainly by rise in temperature and that methane could be released due to gas hydrate dissociation. It is important to underline that several factors mitigate the impact of the liberated CH_4 on the sediment-ocean-atmosphere system. The released CH_4 may dissolve in local pore waters, remain trapped as gas, or rise toward the seafloor as bubbles. Up to 90% or more of the CH_4 that reaches the sulfate reduction zone in the near-seafloor sediments may be consumed by anaerobic CH_4 oxidation [61–65]. Moreover, methane emitted at the seafloor only rarely survives the trip through the water column to reach the atmosphere. In fact, within the water column, oxidation by aerobic microbes is an important sink for dissolved CH_4 over some depth ranges and at some locations (e.g., [66]).

Since authors predict the current and near-future evolution of sea-level rise remain in the range of few decimeters [54, 67], it is mainly the warming of the oceans that has the potential to significantly alter the stable conditions of methane hydrates [68]. So many efforts should be devoted to improve the knowledge about the Arctic gas hydrate reservoirs, because they could play an important role regarding world warming changes, considering that over the past decade the Arctic has been warming up twice as fast as the global average [69]. Moreover, we should improve the knowledge about shallow gas hydrate deposits (water column < 500 m) which could be completely dissociated determining a change of the characteristics of the underground with unpredictable effects. Finally, it is worth to mention that the Arctic is an area logistically difficult to investigate due to the extreme environmental conditions. So, modeling is indispensable for planning and identifying the key areas, where more accurate field analyses should be performed. When more detailed data about gas hydrate occurrence will be available, our modeling can be used to extrapolate local information at global scale.

ACKNOWLEDGMENT

The authors wish to thank Manuela Sedmach for the graphic support.

REFERENCES

1. M. T. Reagan and G. J. Moridis, "Large-scale simulation of methane hydrate dissociation along the West Spitsbergen Margin," Geophysical Research Letters, vol. 36, no. 23, Article ID L23612, 2009.
2. M. T. Reagan, G. J. Moridis, S. M. Elliott, and M. Maltrud, "Contribution of oceanic gas hydrate dissociation to the formation of Arctic Ocean methane plumes," Journal of Geophysical Research C, vol. 116, no. 9, Article ID C09014, 2011.
3. N. Shakhova, I. Semiletov, A. Salyuk, V. Yusupov, D. Kosmach, and Ö. Gustafsson, "Extensive methane venting to the atmosphere from sediments of the East Siberian Arctic Shelf," Science, vol. 327, no. 5970, pp. 1246–1250, 2010.
4. G. K. Westbrook, K. E. Thatcher, E. J. Rohling et al., "Escape of methane gas from the seabed along the West Spitsbergen continental margin," Geophysical Research Letters, vol. 36, no. 15, Article ID L15608, 2009.
5. B. Buffett and D. Archer, "Global inventory of methane clathrate: sensitivity to changes in the deep ocean," Earth and Planetary Science Letters, vol. 227, no. 3-4, pp. 185–199, 2004.
6. E. A. G. Schuur, J. G. Vogel, K. G. Crummer, H. Lee, J. O. Sickman, and T. E. Osterkamp, "The effect of permafrost thaw on old carbon release and net carbon exchange from tundra," Nature, vol. 459, no. 7246, pp. 556–559, 2009.
7. M. T. Reagan and G. J. Moridis, "Oceanic gas hydrate instability and dissociation under climate change scenarios," Geophysical Research Letters, vol. 34, no. 22, Article ID L22709, 2007.
8. E. D. Sloan, Clathrate Hydrates of Natural Gases, Marcel Dekker, New York, NY, USA, 2nd edition, 1998.
9. E. D. Sloan, "Natural gas hydrate phase equilibria and kinetics: understanding the state-of-the-art," Revue de l'Institut Francais du Petrole, vol. 45, no. 2, pp. 245–266, 1990.
10. J. P. Kennett, K. G. Cannariato, I. L. Hendy, and R. J. Behl, "Carbon isotopic evidence for methane hydrate instability during quaternary interstadials," Science, vol. 288, no. 5463, pp. 128–133, 2000.
11. E. J. Brook, T. Sowers, and J. Orchardo, "Rapid variations in atmospheric methane concentration during the past 110,000 years," Science, vol. 273, no. 5278, pp. 1087–1091, 1996.
12. J. P. Severinghaus, T. Sowers, E. J. Brook, R. B. Alley, and M. L. Bender, "Timing of abrupt climate change at the end of the Younger Dryas interval from thermally fractionated gases in polar ice," Nature, vol. 391, no. 6663, pp. 141–146, 1998.
13. R. J. Behl, J. P. Kennett, K. G. Cannariato, and L. L. Hendy, "Methane hydrates and climate change: the clathrate gun hypothesis," AAPG Bulletin, vol. 87, no. 10, article 1693, 2003.
14. J. P. Kennett, K. G. Cannariato, I. L. Hendy, and R. J. Behl, Methane Hydrates in Quaternary Climate Change: The Clathrate Gun Hypothesis, vol. 54, American Geophysical Union, Washington, DC, USA, 2002.
15. K. A. Kvenvolden, "Potential effects of gas hydrate on human welfare," Proceedings of the National Academy of Sciences of the United States of America, vol. 96, no. 7, pp. 3420–3426, 1999.
16. T. Sowers, "Late quaternary atmospheric CH4 isotope record suggests marine clathrates are stable," Science, vol. 311, no. 5762, pp. 838–840, 2006.
17. E. G. Nisbet, "Have sudden large releases of methane from geological reservoirs occurred since the Last Glacial Maximum, and could such releases occur again?" Philosophical Transactions of the Royal Society A, vol. 360, no. 1793, pp. 581–607, 2002.
18. G. R. Dickens, "The potential volume of oceanic methane hydrates with variable external conditions," Organic Geochemistry, vol. 32, no. 10, pp. 1179–1193, 2001.
19. U. Tinivella, "The seismic response to overpressure versus gas hydrate and free gas concentration," Journal of Seismic Exploration, vol. 11, no. 3, pp. 283–305, 2002.
20. M. Giustiniani, D. Accettella, M. F. Loreto, U. Tinivella, and F. Accaino, "Geographic information system—an application to manage geophysical data," in Proceedings of the

71st European Association of Geoscientists and Engineers Conference and Exhibition, pp. 340–344, Society of Petroleum Engineers, Amsterdam, The Netherlands, June 2009. View at Scopus
21. U. Tinivella, M. Giustiniani, and D. Accettella, "BSR versus climate change and slides," Journal of Geological Research, vol. 2011, Article ID 390547, 6 pages, 2011.
22. U. Tinivella and J. M. Carcione, "Estimation of gas-hydrate concentration and free-gas saturation from log and seismic data," The Leading Edge, vol. 20, no. 2, pp. 200–203, 2001.
23. M. F. Loreto, U. Tinivella, F. Accaino, and M. Giustiniani, "Offshore Antarctic Peninsula gas hydrate reservoir characterization by geophysical data analysis," Energies, vol. 4, no. 1, pp. 39–56, 2011.
24. D. Antoniades, P. Francus, R. Pienitz, G. St-Onge, and W. F. Vincent, "Holocene dynamics of the Arctic's largest ice shelf," Proceedings of the National Academy of Sciences of the United States of America, vol. 108, no. 47, pp. 18899–18904, 2011.
25. B. M. Vinther, S. L. Buchardt, H. B. Clausen et al., "Holocene thinning of the Greenland ice sheet," Nature, vol. 461, no. 7262, pp. 385–388, 2009.
26. R. G. Graversen, T. Mauritsen, M. Tjernström, E. Källén, and G. Svensson, "Vertical structure of recent Arctic warming," Nature, vol. 451, no. 7174, pp. 53–56, 2008.
27. ACIA, "Arctic Climate Impact Assessment," ACIA Scientific Report, Cambridge University Press, Cambridge, UK, 2005.
28. J. Overpeck, K. Hughen, D. Hardy et al., "Arctic environmental change of the last four centuries," Science, vol. 278, no. 5341, pp. 1251–1256, 1997.
29. P. R. Gent, "Preface to special issue on Community Climate System Model (CCSM)," Journal of Climate, vol. 19, no. 11, article 2121, 2006.
30. S. Solomon, D. Qin, M. Manning et al., Eds., "Climate change 2007: the physical science basis. Contribution of working group I to the fourth assessment report of the Intergovernmental Panel on Climate Change," Assessment Report, Cambridge University Press, Cambridge, UK, 2007.
31. C. J. Willmott, S. M. Robeson, and K. Matsuura, "Short communication A refined index of model performance," International Journal of Climatology, vol. 32, pp. 2088–2094, 2012.
32. G. A. Cherkashov and T. Matveeva, "Undiscovered Arctic gas hydrates: permafrost relationship and resource evaluation," in Proceedings of the American Geophysical Union, Fall Meeting, San Francisco, Calif, USA, December 2011, abstract #GC52A-01.
33. V. M. Dobrynin, Y. P. Korotajeva, and D. V. Plyuschev, "Gas hydrates-one of the possible energy sources," in Long-Term Energy Resources, R. G. Meyer and J. C. Olson, Eds., pp. 727–729, Pitman, Boston, Mass, USA, 1981.
34. K. A. Kvenvolden, "Methane hydrate—a major reservoir of carbon in the shallow geosphere?" Chemical Geology, vol. 71, no. 1–3, pp. 41–51, 1988.
35. G. J. MacDonald, "Role of methane clathrates in past and future climates," Climatic Change, vol. 16, no. 3, pp. 247–281, 1990.
36. R. D. McIver, "Gas hydrates," in Long-Term Energy Resources, R. G. Meyer and J. C. Olson, Eds., pp. 713–726, Pitman, Boston, Mass, USA, 1981.
37. G. D. Holder, R. D. Malone, and W. F. Lawson, "Effects of gas composition and geothermal properties on the thickness and depth of natural-gas-hydrate zones," Journal of Petroleum Technology, vol. 39, no. 9, pp. 1147–1152, 1987.
38. E. D. Sloan and C. A. Koh, Clathrate Hydrates of Natural Gases, CRC Press, Boca Raton, Fla, USA, 3rd edition, 2008.
39. M. Jakobsson, L. Mayer, B. Coakley et al., "The International Bathymetric Chart of the Arctic Ocean (IBCAO) version 3.0," Geophysical Research Letters, vol. 39, no. 12, Article ID L12609, 2012.
40. Giorgetti, A. Crise, R. Laterza, L. Perini, M. Rebesco, and A. Camerlenghi, "Water masses and bottom boundary layer dynamics above a sediment drift of the Antarctic Peninsula pacific margin," Antarctic Science, vol. 15, no. 4, pp. 537–546, 2003.
41. F. W. Jones, J. A. Majorowicz, J. Dietrich, and A. M. Jessop, "Geothermal gradients and heat flow in the Beaufort-Mackenzie Basin, Arctic Canada," Pure and Applied Geophysics, vol. 134, no. 3, pp. 473–483, 1990.

42. S. Chand, J. Mienert, K. Andreassen, J. Knies, L. Plassen, and B. Fotland, "Gas hydrate stability zone modelling in areas of salt tectonics and pockmarks of the Barents Sea suggests an active hydrocarbon venting system," Marine and Petroleum Geology, vol. 25, no. 7, pp. 625–636, 2008.
43. C. Ruppel, G. R. Dickens, D. G. Castellini, W. Gilhooly, and D. Lizarralde, "Heat and salt inhibition of gas hydrate formation in the northern Gulf of Mexico," Geophysical Research Letters, vol. 32, no. 4, pp. 1–4, 2005.
44. T. Bugge, G. Elvebakk, S. Fanavoll et al., "Shallow stratigraphic drilling applied in hydrocarbon exploration of the Nordkapp Basin, Barents Sea," Marine and Petroleum Geology, vol. 19, no. 1, pp. 13–37, 2002.
45. U. Tinivella and M. Giustiniani, "Variations in BSR depth due to gas hydrate stability versus pore pressure," Global and Planetary Change, vol. 100, pp. 119–128, 2013.
46. J. A. Higgins and D. P. Schrag, "Beyond methane: towards a theory for the Paleocene-Eocene thermal maximum," Earth and Planetary Science Letters, vol. 245, no. 3-4, pp. 523–537, 2006.
47. U. Röhl, T. J. Bralower, R. D. Norris, and G. Wefer, "New chronology for the late Paleocene thermal maximum and its environmental implications," Geology, vol. 28, no. 10, pp. 927–930, 2000.
48. K. A. Farley and S. F. Eltgroth, "An alternative age model for the Paleocene-Eocene thermal maximum using extraterrestrial 3He," Earth and Planetary Science Letters, vol. 208, no. 3-4, pp. 135–148, 2003.
49. J. C. Zachos, M. W. Wara, S. Bohaty et al., "A transient rise in tropical sea surface temperature during the Paleocene-Eocene thermal maximum," Science, vol. 302, no. 5650, pp. 1551–1554, 2003.
50. Tripati and H. Elderfield, "Paleoclimate: deep-sea temperature and circulation changes at the Paleocene-Eocene thermal maximum," Science, vol. 308, no. 5730, pp. 1894–1898, 2005.
51. E. Thomas and N. J. Shackleton, "The latest Paleocene benthic foraminiferal extinction and stable isotope anomalies," in Correlation of the Early Paleogene in Northwest Europe, R. O. Knox, R. M. Corfield, and R. E. Dunay, Eds., vol. 101 of Geological Society Special Publication, pp. 401–441, Geological Society of London, London, UK, 1996.
52. J. C. Zachos, K. C. Lohmann, J. C. G. Walker, and S. W. Wise, "Abrupt climate change and transient climates during the Paleogene: a marine perspective," Journal of Geology, vol. 101, no. 2, pp. 191–213, 1993.
53. J. P. Kennett and L. D. Stott, "Abrupt deep-sea warming, palaeoceanographic changes and benthic extinctions at the end of the Palaeocene," Nature, vol. 353, no. 6341, pp. 225–229, 1991.
54. R. J. Nicholls and A. Cazenave, "Sea-level rise and its impact on coastal zones," Science, vol. 328, no. 5985, pp. 1517–1520, 2010.
55. M. J. Bentley, "Volume of Antarctic ice at the last glacial maximum, and its impact on global sea level change," Quaternary Science Reviews, vol. 18, no. 14, pp. 1569–1595, 1999.
56. E. J. Rohling, M. Fenton, F. J. Jorissen, P. Bertrand, G. Ganssen, and J. P. Caulet, "Magnitudes of sea-level lowstands of the past 500,000 years," Nature, vol. 394, no. 6689, pp. 162–165, 1998.
57. H. W. Menard and S. M. Smith, "Hypsometry of ocean basin provinces," Journal of Geophysical Research, vol. 71, no. 18, pp. 4305–4325, 1966.
58. W. R. Peltier, "Ice age paleotopography," Science, vol. 265, no. 5169, pp. 195–201, 1994.
59. W. Ludwig, P. Amiotte-Suchet, and J.-L. Probst, "Enhanced chemical weathering of rocks during the last glacial maximum: a sink for atmospheric CO2?" Chemical Geology, vol. 159, no. 1–4, pp. 147–161, 1999.
60. T. H. Shipley, M. H. Houston, R. T. Buffler et al., "Seismic reflection evidence for widespread occurrence of possible gas-hydrate horizons on continental slopes and rises," American Association of Petroleum Geologists Bulletin, vol. 63, pp. 2204–2213, 1979.
61. K.-U. Hinrichs and A. Boetius, "The anaerobic oxidation of methane: new insights in microbial ecology and biochemistry," in Ocean Margin Systems, G. Wefer, D. Billet, D.

Hebbeln, B. B. Jorgensen, M. Schlüter, and T. C. E. V. Weering, Eds., pp. 457–477, Springer, Berlin, Germany, 2002.

62. T. Treude, A. Boetius, K. Knittel et al., "Anaerobic oxidation of methane at Hydrate Ridge (OR)," Geochimica et Cosmochimica Acta, vol. 67, article A491, 2003.
63. W. S. Reeburgh, "Oceanic methane biogeochemistry," Chemical Reviews, vol. 107, no. 2, pp. 486–513, 2007.
64. K. Knittel and A. Boetius, "Anaerobic oxidation of methane: progress with an unknown process," Annual Review of Microbiology, vol. 63, pp. 311–334, 2009.
65. C. D. Ruppel, "Methane hydrates and contemporary climate change," Nature Education Knowledge, vol. 3, no. 10, article 29, 2011.
66. S. Mau, D. L. Valentine, J. F. Clark, J. Reed, R. Camilli, and L. Washburn, "Dissolved methane distributions and air-sea flux in the plume of a massive seep field, Coal Oil Point, California," Geophysical Research Letters, vol. 34, no. 22, Article ID L22603, 2007.
67. J. A. Church, N. J. White, L. F. Konikow et al., "Revisiting the Earth's sea-level and energy budgets from 1961 to 2008," Geophysical Research Letters, vol. 38, no. 18, Article ID L18601, 2011.
68. Biastoch, T. Treude, L. H. Rüpke et al., "Rising Arctic Ocean temperatures cause gas hydrate destabilization and ocean acidification," Geophysical Research Letters, vol. 38, no. 8, Article ID L08602, 2011.
69. J. Hansen, M. Sato, R. Ruedy et al., "Dangerous human-made interference with climate: a GISS modelE study," Atmospheric Chemistry and Physics, vol. 7, no. 9, pp. 2287–2312, 2007.

CHAPTER 7

Paleoenvironmental Distribution Patterns of Sequence Stratigraphy in the Paleocene-Eocene Deposits of Sekonj Trough Event (Central Iran)

Massih Afghah*, Masoud Ghafari, Vahid Ahmadi

Geology Department, Islamic Azad University, Shiraz, Iran

ABSTRACT

The Paleocene-Eocene successions were studied in Sekonj and Abaregh sections and located in South of Kerman in order to determine the microfacies, depositional environment and sequence stratigraphy. The deposits in the studied area are mainly composed of limestone, sandstone, conglomerate and shale. The obtained data from the field and laboratory observations led to identification of 11 microfacies which represent the coastal, tidal flat, lagoon, barrier and proximal open marine which were deposited in a carbonate-sili- ciclastic ramp environment. The sedimentology evidences, vertical analysis of microfacies, facies and relative sea level changes represent that the Paleocene- Eocene succession deposits have two 3rd depositional sequences in the studied area. The sequence 2 cannot be recognized completely due to the lack of outcrop.

Sequence stratigraphic studies show an upward shallowing trend corresponding to the global sea level changes from the base to the upward of the succession.

Keywords: Paleocene-Eocene, Facies, Depositional Environment, Sequence Stratigraphy, Sekonj Trough

1. INTRODUCTION

The Paleogene period begins in the most parts of Central Iran and the southern slopes of Alborz with the presence of post-orogenic clastic sediments and with the apparent angular unconformity, but a sedimentary trough was formed at the time of Laramide event and between Sekonj area and Shishtu mountain which predominantly is consists of fine-grained clastic sediments and shows the continuous marine conditions and the continuous, gradual transition of the Cretaceous sediments to Paleogene deposits in the region with its considerable thickness [1] . Laramide event formed the Kerman conglomerate during the late Mesozoic era in most areas of Kerman which represents the marine regression conditions [2] . Laramide orogeny event (late Cretaceous-Paleogene) has an important role in geology of Iran and in the studied area. The initiation, termination and effects of the Laramide orogeny event is not similar in anywhere of Iran and even in an identified structural sedimentary basin [3] , but its effects are visible as folding, metamorphic upland and flysch basins on individual structural zones of Iran. Although few studies have been done in the recent years on the stratigraphy of Paleocene-Eocene carbonate sediments in Kerman areas by [4] - [10] , but the Sekonj sedimentary trough is not considered. In this study, facies, depositional environment and sequence stratigraphy of the Sekonj and Abaregh sections in Sekonj sedimentary trough will be analyzed.

2. MATERIALS AND METHOD

The studied area is located in Central Iran structural zone which its event is associated with the history of this micro-continent. Central Iran micro-continent is considered as a part of central Iran. The sekonj section with a thickness of 414 meter is located at 30°01'48''N and 57°29'22''E, in the north east of sekonj village, in the south ridge of Polvar mountain and in the east flank of anticline, about 25 km of Mahan city and 50 km south East of Kerman city (Figure 1) [11] . The Paleocene-Eocene succession in this section is mainly characterized by laminated to thick-bedded, fine to medium-grained, red to brown sandstones bearing olistolith and sandy limestone with interbedded thin layers of shale and alternating with limestones and sandstones. Abaregh section with a thickness of 325 meter is located at 29°20'51''N and 57°56'34''E, about 54 km of Bam city and in the north west of Abaregh village. The lithostratigraphic units of this section are including: 1) the succession of thick-bedded limestones with sandy limestones, 2) limestone and sandy limestone with interbedded shale, sandstone and conglomerate, 3) sandstone, shale and with interbedded limestone. The Sekonj and Abaregh sections are located in the southeast part of Kerman and 1: 100,000 geological maps (Figure 1) [12] [13] and [14] . The general structure of these sections is a flank of the antiform which the oldest rock unit belongs to the pectin-limestone, shale and marl with the veins of gypsum and is cretaceous in age. The youngest rock unit belongs to the Quaternary sediments in the studied area. The biostratigraphy studies show Thanetian-Ypersian age for the studied successions. In terms of tectonically, the studied area is located in the east of Gowk fault (the old name of Golbaf) and is affected by the performance of the fault. Generally, the general structure of Gowk fault system is as a thrust with a ramp geometry and dextral movement (Figure 2) [15] and [16] . Generally, the function of these structures led to many changes in sedimentary facies at short distances, depositional gaps, discontinuities and sudden resection of geological structures in Sekonj sedimentary trough (Figure 2).

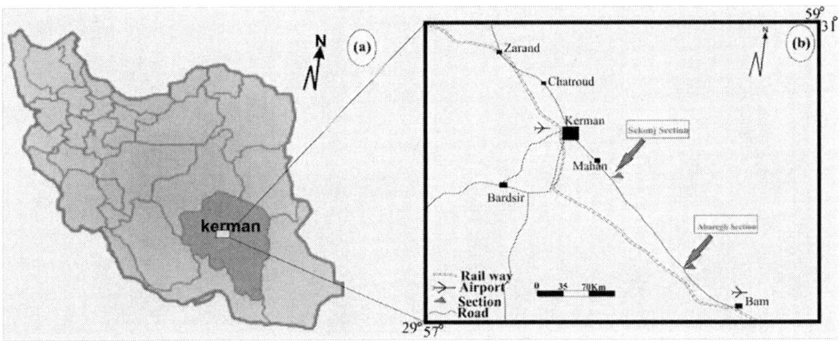

Figure 1. (a) A view of Kerman province position in Iran. (b) Geographical position and access roads of the studied section.

In the field studies, based on the observed changes in genus, color, bedding and grain size, 250 samples were taken from the Sekonj and Abaregh successions and thin sections were prepared from them. Abundance of skeletal and non- skeletal allochems were measured based on the compared charts of [17] . The silisiclastic rocks are named by the classification of [18] and the carbonates rocks are basis on the nomenclature of [19] . The results of field observations and laboratory studies led to reconstruction of the sedimentary paleoenvironment and sedimentary model in the studied sections. Analysis of the sequence stratigraphy in the studied sections is based on the sequence stratigraphy principles of [20] [21] and [22] .

3. SEDIMENTARY FACIES

Field observations, analysis of lithology and microscopic studies led to the identification of 5 facies belts in the form of 11 microfacies and facies (Figure 3 and Figure 4). Thus, the facies and microfcaies which were introduced based on Flugel model (Flugel, 2010) located in the coastal (Mf1, Mf2, Mf3, Mf4), tidal flat (Mf5, Mf7), lagoon (Mf8, Mf9), barrier (Mf10, Mf11) and open marine (Mf13) facies belts.

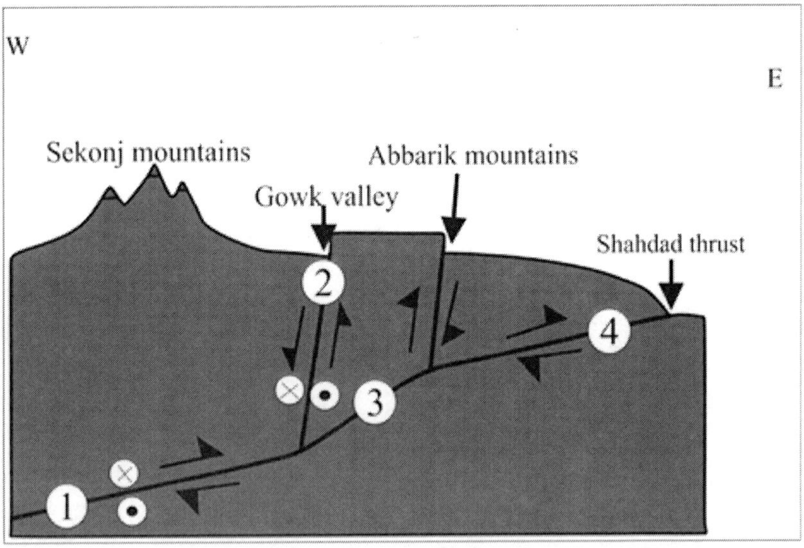

Figure 2. The structure of Gowk fault. The gentle slope parts (1, 4) which are separated with a steep slope stair (3), show the net thrust parts and led to uplift Abbarik mountains and the occurrence of normal oblique fault (2) in the Gowk valley.

3.1. Coastal Facies Belt

Coastal facies belt are located in the middle to upper part of the Sekonj section and in the lower to the upper part of the Abaregh section and are composed of thin to thick layers. The most observed coastal facies belongs to Mf2.

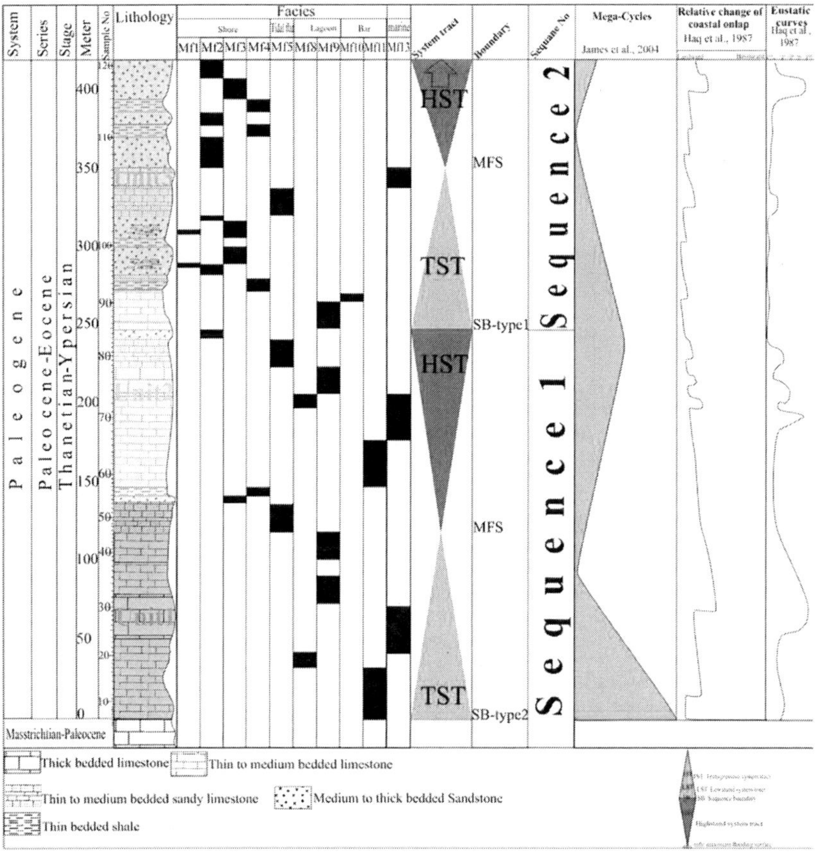

Figure 3. Lithostratigraphy column, facies, sequence stratigraphy and the comparison of global sea level changes in the Sekonj section.

3.2. Conglomerate (Mf1)

The sediments of this facies is visible as lens-shaped between the sandstone successions and has a limited lateral extension in the Sekonj section, while it can be seen as identified successions in the lower part of the Abaregh section. The size of the main components are variable between 5 to 100 mm in this facies which the pebbles and cables are located in the fine to coarse-grained matrix according to the classification of Tucker [23] . The pebbles and cables have contact with together in this lithofacies and show the grain-supported fabric with low matrix according to the classification of Pettijohn. In the field studies, the facies is light to dark gray and can be seen with secondary calcite veins, carbonate, metamorphic and volcaniclastic fragments with poor sorting and moderate to good roundness and probably polymodal (probably olistolith) (Figure 5). The conglomerate is polymictic and has a significant percentage of olistolith grain (granule size, granular and massive bedding). The thickness of fragments in the conglomerate succession are reduced laterally and formed in a channel environment. The conglomerate facies probably have been deposited by the processes of river floods and the slow flow of the river. The poor sorting is attributed to the rapid sedimentation and the high accumulation of grains can be the result of interaction between the grains during the flood flows [24] .

3.3. (Quartzarenite) (Mf2)

This petrofacies is just in the Sekonj section and has been appeared in the various parts of the facies column. The facies probably have been existed in the Abaregh section which has been converted to other sandstone facies during the time. The microfacies is thin to thick-bedded and light brown to white color in the field. The grains of this microfacies are fine to medium-grained, subrounded with good sorting and textural maturity is moderate to good. Alternation of thin layers of shale and sandstone were observed in this microfacies. Its textural maturity is good and monocrystalline quartz grains with parallel extinction and slightly undulatory extinction are with small amount of calcite cement (Figure 5). The contact between quartz grains are straight, sutured and concavo-convex contacts which are caused by pressure solution. The frequency of quartz grains are more than 95% and are monocrystalline with parallel extinction. Based on the cross and parallel laminations, the sandstones were deposited in a high energy environment which represented a foreshore environment. The absence of fairly good roundness and sorting with olistolith grains indicate rapid tectonic activities which prevent the formation of quartz grains with good sorting. The large amounts of monocrystalline quartz with parallel extinction than other types of grains show the inter cratonic origin [25] and the low frequency of feldspars along with rock fragments can be assumed a source area with strong chemical weathering and long transportation [26] . The presence of cross-bedding, trace fossils and its relationship with other facies represent the deposition of this facies in shallow water environment with different energy [27] .

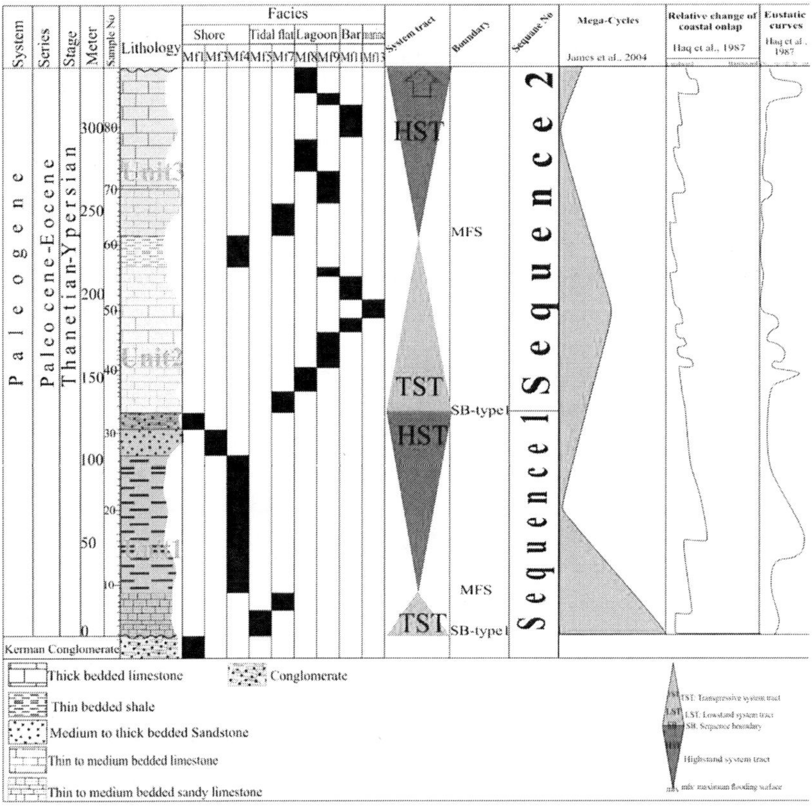

Figure 4. Lithostratigraphy column, facies, sequence stratigraphy and the comparison of global sea level changes in the Abaregh section.

3.4. (Sublitharenite) (Mf3)

This petrofacies was observed in the upper part of the succession in the Sekonj section, but has been limited in the lower part of the Abaregh section. Its mineralogical maturity is sub-mature to mature due to the high percentage of quartz and very small amount of chert debris and also the high proportion of dolomite, muscovite and feldspar than to the total sedimentary rock fragments and feldspar. The mineralogical maturity of the microfacies is sub-mature due to the high percentages of quartz and chert than to the total sedimentary rock fragments and feldspar. In the field observations, the facies is brown to light brown-colored in surface sample, thin to medium bedded and bearing cross bedding with alternating shale facies. The frequency of quartz particles are 70% to 80%, fine to medium-grained and sub-rounded (Figure 5). The grains of the microfacies are fine to medium-grained. In terms of petrological characteristics, is similar to the previous petrofacies which the main difference is the amount of rock fragments. The amounts of rock fragments can be changed between 5% - 15%. The quartz particles are often monocrystalline but some quartz are polycrystalline. The textural maturity of this petrofacies is often immature and sometimes sub-mature. The types of sandstone petrofacies and their textural characteristics such as the types of main grains and the proportion between them, the mature textural maturity and good sorting represent a shore face environment.

Figure 5. (a), (b) Conglomerate facies. The grain size of this facies is pebble to silt. (c), (d) Quartz arenite petrofacies. The quartz grains are aboundant (XPL). (e), (f) Sublitharenite petrofacies. The quartz and feldspar grains are visible in this petrofacies (XPL). (g) The laminations in the sandstone successions. (h) Cross-lamination in the sandstone succession. (i), (j) A view of shale succession in the studied sections.

3.5. Shale (Mf4)

The facies has most extensive in the lower part of the facies column in the Sekonj section which has been exposed with alternating and interbedded with other lithological units. In the field observations, the facies is black to gray-colored, foliated, very soft and non-fossil which can be seen in carbonates with approximately high thickness in compared of alternating with sandstones (Figure 6). The presence of fine-grained quartz grains represent that the facies gradually changed to siltstone in some units. Alternating layers of sandstone and shale indicate deposition of shale facies in a mud plain, in shallow coastal environments. The shales with low thickness were mainly deposited in a mud plain near the sea and with alternating sandstone facies of shallow coastal environment and carbonate rocks. The high lateral extension of shale facies between the semi-restricted deposits indicates reduction of environment mechanical energy in mud plain.

4. TIDAL FLAT FACIES BELT

This facies belt is composed of 2 microfacies which can be seen with thin to thick-bedded layers throughout the clastic-carbonate succession. The most observed coastal facies belongs to Mf5.

4.1. Sandy Mudstone (Mf5)

The microfacies observed in both sections, in the lower part of the Abaregh section and in the different parts of the Sekonj section in the studied succession. The size of the sands is very fine to coarse-grained. The average frequency of components in the microfacies are including quartz particles about 10% - 20%, matrix 45%, rock fragments 8% and calcareous cement 10%.

More quartz grains are rounded and slightly angular, in the form of single crystal and polycrystalline with parallel extinction, although a few of them has undulatory extinction (Figure 6). It seems that the quartz particles are formed during breaking the pre-rounded grains and are located in a carbonate matrix without contact with each others. Textural maturity of the microfacies is immature due to the presence of abundant cement and matrix. The indeterminate microfossils fragments broken and transported and can be seen in calcite cement. The main characteristics of this facies belt are the presence of sand-sized quartz clastic grains, the secondary calcite veins with low variability of fossil assemblages in the tidal flat Mf5 microfacies. The presence of calcite veins and also stylolites in the microfacies is one of the features of intertidal flat and supratidal flat which have displacement with spreading to the land [28] . The detrital loads such as quartz grains in the lime mud matrix represent the tidal flat for this microfacies [29] .

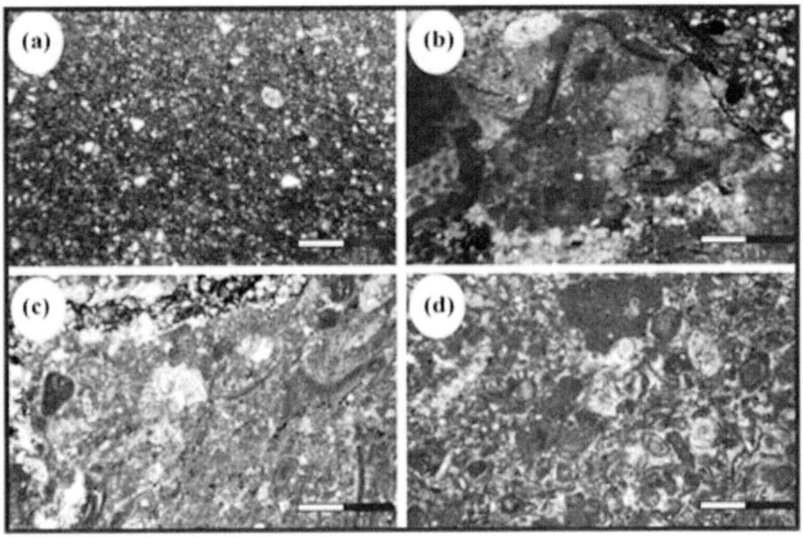

Figure 6. (a), (b) Sandy mudstone microfacies. The quartz grains and bioclasts are visible in the photos (under normal light). (c), (d) Interclast bioclast wackestone/packstone. Bioclasts and interclasts are visible in the photos (PPL).

4.2. Interclast Bioclast Wackestone/Packstone (Mf7)

In the field observations, the microfacies is dark gray color, with medium to thick-bedded layers which interclasts and microfossils can be observed with the naked eyes. The microfacies only are exposed in the end of the middle part of the Abaregh section with medium-bedded limestones. The most allochems of the microfacies are composed of skeletal fragments and bioclasts about 5% - 15%. The bioclasts are including the bivalve and algae fragments or indeterminate fossils which are allochthonnous and have been transported to the environment (Figure 6). The difference between this microfacies with Mf8 is an increase in the amount of interclast (20%). The allochems within the microfacies have strongly micritized. Due to the absence or very low diversity of bioclasts and cooperation with the shallow facies, the sub-environments of tidal flat is recommended for the microfacies [30] . The sedimentation changes from clastic to carbonate occur near the coast areas and inter-ramp where the river eroded the detrital particles and transported to the sea. Also, the residue of bioclasts which observed in the Mf7 microfacies is another feature of tidal flat facies belt.

5. LAGOONAL FACIES BELTS

The microfacies group is located between tidal flat and reef microfacies. The lagoon microfacies has alternating with shale facies. The facies belt consists of two facies, thin to medium bedded and dark limestone and with alternating tidal flat and barrier facies.

5.1. Miliolid Peloid Bioclast Wackestone (Mf8)

The main components of the microfacies are the imperforated larger benthic foraminifera (Miliolids) with high frequency and with a small amount of benthic foraminifera (small Miscellanea). Other bioclasts include green algae, crustose coralline red algae and bivalve fragments are rarely present (Figure 7). The frequency of peloid grains is about 15% with good roundness and in a micritic matrix. The wackestone matrix with fragmentation of the foraminiferal shells is justifiable in the facies. The presence of skeletal grains with low diversity and the algae is one of the features of lagoonal facies belt. The miliolid, green algae and small amount of indeterminate fossils in micritic matrix indicate deposition in a lagoon environment with low energy [31] . The peloid in lime mud matrix and low diversity of fossils represent deposition in restricted shallow sub-tidal environment with low sedimentation rate. The imperforate foraminifera such as miliolids with high frequency indicate hypersaline, shallow restricted lagoon [32] .

5.2. Algae Bioclast Packstone/Wackestone (Mf9)

This microfacies is characterized by abundant crustose coralline red algae (Such as Lithothamnion sp.). Also, the calcareous bioclasts including the fragments of bivalve are rarely present. Also small amounts of detrital quartz are present in this facies (Figure 7). Moreover, the abundant bioclasts (often bivalve fragments) with poor sorting and algae represent an environment with medium energy. Also, the presence of detrital quartz with poor sorting suggests deposition near the source region. According to the studies by Bachman and Hirsch [33] and Penny et al [34] , the presence of green algae shows a lagoon environment.

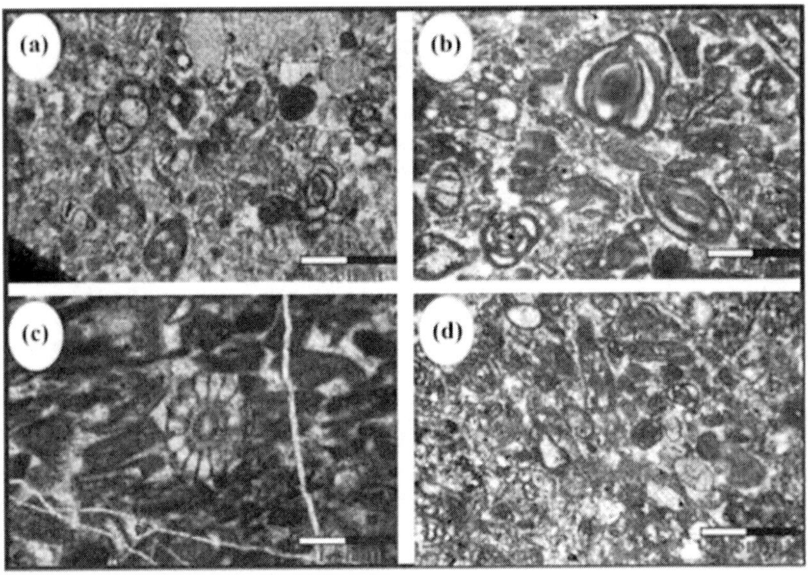

Figure 7. (a), (b) Miliolid peloid bioclast wackestone. The miliolids and peloid grains can be identified in the photos. (c), (d) Algae bioclast packstone/wackestone. The algae, other bioclasts and complete fossils can be observed in the photos (PPL).

6. BARRIER FACIES BELT

The barrier facies belt has abundant fossils in the studied area and based on 2 microfacies are introduced.

6.1. Coral Boundstone (Mf10)

The microfacies with limited lateral extension is in the upper part of the Sekonj section (Figure 8). The main components of the microfacies are only corals (Figure 8). In the field observations, this microfacies is characterized by gray to cream-colored, medium-bedded limestone. Also, small amounts of other bioclasts can be seen in the microfacies. The micritic cements have filled most of the porosity and cavities in the corals. The microfacies are association with Mf9 microfacies in the facies column and its limited extension in the field observations represents deposition as patch reefs in the lagoon toward the carbonate barrier and open marine. The boundstone is interpreted to be formed above the fair- weather wave base and in a high energy environment which the autochthonous carbonate are formed and based on the occurrence of coral boundstone facies suggests a high energy barrier environment.

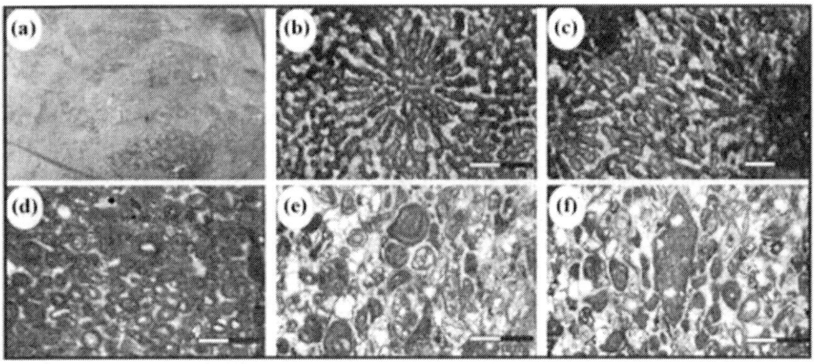

Figure 8. (a) A close-up view of boundstone succession in the field observations. (b), (c). Coral boundstone microfacies (under normal light). (d)-(f) Bioclast peloid ooid packstone/grainstone microfacies. Fossil, peloid and ooid are clearly visible in three images (PPL).

6.2. Bioclast Peloid Ooid Packstone/Grainstone (Mf11)

This facies is in the middle and upper part of the Abaregh section and in the upper part of the Sekonj section (Figure 8). This microfacies is characterized by gray, thin to thick-bedded limestone with alternating shale facies. Peloid and ooild with more than 25% are the most abundant non-skeletal particles. In some places of the microfacies, peloids are micritized which indicate energy fluctuations. Minor constituent is composed of less than 5% bioclast and interclast (Figure 8). Abundance of allochems, presence of ooilds with tangential texture, small amounts of lime mud and abundance of cement in this facies belt show a high energy environment [35] . Also, the presence of peloid and indeterminate ooids suggest a barrier environment and above the fair-weather wave base. The ooids in the Mf11 microfacies represent a shallow, high energy and turbulent environment which suggests deposition in ooidal carbonate barrier.

7. OPEN MARINE FACIES BELT

In the field observations, this facies belt is medium to thick-bedded and darker than barrier facies belt. All of the proximal open marine microfacies have low thickness than other facies belts in the stratigraphic column. The proximal open marine facies belt is composed of one microfacies which can be seen in the stratigraphic column.

7.1 Bioclast Packstone/Wackstone (Mf13)

In the field observations, the apparent evidence of this facies is similar the lagoon facies. This facies can be seen in different parts of the succession in the Sekonj section and in the middle part of the Abaregh section. Allochems of this microfacies are including 40% echinoid fragments and larger benthic foraminifera with 50% lime mud and the other is cement (Figure 9). This facies with althernating Mf8 microfacies has similar fossil fragments. The similarity between the skeletal fragments of this facies with bioclasts of barrier facies indicates deposition in the upper part of the carbonate barrier and in the proximal open marine. Diversity and the bioclasts size in this microfacies show deposition in the open marine sub-environments [35]. The presence of deep sea benthic foraminifera suggests deposition in shallow open marine environment.

8. DEPOSITIONAL PALEOENVIRONMENT

The Paleocene-Eocene carbonate-siliciclastic sedimentary successions in the east of Central Iran have significant development, distribution and thickness which as a result of the wide stratigraphic distortion during the tectonic events were not well studied. Laramide event formed the Kerman conglomerate during the late Mesozoic era in most areas of Kerman which represents the marine regression conditions. According to the obtained information from the analysis microfacies, depositional environment and also the absence of storm facies, the Paleocene-Eocene succession has been deposited in a carbonate-siliciclastic ramp in the studied area. Spreading of the carbonate barrier dunes facies shows the ramp environment [36] . The possibility of the rimmed-shelf existence requires the presence of barrier reef in the area. However, the barrier environment of the studied area is limited to the patch reefs assemblages which are separated the lagoon from the open marine sub-environments. The gradual changes of facies, presence of barrier and the absence of sliding and slumping facies with the macroscopic and microscopic characteristics of carbonate-siliciclastic facies and comparison with the models presented by [37] and indicate that the facies have been deposited in a shallow marine environment, in a mixed carbonate-siliciclastic ramp. By studying thin sections, determining depositional microfacies and comparing with the current depositional environments, a range from lagoon to open marine environment has been identified which separated by barrier and patch reef. Coastal, tidal flat, lagoon, barrier, patch reefs and proximal open marine facies belts of the depositional paleoenvironment were formed the carbonate-siliciclastic ramp (Figure 10).

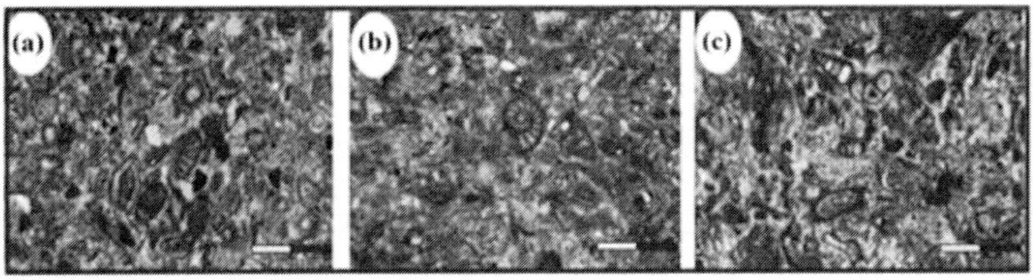

Figure 9. (a)-(c) Bioclast packstone/wackstone with different fossils in the proximal open marine (PPL).

9. SEQUENCE STRATIGRAPHY

According to the sequence stratigraphy studies of Paleocene-Eocene succession in the studied area, there are two important points including: 1) sediments have been deposited in coastal and shallow marine basin. 2) the nature of carbonate and siliciclastic were existed in the time of formation in the basin. The studied sections are composed of two 3rd incomplete depositional sequences which include 4 system tracts (Figure 11).

9.1. Sequence 1

The sequence 1 with a thickness of 249 and 126 meter are respectively identified in the Sekonj and Abaregh sections which represent a carbonate and a little siliciclastic sequence and has been deposited in coastal, tidal flat, barrier and proximal open marine environments. The sequence has high thickness in the Sekonj section and the thickness of siliciclastic layers gradually increase from the bottom to up of the succession. The sequence 1 with low thickness in the Abaregh section is more siliciclastic and gradually converts into the carbonate facies. In addition, the sequence has been deposited during the period of rising sea level and the Paleocene-Eocene carbonate succession is placed on the thick bedded Cretaceous carbonate succession in the Sekonj section. The sediments have been deposited unconformably on the Kerman conglomerate deposits in the Abaregh section. The field and laboratory observations show that the boundary between the Paleocene-Eocene and the Cretaceous carbonate successions is gradual and continuous and there is no evidence of out of the water and regression conditions. Hence, the lower boundary of this sequence is a type 2 (SB2) (Figure 3 and Figure 4). The first depositional sequence has been placed unconformably on the Kerman conglomerate deposits in the Abaregh section. So, the lower boundary of this sequence with Kerman conglomerate is a type 1 (SB1). The conversion of Cretaceous thick-bedded limestone to Paleocene-Eocene thin to medium-bedded limestone shows a small amount of water depth reduction. Due to the absence of sufficient evidences, lowstand system tract (LST) is not identifiable from transgressive system tract (TST) in the both sections.

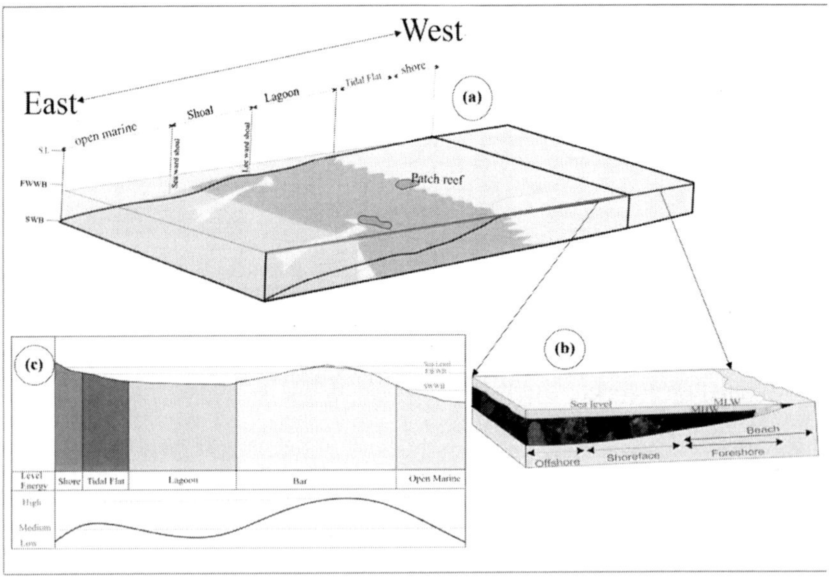

Figure 10. (a) Depositional paleoenvironment model of Paleocene-Eocene succession in the studied area. (b) Subdivision in the coastal environment. (c) Transverse profile of depositional paleoenvironment in the studied area and energy level in the different parts of the studied succession.

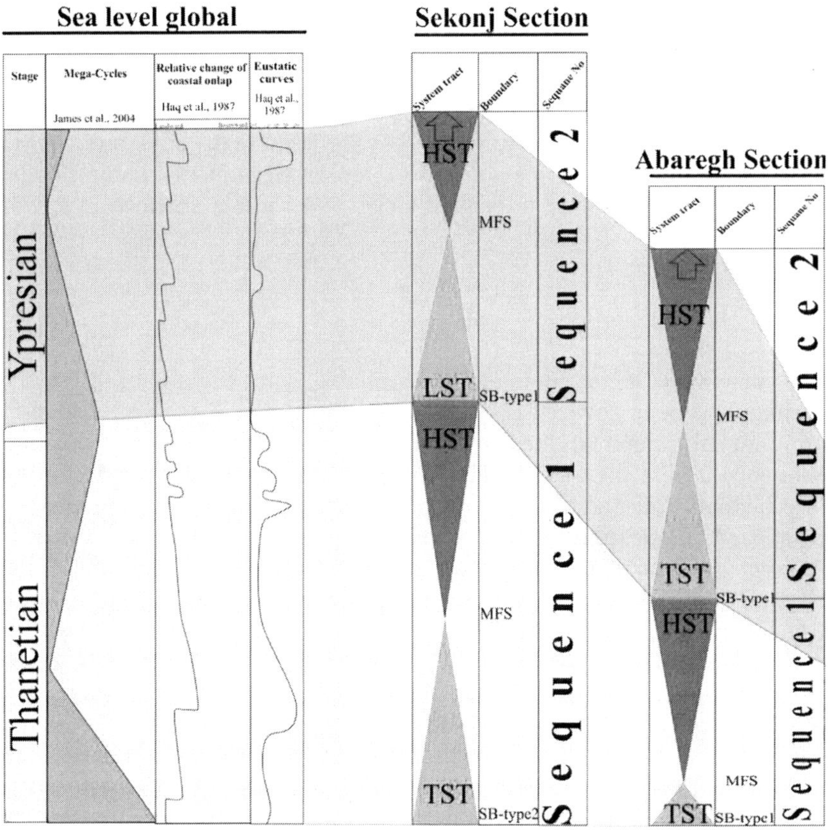

Figure 11. Correlation of sequence stratigraphy in the studied sections with global sea level changes.

So, the sequence is started with the transgressive system tract (TST) on top of the sequence boundary (Figure 3 and Figure 4). Transgressive system tract (TST) has been deposited in proximal open marine, lagoon and barrier facies belts which indicate the rate of the sedimentation is more than the accommodation space. But, the system tract has been deposited in tidal flat facies belt in the Abaregh section. Highstand systems tract (HST) ended to sequence boundary type 1 (SB1) and is characterized by coastal microfacies and the sandstone with cross-lamination in the Sekonj section and with the conglomerate succession in the Abaregh section. The early highstand systems tract in Sekonj section is characterized by a proliferation of grain-supported shoal facies. This part mainly comprises the shoal and shore facies association. The overhead part of sequence 1 (late HST) indicates upward shallowing trend. The shallowing-upward trend from shore and tidal flat is indicative of a progradational stacking pattern during the highstand systems tract. Maximum flooding surface (mfs) is characterized by lagoon microfacies (Mf9) in the Sekonj section and tidal flat microfacies (Mf7) in the Abaregh section. It seems that the sedimentation phase is calmer and accumulation pattern is in the shallow sediments which represent the accommodation space is less than the rate of the sedimentation. The falling sea level was occurred after maximum flooding surface (mfs) and ended to coastal facies and show a shallowing upward parasequence.

9.2. Sequence 2

The sequence 2 is not completely identifiable, since the end of the studied sections has been covered (Figure 11). The sequence 2 with a thickness of 165 and 199 meter are respectively in the Sekonj and Abaregh sections. Sequence 2 is in the Sekonj section with more siliciclastic nature and the interbedded limestone can be seen in some places. But, the sequence is composed of limestone with interbedded sandstone and shale layers in the Abaregh section. Three system tracts have been identified in the Sekonj section, but due to the absence of sufficient evidences, separation of transgressive system tract (TST) from lowstand system tract (LST) was not possible in the both sections. Sequence 2 has lagoonal nature in the Sekonj section and coastal nature in the Abaregh section. Lowstand system tract (LST) is located on the sequence boundary type 1 (SB1) and has a constant trend. The system tract has been just identified in the Sekonj section and includes coral reef boundstone succession. It seems that the system tract was a shallow environment and gradually increased its depth and reached to the shale facies which indicates the maximum flooding surface (mfs). The highstand systems tract (HST) overlying the maximum flooding surface (mfs) shows a regressive trend in the both sections. Also, the rate of sedimentation is more than accommodation space. The end of this system tract is not identifiable due to the lack of outcrop and coverage.

9.3. Correlation of Sequence Stratigraphy in the Studied Sections

Sequence 1 is mainly composed of limestone in the Sekonj section and alternating limestone, sandstone, shale and conglomerate in the Abaregh section. Sequence 1 has abundant algae fragments in the limestone successions in the both sections. The lower boundary of sequence 1 is a type 2 sequence boundary (SB2) which the mentioned limestone succession overlies the Cretaceous thick-bedded limestone succession in the Sekonj section, while the mentioned limestone succession has been deposited on the Kerman conglomerate in the Abaregh section. This difference indicates that the Kerman conglomerate successions have not been deposited in the Sekonj section or have been removed as a result of the tectonic activities. In addition, the maximum thickness of the sequence 1 is belongs to the Sekonj section, while the conditions are reversed in the sequence 2. The thickness of succession has been reduced from East to the West of the basin. This is so that, at first the limestone succession and then siliciclastic succession have been deposited in the north of basin (Sekonj section) and the thickness of limestone succession is more than siliciclastic succession. Generally, the global sea level changes included two transgressions at the time of deposition of this succession which are match with the transgression and regression of sea water in the sequence 1 and 2 of the studied area (Figure 11). The role of tectonic activities become more upward the basin and high rate of the subsidence in the basin led to increase the accommodation space and sediment supply in both sections. A long period of shoal conditions in Sekonj section and lagoon conditions in Abaregh section, reflecting a balanced situation between accommodation and sedimentation, characterizes the aggradational sedimention pattern.

10. CONCLUSION

Generally, the Palecene-Eocene succession is present in some part of the south of Kerman, due to erosion during the Laramide orogeny event but the basal part of Palecene-Eocene succession of Thanetian-Ypersian age in the sections, transgressively overlies the top of late Cretaceous and Kerman Conglomerate. Quaternary sediments are usually the upper border of the above sequence. Palecene- Eocene succession in the Study area basin mainly consists of carbonate-clastic (limestone, dolomite, sandstone, shales, Conglomerate). The other two sections are characterized by a mixed succession of carbonate siliciclastic shelf deposits. The regional differentiation most probably reflects their position on separate tectonic blocks on which different facies conditions developed due to different tectonic movements. Facies, depositional environment and sequence stratigraphy studies of Palecene-Eocene successions in Sekonj sedimentary trough provided the following consequences: Identification of 11 microfacies in 5 facies belts including coastal, tidal flat, lagoon, barrier and proximal open marine environments which have been deposited in a carboanate-siliciclastic ramp. The most observed siliciclastic facies are in the Sekonj section which all of the siliciclastic facies can be seen in it. The observed facies and microfacies are including: Mf1, Mf2, Mf3, Mf4, Mf5, Mf8, Mf9, Mf10, Mf11, Mf13 in the Sekonj section and Mf1, Mf3, Mf4, Mf5, Mf7, Mf8, Mf9, Mf11, Mf13 in the Abaregh section. The lateral trend of facies and microfacies from Sekonj section to Abaregh section is characterized by deposition of carbonate microfacies containing miliolids, algae and fossil fragments in the lower part of the Sekonj section and upper part of the Abaregh section. Sequence stratigraphy analysis of the studied sediments led to identification of two 3rd depositional sequences with 2 maximum flooding surfaces. So, four system tracts were separated from together in the Sekonj section. Sequence 1 has been completely observed in the both sections, but due to the lack of outcrop, the sequence 2 was not completely interpreted at the end of the succession. In addition, the regional tectonic activities caused the differences in the two sequences. The predominant facies associations developed in Palecene-Eocene demonstrate an overall transgression-regression cycle in the study area.

CITE THIS PAPER

Afghah, M., Ghafari, M. and Ahmadi, V. (2017) Paleoenvironmental Distribution Patterns of Sequence Stratigraphy in the Paleocene-Eocene Deposits of Sekonj Trough Event (Central Iran). Open Journal of Geology, 7, 647-665. https://doi.org/10.4236/ojg.2017.75044

REFERENCES

1. Dimitriovic, M.D. (1973) Geology of Kerman Region. Institute for Geological and Mining Exploration of Iran, Report No. Y/52, 335 p.
2. Aghanabati, A. (2006) Geology of Iran. Geological Survey & Mineral Explorations of Iran (GSI), 586 p. (In Persian)
3. Abbasnejad, A. and Dastanpour, M. (1998) Earthquakes and Seismicity of the Kerman Province. Kerman Cultural Services Publication, 255 p. (In Persian)
4. Hajian, J. (1996) Geology of Iran, Paleocene-Eocene in Iran. Geological Survey of Iran, 460 p. (In Persian)
5. Shahmoradi, M. and Faryabi, A. (2011) Biostratigraphy and Paleoecology of the Paleocene-Eocene Deposits Foraminifera in Abaregh Section, NW of Bam, Kerman. 2nd Geological Congress of Iranian Plateau, University of Zarand in Kerman. (In Persian)
6. Samadi Afkham, M. (2012) Biostratigraphy and Microfacies Studies of the Paleocene-Eocene Depositional Succession in the North of Khoosf (Dahanrud Section), East of Iran. M.S.c Geology' Thesis (The Field of Stratigraphy and Paleontology), University of Birjand, 145 p. (In Persian)
7. Hashemiazizi, S.H., Miranshabestari, Q.R. and Khazaei, A.R. (2013) Petrography, Depositional Facies and Diagenesis Analysis of Paleocene-Eocene Carbonate Deposits in Ching Anticline in the West of Birjand, East of Iran. The Journal of Sedimentary Facies, 29, 107-122. (In Persian)
8. Babazadeh, S.A. (2014) Statistical Analysis of Foraminifera in the Eocene Ramp in the Lut Block, East of Iran. Stratigraphic and Sedimentological Journal, 30, 73-88. (In Persian)
9. Shokuhimoghadam, N., Miranshabestari, Q.R. and Khazaie, A.R. (2011) Microfacies and Depositional Environment of Paleocene-Eocene Carbonate Succession in the South East of Birjand, East of Iran. Journal of Sedimentary Facies, 4, 172-187. (In Persian)

10. MoeinAldini, Z., Faryabi, A., Raeis Alsadat, S.N. and Lotf Abad Arab, A. (2013) Analysis the Biogenic Structures of the Paleocene Eocene Flysch Deposits in Chahartagari Section, SE of Kerman. Journal of Paleontology, 1, 103-113. (In Persian)
11. Gitashenasi Cartographic and Geographical Institute (2007) Road Atlas of Iran, Scale of 1/100000.
12. Sahandi, M. R. and Rahim-zadeh, F. (1992) Geological Map of Kerman 1/250000. Geological Survey & Mineral Explorations of Iran (GSI).
13. Djokovic, I., Dimitrijevic, M.N., Cvetic, S. and Dimitrijevic, M.D. (1978) Geological Map of Khaneh Khatun 1/100000. Geological Survey & Mineral Explorations of Iran (GSI).
14. Azizan, H.A., Shahraki, A. and Safoori, S. (1999) Geological Map of Kerman 1/100000. Geological Survey & Mineral Explorations of Iran (GSI).
15. Walker, R. and Jackson, J. (2004) Active Tectonics and Late Cenozoic Strain Distribution in Central and Eastern Iran. Tectonics, 23, TC5010.
16. Shafiebafti, A., Pour Kermani, M., Shahpasand-Zadeh, M. and Iran Manesh, F. (2008) Tectonic Geomorphology and Analysis of the Kuh-Banan Fault Activity by Calculating Geomorphic Indices. Journal of Geology and Environment, No. 7, 1-17. (In Persian)
17. Flugel, E. (2010) Microfacies of Carbonate Rocks, Analysis, Interpretation and Application. Springer-Verlag, Berlin, 976 p.
18. Pettijohn, F.J. (1975) Sedimentary Rocks. Harper and Row, New York, 628 p.
19. Dunham, R.G. (1962) Classification of Carbonate Rocks According to Depositional Texture. In: Ham, W.E., Ed., Classification of Carbonate Rocks, American Association of Petroleum, Geologists Memoir 1, 108-121.
20. Catuneanu, O. (2006) Principles of Sequence Stratigraphy. Department of Earth and Atopospheric Science, University of Alberta Edmonton, Alberta, Canada, 387 p.
21. Catuneanu, O., Abreu, V., Bhattacharya, J., Blum, M., Dalrymple, R., Eriksson, P., Fielding, C.R., Fisher, W., Galloway, W., Gibling, M., Giles, K.A., Holbrook, J.M., Jordanm, R., Kendall, C.G.St.C., Macurda, B., Martinsen, O.J., Miall, A.D., Neal, J.E., Nummedal, D., Pomar, L., Posamentier, H.W., Pratt, B.R., Sarg, J.F. Shanley, K.W., Steel, R.J., Strasser, A., Tucker, M.E. and Winker, C. (2009) Towards the Standardization of Sequence Stratigraphy. Earth-Science Reviews, 92, 1-33.
22. Catuneanu, O., Galloway, W.E., Kendall, C.G.S.C., Miall, A.D., Posamentier, H.W., Strasser, A. and Tucker, M.A. (2011) Sequence Stratigraphy: Methodology and Nomenclature. Newsletters on Stratigraphy, 44, 173-245.
23. Tucker, M.E. (2001) Sedimentary Petrology: An Introduction to the Origin of Sedimentary Rocks. Blackwell, Hoboken, 262 p.
24. Miall, A.D. (2000) Principle of Sedimentary Basin Analysis. Springer-Verlag, New York, 668 p.
25. Wanas, H.A. and Abdel-Maguid, N.M. (2006) Petrography and Geochemistry of the Cambro-Ordovician Wajid Sandstone, Southwest Saudi Arabia: Implications for Provenance and Tectonic Setting. Journal of Asian Earth Sciences, 27, 416-429.
26. Pettijohn, F.J., Potter, P.E. and Siever, R. (1987) Sand and Sandstone. 2nd Edition, Springer, New York, 553 p.
27. Kostic, B., Bech, A. and Aigner, T. (2005) 3-D Sedimentary Architecture of a Quaternary Gravel Delta (SW-Germany): Implication for Hydrostratigraphy. Sedimentary Geology, 181, 147-171.
28. Shinn, E.A. (1983) Tidal Flat Environment. AAPG Mem, 33, 173-210.
29. Palma, R.M., Lopez-Gomez, J. and Piethe, R.D. (2007) Oxfordian Ramp System (La Manga Formation) in the Bardas Blancas Area (Mendoza Province) Neuquén Basin, Argentina: Facies and Depositional Sequences. Sedimentary Geology, 195, 113-134.
30. Jamalian, M., Adabi, M.H., Moussavi, M.R., Sadeghi, A., Baghbani, D. and Ariyafar, B. (2011) Facies Characteristic and Paleoenvironmental Reconstruction of the Fahliyan Formation, Lower Cretaceous, in the Kuh-e Siah area, Zagros Basin, Southern Iran. Facies, 57, 101-122.
31. Geel, T. (2000) Recognition of Stratigraphic Sequence in Carbonate Platform and Slope Deposits: Empirical Models Based on Microfacies Analysis of Palaeogene Deposits in South Eastern Spain. Palaeogeography, Palaeoclimatology, Palaeoecology, 155, 211-238.

32. Wilson, J.L. (1975) Carbonate Facies in Geologic History. Springer Verlag, New York, 471p.
33. Bachmann, M. and Hirsch, F. (2006) Lower Cretaceous Carbonate Platform of the Eastern LEVANT (Galilee and the Golan Heights): Stratigraphy and Second-Order Sea-Level Change. Cretaceous Research, 27, 487-512.
34. Penney, S.J. and Racey, A. (2004) Ecology of Extant Nummulitids and Other Larger Benthic Foraminifera: Applications in Palaeoenvironmental Analysis. Earth-Science Reviews, 67, 219-265.
35. Spalletti, L.A., Poire, D.G.E. and Schwartz, G.V.D. (2001) Sedimentologic of a Neocomian Marine Carbonate-Siliciclastic Ramp: Neuquen Basin Argentina. South America. Journal of Earth Sciences, 14, 609-624.
36. Elrick, M. and Read, J.F. (1991) Cyclic Ramp to Basin Carbonate Deposits, Lower Mississippian, Wyoming and Montana. Journal of Sedimentary Petrology, 61, 1194-1224.
37. Selly, R.C. (1996) Anicient Sedimentary Environments and Their Sub-Surface Diagnosis. 4th Edition, Nelson Thornes (Publisher) Ltd., England, 315 p.

CHAPTER 8

Remote Sensing and Geographic Information System for Fault Segments Mapping a Study from Taiz Area, Yemen

Anwar Abdullah, Shawki Nassr, and Abdoh Ghaleeb

Geology Department, Faculty of Applied Science, Taiz University, Taiz, Yemen

ABSTRACT

GIS and remote sensing data for allowing detection of structural features, such as faults, offer opportunities for improving mapping and identifying the areas that are likely to be locations of faulting areas. Landsat ETM-7 satellite data images were used and band-5 was found as the most suitable band for lineament delineation, based on the ability to identify geological features. Four contributing factors, namely, drainage patterns, faults (previously mapped), lineaments, and lithological contacts layers, were parameters used in this study to produce a fault potential prediction map using the overlay model techniques. The potential map (fault susceptibility map) classifies the study area into five potential zones, namely, very low, low, moderate, high, and very high potential. The areas covered by moderate to the highest potential zones were considered as fault segments (fault lines) in the area. The comparison of the potential map and the published fault map by using GIS matching techniques shows that 75 fault segments (fault lines) in the potential map were not properly identified in the study area. The correlation between fault segments and faults data collected from field work stations shows that there were 39 fault segments which may represent new faults in the area being identified. The presence of these faults is not known from the literature; this leads to updating and revising of existing geological map of the study area.

1. INTRODUCTION

Faults are weakness zones in the brittle part of the lithosphere, along which movement can take place in response to induced stresses. When faults undergo displacement, depending on geological and structural conditions, strain markers can be formed on the fault surface [1]. The presence of faults in any area is based on displacement of rock layers. But also, most of faults are represented by some geological features such as drainage patterns, lineaments (linear features), and lithological contacts between rock units within the rocks of the area. The presence of faults may be indicated by these geological features (factors).

The term lineament was first introduced by [2, 3] who recognized the existence of linear geomorphic features and interpreted them as surface expressions of zones of weakness or structural displacement of the earth's crust. Lineaments are linear features on the Earth's surface, usually related to the subsurface phenomena. Generally, lineaments are related to large fractures and faults where their orientation and number give an idea of fracture pattern of rocks [4]. In the recent years, the lineaments have been defined as natural crustal structures that may represent a zone of structural weakness [5].

The drainage system, which develops in an area, is strictly dependent on the slope, the nature, and attitude of bedrock and on the regional and local fracture pattern [6]. Most stream networks are adapted to regional slope and geological structures, picking out the main fractures in the underlying rocks [7].

The contact between two lithologies can also appear as a linear feature. This contact may appear as a change in drainage pattern across the structural features [8] or the two units may have different spectral properties [9].

Based on the definitions of lineaments, drainage patterns (including pattern the length, spatial distributions), and the lithological contacts between different rock types, the faults could be mapped in the study area. These geological features mostly resemble a fault lines in the area. The most important features in the area are the presence of drainage lines patterns and fractures. Lineaments, drainages, lithological contacts, and previous fault lines data are important data and used in this research for fault segments mapping using GIS technique.

With the advent of remote sensing and computer technology in the geosciences, geological investigation and interpretation have entered a new era. Remote sensing technology is very efficient for collecting data. Computer technology, such as computer-based geographic information system (GIS), supplies a different method for data storage, integration, analysis, and display. The combination of remote sensing and GIS provides an optimum system for various geological investigations such as fault mapping [10].

Several studies have been carried out to enhance geological knowledge and revise existing geological maps by using optical remotely sensed data and discovered new faults that were previously unmapped [11–16]. Furthermore, authors in [17, 18] conclude by suggesting that satellite images and/or geographic information systems (GIS) are useful for tectonic mapping and identifying previously unmapped faults.

The benefit of integrating different data and techniques in structural features such as faults identification and analysis made possible using remote sensing and GIS techniques is the ability to identify faults based on their characteristics. Creation and classification of various types of geological features can be carried based on the purpose of the study. Once the identification, preparation, and processes are complete, geographic information systems (GIS) functionality, such as vector and raster spatial analysis and overlay, can be employed for structural mapping and analysis using powerful software programs [19].

Generally, Yemen is located in the southwestern part of Arabian Peninsula, and it is affected by an active rifting zone of the Gulf of Eden and the Red Sea opening. This active (rifting) zone can cause the reactivation of the old fault systems and may create new faulting lines in the area. The active faults (such as, rock mass movements or sliding on the fault planes and bedrock lithologies cracking) may cause severe damages in some places, especially in urban or settlements areas. For example, bedrock cracking and rock mass sliding happened in the 2003 and 2010 in some places in Taiz city caused severe damages, and many houses, roads, and other infrastructures were destroyed.

The previous works (data) such as geological and structural maps are very old data, showing a few of fault lines in the study area. For this purpose, mapping of faults may lead to updating these data. In this paper, the using of remote sensing (data processing) and GIS (modeling) techniques could be helpful tools for mapping of faults, where the different geological features (factors) are taken in the considerations. The results of the work could be considered from the government or the people before or during the different construction projects, to increase the safety, especially in the urban areas.

The purpose of this study was to test the integration between remote sensing and geographic information system (GIS) for detecting the fault segments (fault lines) over the study area and to investigate the ability of this method in giving real results compared to the previous data with respect to the ground truth (field work).

The study area located in the southwest of Yemen has an area of 176 km^2. It includes Taiz city and extends between Rubayi and Al-Hawban areas (Figure 1).

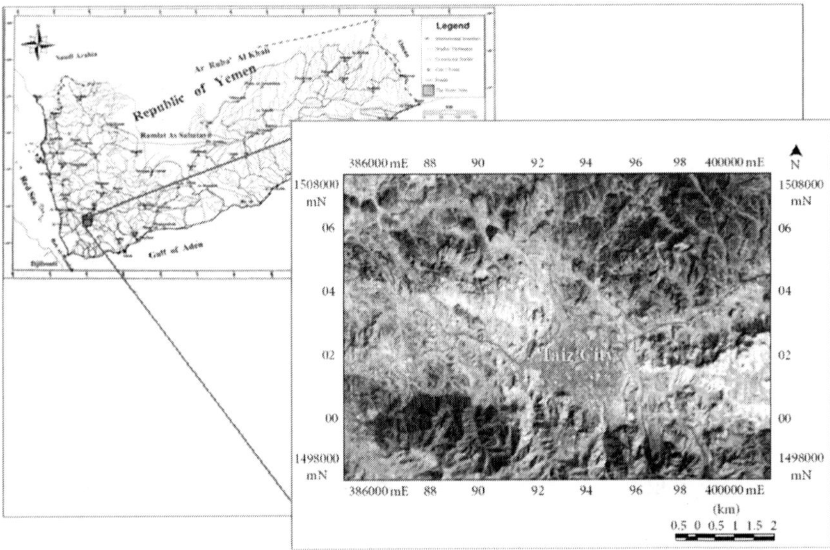

Figure 1: Location of the study area.

According to GSY [20] classification, the rocks in the study area are divided into two groups as follows.(1)Tertiary (granite body intrusive).(2)Tertiary (volcanic rocks such as basalt, rhyolite, dacite, ignimbrite, and ashflow deposits).

The geological map of the study area (Figure 2) was digitized from the geological sheet map of Taiz with scale 1 : 250,000.

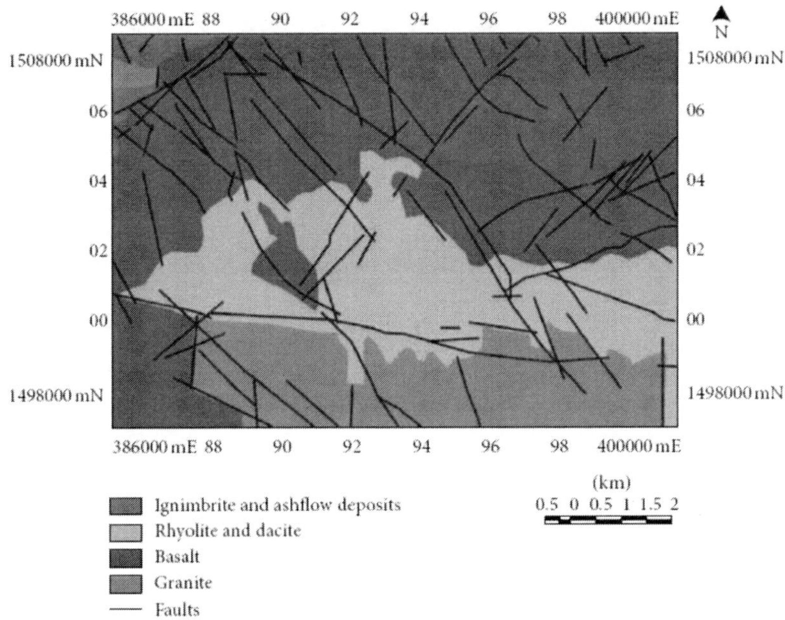

Figure 2: Geological map of the study area [20].

2. MATERIALS AND METHODS

Various data were used for fault segment susceptibility mapping (potential fault zones mapping) in the study area. Various image processing and enhancement techniques were applied on different remote sensing data including the Landsat ETM-7 (Enhanced Thematic Mapper) satellite images and SRTM (The Shuttle Radar Topography Mission) to get maximum information by using different approaches of extraction methods. The other data used in this study were topographic maps, geological maps, and field data.

During conducting this study, many different software packages were used since there was no single software that would process all steps in the analyses. Therefore, the diverse software used for the analysis in the current study is PCI Geomatica (version 9.1), ERDAS (version 8.4), and Ilwis software (version 3.3).

2.1. Data Processing

According to this paper the linear features delineation was based on decision of the most appropriate data, such as drainage patterns, lineament, fault (previously mapped), and lithological contact layers for mapping of fault segments using GIS. Many of researchers such as [21–26] have utilized lineaments, drainage patterns, faults, and lithology factors as important factors to measure and map the susceptibility of geological hazard. Therefore, the input attributes of the structural features of an area should be studied as a first consideration.

The previous data about the study area including geological, structural, and topographic maps are digitized. To prepare these maps, first, the maps are converted into a digital format by using the scanner. There are some data which were extracted from data in digital formats, such as Landsat ETM satellite images and SRTM.

Generally, four digital layers (maps) such as drainage patterns, fault (previously mapped), lineament, and lithological contact were prepared and converted into secondary data such as drainage buffer, fault buffer, lineament buffer, and lithological contacts buffer layers.

These layers were converted into slicing layers with five classes. Each class was determined as 30 m. This is because these layers were generated from different data (ETM, SRTM, and previous maps) with different spatial resolutions and scales. Due to this, these layers should be enhanced and resampled (rescaled) to be suitable in resolution for GIS model. The resolution was selected as 30 m for this process based on the ability of satellite image of 30 m spatial resolution to identify the geological features in the area. Then, slicing layers were converted into weight layers (thematic layers). These procedures are illustrated in Figure 3.

Arbitrary classifications are still common; however, the main classification approaches are ranking, natural breaks, equal interval classes, equal area classes, and mean value and standard deviation intervals [25, 27–29]. Hence, the classification determines the spatial distribution of buffering zones or susceptibility (classifying susceptibility) based on equal area classes.

Classes (five classes of each weight layer) and their weight values are given in Table 1. Each class has a weight to express the contribution to the occurrence of fault segments. The weight value was assigned to be between 1 to 5 (Table 1). These weight layers (maps) were input into GIS model. The resultant map (susceptibility map) was classified into different zones, very low, low, moderate, high, and very high fault susceptibility zones or equal area classes using ranking approach.

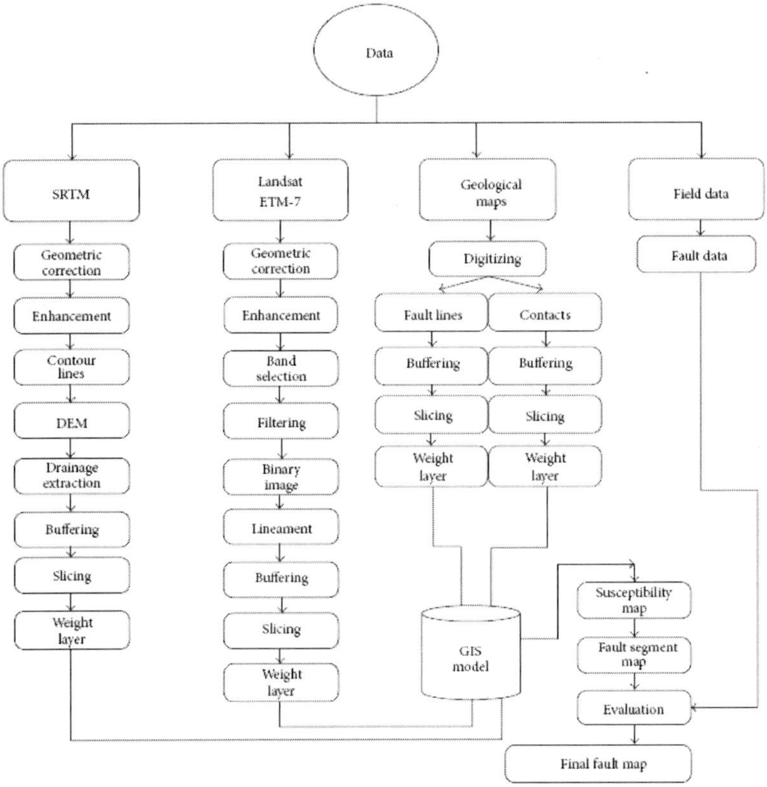

Figure 3: Chart of the methodology in this study.

Table 1: Parameters and the weight values are given to different factors in the study area.

Classes	Weight value
<30 m	5
30–60 m	4
60–90 m	3
90–120 m	2
>120 m	1

2.1.1. Lineament Factor

Lineaments are related to large structural fractures [4] and it may represent zones of weakness [22]. Also, the presence of tectonic structures such as linear features could be important factor in geological hazard occurrences [30, 31].

Filtering is useful in image enhancement. An edge-enhancing filter can be used to highlight any changes of gradient within the image features, such as structural lines [17, 28, 32–37]. They generated lineament maps and determined several significant structural features.

In the recent years, the lineaments have been defined as natural crustal structures that may represent a zone of structural weakness [38]. Landsat ETM-7 satellite data were used and the first step was to select the band that should be used for lineament extraction [39]. Visual inspection of the individual bands was carried out, based on the ability to identify features, and band-5 (1.55–1.75 μm) (SWIR) was selected as shown in Figure 4. And it was stretched linearly to output range from 0 to 255 (Figure 5), because it is the least affected band by the scattering, through the travel path of the long wavelength in the atmosphere. Therefore, it shows a good contrast and better display of geological features compared to other bands.

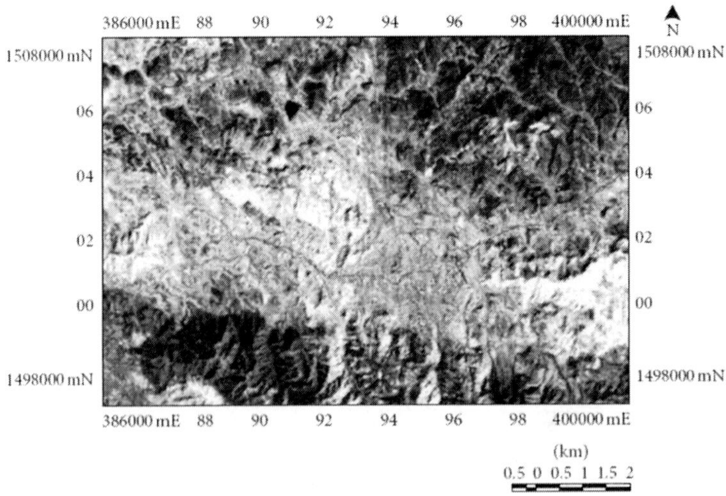

Figure 4: Original band-5 of Landsat ETM-7.

The second step was to select the filter type. For this purpose, different types of filters are tested such as 3 by 3 sobel kernel filter, 5 by 5 edge kernel filter, and 7 by 7 edge kernel filter. The 5 by 5 and 7 by 7 edge enhancement filters give thicker and less linear areas after threshold application; sobel results in thinner and more linear zones that give the best result compared with other two filters.

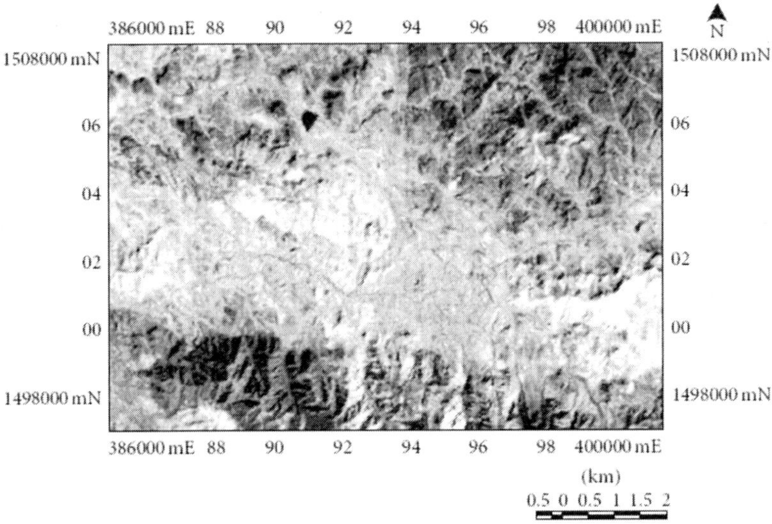

Figure 5: Band-5 of Landsat ETM-7 after contrast enhancement.

The sobel filter was the best which convolved over band-5 in four principal directions and the filtered image was added back to the original image. The sobel filters in four principal directions are given in Table 2.

Table 2: Sobel kernel filters in four main directions.

Four main principal directions											
NW-SE			E-W			NW-SE			N-S		
−2	−1	0	−1	−2	−1	0	1	2	1	0	−1
−1	0	1	0	0	0	−1	0	1	2	0	−2
0	1	2	1	2	1	−2	−1	0	1	0	−1

Then, the four enhanced filtered images combined together to produce one threshold filtered image (Figure 6), and this image was converted to binary image by applying adaptive thresholding. This adaptive threshold used to create binary images is controlled by two parameters, and the suitable values of adaptive thresholding parameters are windows size of 3, and this value is equal to thinning of pixels and the mean offset of 3. The final binary image of the study area is shown in Figure 7.

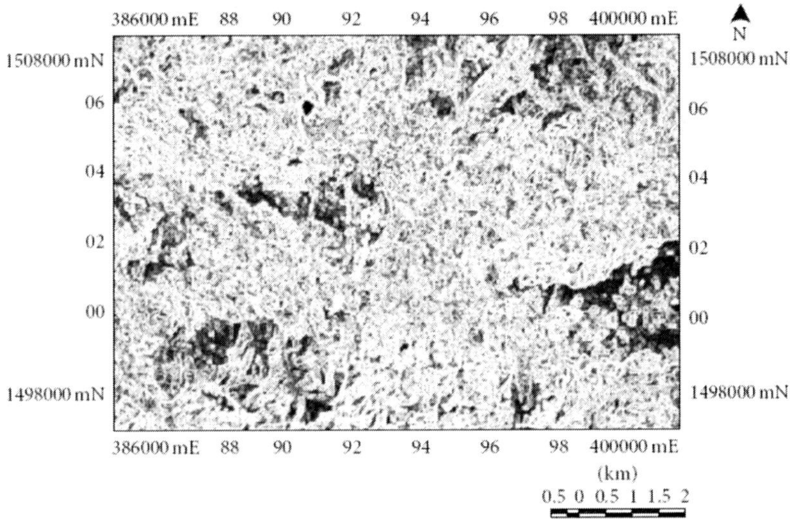

Figure 6: Threshold image produced by combining four filtered images.

The lineaments detected during the interpretation process were digitized directly on the image on the screen, and the final lineament map was recorded and stored in vector files (Figure 8).

There are important points which need to be mentioned and should be followed in relation to the above procedure for lineament mapping. First, care is needed in the interpretation of the result of such analysis to exclude man-made features; therefore, the final map was screened to remove lineaments related to cultural features such as roads, canals, and field plantation boundaries by comparing it with the topographic maps and geological map of the area. All procedures mentioned above have been followed during trace lineaments from binary image.

The lineament map was prepared and converted into lineament buffer layer and then converted into slicing layer with five classes. Finally, this layer was converted into weights layer (Figure 9). The weight value was assigned to be between 1 and 5 (Table 2). And, this layer was ready to be input into GIS modeling.

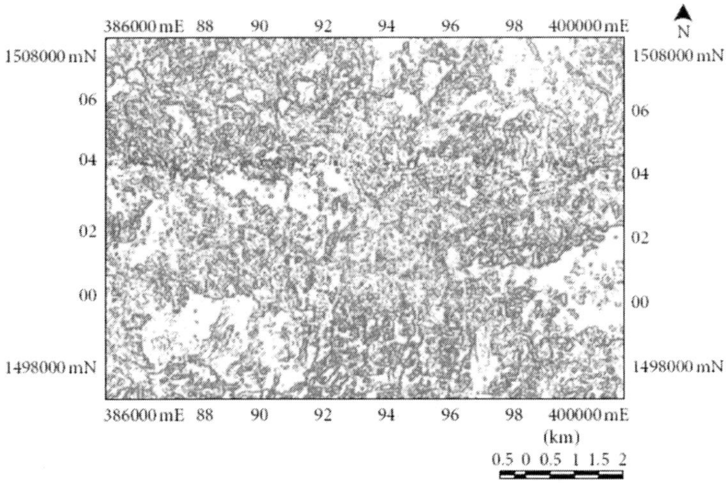

Figure 7: Binary image produced by using sobel filters.

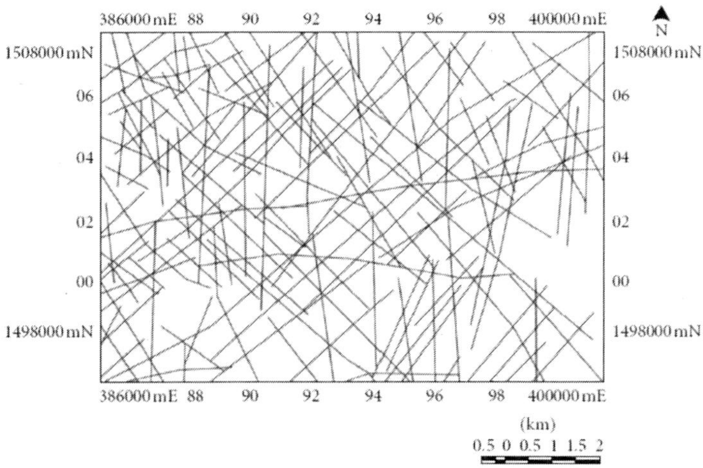

Figure 8: Lineaments extracted from binary image.

2.1.2. Drainage Patterns Factor

The drainage pattern is apparently being controlled by structure and lithology in the study area. The lithologic variation has given a rise to different drainage patterns. For example, radial and dendritic drainage patterns are developed over granitic rocks. Moreover, the most important feature in the area is the presence of drainage lines patterns and fault lines. It is clearly to see that there is a good relationship between these fault lines and drainage pattern system distribution especially with third and fourth river orders in the area.

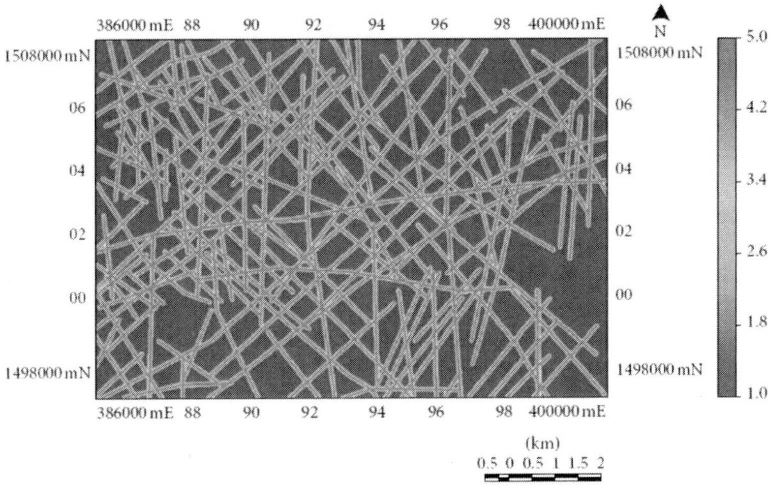

Figure 9: Lineament weight layer.

Geological features are any alignment of features on satellite images such as the various types recognized including topographic, drainage, vegetative, and color alignments. Digital elevation models (DEMs) are very useful in aiding the classification of landforms for many purposes such as recognition of drainage patterns. Extraction of topographic feature information from DEMs has become increasingly popular in structural analysis [40]. Digital elevation models (DEMs) data were used to trace tectonic features and mapping geologically and topographically defined structures in many areas [41]. The Shuttle Radar Topography Mission (SRTM) data were also used to trace drainage pattern system and mapping structures features [42]. The extraction of negative system lines (fault lines) from DEM data mostly resembles drainage lines patterns [43].

SRTM radar images with 90 m resolution were used to obtain contour line data. This contour data was used to create the DEM with 30 m resolution of the study area. The interpolation and resampling processing techniques were used to convert the contour line data (polygon) into 30 m grid layer and the grid layer was converted into a DEM with 30 m spatial resolutions (Figure 10).

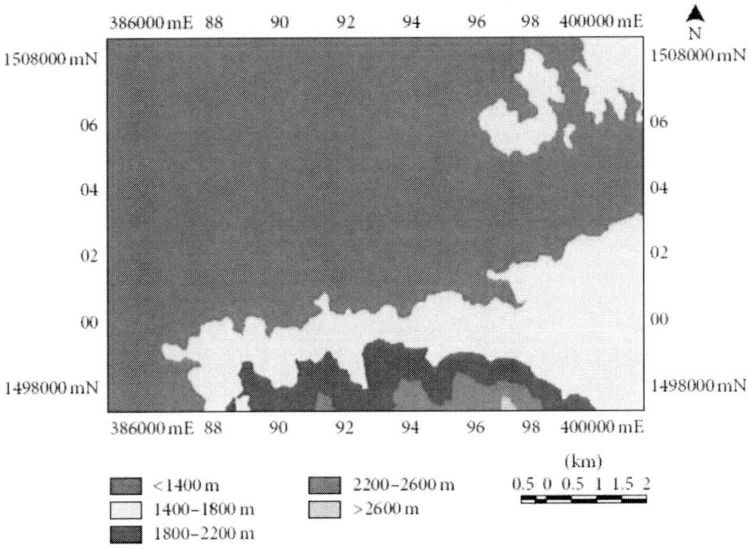

Figure 10: Digital elevation model (DEM) of the study area.

This DEMs data was used for automatic extraction of drainage patterns over the study area using drainage basin from elevation data function (DRAIN) of PCI Geomatica software version 9.1. The drainage basin from elevation data function (DRAIN) locates the drainage networks (river or stream) within the DEM. The output data channel will appear as lines representing the rivers and streams [44].

The result of the drainage pattern map using automatic extraction technique is shown in Figure 11. This technique helps us to extract the drainage patterns over the settlement area, and the topographic maps were used as reference data.

This map was converted into vector layer. The drainage pattern layer (vector layer) was converted into drainage buffer layer and then converted into slicing layer with five classes. This map was converted into weights layer (Figure 12). The weight value was assigned to be between 1 and 5 (Table 1). This layer was ready to be input into GIS modeling.

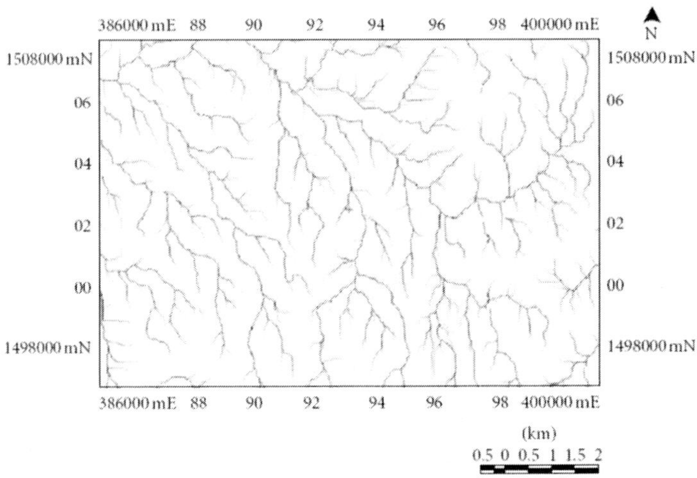

Figure 11: Drainage patterns map showing the nature of drainage system in the area.

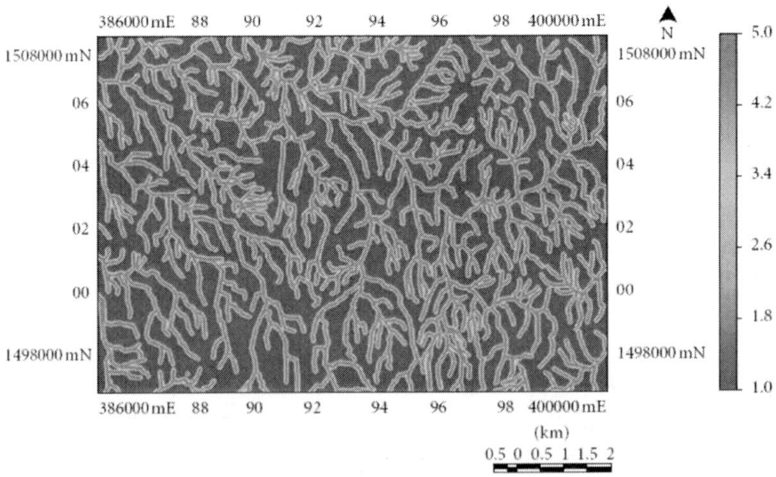

Figure 12: Drainage weight layer.

2.1.3. Lithological Contact Factor

Lithological contacts were found in the field, such as bedding planes and lithological contacts between different rock types. The lithological contact layer of the study area (Figure 13) was digitized from the geological sheet map of Taiz with scale 1 : 250,000. The lithological contact layer (vector layer) was converted into lithological buffer layer and then converted into slicing layer with five classes. And this map was converted into weights layer (Figure 14). The weight value was assigned to be between 1 and 5 (Table 1). This map was ready to be input into GIS modeling.

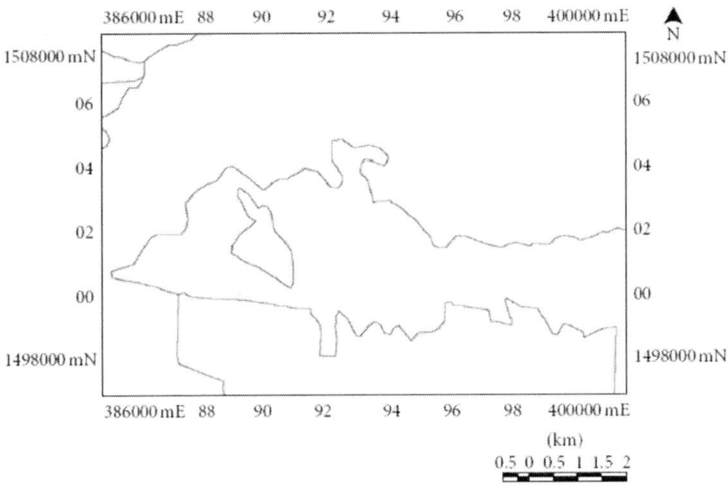

Figure 13: Lithological contacts layer [20].

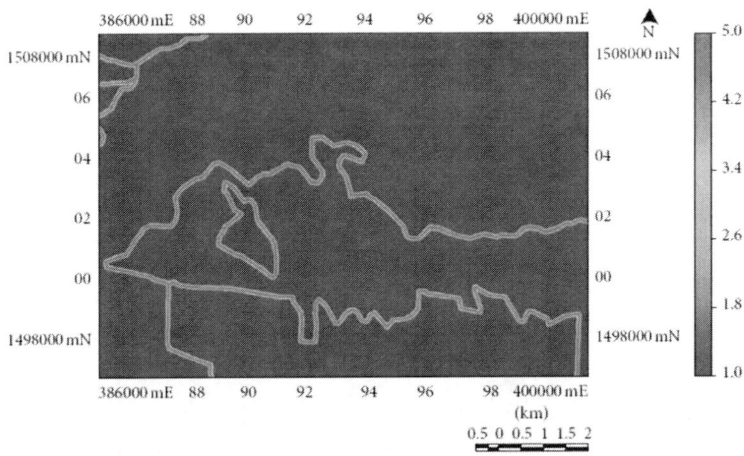

Figure 14: Lithological contacts weight layer.

2.1.4. Fault Factor

Geological structures are patterns such as faults, folds, joints, and others found in the rocks resulting from natural deformation [45]. The fault map of the study area (Figure 15) was digitized from the geological and the hydrogeological sheet maps of Taiz area with scales 1 : 250,000 and 1 : 50,000. The first sheet was prepared by geological survey of Yemen (GSY) and the second was prepared by National Water Resources Authority, Dar El-Yemen Hydro consultant [46]. This layer (fault layer) was converted into fault buffer layer and then converted into slicing layer with five classes. And this map was converted into weights layer (Figure 16). The weight value was assigned to be between 1 and 5 (Table 1). This layer was ready to be input into GIS modeling.

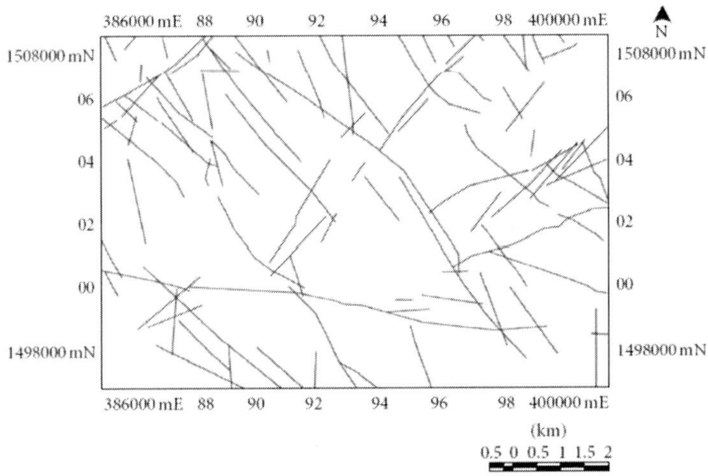

Figure 15: Fault map of the study area [20, 46].

2.2. GIS Model

The major goal of this study is to produce the fault susceptibility and the fault segment maps in order to get information about the possible locations of new faults existing in the area.

Generally, four digital weight layers such as drainage patterns, fault (previously mapped), lineament, and lithological contact layers were prepared from different data. The first step of the GIS modeling involves the input data (weight layers), and the second step includes selection of the suitable equation for this model to determine the susceptibility of faulting segments in the study area.

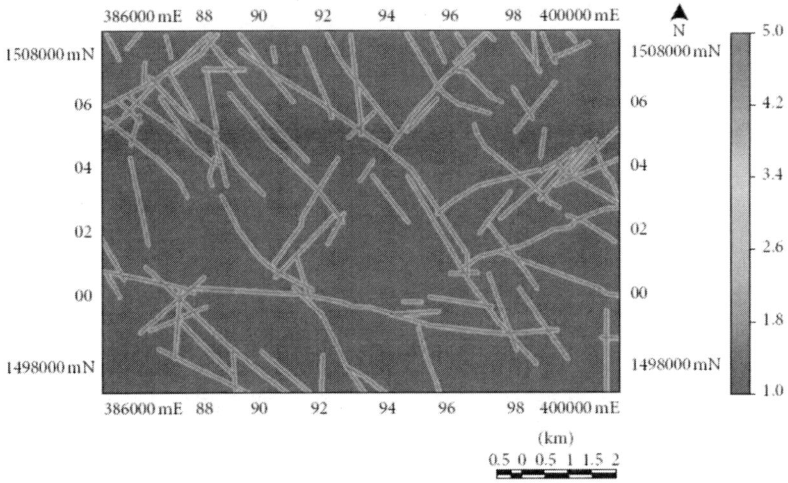

Figure 16: Fault weight layer.

In this study, the overlay layers model (spatial multilayers analysis) was used preformed on these thematic maps with different weights by combining and aggregating these maps together with considering all different parameters. The fault susceptibility model produced with a modified linear equation of drastic model [47], using the following equation:where, DRN_W is the drainage weight layer, FLT_W the fault weight layer (previously mapped), LIN_W the weight layer, and LCT_W the lithological contact weight layer.

Researchers such as [48–50] classified the susceptibility into five potential classes, very low (VL), low (L), medium (M), high (H), and very high (VH).

The resultant map (susceptibility map) was classified into different zones, including very low, low, moderate, high, and very high fault susceptibility zones or equal area classes using ranking values.

2.3. Evaluation of the Susceptibility Map

The above-mentioned technique was used to produce the fault segment susceptibility map from the different data. The result obtained from this technique needs to be evaluated in some manner. The first evaluation was made with respect to the fault map (previous fault map) using GIS overlay technique to determine where the fault segments and true faults are matched over the study area.

During fieldwork, structural data (strike/dip of faults) were collected from 75 stations in the study area. These data were used for the fault segment map evaluation. These stations were located in the fault segments map to make whether there is any correlation between the fault segments and the faults collected from the field.

3. RESULTS AND DISCUSSION

The final susceptibility map of faulting (potential fault zones) was produced as a result of the integration of all parameters of weight layers (drainage, fault (previously mapped), lineament, and lithological contact). This map was classified into five zones, namely, very low potential zone (VL), low potential zone (L), medium potential zone (M), high potential zone (H), and very high potential zone (VH), as shown in Figure 17.

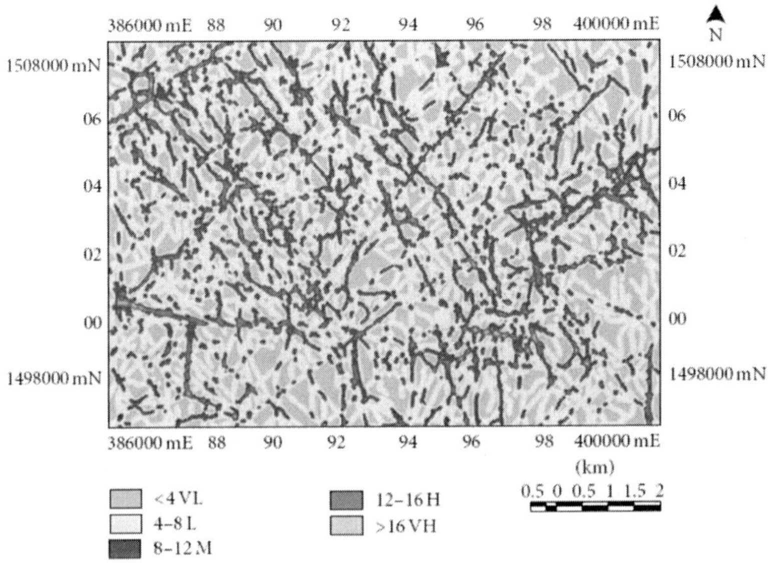

Figure 17: Susceptibility map of fault zones over the study area, classified into five zones, namely, very low potential zone (green), low potential zone (yellow), medium potential zone (blue), high potential zone (pink), and very high potential zone (cyan).

The potential zones in the fault susceptibility map show that the very low potential zone has a total area of 22.87%, and the low potential zone has a total area of 51.89%. More than half of the study area are dominated by very low and low potential zones. And this area does not reflect faults (faults in the previous map), while the 22.40% of moderate, 2.73% of high, and 0.11% of very high potential zones areas almost cover all the faults.

The evaluation of the susceptibility map has been computed by using GIS overlay capability techniques with the fault map of the study area to determine where the potential zones and faults are matched. In order to evaluate our susceptibility map for faulting, a comparison with the known structural fault lines has been done. The result of this technique shows that all of the faults were matched with the moderate to high and very high potential zones (continuous areas from moderate to high and very high).

There were some of moderate to high and very high potential areas in the susceptibility map that could not match any fault lines in the output map of the fault map. These fault segments identified by this technique may be surface reflections of the new fault locations in the study area which were not identified by previous works. This means that the use of this applied model worked successfully to identify linear features (e.g., fault lines) in the area of interest. The results suggest an additional 75 of new fault segment lines, which were not recognized by the fault map over the study area (Figure 18).

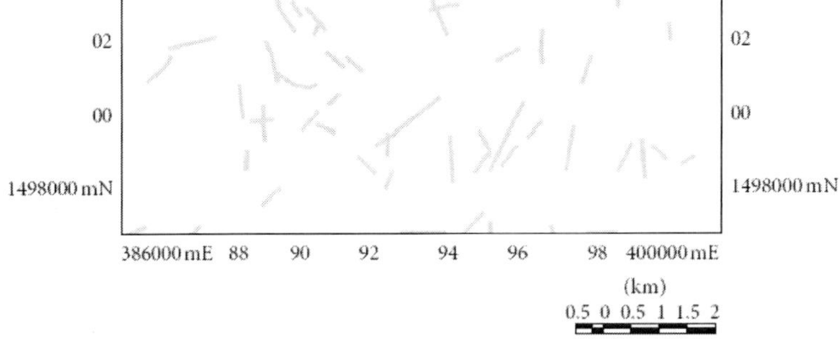

Figure 18: The 75 fault segments (faults lines) as a vector layer extracted from the susceptibility map over the study area.

Moreover, the fault segments obtained from this model were shorter in length compared with faults that were observed in the field. For this reason, this paper attempted to determine the accuracy of this technique for fault segments mapping and the relationship between these segments and ground truth (faults) in terms of the length and the spatial distribution in the study area.

Based on the pattern of fault segments including the length, spatial distributions, and visual interpretation with a topographic map, a drainage pattern map, and three dimensions of DEM and Landsat ETM, the true length of fault segments could be recognized from this analysis and could be implemented in order to assess the extraction accuracy of every true fault length in the area.

Validation of result was made with the geological truth (field work) in the study area. During fieldwork, many faults were identified in different field work outcrops. Strike/dip readings of faults have been collected during the field work from different rock outcrops in the area. Based on the field data, these linear features identified on Landsat imagery and DEM are interpreted to be faults (Figure 19).

Figure 19: The 39 fault segments identified as faults in the study area showing true fault length and points of field work stations, normal fault segments indicated with red points, reverse fault segments indicated with green points, and strike slip fault segments indicated with blue points.

Table 3: Structure reading of faults from the field.

Location	North	East	Strike/dip	Name of fault	Location	North	East	Name of fault
1	1506721	387493	N224E/43°	Normal	21	1504706	392729	Normal
2	1505753	387024	N223E/56°	Normal	22	1504886	392998	Normal
3	1507429	386266	N308E/78°	Reverse	23	1502642	386915	Reverse
4	1503679	386486	N311E/16°	Reverse	24	1501106	385718	Reverse
5	1505813	389149	N312E/79°	Reverse	25	1499311	395042	Shear
6	1500837	396967	N329E/58°	Shear	26	1500278	390156	Reverse
7	1503011	398523	N184E/93°	Shear	27	1500228	392968	Reverse
8	1505863	390066	N313E/96°	Reverse	28	1503100	393357	Normal
9	1499101	398982	N4E/12°	Shear	29	1502003	388032	Normal
10	1507229	390136	N228E/66°	Normal	30	1505354	390315	Shear
11	1505334	389717	N315E/8°	Reverse	31	1502412	388301	Shear
12	1506551	390625	N15E/65°	Normal	32	1499999	393786	Shear
13	1502303	389926	N219E/45°	Normal	33	1500358	393906	Shear
14	1501624	390674	N225E/15°	Normal	34	1502293	394434	Shear
15	1500837	391781	N226E/11°	Normal	35	1498832	398294	Reverse
16	1502632	390305	N227E/24°	Normal	36	1503579	385947	Shear
17	1504397	389448	N305E/18°	Reverse	37	1502143	387334	Shear
18	1499061	397037	N352E/47°	Shear	38	1504347	398483	Reverse
19	1501994	396239	N360E/50°	Shear	39	1505903	397376	Reverse
20	1502861	389757	N217E/47°	Normal				

The fieldwork stations were located in the fault segments map to make whether there is any correlation between the fault segments (faults lines identified by this technique) and the faults data collected from field work stations. There was good correlation between a few fault segments lines and faults, where 39 locations came in the same or nearby the location of fault segments (faults lines). Therefore, these 39 fault segments (faults lines) were identified as new faults (Figure 19). The structural reading (strike/dip) of faults from 93 fieldwork stations are given in Table 3.

Three fault systems were identified. These are the E-W, NW-SE, and NE-SW fault systems [46]. The area was subjected to intensive tectonic activity which resulted in a series of grabens, horsts, and tilted fault blocks. The tectonic activity deformed the area before the Tertiary volcanic activity, especially through the NE-SW faults which run parallel to the rift system of the Gulf of Aden, and it is considered the oldest fractures in the area. The NW-SE faults are run parallel to the opening of the Red Sea and occurred in the area during the Tertiary volcanic activity. The N-S and E-W fault system are the youngest major faults in the area. Smaller faults (as well as joints) also follow the same trends as the major faults. However, more intensive faulting is in the NW-SF direction [46].

Conjugate faults completely defined the principal stress direction. Strike and dip of the faults were plotted in equal-area projection of Schmidt net to obtain and calculate the directions of principal stresses which acted on those faults.

Totally 10 strike-dip readings of the conjugate faults from different rock units of the area were analyzed with the Stereonet program (Stero32 Version 0.9). The stereographic plot was prepared (Figure 20). According to the result of the conjugate faults analysis, compressional principal stress () is measured to be in the NE-SW direction.

This analysis indicates that the N-S and E-W orientated fault sets are sets of shear fractures which could be typical for strike slip faults. The NW-SE and NE-SW orientated sets are sets of tensional and extensional fractures, which could be typical for normal and reverse faults, respectively. Different field observations have also indicated that many damages were found or occurred along the NW-SE faults.

4. CONCLUSIONS

This study shows that remote sensing data and GIS technique are powerful tools in identifying geological structures such as fault segments in the urban area such as the Taiz city, Yemen. The use of different types of data such as drainage patterns, faults (previously mapped), lineament, and lithological contacts with spatial resolution of 30 m was combined through a process of integration and index overlay modeling technique for producing the susceptibility map (potential fault zones) of fault segments in the study area. GIS spatial overlay technique was used to determine the spatial relationships of all the criteria (factors) and subcriteria (classes) within layers (maps) to classify and map the potential area of fault zones of the study area. The fault segments map shows that there is good correlation and distribution between fault segments and faults (previous fault lines) in the study area. There are 39 fault segments which may represent new faults in the area being identified. The presence of these faults is not known from the literature. These faults could be active due to the rift system of the Gulf of Eden and opening Red Sea. Active or reactivated faults can cause enormous damage to settlement or urban (houses or construction project) areas by rock mass movement along the fault surfaces or bedrock cracks, damage which refers to loss of life or injury, or damage to roads. Furthermore, geological survey of Yemen is requested to update the maps and create new ones using new data and techniques and advise the people about the proper sitting of houses and restricting settlement activities in danger areas. GIS models can then yield more accurate estimates of the fault mapping using more data such as geophysical data.

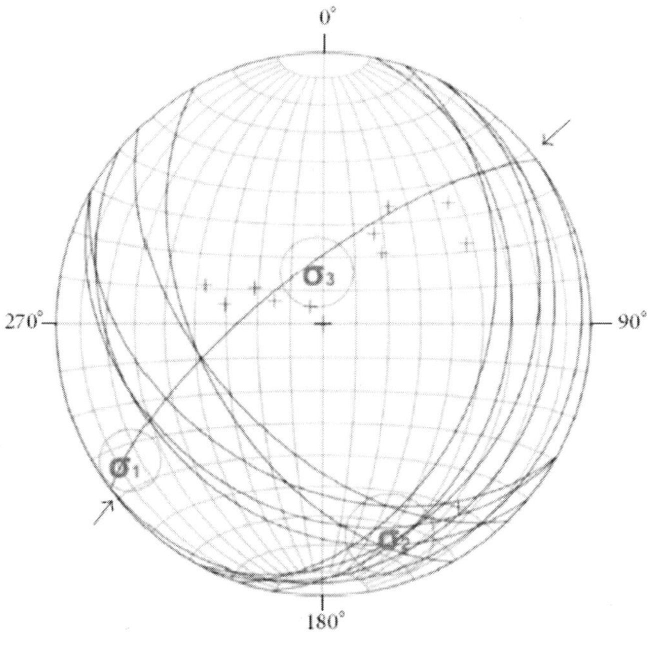

Figure 20: Result of conjugate faults using stereographic analysis.

ACKNOWLEDGMENT

The authors would like to thank the Geology Department, Faculty of Applied Science, Taiz University, Taiz, Yemen for the support to this research and staff for their valuable helps and comments.

REFERENCES

1. B. Dehandschutter, Study of the structural evolution of continental basins in Altai, central Asia [Ph.D. thesis], 2001.
2. W. H. Hobbs, "Lineaments of the Atlantic Border region," Geological Society, vol. 15, pp. 483–506, 1904.View at Google Scholar
3. W. H. Hobbs, "Repeating patterns in the relief and in the structure of the land," Geological Society, vol. 22, pp. 123–176, 1911. View at Google Scholar
4. L. E. Arlegui and M. A. Soriano, "Characterizing lineaments from satellite images and field studies in the central Ebro basin (NE Spain)," International Journal of Remote Sensing, vol. 19, no. 16, pp. 3169–3185, 1998.
5. R. T. Walker, "A remote sensing study of active folding and faulting in southern Kerman province, southeast Iran," Journal of Structural Geology, vol. 28, no. 4, pp. 654–668, 2006.
6. C. Travaglia and N. Dainelli, Groundwater Search by Remote Sensing: A Methodological Approach, Environment and Natural Resources, FAO, Rome, Italy, 2003.
7. M. Morisawa, Rivers, Longman, New York, NY, USA, 1985.
8. C. E. Brockmann, A. Fernandez, R. Ballon, and I. I. Claure, "Analysis of geological structures based on landsat-1 images," in Remote Sensing Applications for Mineral Exploration, W. L. Smith, Ed., pp. 292–317, Dowden, Hutchinson and Ross, Strondsberg, Pa, USA, 1977. View at Google Scholar
9. P. T. Nguyen and D. Ho, "Multiple source data processing in remote sensing," in Digital Image Processing in Remote Sensing, J. P. Muller, Ed., pp. 153–176, Taylor and Francis, Philadelphia, Pa, USA, 1988. View at Google Scholar
10. X. Chen, Application of remote sensing and GIS techniques for environmental geologic investigation, northeast Iowa [Ph.D. thesis], University of Iowa, Iowa, Iowa, USA, 1992.
11. A. Ü. Akman and K. Tüfekçi, "Determination and characterization of fault systems and geomorphological features by RS and GIS techniques in the WSW part of Turkey," in Proceedings of the 20th ISPRS Congress, pp. 899–904, Istanbul, Turkey, 2004.
12. P. R. E. Guerra, Faulting evidence of isostatic uplift in the rincon mountains metamorphic core complex, an image processing analysis [Ph.D. thesis], 2000.
13. M. A. Juhari and A. Ibrahim, "Geological applications of Landsat TM imagery: mapping and analysis of lineaments in NW Peninsula Malaysia," in Proceedings of the 18th Asian Conference on Remote Sensing, pp. J-1-1–J-1-8, Kuala Lumpur, Malaysia, 1997.
14. W. F. P. G. Micheal, "Lineaments analysis South Florida region, aquifer storage and recovery regional study," Draft Technical Memorandum, Central and Southren Florida Project, Army Corps of Engineers, Jacksonville, Fla, USA, 2004. View at Google Scholar
15. I. D. Novak and N. Soulakellis, "Identifying geomorphic features using LANDSAT-5/TM data processing techniques on Lesvos, Greece," Geomorphology, vol. 34, no. 1-2, pp. 101–109, 2000.
16. S. Solomon and W. Ghebreab, "Lineament characterization and their tectonic significance using Landsat TM data and field studies in the central highlands of Eritrea," Journal of African Earth Sciences, vol. 46, no. 4, pp. 371–378, 2006.
17. K. S. Kavak and H. Cetin, "A detailed geologic lineament analysis using landsat TM data of Gölmarmara/Manisa region, Turkey," Online Journal of Earth Sciences, vol. 1, no. 3, pp. 145–153, 2007.View at Google Scholar

18. L. A. Rutty, The basement fracture pattern of sothern Ontario: a tectonic interpretation based on landsat TM imagery, airphotos and field data [M.S. thesis], National Library of Canada, Ottawa, Canada, 1993.
19. K. M. M. Elias, "Multiple data set integration for structural and stratigraphic analysis of Oil and Gas bearing formation using GIS," in Proceedings of the Map India Conference, Geology & Mineral Resource, 2003.
20. Geological Survey of Yemen (GSY), Geological Map of Taiz Area (1:250, 000), 1990.
21. S. Bai, J. Wang, G. N. Lu, P. G. Zhou, S. S. Hou, and S. N. Xu, "GIS-based logistic regression for landslide susceptibility mapping of the zhongxian segment in the three gorges area, China," Geomorphology, vol. 115, pp. 23–31, 2010. View at Google Scholar
22. R. L. Bates and J. A. Jackson, Glossary of Geology, American Geological Institute, Alexandria, Va, USA, 1987.
23. M. L. Süzen and V. Doyuran, "Data driven bivariate landslide susceptibility assessment using geographical information systems: a method and application to Asarsuyu catchment, Turkey," Engineering Geology, vol. 71, no. 3-4, pp. 303–321, 2004.
24. F. Guzzetti, A. Carrara, M. Cardinali, and P. Reichenbach, "Landslide hazard evaluation: a review of current techniques and their application in a multi-scale study, Central Italy," Geomorphology, vol. 31, no. 1–4, pp. 181–216, 1999.
25. C. F. Chung and A. G. Fabbri, "Probabilistic prediction models for landslide hazard mapping," Photogrammetric Engineering and Remote Sensing, vol. 65, no. 12, pp. 1389–1399, 1999.
26. F. C. Dai and C. F. Lee, "Landslide characteristics and slope instability modeling using GIS, Lantau Island, Hong Kong," Geomorphology, vol. 42, no. 3-4, pp. 213–228, 2002.
27. L. Ayalew, H. Yamagishi, and N. Ugawa, "Landslide susceptibility mapping using GIS-based weighted linear combination, the case in Tsugawa area of Agano river, Niigata prefecture, Japan," Landslides, vol. 1, pp. 73–81, 2004. View at Google Scholar
28. T. B. Minor, J. A. Carter, M. M. Chesley, and R. B. Knowles, "An integrated approach to groundwater exploration in developing countries using GIS and remote sensing," in Proceedings of the International American Congress on Surveying and Mapping/American Society for Photogrammetry and Remote Sensing (ACSM/ASPRS '94), pp. 418–428, 1994.
29. S. Lee and K. Min, "Statistical analysis of landslide susceptibility at Yongin, Korea," Environmental Geology, vol. 40, no. 9, pp. 1095–1113, 2001.
30. L. Donati and M. C. Turrini, "An objective method to rank the importance of the factors predisposing to landslides with the GIS methodology: application to an area of the Apennines (Valnerina, Perugia, Italy)," Engineering Geology, vol. 63, no. 3-4, pp. 277–289, 2002.
31. A. Günther, "SLOPEMAP: programs for automated mapping of geometrical and kinematical properties of hard rock hill slopes," Computers and Geosciences, vol. 29, no. 7, pp. 865–875, 2003.
32. S. A. Ali and S. Pirasteh, "Geological applications of Landsat Enhanced Thematic Mapper (ETM) data and Geographic Information System (GIS): mapping and structural interpretation in south-west Iran, Zagros Structural Belt," International Journal of Remote Sensing, vol. 25, no. 21, pp. 4715–4727, 2004.
33. G. Jordan, B. M. L. Meijninger, D. J. J. V. Hinsbergen, J. E. Meulenkamp, and P. M. V. Dijk, "Extraction of morphotectonic features from DEMs: development and applications for study areas in Hungary and NW Greece," International Journal of Applied Earth Observation and Geoinformation, vol. 7, no. 3, pp. 163–182, 2005.
34. A. Anwar, M. A. Juhari, and A. Ibrahim, "The extraction of lineaments using slope image derived from digital elevation model: case study of Sungai Lembing—Maran area, Malaysia," Journal of Applied Sciences Research, vol. 6, no. 11, pp. 1745–1751, 2010.
35. K. Koike, S. Nagano, and M. Ohmi, "Lineament analysis of satellite images using a Segment Tracing Algorithm (STA)," Computers and Geosciences, vol. 21, no. 9, pp. 1091–1104, 1995.

36. A. Mah, G. R. Taylor, P. Lennox, and L. Balia, "Lineament analysis of Landsat Thematic Mapper images, Northern Territory, Australia," Photogrammetric Engineering & Remote Sensing, vol. 61, no. 6, pp. 761–773, 1995.
37. A. Anwar, M. A. Juhari, and A. Ibrahim, "A comparison of landsat TM and SPOT data for lineament mapping in Hulu Lepar area, Pahang, Malaysia," European Journal of Scientific Research, vol. 34, no. 3, pp. 406–415, 2009.
38. A. Masoud and K. Koike, "Tectonic architecture through Landsat-7 ETM+/SRTM DEM-derived lineaments and relationship to the hydrogeologic setting in Siwa region, NW Egypt," Journal of African Earth Sciences, vol. 45, no. 4-5, pp. 467–477, 2006.
39. M. L. Süzen and V. Toprak, "Filtering of satellite images in geological lineament analyses: an application to a fault zone in Central Turkey," International Journal of Remote Sensing, vol. 19, no. 6, pp. 1101–1114, 1998.
40. A. Ganas, S. Pavlides, and V. Karastathis, "DEM-based morphometry of range-front escarpments in Attica, central Greece, and its relation to fault slip rates," Geomorphology, vol. 65, no. 3-4, pp. 301–319, 2005.
41. S. Sarapirome, A. Surinkum, and P. Saksutthipong, "Application of DEM data to geological interpretation: Thong Pha Phum area, Thailand," in Proceedings of the 23rd Asian Conference on Remote Sensing (ACRS '02), Kathmandu, Nepal, 2002.
42. G. Sarp and V. Toprak, "An Integrated Lineament Analysis from Satellite Images," in Proceedings of the 28th Asian Conference on Remote Sensing (ACRS '07), Kuala Lumpur, Malaysia, 2007.
43. A. Anwar, M. A. Juhari, and A. Ibrahim, "Automatic mapping of lineaments using shaded relief images derived from digital elevation model (DEMs) in the Maran—Sungi Lembing area, Malaysia," Electronic Journal of Geotechnical Engineering, vol. 15, pp. 949–957, 2010.
44. P. C. I. Geomatica, PCI Geomatica User's Guide Version 9. 1, Richmond Hill, Ontario, Canada, 2001.
45. A. Ibrahim and M. A. Juhari, Dictionary of Geological the Basic Terms, Malaysia, National University of Malaysia, Selangor, Malaysia, 1990.
46. National Water Resources Authority, Hydrogeologic Map of Taiz Area (1: 50, 000), Dar El-Yemen Hydro Consultant, 1997.
47. L. Aller, T. Bennett, J. H. Lehr, and R. J. Petty, "DRASTIC: a standard system for evaluating groundwater pollution potential using hydrogeologic settings," Tech. Rep. EPA/600/2 85/018 R.S., Kerr Enviromental Research Laboratory, Enviromental Protection Agency, Ada, Okla, USA, 1995. View at Google Scholar
48. M. J. Crozier, "Field Assessment of Slope Instability," in Slope Instability, D. Brunsden and D. Prior, Eds., pp. 103–142, John Wiley and Sons, New York, NY, USA, 1984. View at Google Scholar
49. Q. Zaruba and V. Mencl, Landslides and Their Control, Elsevier, Amsterdam, The Netherlands, 1982.
50. I. Das, S. Sahoo, C. van Westen, A. Stein, and R. Hack, "Landslide susceptibility assessment using logistic regression and its comparison with a rock mass classification system, along a road section in the northern Himalayas (India)," Geomorphology, vol. 114, no. 4, pp. 627–637, 2010.

CHAPTER 9

Petroleum geology features and research developments of hydrocarbon accumulation in deep petroliferous basins

Xiong-Qi Pang 1,2, Cheng-Zao Jia 3,4, Wen-Yang Wang1,2

1.State Key Laboratory of Petroleum Resources and Prospecting, Beijing, China

2.Basin and Reservoir Research CenterChina University of Petroleum, Beijing, China

3.PetroChina Company Limited. Beijing, China

4.Research Institute of Petroleum Exploration and Development, Beijing, China

ABSTRACT

As petroleum exploration advances and as most of the oil–gas reservoirs in shallow layers have been explored, petroleum exploration starts to move toward deep basins, which has become an inevitable choice. In this paper, the petroleum geology features and research progress on oil–gas reservoirs in deep petroliferous basins across the world are characterized by using the latest results of worldwide deep petroleum exploration. Research has demonstrated that the deep petroleum shows ten major geological features. (1) While oil–gas reservoirs have been discovered in many different types of deep petroliferous basins, most have been discovered in low heat flux deep basins. (2) Many types of petroliferous traps are developed in deep basins, and tight oil–gas reservoirs in deep basin traps are arousing increasing attention. (3) Deep petroleum normally has more natural gas than liquid oil, and the natural gas ratio increases with the burial depth. (4)

The residual organic matter in deep source rocks reduces but the hydrocarbon expulsion rate and efficiency increase with the burial depth. (5) There are many types of rocks in deep hydrocarbon reservoirs, and most are clastic rocks and carbonates. (6) The age of deep hydrocarbon reservoirs is widely different, but those recently discovered are predominantly Paleogene and Upper Paleozoic. (7) The porosity and permeability of deep hydrocarbon reservoirs differ widely, but they vary in a regular way with lithology and burial depth. (8) The temperatures of deep oil–gas reservoirs are widely different, but they typically vary with the burial depth and basin geothermal gradient. (9) The pressures of deep oil–gas reservoirs differ significantly, but they typically vary with burial depth, genesis, and evolution period. (10) Deep oil–gas reservoirs may exist with or without a cap, and those without a cap are typically of unconventional genesis. Over the past decade, six major steps have been made in the understanding of deep hydrocarbon reservoir formation. (1) Deep petroleum in petroliferous basins has multiple sources and many different genetic mechanisms. (2) There are high-porosity, high-permeability reservoirs in deep basins, the formation of which is associated with tectonic events and subsurface fluid movement. (3) Capillary pressure differences inside and outside the target reservoir are the principal driving force of hydrocarbon enrichment in deep basins. (4) There are three dynamic boundaries for deep oil–gas reservoirs; a buoyancy-controlled threshold, hydrocarbon accumulation limits, and the upper limit of hydrocarbon generation. (5) The formation and distribution of deep hydrocarbon reservoirs are controlled by free, limited, and bound fluid dynamic fields. And (6) tight conventional, tight deep, tight superimposed, and related reconstructed hydrocarbon reservoirs formed in deep-limited fluid dynamic fields have great resource potential and vast scope for exploration. Compared with middle–shallow strata, the petroleum geology and accumulation in deep basins are more complex, which overlap the feature of basin evolution in different stages. We recommend that further study should pay more attention to four aspects: (1) identification of deep petroleum sources and evaluation of their relative contributions; (2) preservation conditions and genetic mechanisms of deep high-quality reservoirs with high permeability and high porosity; (3) facies feature and transformation of deep petroleum and their potential distribution; and (4) economic feasibility evaluation of deep tight petroleum exploration and development.

Keywords: Petroliferous basin, Deep petroleum, geology features, Hydrocarbon accumulation, Petroleum exploration, Petroleum resources

1. INTRODUCTION

As the world demands more petroleum and petroleum exploration continues, deep petroleum exploration has become an imperative trend. As it is nearly impractical to expect any major breakthrough in middle or shallow basins (Tuo 2002), petroleum exploration turning toward deep basins has become inevitable. After half a century's exploitation in major oilfields across the world, shallow petroleum discoveries tend to be falling sharply (Simmons 2002). Nor are things optimistic in China, where the rate of increase of mid-and-shallow petroleum reserves is increasingly slowing down (Wang et al. 2012). At the same time, the world's petroleum consumption continues to increase. According to BP Statistical Review of World Energy 2014, from 2002 to 2012, the world's petroleum consumption increase was virtually the same as its petroleum output increase, with annual average oil and natural gas consumption increases of 1.35 % and 3.14 %, respectively, compared to the annual average output increases of 1.47 % and 3.23 %. In China, however, the petroleum consumption increase is far greater than its output increase. According to National Bureau of Statistic of China statistics in 2013, from 2000 to 2013, China's average annual oil and natural gas consumption increases were 6.1 % and 14.6 %, respectively, compared to the annual average output increases of 1.93 % and 11.9 %. Deep petroleum, as one of the strategic "three-new" fields for the global oil industry (Zou 2011) as well as one of the most important development areas for China's oil industry, forms the most important strategically realistic area for China's oil industry to lead future petroleum exploration and development (Sun et al. 2013). All indicate that deep hydrocarbon exploration is an inevitable choice toward ensuring energy supply and meeting market demands.

After half a century's effort, gratifying achievements have been made in deep petroleum exploration throughout the world, despite being faced with challenges and problems today. The Former Soviet Union discovered four 6,000 m or deeper industrial oil–gas reservoirs out of its 24 petroliferous basins (Tuo 2002). The oil discovered in deep basins in Mexico, the USA, and Italy contributes more than 31 % of their present recoverable oil reserves (Kutcherov et al. 2008) and the natural gas discovered there makes up approximately 47 % of their total proved natural gas reserves (Burruss 1993). China, too, has appreciable achievements in deep petroleum exploration. Compared with 2000, the deep reserves discovered in West China in 2013 increased an average of 3.5 times. The ratio of deep petroleum reserves increased from 40 % in 2002 to 80 % in 2013. Of the 156 well intervals in the Tarim Basin that have been tested so far, 58 have gone deeper than 5,000 m. The deep drillhole success rate in the Jizhong Depression is as high as 21.4 %. A

5,190-m-deep “Qianmiqiao buried hill hydrocarbon reservoir” was discovered in the Huanghua Depression (Tuo 2002). Despite these achievements, however, a lot of problems have also emerged in deep petroleum exploration. These include (a) the difficulty in understanding the conditions of deep oil–gas reservoirs and evolution due to the multiple tectonic events having taken place in deep basins (Zhang et al. 2000; He et al. 2005), (b) the difficulty in evaluating the resource potential and relative contribution due to the complex sources and evolution processes of deep petroleum (Barker 1990; Mango 1991; Dominé et al. 1998; Zhao et al. 2001; Jin et al. 2002; Zhao et al. 2005; Darouich et al. 2006; Huang et al. 2012; Pang et al. 2014a), (c) the difficulty in predicting and evaluating favorable targets due to the complex genesis and distribution of deep, relatively high-porosity and high-permeability reservoirs (Surdam et al. 1984; Ezat 1997; Dolbier 2001; Rossi et al. 2001; Moretti et al. 2002; Lin et al. 2012), and (d) the difficulty in predicting and evaluating the petroleum possibility in deposition targets due to the complex deposition mechanism and development pattern of deep petroleum (Luo et al. 2003, 2007; Ma and Chu 2008; Ma et al. 2008; Pang et al. 2008). All these problems provide a tremendous challenge to deep petroleum exploration.

With abundant resource bases and low proved rates, deep petroliferous basins are important for further reserve and output increases (Tuo 2002; Zhao et al. 2005; Dai 2006, Pang et al. 2007a; Zhu and Zhang 2009; Sun et al. 2010; Pang et al. 2014a). According to Dai (2006), the proved rate of the exploration concessions of PetroChina is 17.6 % for deep oil and 9.6 % for deep natural gas, far lower than their mid-and-shallow counterparts of 39.6 % for oil and 14.6 % for natural gas. Pang et al. (2007a, b) suggest that West China contains around 45 % of the residual petroleum resources of China, and 80 % of these residual resources are buried in deep horizons more than 4,500 m below the surface, yet the present proved rate is less than 20 %. As such, implementing deep petroleum resource research, tapping deep petroleum and increasing petroleum backup reserves are urgently needed if we ever want to relieve the nation’s petroleum shortage and mitigate energy risks. Many scholars have investigated deep petroleum geologies and exploration (Perry 1997; Dyman et al. 2002; Pang 2010; Ma et al. 2011; He et al. 2011; Wang et al. 2012; Wu et al. 2012; Bai and Cao 2014). Our study in connection with the national “973 Program” (2011CB201100) involves a summary and description of the development and orientation of research by scholars in China and elsewhere with respect to petroleum geology and hydrocarbon accumulation in deep petroliferous basins.

2. CONCEPT AND DIVISION CRITERIA OF DEEP BASINS

Deep basins are also called deep formations by some scholars. The definition and criteria of deep petroliferous basins differ from country to country, from institution to institution and from scholar to scholar.

2.1 Concept and division criteria of deep basins proposed by overseas scholars

So far, there are two sets of definition and criteria for deep petroliferous basins outside China. One is according to the formation depth, i.e., formations within a certain limit of depth are called deep formations. However, the criteria for classifying deep basins also differ from scholar to scholar. Representative criteria include 4,000 m (Rodrenvskaya 2001), 4,500 m (Barker and Takach 1992), 5,000 m (Samvelov 1997; Melienvski 2001), and 5,500 m (Manhadieph 2001; Bluokeny 2001). Another is according to the formation age, i.e., for a given basin, formations older in age and deeper are called deep formations (Sugisaki 1981). Table 1 summarizes the criteria used by different institutions and scholars for deep basins from which it is easy to see that 4,000 and 4,500 m are the criteria accepted by more institutions and scholars.

Table 1. Criteria for deep basins proposed by scholars outside China

Basis for deep basins	Criteria for deep basins	Targeted area	Researcher and year
Formation depth	>4,000 m	In former Soviet Union	Rodrenvskaya (2001)
	>4,500 m	Caspian Basin	
	>4,500 m	In the USA	
		Gulf of Mexico, USA	Barker and Takach (1992)
	>5,000 m		Samvelov (1997)
		West Siberia Basin, East Siberia Basin	Melienvski (2001)
	>5,500 m	South Caspian Basin	Manhadieph (2001)
		Timan-Pechora Basin	Bluokeny (2001)
Formation age	Stratigraphically old formations with large buried depths	In the USA	Sugisaki (1981)

2.2 Concept and division criteria of deep basins proposed by Chinese scholars

Chinese scholars use roughly the same criteria for deep petroliferous basins as their overseas counterparts. Most of them use three indicators: (1) formation depth (Wang et al. 1994; Li and Li 1994; Tuo et al. 1994; Zhou et al. 1999; Hao et al. 2002; Shi et al. 2005; Dai et al. 2005); (2)

formation age (Kang 2003; Ma et al. 2007; Ma and Chu 2008); and (3) formation characteristics (Tuo et al. 1999a; Wang et al. 2001; Wang 2002; Pang 2010). Table 2 summarizes the criteria used for deep basins. Obviously, the concept of deep basins does not only differ from scholar to scholar, it also varies with the basin position and formation characteristics.

Table 2. Criteria of deep basins proposed by Chinese scholars

Basis for deep basins	Criteria for deep basins	Targeted area or parameter features	Researcher and year
Formation depth	>2,500 m	Bohai Bay Basin	Qiao et al. (2002)
	>2,800 m	Songliao Basin	Wang et al. (1994)
	>3,500 m	Liaohe Basin	Li et al. (1999)
		Bohai Bay Basin	Tuo (1994)
		Bohai Bay Basin	Zhou et al. (1999)
		Yinggehai Basin	Hao et al. (2002)
	>3,500 m	East China basins	Ministry of Land and Resources (2005)
	>4,500 m	Junggar Basin	Shi et al. (2005)
		Tarim, Junggar, Sichuan basins	Dai (2003)
		Sichuan Basin	Zhao et al. (2005)
	>4,500 m	West China basins	Ministry of Land and Resources (2005)
Formation age & depth	Stratigraphically old with large buried depths	Varies from basin to basin	Kang (2003), Ma et al. (2007), Ma and Chu (2008)
Formation characteristics	Formation thermal evolution level	$R_o \geq 1.35$ %	Tuo et al. (1999b), Tuo (2002)
	Formation thermal evolution level or formation pressure	$R_o \geq 1.35$ % or formation depth overpressure	Wang et al. (2001), Tuo et al. (1999b), Wang et al. (2002)
	Formation thermal evolution level and tightness level	$R_o \geq 1.35$ % or sandstone formation $\Phi \leq 12$ %, $K \leq 1$ mD, $Y \leq 2$ μm	Pang (2010)

2.3 Importance of using the same concept and criteria in deep basins

No uniform concept or criteria have been agreed upon by scholars either in or out of China with respect to deep petroliferous basins, hence preventing further development and mutual promotion on science research. For this reason, we suggest using 4,500 m as the criterion for deep basins on grounds of the following considerations:

First, this classification represents a succession to previous findings. The U.S. Geological Survey and some former Soviet Union scholars used 4,500 m as the criteria for deep basins (Barker and Takach 1992). Chinese scholars, represented by Shi, Dai, and Zhao et al., also used 4,500 m to demarcate deep formations (Shi et al. 2005; Dai et al. 2005; Zhao et al. 2005). Chinese administrations like Ministry of Land and Resources even issued public documents that define deep petroliferous basins in West China to 4,500 m. Second, 4,500 m represents the general depth at which the hydrocarbon entrapment mechanism of a petroliferous basin transits from buoyancy accumulation to non-buoyancy accumulation. Above this depth, the porosities of the sand reservoirs are generally above 12 %; the permeabilities are higher than 1 mD; and the pore throat radii are larger than 2 μm. "High-point accumulation, high-stand closure, high-porosity enrichment, high-pressure accumulation" (Pang et al. 2014a) normal oil–gas reservoirs generally formed under the action of buoyant forces. Below this depth, to the contrary, "low-depression accumulation, low-stand inversion, low-porosity enrichment, low-pressure stability" unconventional oil–gas reservoirs generally formed. To make things easier, we divide a petroliferous basin into four parts according to the buried depth, using the criteria accepted by previous scholars: shallow (<2,000 m), middle (2,000–4,500 m), deep (4,500–6,000 m), and ultra-deep (>6,000 m). According to the maximum depths of basins, we divide them into shallow basins (<2,000 m), middle basins (2,000–4,500 m), deep basins (4,500–6,000 m), and ultra-deep basins (>6,000 m). Third, deep basins should be classified according to the depth rather than incorporating the geological aspects that constrain the depth distribution of hydrocarbon entrapment. For example, the fact that oil–gas reservoirs in East China basins are commonly shallow while those in West China basins are commonly deep is attributable to their respective unique basin evolution geologies such as the geothermal gradient, reservoir rock type, formation age, and evolution history. These should not form the basis for diverging the criteria for basin depths.

3. EXPLORATION FOR DEEP OIL–GAS RESERVOIRS

3.1 Exploration for deep oil–gas reservoirs across the world

Following the discovery of the first deep hydrocarbon field below 4,500 m in the USA in 1952, deep petroleum exploration boomed in many countries. Seventy countries tried deep exploration (Wu and Xian 2006). Echoing breakthroughs in deep well drilling and completion techniques, a succession of major breakthroughs have been made in deep hydrocarbon reservoir exploration (Dyman et al. 2002). First, major breakthroughs in drilling operation led to the discovery of a number of oil–gas reservoirs including a gas reservoir in the Cambrian–Ordovician Arbuckle Group dolomites at 8,097 m depth in the Mills Ranch gas field in the Anadarko Basin in 1977 (Jemison 1979). From 1980, deep petroleum exploration started to extend from onshore to offshore. Examples include a gas field discovered in Permian Khuff Formation limestones at 4,500 m in the Fateh gas reservoir in the Arabian-Iranian Basin in 1980, and an oil reservoir at a depth of 6,400 m was discovered in the Triassic dolomites of the Villifortuna-Trecate oilfield in Italy in 1984. Recently, major breakthroughs in deep oil exploitation have been reported in the deep and ultra-deep waters of the Gulf of Mexico, East Brazil, and West Africa (Bai and Cao 2014). According to IHS data as of 2010, 171 deep basins and 29 ultra-deep basins had been discovered out of the 1,186 petroliferous basins in the world. These deep basins are predominantly situated in the former Soviet Union, Middle East, Africa, Asia-Pacific, North America, and Central and South America (Fig. 1). A total of 1,290 oil–gas reservoirs have been discovered in deep basins and 187 oil–gas reservoirs in ultra-deep basins across the world. Breakthroughs are continuously reported around the world in the course of deep exploration. First, the drilling depth continues to increase, the maximum being deeper than 10,000 m, as exemplified by the deepest well with 12,200m drilling depth, SG-3 exploratory well. The deepest oil reservoir discovered so far is the Tiber clastic rock oil reservoir (1,259 m underwater and 8,740 m underground). The depth of gas wells continues to increase, and the deepest gas reservoir discovered so far (8,309–8,322 m) is a Silurian basin gas reservoir in the Anadarko Basin. Second, the manageable formation temperature and formation pressure in drilling operations are also continuously increasing. So far, the highest temperature encountered is 370 °C and the highest pressure encountered is 172 MPa (Table 3).

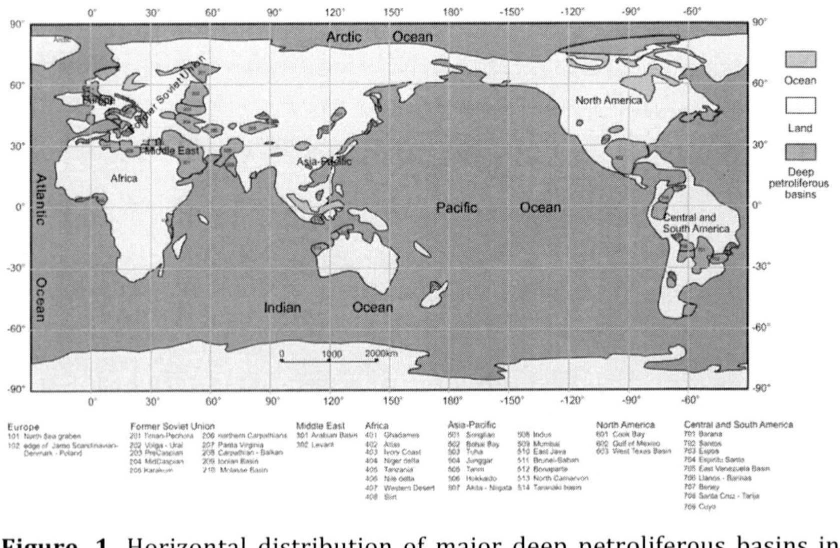

Figure. 1. Horizontal distribution of major deep petroliferous basins in the world

Table 3. Geological characteristics of world representative deep oil–gas reservoirs known so far

Feature	Name	Year	Parameters	Region
Deepest well	SG-3 exploratory well	1992	Completion depth 12,200 m	Kola Peninsula, Russia
Deepest oil reservoir	Tiber clastic rock oil reservoir	2009	Buried depth 8,740 m	Gulf of Mexico abyssal basin, USA
Deepest gas reservoir	Mills Ranch gas reservoir	1977	Buried depth 7,663–8,083 m	Western Interior Basin, USA
Deep hydrocarbon reservoir with highest porosity	Gaenserndorf Ubertief oilfield Hauptdolomit Formation gas reservoir	1977	Porosity 35 %–38 %	Vienna Basin, Austria
Deep hydrocarbon reservoir with lowest porosity	Mora hydrocarbon reservoir	1981	Porosity 2.6 %–4 %	Sureste Basin, Mexico
Deep hydrocarbon reservoir with highest permeability	Platanal oilfield 4830–Cretaceous hydrocarbon reservoir	1978	Permeability 7,800 mD	Sureste Basin, Mexico
Deep hydrocarbon reservoir with lowest permeability	Wolonghe Huanglong structural belt gas reservoir	1980	Permeability 0.01 mD	Sichuan Basin, China
Deep gas reservoir with highest temperature	Satis hydrocarbon reservoir, Tineh Formation gas reservoir	2008	Temperature 370 °C	Nile Delta Basin, Egypt
Deep oil reservoir with lowest temperature	Sarutayuskoye oilfield Starooskolskiy Group oil reservoir	2008	Temperature 47 °C	Pechola Basin, Russia
Deep gas reservoir with highest pressure	Zistersdorf Ubertief 1 oilfield Basal Breccia gas reservoir	1980	Pressure 172 MPa	Vienna Basin, Austria
Deep hydrocarbon reservoir with lowest pressure	Akzhar East oilfield Asselian VIII (PreCaspian) Unit hydrocarbon reservoir	1988	Pressure 8.4 MPa	Caspian Basin, Kazakhstan

According to USGS and World Petroleum Investment Environment Database, from 1945 to 2014, the world's normal petroleum resource has increased from 96 billion ton in 1945 to 630 billion ton in 2014, the annual average increase being as high as 8.06 % (Fig. 2a), and the natural gas resource has also increased from 260 trillion m^3 in 1986 to 460 trillion m^3 in 2013, the annual average increase being as high as 2.85 % (Fig. 2b). Over the past years, the world has shown robust momentum for deep petroleum exploration. The number of oil–gas reservoirs discovered keeps growing fast (Fig. 3). According to data provided by Kutcherov et al. (2008), more than 1,000 hydrocarbon fields have been developed at depths of 4,500–8,103 m, the original recoverable oil reserve of which contributes 7 % of the world's total amount and the natural gas reserve makes up 25 %. According to IHS data, as of 2010, for the 4,500–6,000 m deep hydrocarbon fields in the world, the proved recoverable residual oil reserve is 83.8 billion ton or 35.5 % of the total recoverable oil reserve, and the natural gas is 65.9 billion ton oil equivalent or 44.4 % of the total productive natural gas reserve; for the 6,000 m or deeper hydrocarbon fields in the world, the proved recoverable residual oil reserve is 10.5 billion ton or 4.45 % of the total productive oil reserve, and the natural gas is 7 billion ton oil equivalent or 4.7 % of the total productive natural gas reserve (Fig. 4).

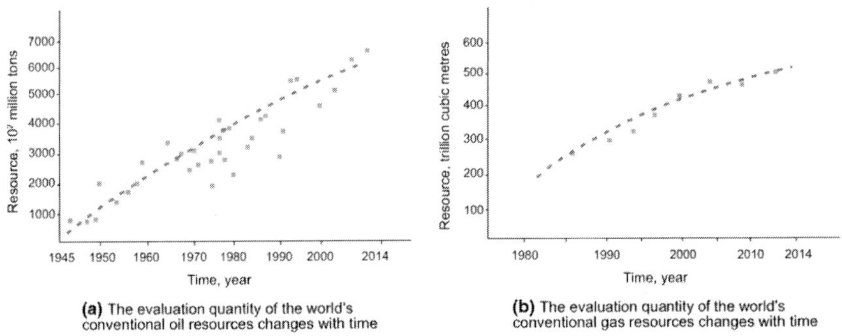

Figure. 2. World petroleum evaluation result and variation as a function of time

3.2 Exploration of deep oil–gas reservoirs in China

China started deep petroleum exploration from the late 1970s, having discovered a number of large deep oil–gas fields in the deep parts of some large sedimentary basins including Tarim, Erdos, and Sichuan basins, and made important progresses in the deep parts of the Daqing, Zhongyuan,

Dagang, and Shengli fields in East China's petroleum region (Feng 2006; Song et al. 2008; Wu and Xian 2006). With the nation's breakthroughs in deep and ultra-deep well drilling techniques and equipment, onshore petroleum exploration has continued to extend toward deep and ultra-deep levels (Sun et al. 2010); petroleum exploration has also undergone a transition from shallow to deep and further to ultra-deep levels. On July 28, 1966, China's first deep well, Songji-6 of Daqing (4,719-m well depth), was completed, marking the transition of China's drilling operation from shallow wells to middle and deep wells, and signaling that China's petroleum exploration was turning from shallow toward deep levels. From 1976, China's petroleum exploration marched toward ultra-deep levels. On April 30, 1976, China's first ultra-deep well, Nuji well in Sichuan (6,011 m well depth), was completed, marking the entry of China's petroleum exploration into ultra-deep levels (Wang et al. 1998). So far, China has drilled deep wells in 15 large basins with sedimentary thicknesses larger than 5 km (Pang 2010). Of the 176 deep exploratory wells drilled in the Jizhong Depression (the average well depth is 4,521 m), 37 have yielded industrial petroleum flows. The exploratory well success rate has reached 21.4 % (Tuo 2002). Of the 156 pay zone well intervals tested in the Tarim Basin, 58 have their bottom boundaries deeper than 5,000 m (Pang 2010). According to statistics, as of 2010, of the 47 petroliferous basins in China, seven deep basins have been discovered, out of which 210 deeper than 4,500 m oil–gas reservoirs have been identified. Shallow basins at a depth of 2,000 m or shallower are predominantly found in China's Inner Mongolia and Tibet; 2,000–4,500 m middle deep basins are typically located in the seas of East China; 4,500–6,000 m deep basins are distributed in Central China and Southern North China. China also has a huge stock of ultra-deep petroleum resources, having discovered some ultra-deep basins with buried depths of more than 6,000 m, including the Tarim Basin and Songliao Basin. These are mostly located in Northwest and Northeast China (Fig. 5).

China is rich in deep petroleum resources with vast room for further exploration. According to a 2005 statistics of Shi et al., the deep oil resource within the mineral concession of CNPC is approximately 51.5×10^8 t or 12 % of the total; the deep natural gas reserve is 4.25×10^{12} m^3 or 19 % of the total. Zhu and Zhang (2009) suggest that China's deep petroleum resource reserves are extremely non-uniform and mostly found in Xinjiang. In the Junggar Basin, the middle–shallow and deep oil geological resources are 9.7×10^8 t or approximately 18 % of the basin's total amount; the deep natural gas resource is $2{,}081 \times 10^8$ m^3 or approximately 32 % of the basin's total amount. Pang (2010) discovered after studies that the Tarim Basin has the richest oil resources in the deep part at 33.7×10^8 t or 56 % of the basin's total oil resource, natural gas resources in the deep part at $29{,}244 \times 10^8$ m^3 or 37 % of the basin's total natural gas resource. Statistics indicate that China's deep petroleum resource is 30,408 million ton, which is 27.3 % of the nation's total oil resource (Fig. 6a); its deep natural gas resource is 29,120 billion m^3, which is 49.2 % of the nation's total natural gas resource (Fig. 6b). Since 2000, China's petroleum exploration has continued to extend toward deep and ultra-deep levels. In the Junggar Basin, the ratio of deep exploratory wells increased from 3 % in 2000 to 15 % in 2013 (Fig. 7a). In the Tarim Basin, this ratio increased from 65 % in 2000 to 92 % in 2013 (Fig. 7b). The ratio of newly increased petroleum reserves in deep formations has also continued to rise. In the Tarim Basin, the ratio of deep oil increased from 66 % in 2000 to 92 % in 2013 (Fig. 8a); the ratio of deep natural gas also increased from 66 % in 2004 to 92 % in 2013 (Fig. 8b).

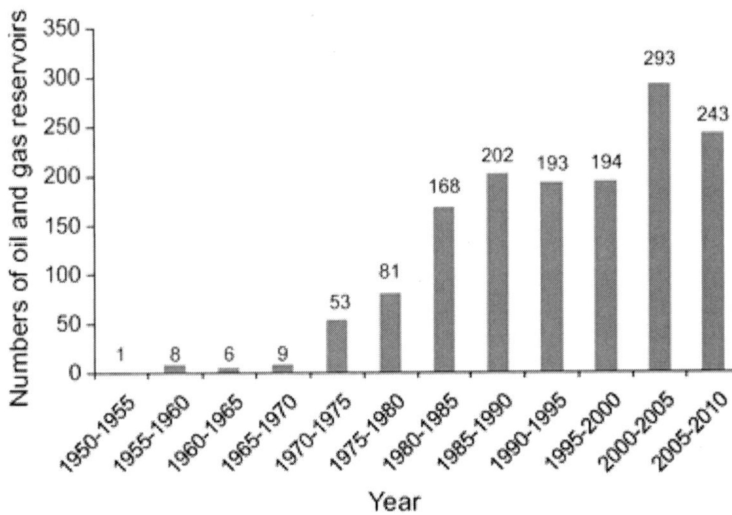

Figure. 3. Number of deep oil–gas reservoirs discovered in the world and its variation as a function of year

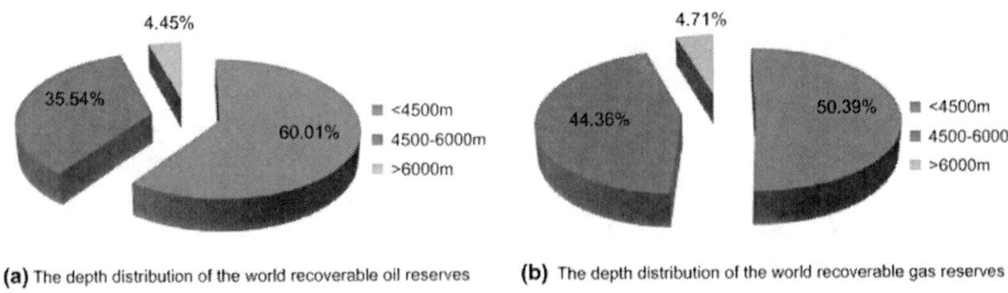

Figure. 4. Distribution of world recoverable oil and gas reserves at middle and deep depths of petroliferous basins

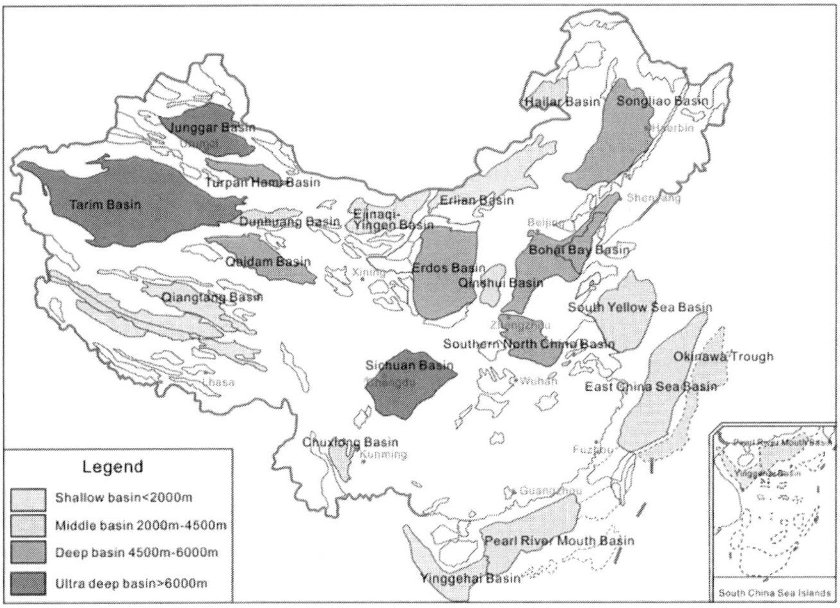

Figure. 5. Depth classification and horizontal distribution of petroliferous basins in China

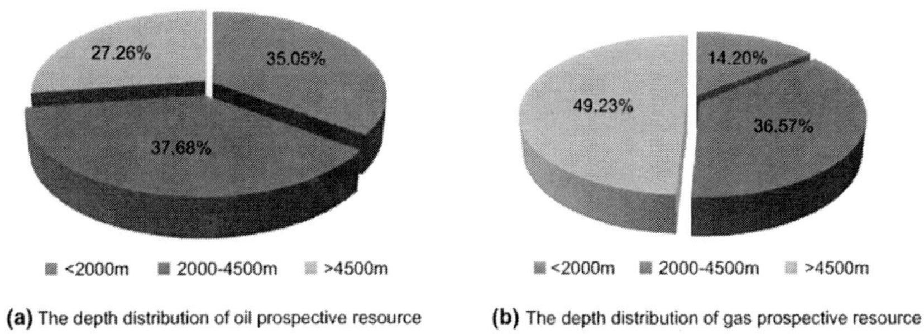

Figure. 6. Depth distribution of petroliferous basin resources in China

4. GEOLOGICAL FEATURES OF DEEP OIL–GAS RESERVOIRS

Compared with middle or shallow petroliferous basins, deep basins have large buried depths and the features of high temperature, high pressure, low porosity, low permeability, complex structural styles, and highly variable sedimentary forms. These special properties have been responsible for the unique characteristics of deep basin oil–gas reservoirs compared with their middle or shallow counterparts. Many scholars (Zappaterra 1994; Dyman and Cook 2001; Liu et al. 2007a, b; Wang et al. 2012) have examined deep oil–gas reservoirs. Table 4 lists some typical hydrocarbon fields (reservoirs) discovered in deep petroliferous basins in the world, from which we can observe their differences and varieties in terms of formation age, lithology, buried depth, porosity, pressure, petroleum phase, trap type, and basin type.

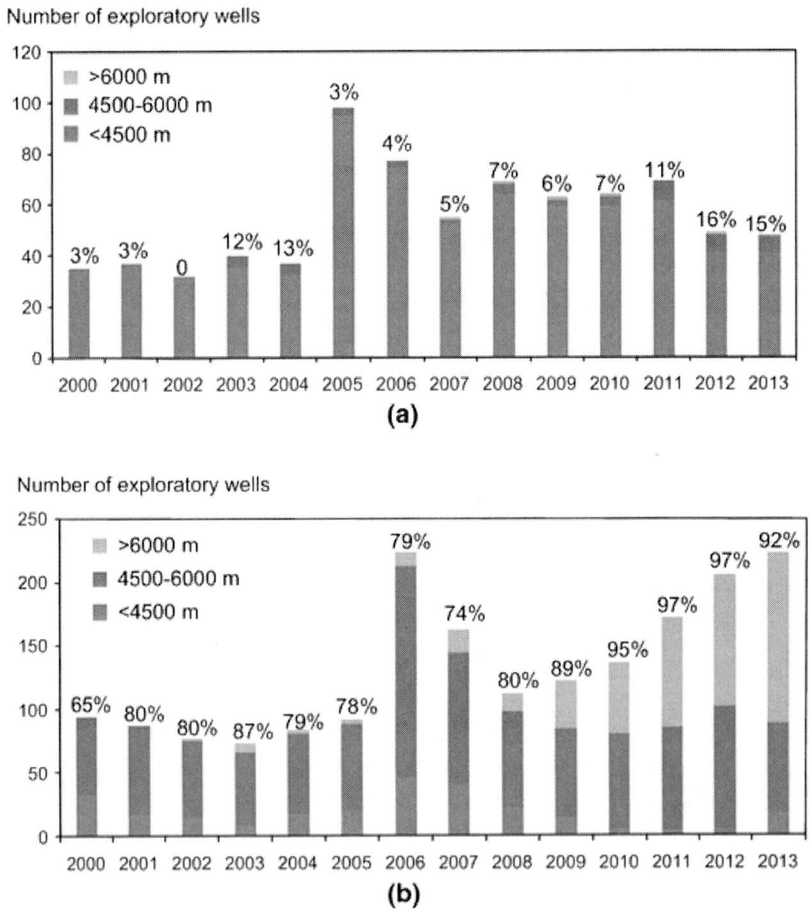

Figure. 7. Depths of exploratory wells drilled in the Junggar and Tarim basins. **A** Number and depth of exploratory wells drilled in the Junggar Basin over time, **B** Number and depth of exploratory wells drilled in the Tarim Basin since 2000

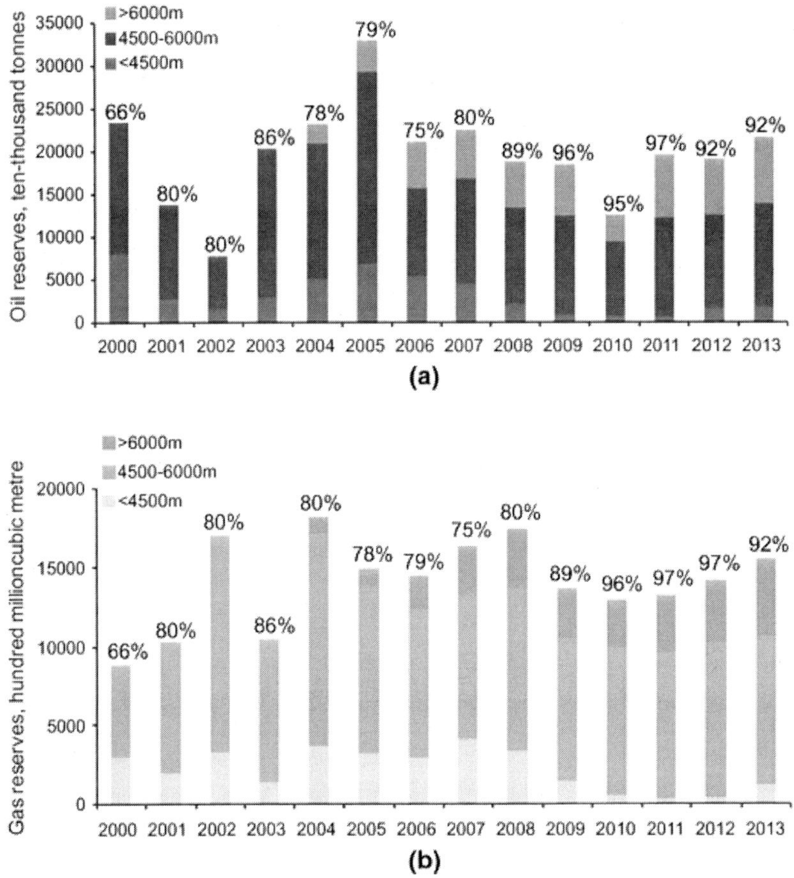

Figure. 8 . Proved deep oil and gas reserves in the Tarim Basin discovered each year since 2000. **A** Depth distribution of proved oil reserves in the Tarim Basin discovered each year since 2000, **B** Depth distribution of proved natural gas reserves in the Tarim Basin

4.1 While oil–gas reservoirs have been discovered in many types of deep petroliferous basins, most have been discovered in low heat flux deep basins

Oil–gas reservoirs have been discovered in all types of deep petroliferous basins (Fig. 9). Based on the basin classification system of Ingersoll (1995), Bai and Cao (2014) classified 87 deep petroliferous basins into seven groups: continental rift, passive continental marginal, foreland, interior craton, fore-arc, back-arc, and strike-slip basins, of which the passive continental marginal basins (25) and foreland basins (41) are the richest in deep petroleum, followed by the rift basins (12). These three types contribute 47.7, 46.4, and 5.6 % of the world's deep proven and probable (2P) recoverable petroleum reserves. The deep 2P recoverable petroleum reserves in the back-arc basins (2), strike-slip basins (3), and interior craton basin (1) contribute merely 0.3 % of the world's total (Fig. 10).

Table 4. Geological features of representative deep hydrocarbon fields (reservoirs) in the world

Country/Area	Oil/gas field/reservoir name	Reservoir lithology	Target formation age	Buried depth, km	Porosity, %	Oil/gas phase	Fluid pressure	Trap type	Basin type
United States	Murphy Creek oilfield	Clastic rock, clastic limestone	J–K	>6.2	5–15	Gas/condensate gas/heavy oil	Constant pressure	Structural	Foreland basin
	Harrisville oilfield	Sandstone	J–E	>7.0	10–15	Gas/oil	Overpressure	Structural	Passive continental marginal basin
Former Soviet Union	Shebelinka oilfield	Clastic rock	C–P	>7.0	14–17	Gas/condensate gas	Overpressure	Structural	Rift basin
	Shebsh oil and gas field	Clastic rock	E	>6.5	12–18	Gas/condensate gas	Overpressure	Structural	Foreland basin
Central & South America	Ceuta oilfield	Limestone	K	>6.0	4–15	Oil	Overpressure	Structural	Foreland basin
	Luna oilfield	Dolomite	J–K	>6.5	8–12	Oil/gas	Overpressure	Structural	Passive continental marginal
West Europe	Malossa oilfield, Villafortuna-Trecate oilfield	Platform limestone	T	6.3	8–15	Gas/condensate gas/oil	Overpressure	Structural	Foreland basin
Mideast	Omani oilfield	Clastic limestone	O	6.0	5–10	Gas	Constant pressure	Structural	Passive continental marginal
Japan	Niigata Basin	Rhyolite, andesite and volcanoclastic rock	N_1	>4.7	4.3	Gas/condensate oil	Overpressure	Structural	Back-arc basin
Africa	Serir oilfield	Clastic rock	E	7.0	8–22	Gas	Overpressure	Structural	Rift basin
China	Puguang gas field	Dolomite	Z–P	7.2	?	Gas	Constant pressure	Structural/lithological	Craton-foreland basin
	Moxi Longwangmiao gas field	Dolomite	Z–∈	>4.8	4.8	Gas	Overpressure	Structural-lithological	Craton-foreland basin
	North Tarim gas field	Carbonate	O	7.1	4.8	Gas	Constant pressure	Structural	Craton-foreland basin
	Mosuowan oilfield	Volcanic rock	C	7.5	3.3	Gas	Overpressure	Structural	Foreland-rift basin
	Xinglongtai buried hill hydrocarbon reservoir	Metamorphic rock	γ	>4.5	3–8	Overpressure		Structural	Rift basin

As a matter of fact, the distribution divergences of deep oil–gas reservoirs in petroliferous basins are essentially decided by the geothermal gradients of the sedimentary basins. Compared with the higher geothermal gradient counterparts, lower geothermal gradient sedimentary basins contain far more deep petroleum resources since when they reached the same buried depth, they had lower pyrolysis temperatures, and source rocks were richer in residual organic matter and hence had greater ability to generate and preserve hydrocarbon. Figure 11 compares the deep hydrocarbon potentials of source rocks in basins with different geothermal gradients in China as a function of depth, from which we can observed that "hot" basins expelled less hydrocarbon indicating they make up a smaller proportion of deep petroleum resources than their "cold" counterparts. As the geothermal gradient increases, the ratio of deep petroleum resources reduces. According to geothermal gradient records of 405 deep oil–gas reservoirs across the world, 318 % or 78.5 % were discovered in deep basins with geothermal gradients of 1–2 °C/100 m; 79 % or 19.5 % were discovered in deep basins with geothermal gradients of 2–3 °C/100 m; and 8 % or 2 % were discovered in deep basins with geothermal gradients larger than 3 °C/100 m (Fig. 12). In China, the geothermal gradient increases from the west toward the east. The number and reserves of deep oil–gas reservoirs discovered in West China basins is far larger than that discovered in East China basins (Fig. 13).

4.2 While many types of petroliferous traps are developed in deep basins, oil–gas-bearing features in deep basin traps are arousing increasing attention

Like their middle–shallow counterparts, deep basins also contain a variety of traps which, according to the conventional trap classification, include tectonic traps, stratigraphic traps, lithological traps, structural-lithological traps, structural-stratigraphic traps, and lithological-stratigraphic traps. Different types of traps differ significantly in terms of their reserves. After a statistical analysis on 837 deep oil–gas reservoirs in the USA, Dyman et al. (1997) discovered that structural traps and combination traps make up as much as 66.9 %. Only in Anadarko and California basins are there more lithological traps than structural ones (Fig. 14). Bai and Cao (2014), after summarizing the trap types and reserves of the world's deep oil–gas reservoirs, discovered that structural traps have 73.7 % of the world's deep recoverable 2P petroleum reserves, with structural-lithological traps and stratigraphic traps contributing 21.9 % and 4.4 %, respectively (Fig. 15).

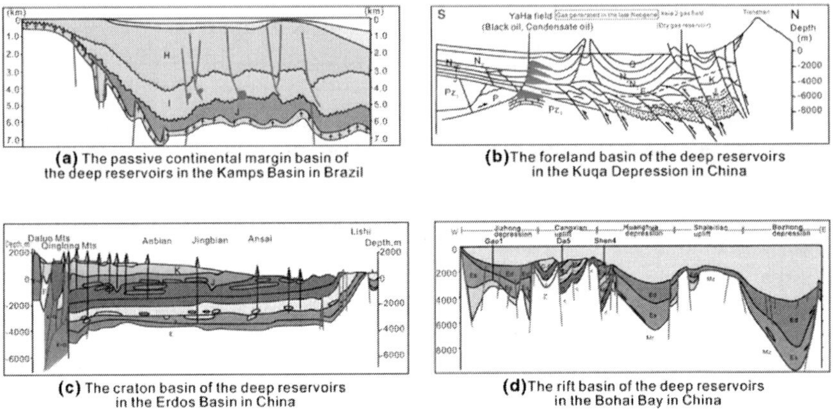

Figure. 9. Deep oil–gas reservoirs developed in different types of basins (Li and Lü 2002; Li 2009)

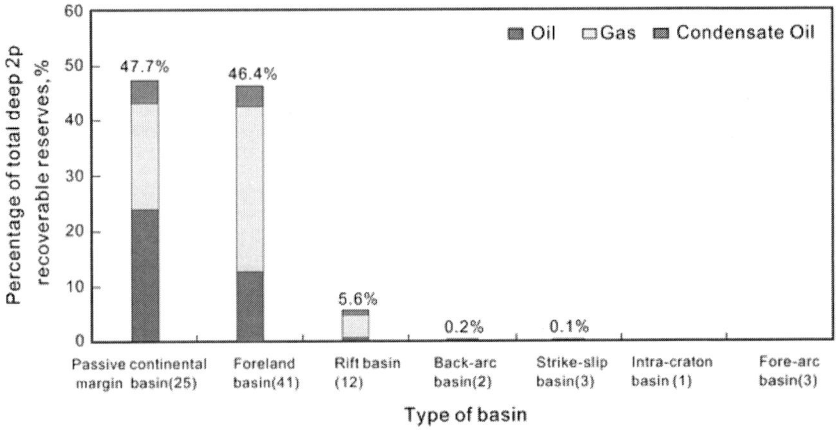

Figure. 10. Deep hydrocarbon reservoir distribution in different types of basins in the world (Bai and Cao 2014)

Recently, with the discovery of tight, continuous oil–gas reservoirs in Canada's Alberta Basin, the USA's Red Desert Basin and Green River Basin, and China's Erdos, Sichuan, and Songliao basins, people have become more interested in these unconventional oil–gas reservoirs. First, these reservoirs have completely different genesis compared with conventional reservoirs, and their discovery has brought on a novel petroleum exploration field. Second, these reservoirs are widely and continuously distributed with vast resource potentials and great scope for petroleum exploration. Third, these reservoirs formed inside deep basin traps between the buoyancy accumulation threshold of a petroliferous basin and the basement of the basin. Their buried depths were quite large, but can be very shallow at present as a result of subsequent tectonic events in the basin. Deep basin traps are a special type of hydrocarbon trap in which the reservoir media have porosities smaller than 12 %, permeabilities smaller than 1×10^{-3} μm^2, and throat radii smaller than 2 μm. Hydrocarbon was not subject to buoyancy in its accumulation, thus making it possible to spread continuously. The more developed the sources rocks were in a deep basin trap, the more continuous the reservoirs were distributed close to the source rocks and the richer the petroleum resources they provide. Figure 16 gives a typical conceptual model and shows the difference about the development and distribution of a deep basin trap-controlled hydrocarbon reservoir in a petroliferous basin and conventional traps.

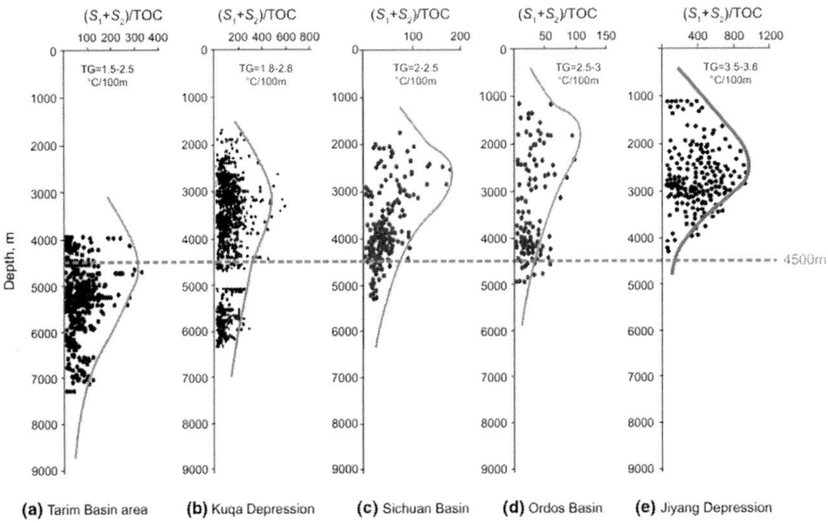

Figure. 11. Comparison of hydrocarbon potentials of deep source rocks in basins in China with different geothermal gradients

4.3 The composition of deep petroleum is widely different with more natural gas than liquid oil and the natural gas ratio increases with the buried depth

The composition of deep petroleum in petroliferous basins varies and includes gaseous hydrocarbon, condensate gas, condensate oil, liquid hydrocarbon, and oil–gas coexistence. Phase statistics of 1,477 deep oil–gas reservoirs in the world demonstrate that oil–gas miscible phases contribute 54 % and gas phases contribute 40 %. The oil phase makes up a very small proportion of 6 % (Fig. 17). Generally, as the formation depth increases, natural gas makes up a larger proportion in deep petroleum and overtakes liquid hydrocarbon as the prevailing type of petroleum resources. Figure 18 shows how the oil and natural gas reserves discovered from different formations of East China's Bohai Bay Basin and West China's Tarim and Junggar basins vary as a function of depth. By and large, the ratio of older deep basin gas reservoirs increases due to pyrolysis of crude oil as a result of extended high temperature exposure of the oil reservoir in deep basins, or thermal cracking in the source rocks when they reached high maturity (Dyman et al. 2001). The Hugoton gas field and Mills Ranch in the USA's Anadarko Basin, for example, are pure gas fields. Many deep gas reservoirs in China's petroliferous basins also originated from earlier oil or oil–gas reservoirs that were cracked into gas under high temperatures, as exemplified by the large Puguang carbonate gas field (Du et al. 2009), the large Kela-2 gas field (Jia et al. 2002), and the Hetian gas field (Wang et al. 2000). Huge liquid or condensate oil reservoirs have also been discovered in a number of deep basins such as the USA's Rocky Mountain Basin, where gas wells make up only 34 % of the deep exploratory wells while most are oil wells and no pyrolysis has taken place deeper than 6,000 m. This is often because the low subsurface temperature or high pressure of the formation had prevented the crude oil from coming to its threshold pyrolysis temperature (Svetlakova 1987). A lot of factors can be related to deep basin petroleum phases. These include (1) the type of the original organic matter; (2) temperature and pressure; and (3) subsequent adjustment or reformation.

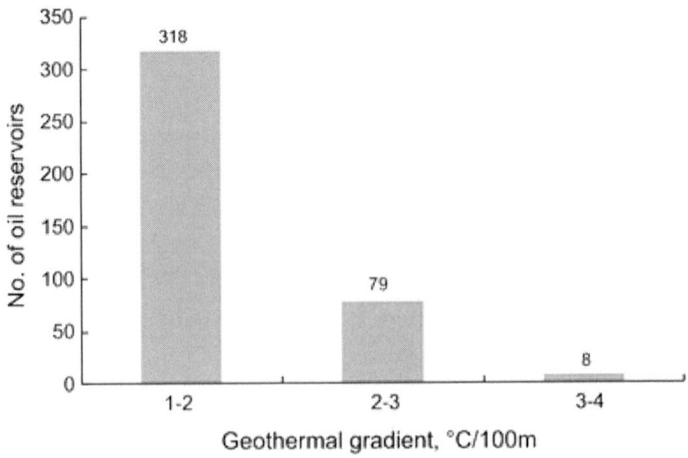

Figure. 12. Number of deep oil–gas reservoirs discovered in basins in the world with different geothermal gradients

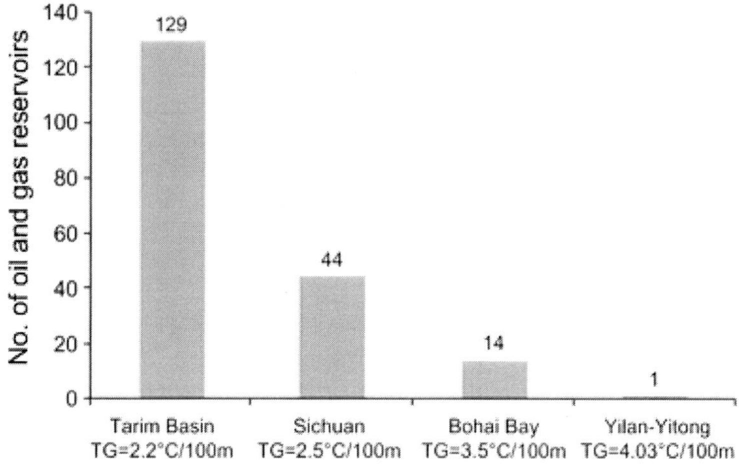

Figure. 13. Number of deep oil–gas reservoirs discovered in basins in China with different geothermal gradients

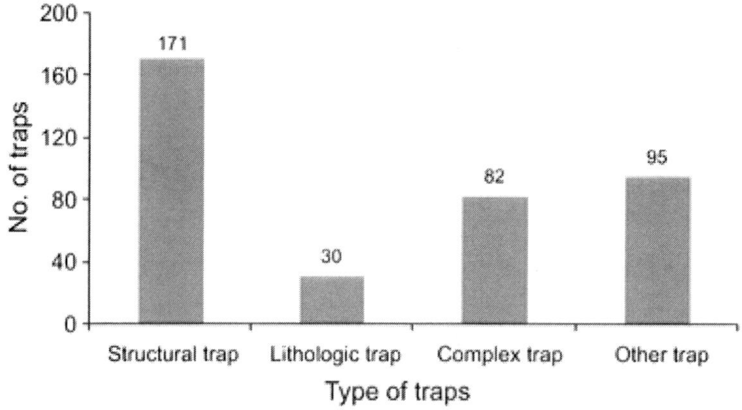

Figure. 14. Trap types of deep oil–gas reservoirs in the USA (Dyman et al. 1997)

4.4 Residual organic matter in deep source rocks reduces but the hydrocarbon expulsion rate and efficiency continuously increase with the buried depth

As far as deep source rocks are concerned, they have much smaller measured residual hydrocarbon amount and hydrocarbon potential than their middle–shallow counterparts, and both reduce with the increase of the buried depth. They appear to have the following characteristics: (1) the residual hydrocarbon amount per unit parent material in the source rocks (total organic carbon, TOC) is represented by S_1/TOC or "A"/TOC (Dickey 1975; Hao et al. 1996; Durand 1988). The residual hydrocarbon amount in deep source rocks first increases then reduces with the increase of the depth or R_o and was already very small in deep basins (Fig. 19); (2) the hydrocarbon potential is represented by the H/C atomic ratio, O/C atomic ratio, and hydrogen index (Tissot et al. 1974; Tissot and Welte 1978; Jones and Edison 1978; Baskin 1997; Zhang et al. 1999). The H/C atomic ratio and O/C atomic ratio (Fig. 20) and hydrocarbon index HI (Fig. 21) reduce with the increase of the buried depth; and (3) the hydrocarbon potential index of source rocks is represented by $(S_1 + S_2)$/TOC (Zhou and Pang 2002; Pang et al. 2004). When the TOC is more than 0.1 %, the hydrocarbon potential index of deep source rocks shows a "big belly" profile of increasing followed by reducing with the increase of the depth or R_o (Fig. 22). In a word, as the buried depth increases and the hydrocarbon potential of source rocks gradually reduces, the accumulated oil–gas volume and hydrocarbon expulsion efficiency of source rocks appear to increase gradually with the increase of the buried depth (Fig. 23), reflecting the increase in the contribution made by source rocks to oil–gas reservoirs. This indicates that the quality and effectiveness of source rocks should be judged by investigating how much hydrocarbon was generated and expelled by source rocks rather than by relying on how much hydrocarbon or how much hydrocarbon potential is left of the source rocks.

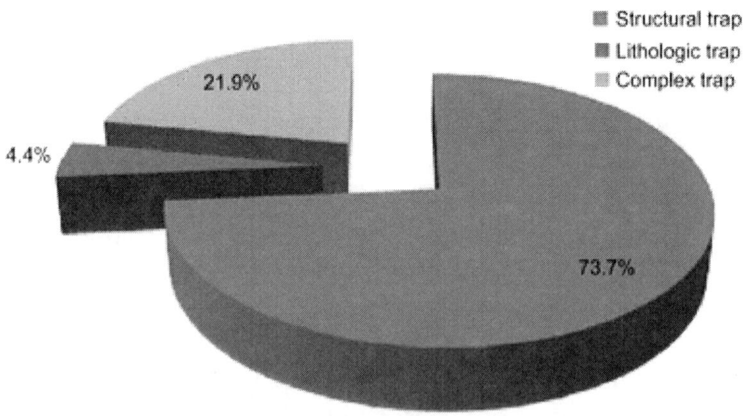

Figure. 15. Deep petroleum reserves in petroliferous basins in the world with different trap types

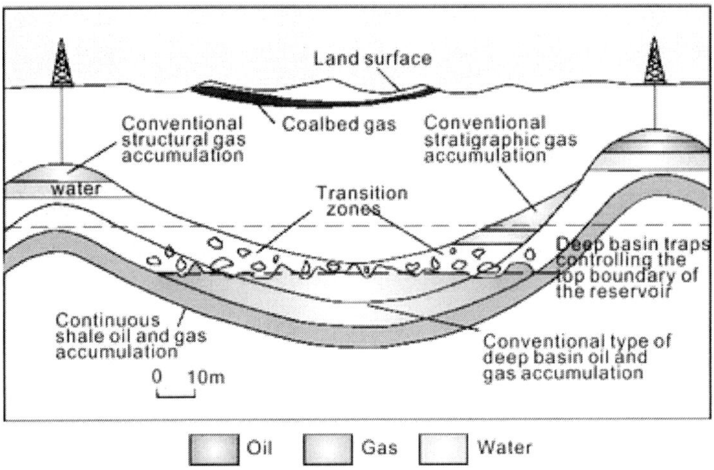

Figure. 16. Difference and comparison of conventional traps versus deep basin traps in petroliferous basins and their petroleum-control features

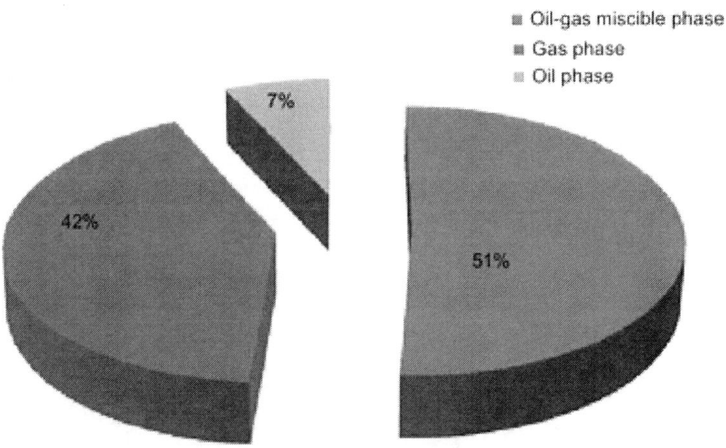

Figure. 17. Phase distribution of deep petroleum in petroliferous basins in China

4.5 While there are many types of rocks in deep hydrocarbon reservoirs, most of them are clastic rocks and carbonate

Deep target formations in petroliferous basins contain a variety of rocks, though clastic rocks and carbonate are the predominant types of deep hydrocarbon reservoirs discovered so far, with certain amounts of volcanic and metamorphic rocks too. Among clastic reservoirs, fractured sandstone reservoirs are the most favorable. Carbonate reservoirs include limestone and dolomite, with carbonate reservoirs extensively found in brittle fractures and karst caves taking the largest proportion. As of 2010, of the 1,477 deep oil–gas reservoirs discovered across the world, 1,035 % or 70.1 % were located in clastic reservoirs, 429 % or 29.0 % were in carbonate reservoirs, and 13 % or 0.88 % were in magmatic and metamorphic reservoirs (Fig. 24).

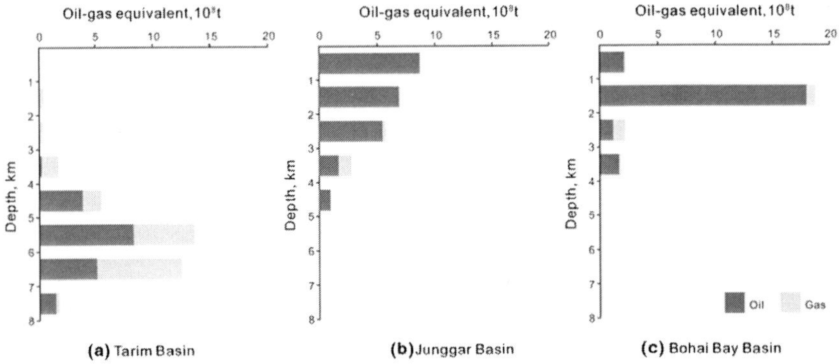

Figure. 18. Petroleum reserves of petroliferous basins in China as a function of buried depth

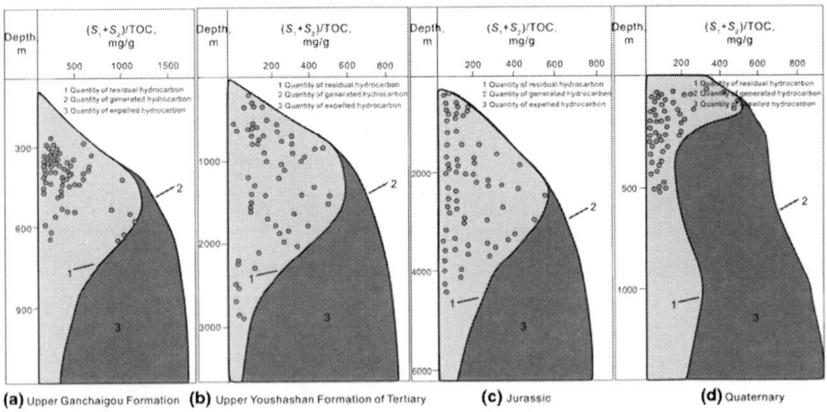

Figure. 19. Hydrocarbon generation, residual hydrocarbon, and hydrocarbon expulsion of source rocks in the Qaidam Basin as a function of buried depth

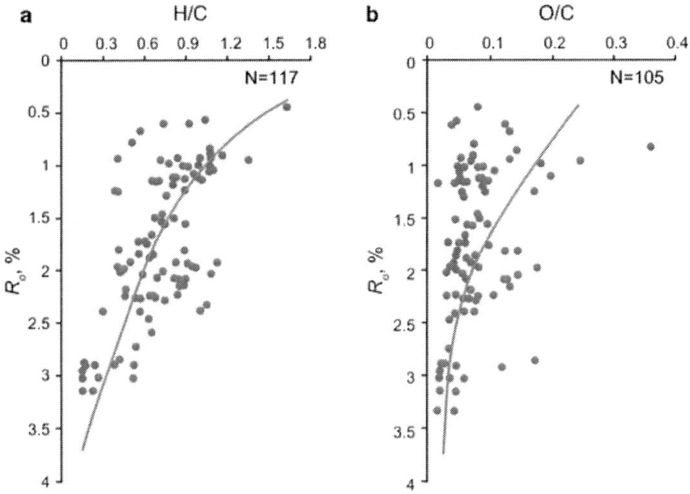

Figure. 20. Kerogen H/C atomic ratio and O/C atomic ratio of source rocks in the Tarim Basin platform area. **A** Change of H/C atomic ratio, **B** change of O/C atomic ratio

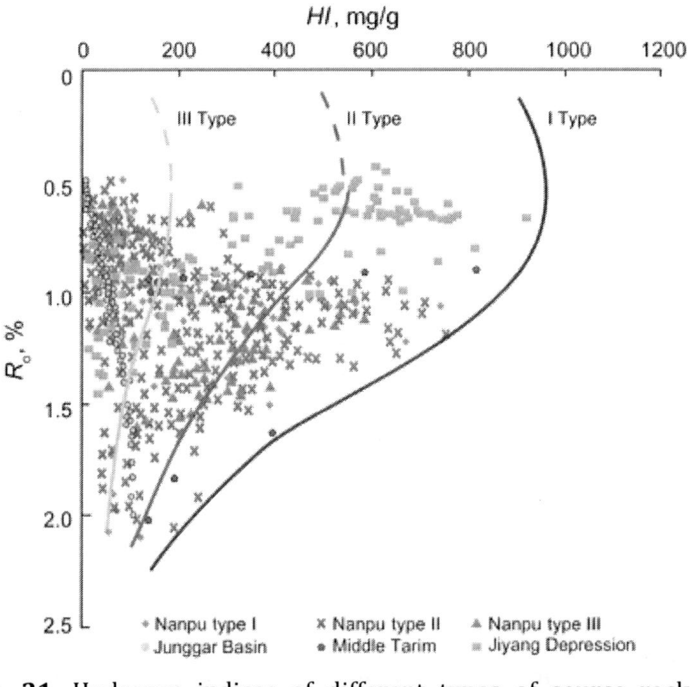

Figure. 21. Hydrogen indices of different types of source rocks as a function of thermal evolution level of the parent material

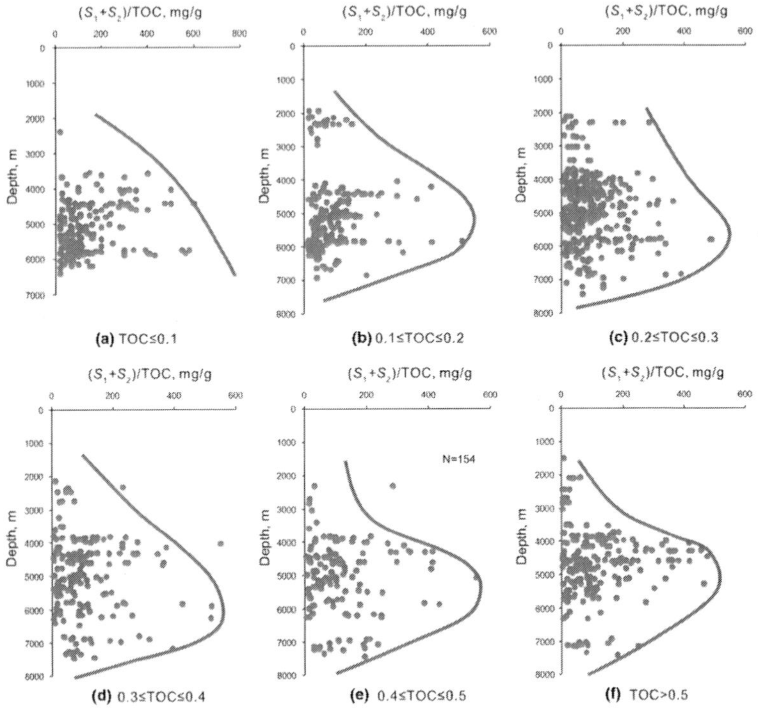

Figure. 22. Hydrocarbon potential of Cambrian–Ordovician carbonate source rocks in the Tarim Basin

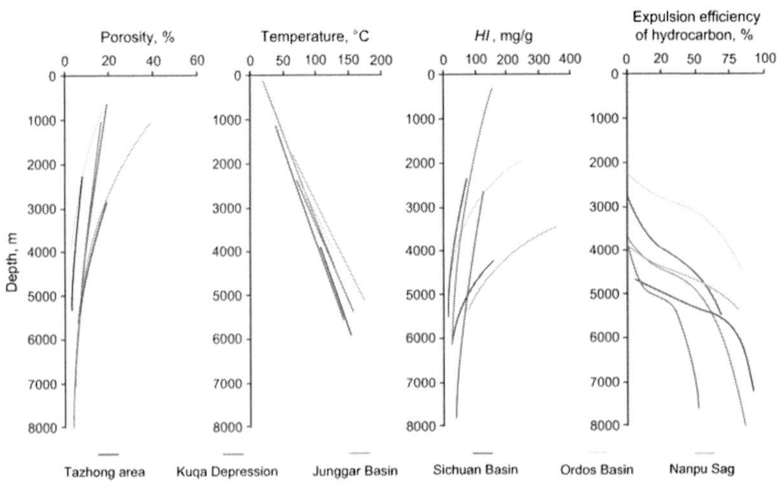

Figure. 23. Hydrocarbon expulsion efficiency of representative petroliferous basins in China as a function of depth

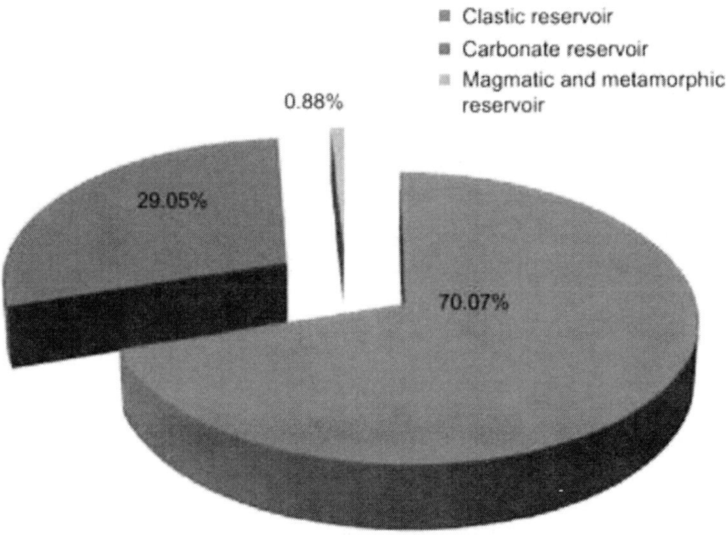

Figure. 24. Distribution of reservoir lithologies of deep oil–gas reservoirs in the world

According to the latest nationwide petroleum resource evaluation made by the Ministry of Land and Resources (2005), deep hydrocarbon reservoirs discovered in China are predominantly carbonate and sandstone reservoirs (Fig. 25). Widespread marine carbonate reservoirs occur in Central West China basins and are responsible for a series of large marine carbonate hydrocarbon fields represented by Central and North Tarim hydrocarbon fields in the Tarim Basin, East and Central Sichuan reef-flat carbonate hydrocarbon fields. In the northwestern flank of the Junggar Basin, the Kuqa Depression of the Tarim Basin and the Xujiahe Formation of the Sichuan Basin, sandstone oil–gas reservoirs in clastic rocks are the predominant type. Besides, a lot of deep volcanic oil–gas reservoirs are also contained in the deep part of the Junggar and Songliao basins. Deep oil–gas reservoirs in bedrock metamorphic rocks have been discovered in places like the Liaohe Depression of the Bohai Bay Basin.

4.6 The age of deep hydrocarbon reservoirs is widely different, but those recently discovered are predominantly Paleogene and Upper Paleozoic

Oil–gas reservoirs in deep petroliferous basins are similar to their middle or shallow counterparts in terms of formation distribution. The ages of the reservoirs cover a wide range. Most of the deep oil–gas reservoirs discovered so far, however, are in five formation systems: Neogene, Paleogene, Cretaceous, Jurassic, and Upper Paleozoic, the deep 2P recoverable petroleum reserves of which account for 12.8 %, 22.3 %, 18.3 %, 12.8 %, and 22.2 % of the world's totals, respectively (Fig. 26). This suggests that the deep petroleum is mainly in Neogene and Upper Paleozoic formations. Also, as the reservoir ages become older, the ratio of deep natural gas in the total deep petroleum reserve tends to increase accordingly.

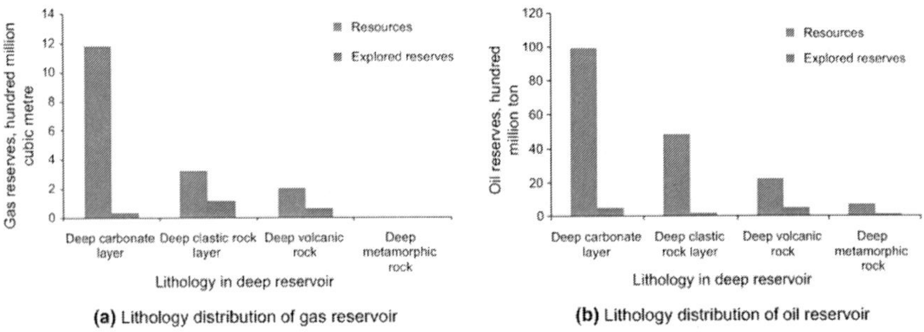

Figure. 25. Distribution of reservoir lithologies of deep oil–gas reservoirs in China

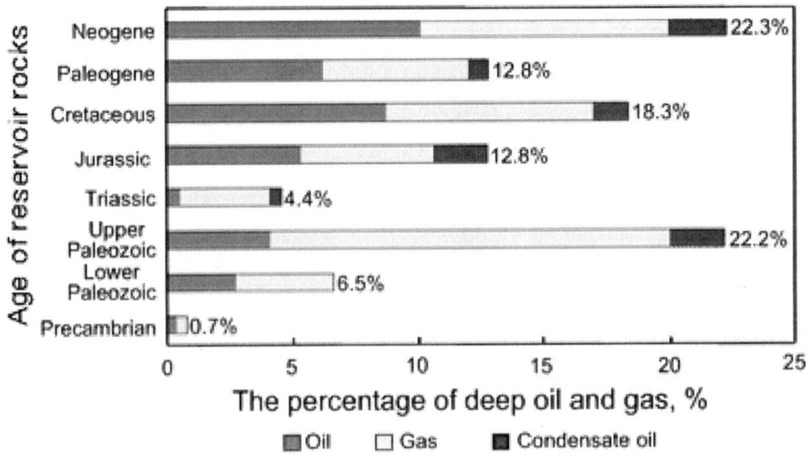

Figure. 26. The age distribution of deep hydrocarbon reservoirs in the world (Bai and Cao 2014)

Table 5. Chronological distribution of deep hydrocarbon reservoir strata in China's petroliferous basins

Basin	Oil/gas field name	Target formation depth, m	Formation lithology	Proved oil reserve, 10^4 t	Proved natural gas reserve, 10^8 m^3	Target formation age
Tarim Basin	Central Tarim gas field	4,500–6,200	Carbonate	38,600	1,020	Ordovician
	Halahatang oilfield	5,900–7,100	Carbonate	20,812	–	Ordovician
	Donghetang oilfield	>6,000	Clastic rock	3,323	–	Carboniferous
	Kuche deep gas region	5,000–8,000	Clastic rock	–	6,448	Cretaceous
Junggar Basin	Xiazijie oilfield	4,800	Volcanic rock	1,548	34	Permian
Sichuan Basin	Longgang gas field	2,800–7,100	Carbonate	–	730	Permian
	Moxi Longwangmiao gas field	4,500–5,500	Dolomite	–	4,404	Sinian, Cambrian
Bohai Bay Basin	Qishan gas field	4,500–5,500	Clastic rock	–	95	Paleogene
	Xinglongtai gas field	4,500–5,500	Metamorphic rock	0.75	–	Archaeozoic
Songliao Basin	Changling gas field	4,500–5,000	Volcanic rock	–	640	Cretaceous

After summarizing the reservoir ages of oil–gas reservoirs in China's deep petroliferous basins (Table 5), we discovered that the reservoir ages of deep oil–gas reservoirs in Central and West China basins are predominantly Paleozoic, meaning the reservoirs are quite old; those in East China basins are predominantly Paleogene or Cretaceous, and reservoir ages of oil–gas reservoirs in the bedrock are predominantly Precambrian.

4.7 The porosity and permeability of deep hydrocarbon reservoirs are widely different, but they vary with the lithology and buried depth

The porosities and permeabilities of target formations for deep oil–gas reservoirs in petroliferous basins vary widely, ranging from high-porosity, high-permeability (with porosity of 38 % and permeability of 7,800 mD) high-quality reservoirs to low-porosity, low-permeability (with porosity lower than 5 % and permeability less than 0.1 mD) tight reservoirs. High-porosity, low-permeability or low-porosity, high-permeability petroliferous reservoirs have also been identified. Deep drilling records across the world demonstrate that as the reservoir depth increases, the compaction effect and consequently the diagenesis intensify, so the porosity of deep rocks tends in general to decrease. The porosities of the world's deep petroliferous basins are mostly in the 10 %–12 % range (Wang et al. 2012). A summary of the porosities and permeabilities of 20,717 oil–gas reservoirs across the world as a function of depth revealed that the porosities and permeabilities of these reservoirs tend to decrease with the increase of depth overall (Fig. 27), though this rule varies from one lithology to another in different areas. Clastic reservoirs, for example, show obvious porosity and permeability decreases in some reservoirs, but the porosity of other reservoirs varies little in deep or ultra-deep formations (Fig. 28), while carbonate reservoirs do not show obvious decreases in their reservoir properties as the depth increases (Fig. 29) due to their high rock brittleness, high compaction stability, and good solubility. The porosities of volcanic reservoirs do not vary much with depth (Fig. 30). The various petroliferous basins in China do not present the same characteristics. Figure 31 compares the reservoir porosity variations of different petroliferous basins in China as a function of depth. It shows that the porosity of sandstone reservoirs decrease with an increase in buried depth, while the porosity is in general retained in carbonate and volcanic rocks until below 6,500 m.

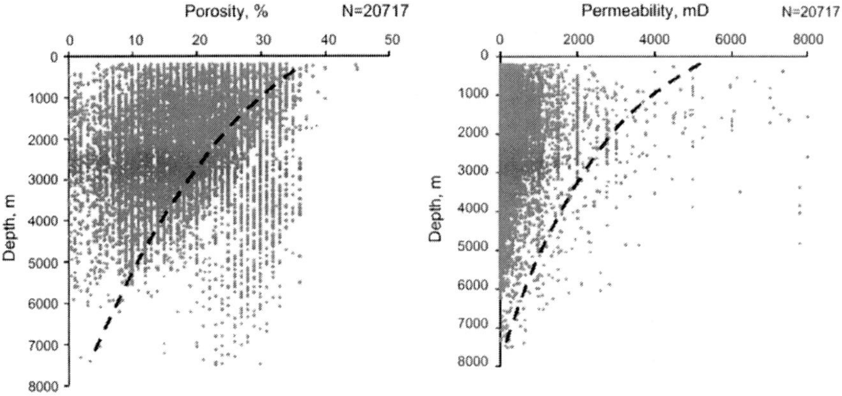

Figure. 27. Reservoir properties as a function of depth of petroliferous basins in the world

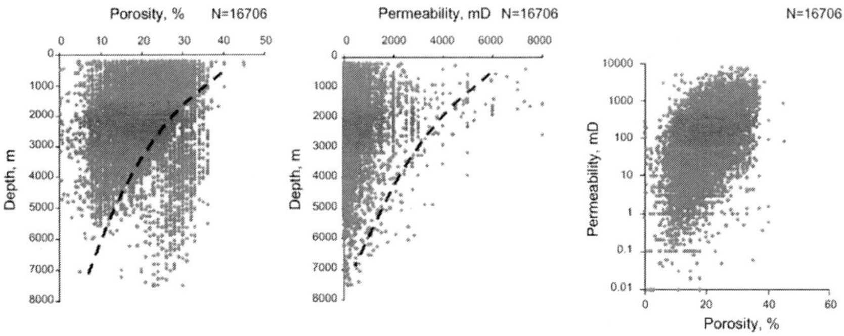

Figure. 28. Reservoir properties of clastic rocks as a function of buried depth of petroliferous basins in the world

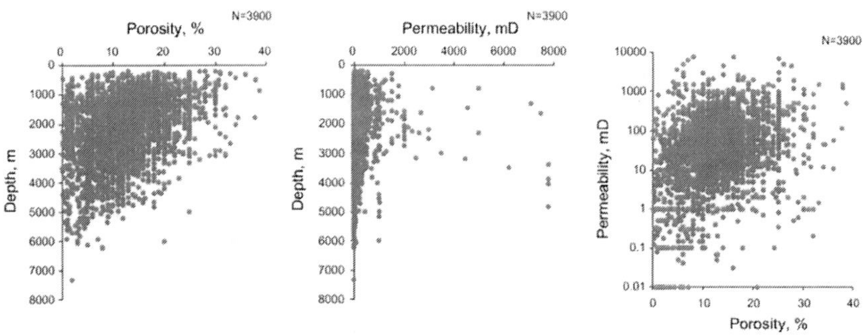

Figure. 29. Reservoir properties of carbonate as a function of buried depth of petroliferous basins in the world

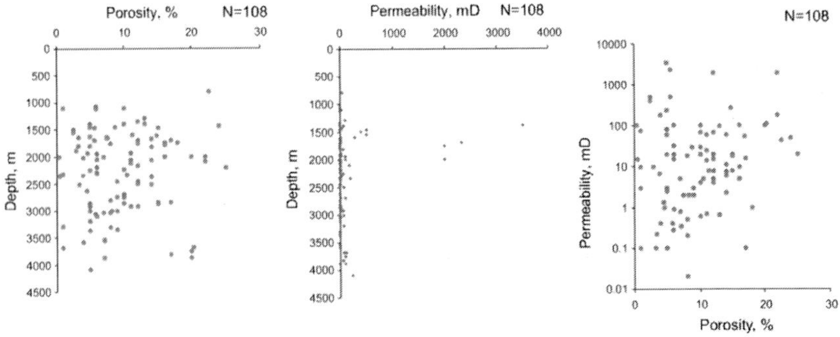

Figure. 30. Reservoir properties of volcanic and metamorphic rocks as a function of buried depth of petroliferous basins in the world

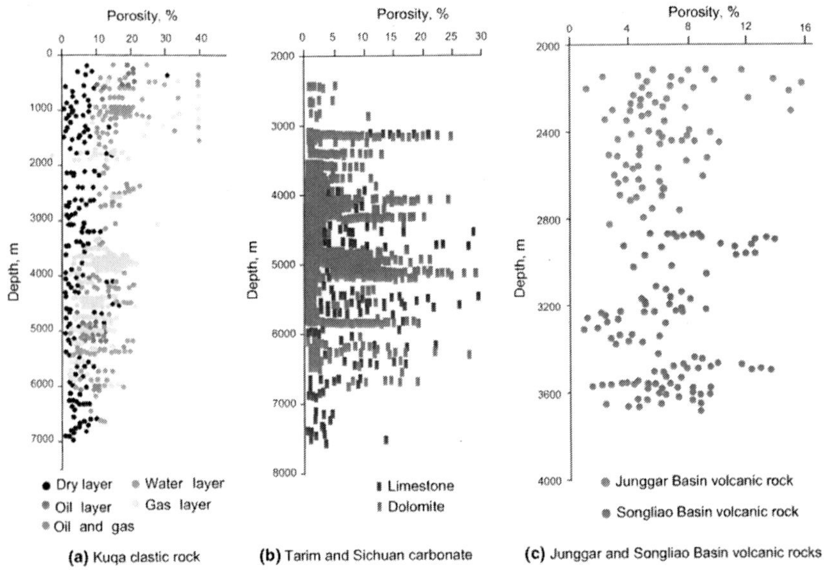

Figure. 31. Porosities of target strata with different lithologies as a function of depth of petroliferous basins in China

4.8 The temperatures of deep oil–gas reservoirs differ widely, but they typically vary with the buried depth and geothermal gradient

The temperature range of oil reservoirs in deep petroliferous basins has exceeded that of liquid hydrocarbon (windows) supposed by traditional kerogen theory (60–120 °C, R_o = 0.6 %–1.35 %): the highest oil reservoir temperature discovered in the world so far is more than 200 °C. Compared with their middle–shallow counterparts, deep oil–gas reservoirs have even higher temperatures which vary even more widely. Statistics of the temperatures and pressures of 428 deeper-than-4,500 m oil–gas reservoirs in the world (Fig. 32) show that the temperatures of deep oil–gas reservoirs can be 200 °C maximum and those of a couple of gas reservoirs are more than 370 °C, compared with the lowest hydrocarbon reservoir temperature of 47 °C. Even at the same depth, the temperatures of deep oil–gas reservoirs vary from one type to another, such as the petroliferous basins in China, those in the east are mostly extensional basins that are typically hot basins with an average geothermal gradient of approximately 4 °C/100 m and oil reservoirs deeper than 4,500 m being hotter than 180 °C; the extrusion basins in the west and the craton basins in the center, to the contrary, have lower geothermal gradients and are typical cold basins (Liu et al. 2012) with an average geothermal gradient of approximately 2.5 °C/100 m. At the same depth of 4,500 m, the temperature in the center and west is less than 120 °C. The temperature difference is approximately 60 °C. Figure 33 compares the geothermal gradients of some of the representative basins in China as a function of time, from which we can observe that, at the same depth, the formation temperature tends to increase from west toward east, reflecting the eastward increase of the geothermal gradient or heat flux.

4.9 The pressure of deep oil–gas reservoirs typically varies with the buried depth, genesis, and evolution period

Oil–gas reservoirs formed under high porosities and high permeabilities in middle–shallow petroliferous basins are typically buoyancy controlled, thereby generally displaying high pressures. Oil–gas reservoirs discovered in deep basins, however, have complex genesis, thereby displaying diverse pressures (Fig. 34). Statistics of the pressure records of 16,552 oil–gas reservoirs in the world revealed significant pressure differences among deep oil–gas reservoirs. The highest of the majority is 130 MPa, with a few outliers as high as 172 MPa, while the lowest is merely 8.4 MPa (Fig. 35).

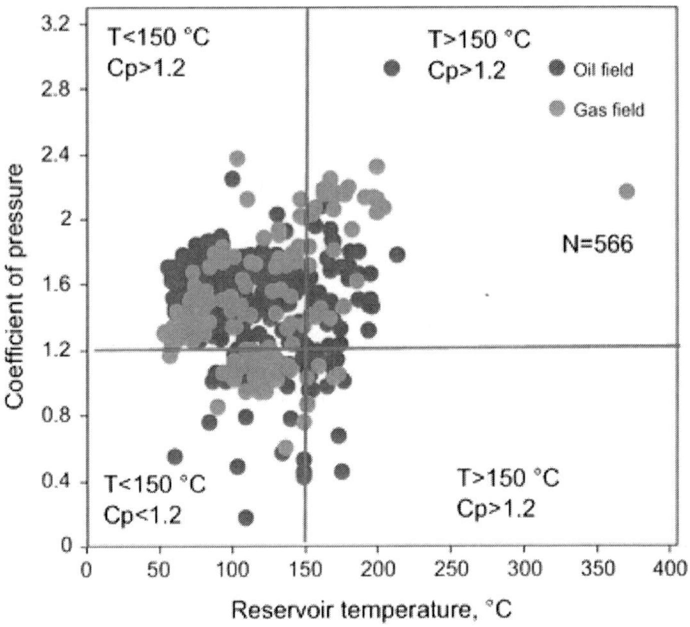

Figure. 32. Scatter diagram of temperatures and pressure coefficients of deep hydrocarbon fields of petroliferous basins in the world

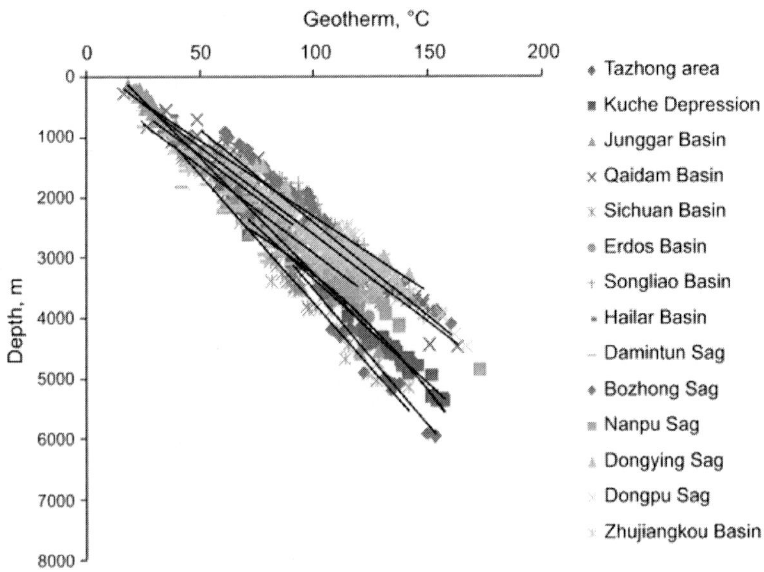

Figure. 33. Formation temperatures of petroliferous basins in different parts of China as a function of buried depth

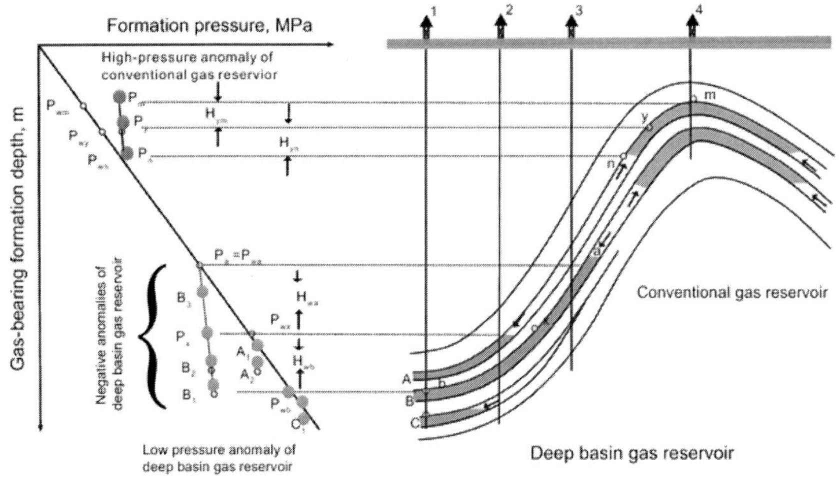

Figure. 34. Comparison of pressures and genesis of conventional gas reservoirs versus deep basin oil–gas reservoirs (Pang et al. 2006)

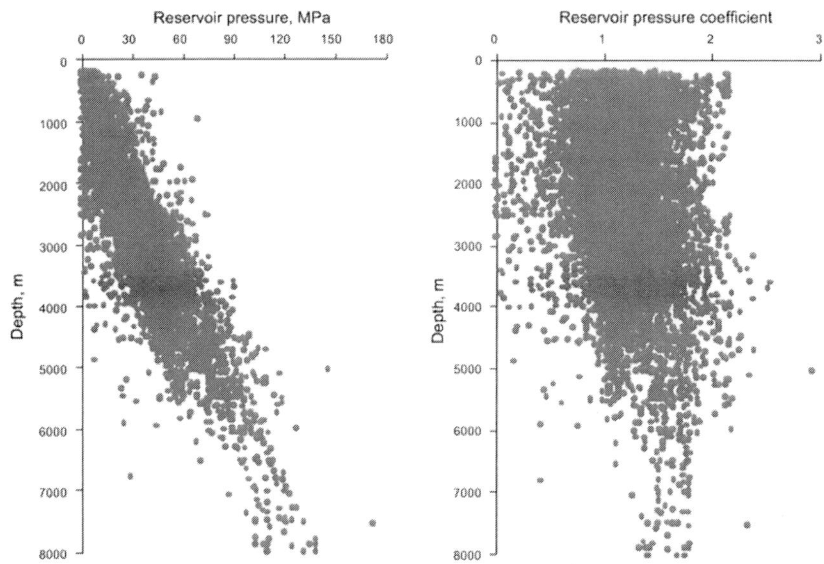

Figure. 35. Hydrocarbon reservoir pressures of petroliferous basins in the world as a function of buried depth

Abnormal high pressures are generally contained in deep tight structural gas reservoirs and tight lithological gas reservoirs, like the Sichuan Xiaoquan gas reservoir which has abnormally high pressures with a pressure coefficient of 1.6–2.0 (Guan and Niu 1995), and the Fuyang tight oil reservoir in Songliao which also displays abnormally high pressures with a pressure coefficient larger than 1.6. Deep oil–gas reservoirs with normal pressures also exist, like the 13 Lunnan buried hill oil reservoirs in the Tarim Basin that have the pressure coefficients between 1.03 and 1.14; and the Lunnan-17 well, Lunnan-30 well, Lunnan-44 well, and Jiefang-123 well oil reservoirs that have coefficients of 1.137, 1.130, 1.143, and 1.148, defining them as normal pressure oil reservoirs (Gu et al. 2001). Low-pressure oil–gas reservoirs are typically found in tight syncline sandstone gas reservoirs, like the tight sandstone hydrocarbon reservoirs discovered in Canada's Alberta Basin (Masters 1979), the tight sandstone hydrocarbon reservoirs discovered in the USA's Red Desert (Spencer 1989), and the tight sandstone hydrocarbon reservoirs in the USA's Green River Basin. The same occurs in the upper Paleozoic tight sandstone reservoirs in China's Erdos Basin and the tight sandstone gas reservoirs discovered in the Jurassic of the Tuha Basin (Fig. 36). In the main, deep oil–gas reservoirs have complex pressures; abnormally high, normal, and abnormally low-pressure hydrocarbon reservoirs can all be found in deep oil–gas reservoirs. So far, the complex hydrocarbon accumulation areas formed by coexistence of the three pressure categories are increasingly found in basins. Research (Pang et al. 2014a) shows that the pressure of deep oil–gas reservoirs is decided by their genesis mechanism and genesis process. Normal oil–gas reservoirs formed in free fluid dynamic fields in high-porosity, high-permeability media generally have high pressure, while unconventional oil–gas reservoirs formed in limited fluid dynamic fields in low-porosity, low-permeability media generally end up with negative pressure, though they appeared to show high pressure during hydrocarbon accumulation into reservoirs. Normal oil–gas reservoirs formed in the early years in petroliferous basins, as the depth increased, superimposed or compounded with the unconventional oil–gas reservoirs originated from the deep part before, eventually giving rise to superimposing or compounded, continuous oil–gas reservoirs, among which there can be high-pressure petroliferous reservoirs coexisting with low-pressure petroliferous reservoirs.

4.10 Deep oil–gas reservoirs may exist with or without a cap, and those without a cap are typically of unconventional genesis

Forming a hydrocarbon reservoir without relying on a cap is the unique geological feature of deep hydrocarbon entrapment. Conventional petroleum geological theory assumes that a cap is an indispensable geological element for the formation and preservation of any hydrocarbon reservoir; without a cap, it would be impossible for hydrocarbon to gather into a reservoir in any high-porosity, high-permeability media, since buoyancy would cause the hydrocarbon to percolate upward to the basin surface. For deep hydrocarbon entrapment, however, as far as the reservoir media were commonly tight, a cap is not indispensable (Fig. 37). The realities underlying this phenomenon are the particular fluid dynamic fields and material equilibrium conditions of deep oil–gas reservoirs: (1) deep reservoirs, if commonly tight, had poor porosities and permeabilities and limited pore throat radii, so hydrocarbon had to overcome great capillary pressures and the hydrostatic pressures of the overlying water columns when it tried to charge into the reservoir; (2) for deep oil–gas reservoirs, buoyancy was no longer the primary drive for hydrocarbon migration; they expelled pore water and expanded their own area by relying on the hydrocarbon volume increase and finally became continuous oil–gas reservoirs; and (3) the sources of tight continuous oil–gas reservoirs formed from deep basins were close to the reservoir, keeping the hydrocarbon under constant diffusion-accumulation equilibrium, thus enabling them to be preserved for a long time under structural stability.

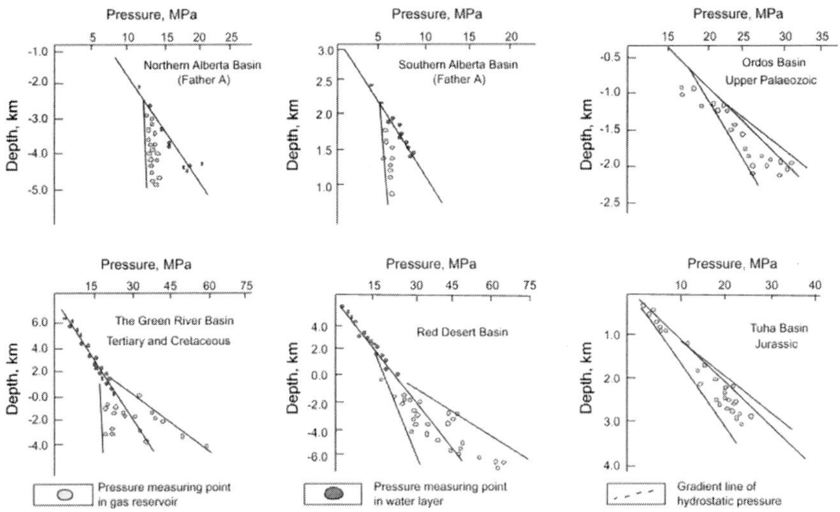

Figure. 36. Reservoir pressures of deep tight sandstone oil–gas reservoirs in representative petroliferous basins in China and elsewhere as a function of depth

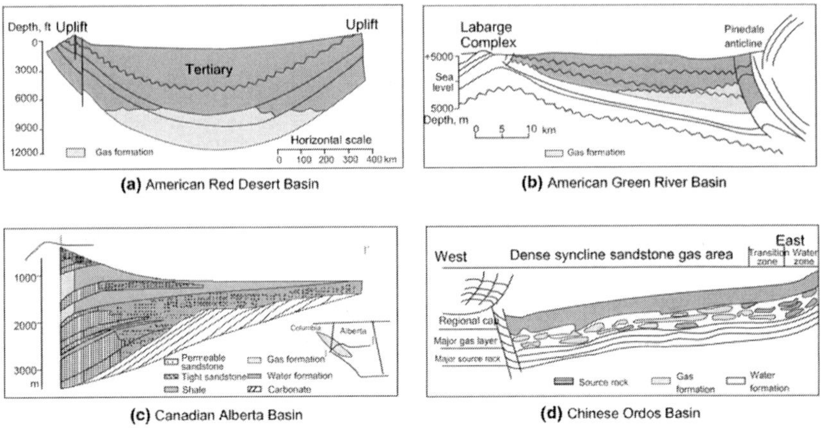

Figure. 37. Reservoir pressures of a few known tight sandstone oil–gas reservoirs in China and worldwide

Deep oil–gas reservoirs can be developed without a cap, but oil–gas reservoirs without a cap are generally found in areas with tight reservoirs and stable tectonism, like deep depressions and slope areas of the basin.

5. MAJOR PROGRESS IN DEEP HYDROCARBON RESERVOIR RESEARCH

Deep hydrocarbon reservoirs are attracting the attention of more and more companies and scholars at home and abroad and a series of research achievements have been made in the past ten years. The major research achievements and progresses are described as follows:

5.1 Multiple deep hydrocarbon sources in petroliferous basins and the formation mechanisms

Traditionally, hydrocarbon researchers believed that hydrocarbon was formed from thermal degradation of organic matter under appropriate temperatures (60–135 °C) and pressure (burial depth 1,500–4,500 m) in sedimentary basins (Hunt 1979; Tissot and Welte 1978). But the formation mechanism of deep hydrocarbons in a high-temperature and high-pressure environment is far more complicated than it was thought to be. So far, four hydrocarbon generation modes have been proposed by scholars.

5.1.1 Successive generation of gas from deep organic matter

The theory of successive generation of gas from organic matter solved the problem of generation and expulsion of natural gas from highly and over-mature source rocks deep down in hydrocarbon basins (Zhao et al. 2005). The "successive generation of gas" mechanism means the conversion of gas-generating matrix and succession of gas-generating time and contribution in the gas-generating process (Fig. 38), including two aspects: 1) the generation of gas from thermal degradation of kerogen and cracking of liquid hydrocarbon and soluble organic matter in coal, which occurs successively in terms of gas generation and contribution; and 2) part of the liquid hydrocarbon generated in the thermal degradation of kerogen is expelled from source rocks to form oil reservoirs, but most of the liquid hydrocarbon remains dispersed in the source rocks, where it is thermally cracked in the highly to over-mature areas so that the source rocks still have great gas-generating potential.

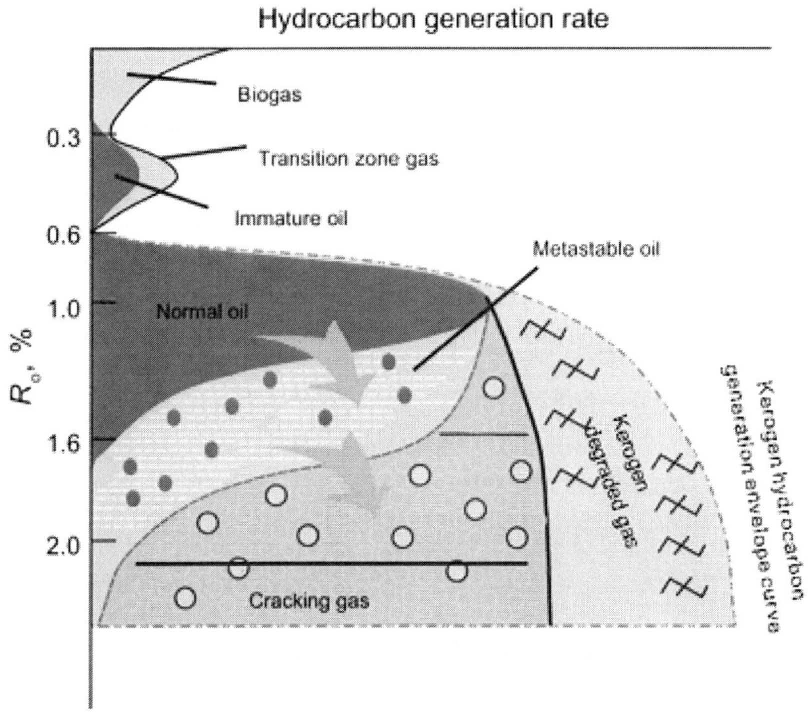

Figure. 38. Successive generation of gas from organic matter (Zhao et al. 2005)

5.1.2 Hydrocarbon generation from hydrogenation of deep organic matter

The generation of hydrocarbons from thermal evolution of organic matter not only needs heat but also hydrogen. The injection of hydrogen-rich fluids into a sedimentary basin will definitely have a great impact on the generation of hydrocarbon. Jin et al. conducted a simulation test with olivine, zeolite, and source rocks, finding that the methane yield increased by 2–3 times after the source rocks interacted with the zeolite and olivine. According to analysis of hydrocarbon generated in low-maturity source rocks in the Dongying Sag and the Central Tarim region by Jin et al. (2002) by means of hydrogenation thermal simulation, the effect of hydrogenation on hydrocarbon generation for Type II_2 kerogen is distinct after the peak hydrocarbon-generating period, while the effect on humic-type kerogen is distinct in all periods. The effect of hydrogenation on source rocks with poor hydrogen kerogen is more distinct (Fig. 39).

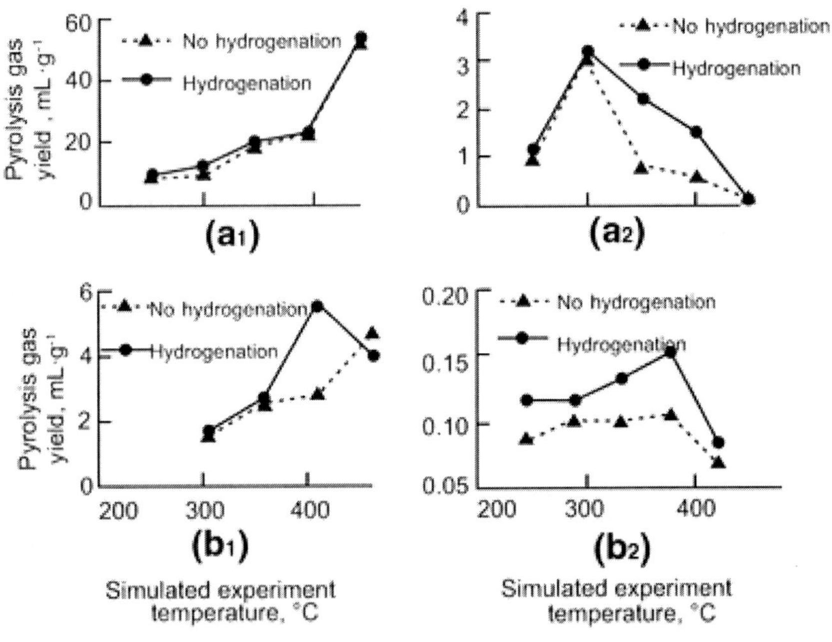

Figure. 39. Simulation test results comparison of different types of source rocks under hydrogenated and non-hydrogenated conditions (Jin et al. 2002). **A** Source rock from Sha-3 member at well Fan15 in the Dongying Sag (Type II_2 kerogen, water added) and **B** Carboniferous source rocks from Central Tarim well He6 (Type III kerogen, water not added)

5.1.3 Hydrocarbon generation and expulsion from deep asphalt cracking

Paleo-reservoirs, if damaged, may produce asphalt, and asphalt may be cracked under high temperature to generate lighter hydrocarbons.

Gong et al. (2004) collected asphaltic sand formed by bio-degradation in the Silurian from the Tarim Basin, and used a high-pressure reaction vessel to investigate the effects of heat on the compositions, isotopes, and physical properties of the sand (Fig. 40). The results indicated that heat had an effect on the compositions and structure of the Silurian asphaltic sand during later-stage burial in the Tarim Basin. The asphaltic sand produces gas at high temperature, and the cracked gas has lighter carbon isotopes: there are less compositions of C_{6+} and above, and it is dominated by light oil; the gas yield is low at low temperature but increases substantially after 400 °C and reaches its peak at 550 °C.

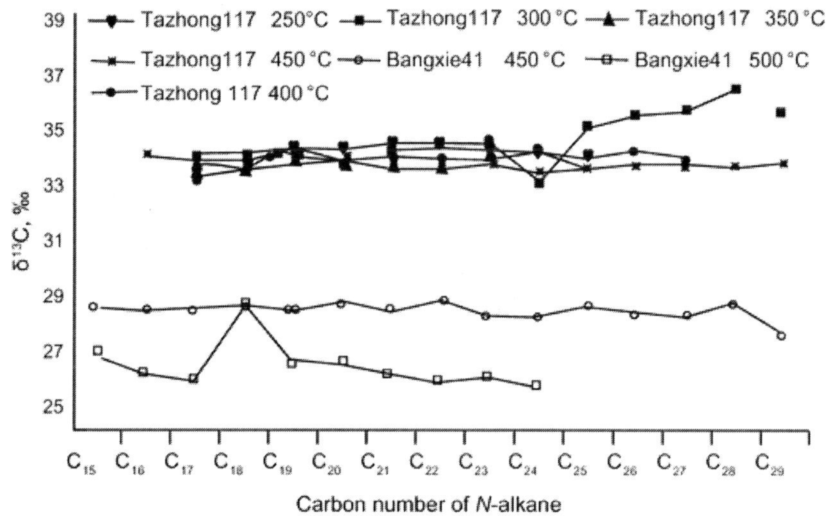

Figure. 40. Curves of *n*-alkane carbon isotopes under different experimental temperatures (Gong et al. 2004)

Huang et al. (2012) obtained the yield curves of the asphalt degradation products (oil and gas) under different temperatures and further evolution process by using artificial and geological samples (Fig. 41). The hydrocarbon-generating process of asphaltic sandstone was divided into three stages: (1) evaporative fractionation stage (R_o<0.9 %), during which the light components of crude oil escaped from damaged paleo-reservoirs to form heavy oil (or asphalt), which could not provide sufficient supply to new hydrocarbon reservoirs; (2) Cracking stage 1 (R_o = 0.9 %–1.8 %), during which oil precipitation increased abruptly and reached a peak, and the yield of hydrocarbon gas increased. It was an important hydrocarbon supply stage in the burial process of asphaltic sandstone (heavy oil); (3) Cracking stage 2 (R_o>1.8 %), during which large amounts of large molecule hydrocarbons were decomposed and hydrocarbon gas was given off. With the rise of temperature, the amount of liquid hydrocarbon in the expelled hydrocarbon components decreased substantially and large amounts of hydrocarbon gas was generated, followed by a substantial decrease of the residual hydrocarbon, and the generation of large amounts of methane. It was an important gas supply stage.

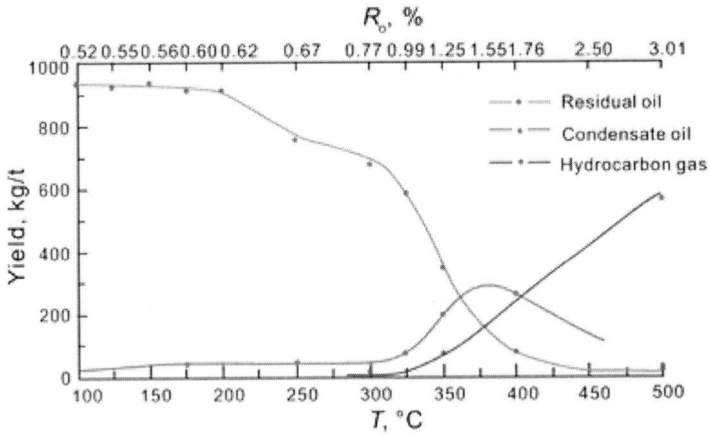

(a) Artificial samples

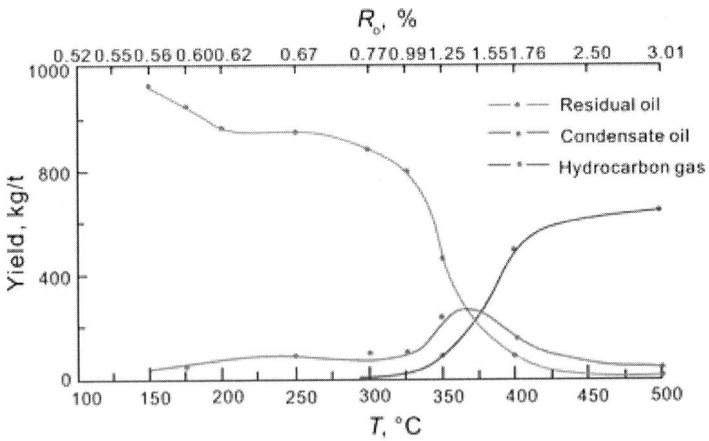

(b) Geological samples

Figure. 41. Hydrocarbon yield curves of asphaltic sand under different temperatures (Huang et al. 2012)

5.1.4 Hydrocarbon generation and expulsion from deep source rocks with low TOC

Pang et al. (2014b) believed that deep-buried poor source rocks with low total organic carbon (TOC) concentration can be regarded as effective source rocks. They generate mass hydrocarbons in the long evolution of geological history, which makes a great contribution to hydrocarbon accumulation. Pang et al. (2014b) obtained the evolutionary charts of TOC of carbonate source rocks based on the hydrocarbon expulsion threshold theory through simulation by using the material balance method (Fig. 42). As shown in the charts, a) the original TOC of source rocks decreases with an increase of R_o. The TOC has a minor increase first and then decreases substantially until a balance is reached. The sharp decrease occurs when R_o = 0.5 %–2.0 %, and the TOC of Type I, II, and III source rocks decreases by 62 %, 48 %, and 25 %, respectively. This is consistent with the period of generation and expulsion of large amounts of hydrocarbon during the thermal evolution of organic matter. b) The abrupt TOC decrease of different types of source rocks occurs at different times. For source rocks of Type I, II, and III organic matter, the abrupt TOC decrease occurs when R_o = 0.5 %, 0.7 %, and 0.9 %, respectively. This may be related to the increasing hydrocarbon expulsion threshold depth of the sources rocks. Many scholars studied the hydrocarbon yield of poor hydrocarbon carbonate source rocks through thermal experiments (Qin et al. 2005; Hao et al. 1993; Cheng et al. 1996; Fan et al. 1997; Xie et al. 2002; Hu et al. 2005; Liu et al. 2010b). According to the test results, the source rocks yielded large amounts of hydrocarbon. The maximum oil and gas yield was 40.4–482.6 kg/t TOC and 115–4,226 m^3/t TOC, respectively. This indicates the source rocks in the deep basins with low TOC concentration are the result of the mass generation and expulsion of hydrocarbons and can be taken as effective source rocks under certain conditions. Pang et al. (2014b) pointed out, for deep source rocks that have yielded large amounts of hydrocarbon in basins, if the regional hydrocarbon prospects are identified and evaluated by using residual organic abundance indices (TOC), it will lead to errors; for highly to over-matured source rocks buried deep in basins, more source rocks are poor in hydrocarbon and the errors will be more obvious. Therefore, residual hydrocarbon indices cannot be used directly to identify and evaluate deep-buried effective source rocks. If TOC is used as a basic index to identify and evaluate source rocks, comparisons should be made in the same geological conditions (Pang et al. 2014b). Figure 43 shows the charts of TOC recovery coefficient under different geological conditions in accordance with the material balance theory, we can conclude that the TOC recovery coefficient of argillaceous source rocks (R_o > 2.0 %) varies with different parent material types; the TOC recovery coefficient of parent material I, II, III type reaches 2.6, 1.75, 1.25 respectively; while for carbonate source rocks the TOC recovery coefficient can reach 2.65, 7.80, 1.30 respectively. The chart allows comparison of TOC of source rocks under different geological conditions on an equal footing. The criteria for evaluation of effective source rocks based on hydrocarbon expulsion thresholds of deep source rocks in basins are listed in Table 6. With the criteria, Pang et al. (2014b) identified and evaluated an effective Cambrian–Ordovician source rock in the Central Tarim region. According to the results, the thickness of the effective source rocks is 47 to 129 meters more than that evaluated using the TOC = 0.5 % criterion, and the scope of distribution is increased by 7 km^2, total oil yield increased by 182,500 million tons, and total reserves increased by 9,300 million tons, accounting for 38.1 % of the total resource. This reflects the importance of research on effective source rocks with low TOC.

5.2 Deep-buried high-porosity and high-permeability reservoirs, the formation and distribution of which are related to structural changes and underground fluid activities

The porosity and permeability of deep hydrocarbon reservoirs in basins are usually low. The deeper the reservoirs are, the smaller the porosity and permeability will be. In actual geological conditions, the porosity and permeability of deep effective reservoirs vary significantly. Research results indicate that the formation of deep-buried high-porosity and high-permeability reservoirs is closely related to structural changes and evolution as well as fluid activities. In general, more faults and disconformities are developed in regions with stronger structural activities. Faults not only serve as pathways for hydrocarbon migration, but also improve the reservoir quality. Fault developed regions generally have favorable porosity and permeability. As shown in Fig. 44, most hydrocarbon reservoirs so far discovered in the Tarim region are distributed in fault regions. This is particularly true for carbonate reservoirs, which are developed near faults (Fig. 44a). Disconformities also serve as pathways for hydrocarbon migration, and improve the geological conditions of reservoirs by weathering and leaching. As shown in Fig. 45, reservoirs near disconformities have high porosity and permeability. High-quality carbonate reservoirs are usually distributed within 200–300 m from an unconformity. Sedimentary fluids have decisive effects on the granular structure and composition of reservoirs, and influence the strain response and the formation of fractures and cavities in the rock-forming process. Surface fluids may change the physical properties of surface rocks, enabling the rocks to accept external fluids and providing conditions for solution-pore type reservoirs reconstructed by external fluids after being buried. Underground hot fluids are favorable for secondary pore formation in deep reservoirs and the accumulation of gas and oil may take place in the dissolution pores and cavities, fractures, and cracks.

The formation of deep effective reservoirs is usually affected by a range of factors. For example, organic matter generates organic acids in the process of thermal evolution. The organic acids dissolve some minerals in the surrounding rocks to form secondary pores and improve the porosity of deep reservoirs. In the fast subsidence–sedimentation of sedimentary basins, overpressure systems may develop in shallow and high-porosity strata to reduce the effective stress, and thus reducing the compaction and restraining the pressure dissolution. At the same time, the fluids in the overpressure systems have poor fluidity, which retards the formation and cementation of rocks. That is why deep-buried overpressure reservoirs usually have high porosities. In summary, the high porosities of deep-buried overpressure reservoirs in basins are the result of the following combined effects: reduced compaction under low effective stress, retarded cementation due to low-fluidity fluids, and dissolution of minerals by organic acids. Therefore, multi-factor integrated research is required for the evaluation of deep-buried effective reservoirs.

Deep-buried high-quality clastic reservoirs developed over different periods in basins in China are related to such factors as abnormally high pressure, early hydrocarbon charging, thermal convection, gypsum-salt effects, and sandstone and mudstone interbeds (Li and Li 1994; Gu et al. 1998; Zhong and Zhu 2003). Dissolution, early long-term shallow burial and late short-term deep burial, abnormally high pressure, and early hydrocarbon charging are key factors influencing the formation of deep-buried high-quality clastic reservoirs in China (Shi and Wang 1995; Yang et al. 1998; Li et al. 2001; Zhong et al. 2008). Many researchers realized that the formation and evolution of high-porosity and high-permeability carbonate reservoirs are related to the epidiagenesis, dissolution of organic acids, dolomitization, abnormally high pressure, modification of thermal fluids, and hydrocarbon charging (Surdam et al. 1989; Davies and Smith 2006; Fan 2005; Li et al. 2006; Zhu et al. 2006).

Table 6. Revised TOC thresholds criterion for evaluation of effective source rocks

Evolutionary period	Muddy source rock			Carbonate source rock		
	I	II	III	I	II	III
Immature ($R_o < 0.5$ %)	0.50	0.50	0.50	0.50	0.50	0.50
Mature ($R_o = 0.5$ %–1.2 %)	0.35	0.45	0.47	0.30	0.40	0.45
Highly mature ($R_o = 1.2$ %–2.0 %)	0.25	0.35	0.45	0.20	0.30	0.40
Over-mature ($R_o > 2.0$ %)	0.20	0.30	0.40	0.15	0.25	0.35

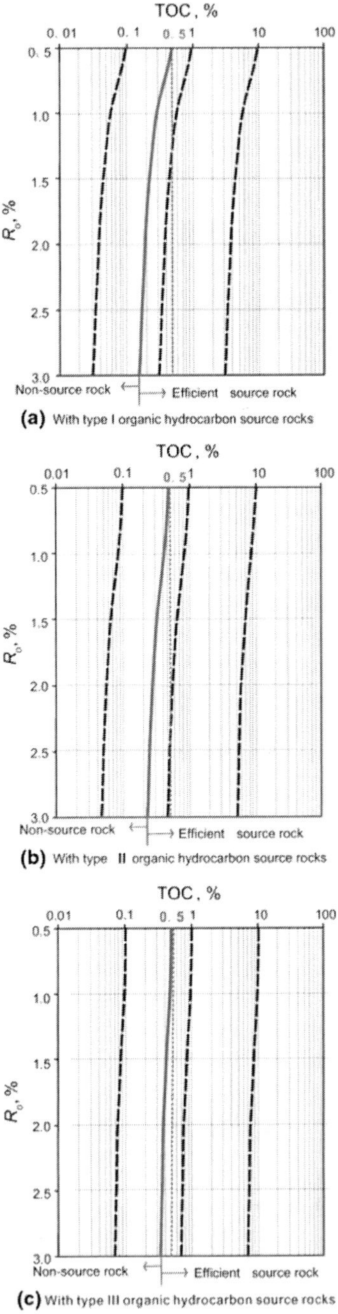

Figure. 42. Evolutionary charts of TOC of different carbonate source rocks of deep petroliferous basins (Pang et al. 2014b)

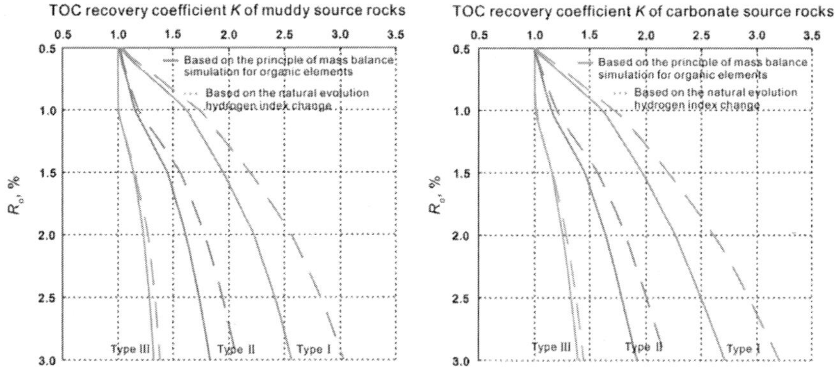

(a) The TOC recovery coefficient plot of different kerogen in muddy source rocks

(b) The TOC recovery coefficient plot of different kerogen in carbonate source rocks

Figure. 43. Charts of restored TOC of source rocks under different geological conditions of petroliferous basins (Pang et al. 2014b)

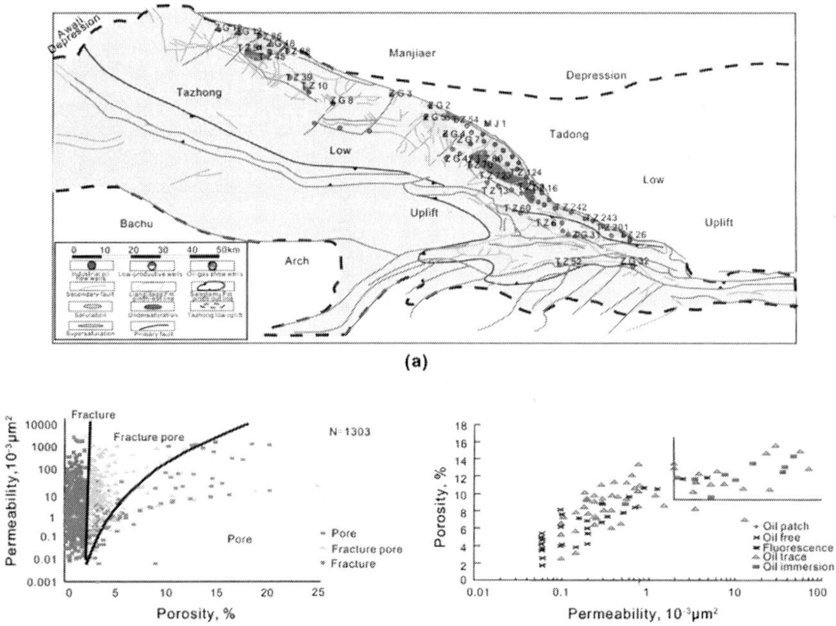

(b) The relationship between porosity and permeability in middle Tarim carbonate

(c) The relationship between porosity and permeability in middle Tarim clastic rocks

Figure. 44. Faults and hydrocarbon distribution in the Central Tarim Basin and the porosity–permeability features in reservoirs

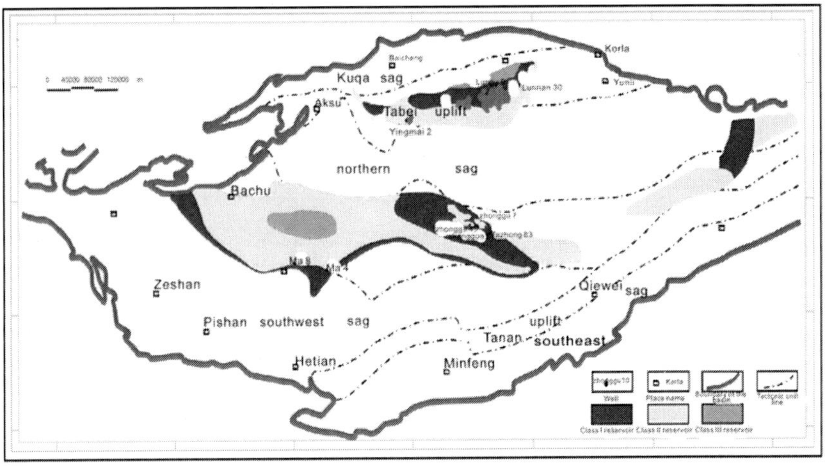

(a) Relationship between porosity distribution and distance to the weathering crust in the Yingshan Formation, Tazhong region

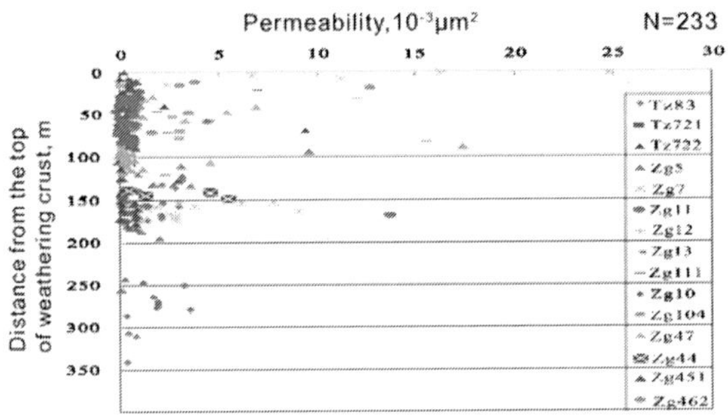

(b) The relationship between permeability of the Yingshan Formation and the distance from the top of weathering crust in Tazhong

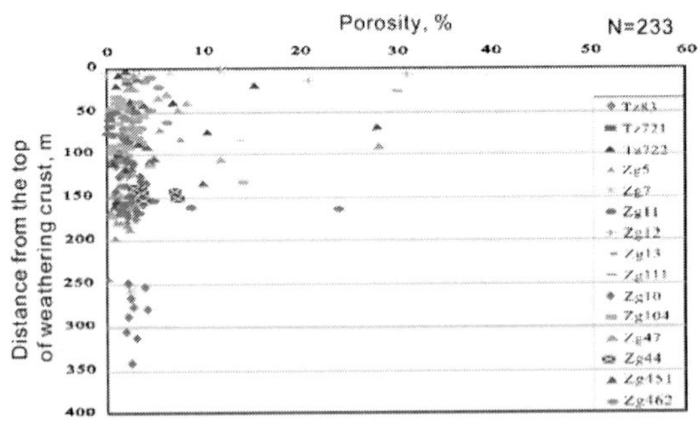

(c) Relationship between porosity distribution and the distance to the weathering crust in Yingshan Formation, Tazhong region

Figure. 45. Weathering crust and hydrocarbon distribution in the basin-platform region, Tarim Basin and the porosity–permeability features in reservoirs (Du et al. 2010)

As research on deep-buried marine facies carbonate reservoirs goes on and exploration breakthroughs are continuously made in China, geologists have achieved a better understanding of the formation mechanisms of secondary pores in carbonate reservoirs. Previous researchers insisted that secondary pores in deep-buried carbonate reservoirs were formed as a result of the dissolution of the carbonate rock exposed to the air, and many scholars focused on paleo-dissolution reservoirs, ignoring the role of karstification (dissolution) in improving the permeability and storage properties of deep-buried carbonate reservoirs (Zhu et al. 2006). In recent years, the fluid-rock interaction in deep-buried hydrocarbon-bearing carbonate reservoirs has attracted the attention of more and more scholars (Land and MacPherson 1992; Fisher and Boles 1990; Williams et al. 2001; Zhang et al. 2005b). As deep-buried karst is not controlled by the base level of erosion on the surface, many substances could induce water–rock reactions, such as acidic water and gases produced in the process of thermal evolution of organic matter; hot water produced by magmatic activities, compaction or diagenesis; acidic gases from deep strata in basins; and hydrogen sulfide gas produced in thermochemical or microbe reduction of sulfate-bearing carbonate rocks (Luo 2003). Pan et al. (2009) believed, based on samples from outcrops and drilling in the Central Tarim region, that hydrothermal karst reservoirs may be an important type of reservoir, which has been ignored in the exploration for Lower Paleozoic carbonate hydrocarbon reservoirs in the Tarim Basin, and that there might be high-quality reservoirs along the faults, the main pathways for the migration of hydrothermal fluids, and fault-associated dissolution fractures and cavities. Li et al. (2010) studied the Mid-Lower Ordovician carbonate reservoirs in the Tahe region in the Tarim Basin. According to the research results, carbonate reservoirs might have experienced strong cementation in the normal process of deep burial diagenesis, but there were hardly any signs of dissolution. The modification of the reservoirs might be related to late epikarstification and structural-thermal fluid processes (Fig. 46). Generally, the epikarstification was followed by the structural-thermal fluid processes. The development and distribution of the latter might be related to the fault-fissure system and early epikarst system. According to the research results on the Ordovician carbonate reservoirs in the Tarim Basin by Lin et al. (2012), four dynamic mechanisms were involved in the structural modification of the reservoirs and the formation of fissures and cavities: (1) penecontemporaneous surface water karstification of carbonate rocks to form dissolution cavities, (2) raised surface fresh water karstification of the reservoir to form dissolution cavities, (3) deep hydrothermal fluid karstification of reservoir to form dissolution cavities, and (4) stress-induced faulting to form fissures (Fig. 47). Zhu et al. (2006) discovered in studying the deep-buried high-quality carbonate reservoirs in the Sichuan Basin that, besides the dolomitization and deep dissolution that controlled the formation of the porous oolitic dolomite in the Feixianguan Formation, strongly corrosive materials produced in the TSR (thermochemical sulfate reduction) boosted the dissolution of dolomite, leading to the formation of porous permeable spongy oolitic dolomite dissolution bodies, which played a constructive role in the formation of the reservoirs. The above research achievements indicated that the deep fluids in basins might be closely associated with the formation of high-quality carbonate reservoirs.

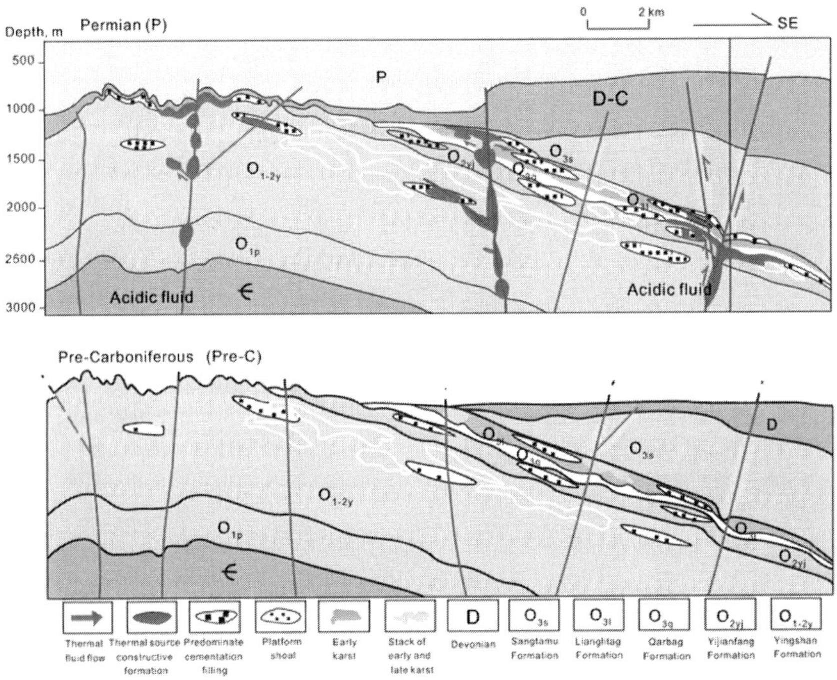

Figure. 46. Structures–fluid effect of Ordovician carbonate and controlling mode of reservoir distribution in the Tahe region, Tarim Basin (Li et al. 2010). Early karst: Mid-Late Ordovician–Silurian (Caledonian); Superposition of early and late-period karst: superposition of Mid-Late Ordovician–Silurian karst (Caledonian) + Late Devonian karst (Hercynian)

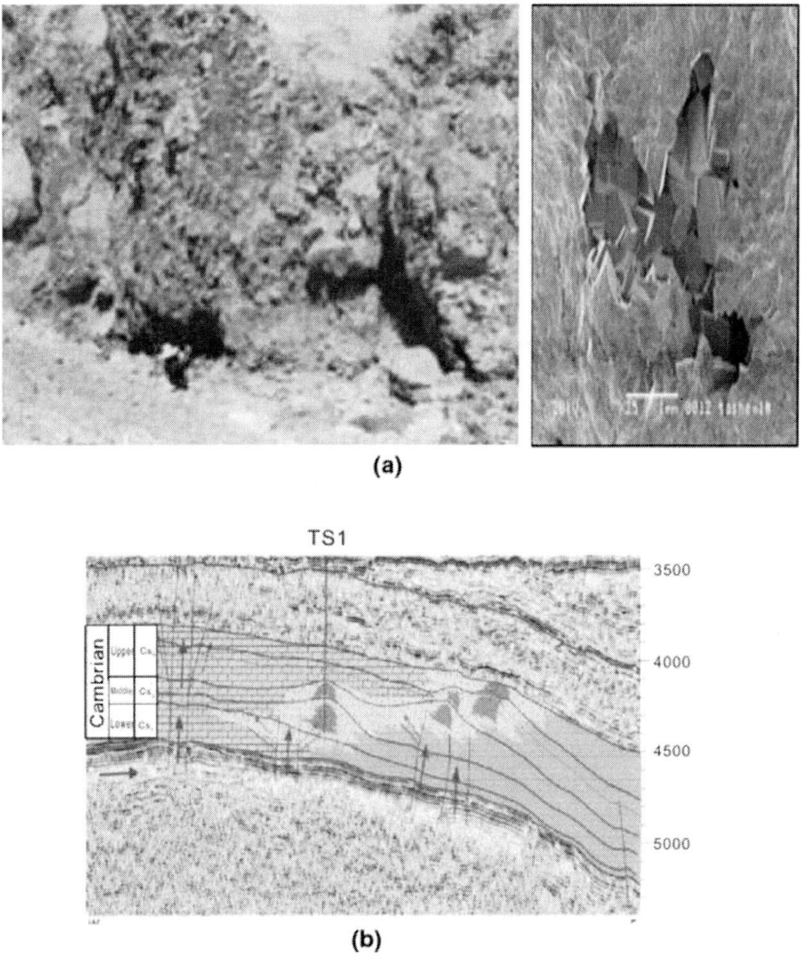

Figure. 47. Cavities and fissures formed as a result of structural changes of carbonate reservoirs in the Tarim Basin (Lin et al. 2012). **A** Dissolution cavities, **B** Fissures

5.3 Capillary pressure difference between inside and outside the target formation is the main force resulting in hydrocarbon accumulation in deep reservoirs

Forces involved in the formation of hydrocarbon reservoirs in hydrocarbon basins mainly include buoyancy, fluid pressure and dynamic force, expansive force of molecular volume, capillary force, molecule adsorption force, and intermolecular binding force (Pang et al. 2000). Under the action of buoyancy, natural gas migrates along migration pathways into traps in the upper part of a structure (White 1885). Under the action of the expansive force of gaseous molecules, natural gas migrates in a piston-type and accumulates in the lower part of a structure, with gas topped by water (Pang et al. 2003). Due to the capillary pressure difference between sandstone and mudstone, natural gas migrates into lens-type sandstones with larger pores, and accumulates at the top of the lens under the action of buoyancy to form a reservoir (Chen et al. 2004).

The temperature and pressure in the shallow strata of sedimentary basins are low and the rock porosity and permeability are relatively high, so the accumulation and migration of hydrocarbon are primarily controlled by buoyancy. But in the deep strata, the temperature and pressure are high and the rocks are relatively tight, so the accumulation and migration of hydrocarbon are controlled by a combination of forces. There are a variety of hydrocarbon reservoirs in the deep strata of sedimentary basins. In addition to conventional reservoirs in high-porosity and high-permeability strata, there are a large number of unconventional reservoirs in tight strata. Many hydrocarbon reservoirs have been discovered in the low-porosity carbonate and sandstone strata of the Junggar Basin and Tarim Basin (Fig. 48). With the ongoing research, geologists are gaining a better understanding of the formation mechanisms of deep hydrocarbon reservoirs. Capillary pressure that causes surface potentials has been considered to resist the flow of underground fluids. But since Magara (1978) puts forward the idea that capillary pressure is the primary power for hydrocarbon migration from source rocks to reservoirs, more and more scholars have realized that capillary pressure difference between mudstone and sandstone is an important factor in the formation of hydrocarbon reservoirs (Barker 1980; Magara 1978; Pang et al. 2000; Chen et al. 2004; Zhao et al. 2007; Li et al. 2007), and particularly that lithological traps such as sandstone lenses and pinch-outs are in a direct contact with deep mudstones in the part of basins. Therefore, capillary pressure difference might be one of the key forces for the formation of lithological hydrocarbon reservoirs included in sources rocks or in contact with source rocks. Surface potential is more significant to the formation of deep-buried lithological hydrocarbon reservoirs (Huo et al. 2014a). In the sand–mud contact zone, there is capillary pressure difference between sandstone and mudstone as former has bigger pores than the latter. Pang et al. (2006) and Wang et al. (2013) through the physical experiments concluded that the greater the capillary pressure difference is between the inner sandstone and outer mudstone in a trap, the better hydrocarbon accumulation conditions will be in the sandstone (Fig. 49). Under strata conditions, as the wall rock is compacted, the pore throat radius of the wall rock is much smaller than that of isolated sand bodies. Due to the capillary pressure difference between them, hydrocarbon migrates inwards. So, capillary pressure difference might be one of the key factors that led to the migration of hydrocarbon from the wall rock into the isolated sand bodies. A large number of tight reservoirs were developed in the deep strata of basins. The research of Pang et al. (2006) showed that hydrocarbon was not affected or less affected by buoyancy in tight media. The surface potential or capillary pressure difference between reservoir and external strata is a key factor affecting the accumulation of hydrocarbon in tight media (Fig. 50). The pressure difference has the following effects: (1) hydrocarbon migrates from the source rock into the reservoir due to capillary pressure difference. The greater the difference between the inner and outer capillary pressure or between the inner and outer surface potential, the higher the saturation of hydrocarbon in the reservoir, and the better the gas-bearing properties. (2) Due to capillary pressure difference, hydrocarbon in the tight reservoir migrates from a low-porosity and low-permeability rock to a high-porosity and high-permeability rock to form hydrocarbon-rich "sweet spots". (3) When faulting or fissuring occurs in a tight reservoir, the capillary force will decrease and the surface potential will drop, causing capillary pressure difference and the accumulation of hydrocarbon to fault zones and formation of "sweet spots". Exploration practice has proved that there are large numbers of hydrocarbon-rich "sweet spots" in the extensive, continuous tight sandstone reservoirs in the deep strata of basins, and most of them were formed due to faulting and fissuring in late periods.

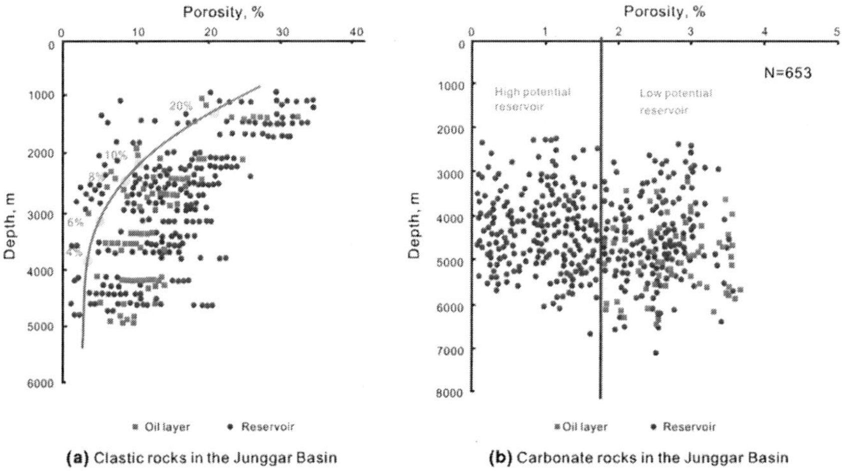

Figure. 48. Diagram of effective porosity in different reservoirs changes as a function of burial depth

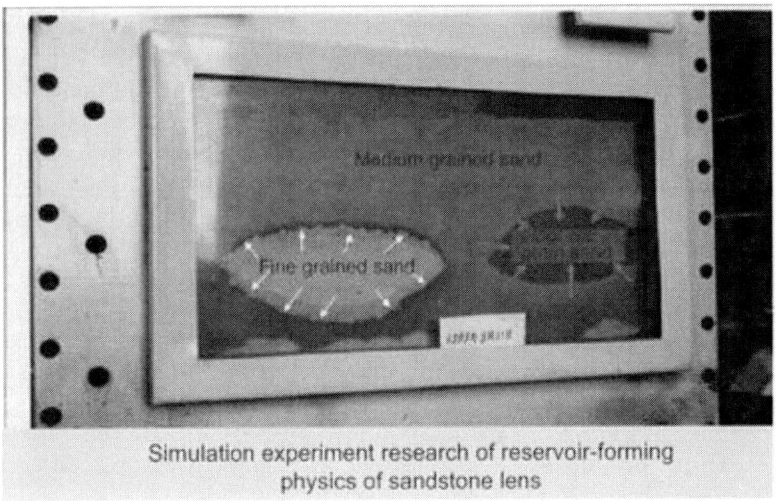

Figure. 49. Physical experiments simulating the formation of a sandstone lens reservoir (Pang et al. 2006; Wang et al. 2013)

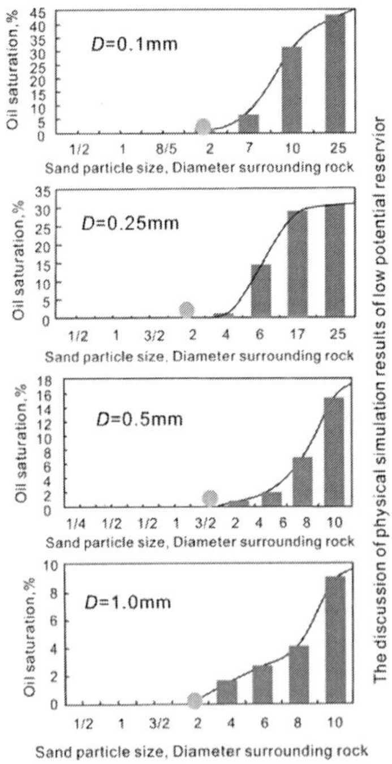

Petroliferous properties of sandstone under various conditions of wall rock (*D*=0.5)

No.	Sample No.	D/d	The sample size	Extraction of oil	Oil content	Oil saturation
1	30mesh-10mesh	1/4	125.1	5.3	0.0042	0.0261
2	30mesh-20mesh	1/2	113.8	4.7	0.0041	0.0255
3	30mesh-30mesh	1	82.4	4.9	0.0059	0.0367
4	30mesh-60mesh	2	167.6	21.2	0.0126	0.0780
5	30mesh-120mesh	4	140.3	180.4	0.1286	0.7929
6	30mesh-200mesh	6	152.4	452.0	0.2966	1.8290
7	30mesh-240mesh	8	142.2	1554.0	1.0905	6.7249
8	30mesh-300mesh	10	126.6	3143.8	2.4833	15.3134

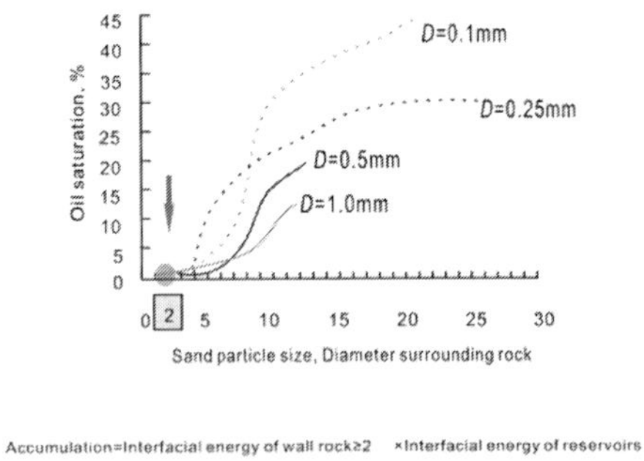

Figure. 50. Physical experiment results of capillary pressure difference (surface potential) controlling hydrocarbon (Pang et al. 2006)

5.4 Three dynamic boundaries of deep hydrocarbon reservoirs

The accumulation and migration of gas and oil are controlled by many geological conditions, including temperature, pressure, source of oil and gas, migration path, traps, etc. The formation and distribution of deep oil–gas reservoirs are mainly constrained by three force balance boundaries in terms of the dynamic mechanics. The first dynamic boundary is buoyancy-controlled threshold. It is the upper boundary of formation and distribution of unconventional tight reservoirs. The buoyancy-controlled threshold is the maximum depth of non-buoyancy-driven hydrocarbon migration in a hydrocarbon basin. The formation and distribution of hydrocarbon reservoirs in hydrocarbon basins are controlled by the dynamic boundary formed by different thresholds (Pang 2010; Pang et al. 2014a). The buoyancy-driven hydrocarbon accumulation threshold is a new geological concept in relation to buoyancy-driven hydrocarbon accumulation (White 1885). Pang et al. (2014a) maintained that the buoyancy-driven hydrocarbon accumulation threshold is the threshold beyond which buoyancy will have less effect on the accumulation of hydrocarbon in highly compacted strata as the burial depth increases. Generally, it is characterized by porosity, pore throat radius, and permeability geological parameters of strata at certain burial depths (Fig. 51).

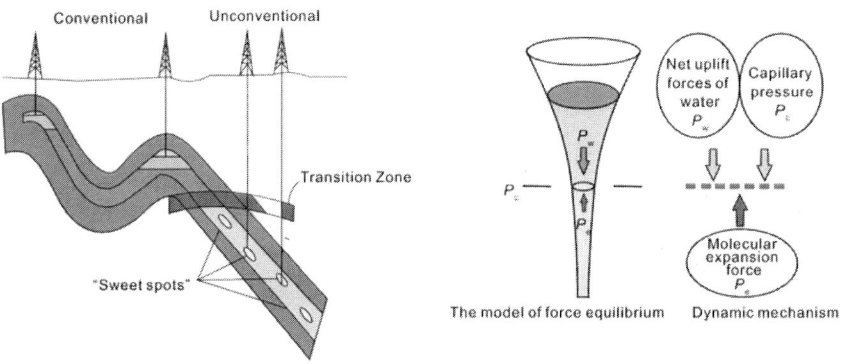

Figure. 51. Conceptual model of buoyancy-controlled thresholds in deep petroliferous basins and controlling the distribution of oil–gas

With respect to the causes of the buoyancy-controlled threshold, scholars have different views. Masters held that buoyancy-controlled threshold is caused by the relative variation of permeability of strata (Masters 1979);

other scholars argued that geological conditions such as diagenetic difference (Cant 1986), fault seals (Cluff and Cluff 2004) and force equilibrium mechanisms (Berkenpas 1991) are the causes. These views may be used to explain the phenomena of non-buoyancy-controlled hydrocarbon accumulation in some basin areas, but cannot explain it in a broader sense. Through physical simulation experiments (Fig. 52), Pang et al. (2013, 2014a) believed that there exists an equilibrium of forces affecting the migration of natural gas at the buoyancy-controlled threshold. Charge pressure is the dynamic force and capillary force and overlying water pressure are the resistances. The equilibrium between the driving force for upward migration of hydrocarbon (P_e) and capillary force in tight media (P_c) and overlying static water pressure (P_w) is the dynamic mechanism of buoyancy-driven hydrocarbon accumulation thresholds. Figure 52 shows the physical simulation experiment results of buoyancy-controlled threshold in a tapered glass tube; it can be quantified by the dynamic equilibrium equation $P_e = P_w + P_c$. Similar results have been obtained based on the physical simulation experiment using a thick tube filled with sandstone of different sizes (Pang et al. 2014a).

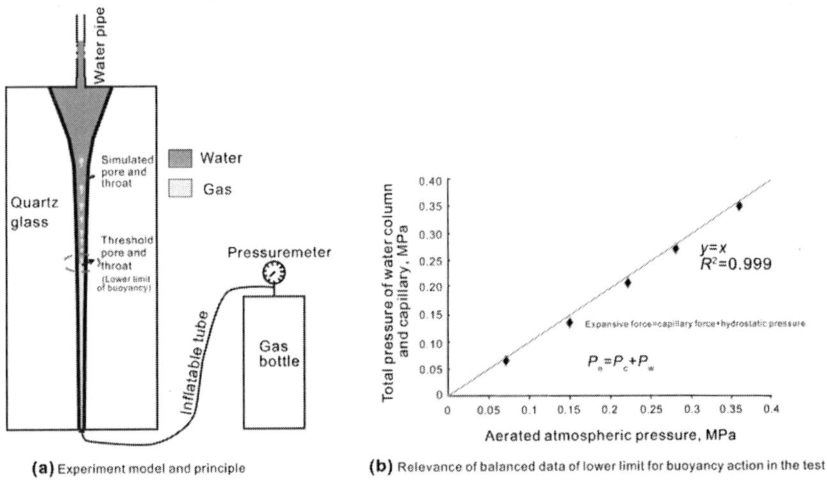

Figure. 52. Physical simulation of buoyancy-controlled threshold (Pang et al. 2013)

The fundamental reason why hydrocarbon in deep tight strata of hydrocarbon basins is not controlled by buoyancy is that the sum of rock capillary force and overlying static water pressure is greater than the pressure in the hydrocarbon reservoir. When the pressure in the hydrocarbon reservoir is greater than the sum of the two forces, buoyancy will cause hydrocarbons to migrate upwards to the surface or scatter. The equilibrium of forces is the dynamic mechanism of buoyancy-controlled threshold, which can be expressed in Eqs. (1)–(3):

$$P_e = P_w + P_c, \quad (1)$$

where P_e is oil reservoir pressure, which can be expressed as follows:

$$P_{eg} = \frac{z \times \rho_g}{M_g} \times R \times T \times 1.01 \times 10^2, \quad (2)$$

$$P_{eo} = \frac{RT}{V - b} - \frac{a}{V^2} = \frac{\rho_o RT}{M_o - \rho_o b} - \frac{\rho_o^2 \times a}{M_o^2}, \quad (3)$$

where P_{eg} is the gas reservoir pressure, MPa; z is the gas deviation coefficient (compressibility factor), dimensionless; R is the universal gas constant, 0.008314 MPa m^3/(kmol K); T is the absolute temperature of natural gas, K; M_g is natural gas molar mass, kg/kmol; ρgρg is the natural gas density under strata conditions, kg/m^3; P_{eo} is the oil reservoir pressure, MPa; ρoρo is the oil density under strata conditions, kg/m^3; M_o is the molar mass of oil, kg/kmol; and a, b are van der Waals constants.

The change of any geological parameter in the force equilibrium equations of the buoyancy-controlled threshold will lead to changes of the critical conditions of the buoyancy-controlled threshold. Factors include different driving forces, fluid physical and chemical properties, strata conditions, and the structural environment of basins. Driving forces include hydrocarbon reservoir pressure (P_e), overlying static water pressure (P_w), and reservoir media capillary forces (P_c). The change of any force will lead to change of the buoyancy-driven hydrocarbon accumulation threshold; fluid physical and chemical properties include hydrocarbon-water interface tension, contact angle, density, and temperature; strata media conditions include porosity, permeability, and pore throat radius; the structural environment of basins refers to structural changes of a basin that affect the distribution scope of the buoyancy-driven hydrocarbon accumulation threshold.

Under actual geological conditions, the buoyancy-controlled threshold is affected by a combination of the above factors. Research results indicate that the depth of the buoyancy-controlled threshold increases and the corresponding porosity, permeability, or pore throat radius decreases as the sand grain size increases under the circumstances that all the conditions are favorable, and that the corresponding burial depth, porosity, permeability, or pore throat radius of the buoyancy-controlled threshold force equilibrium decreases as the sand grain sorting difficulty increases. According to statistics, generally, the buoyancy-controlled threshold of hydrocarbon basins is as follows: porosity <12 %, permeability <1 mD, and pore throat radius <2 μm. Although porosity is often considered a characteristic parameter of buoyancy-controlled threshold, in fact, whether hydrocarbon is controlled by buoyancy in a reservoir or not is mostly decided by the pore throat radius, because it has a direct effect on the capillary force affecting the migration of hydrocarbon. For carbonate reservoirs with smaller porosities, the migration of hydrocarbon is also affected by buoyancy in fissure developed areas. The general relations between porosity, permeability, and pore throat radius are shown in Fig. 53.

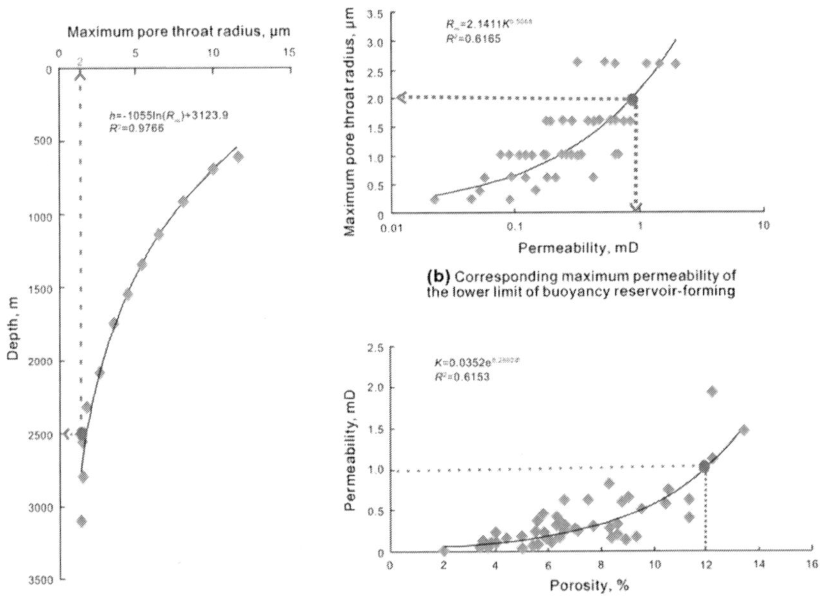

(a) Corresponding maximum pore throat radius of the lower limit of buoyancy reservoir-forming

(b) Corresponding maximum permeability of the lower limit of buoyancy reservoir-forming

(c) Corresponding maximum porosity of the lower limit of buoyancy reservoir-forming

Figure. 53. Key elements of the equilibrium of forces at buoyancy-controlled threshold and their relations

Buoyancy-controlled threshold distribution under actual geological conditions can be predicted based on the equilibrium equation (Eqs. (1)–(3)). Figure 54 shows the predicted buoyancy-controlled threshold of tight sandstone reservoirs of the Upper Paleozoic in the Ordos Basin in China based on the force equilibrium equations. The results indicated that they correspond to geological conditions of porosity <10 % and permeability <1 mD. Below the dynamic boundary, gas is filled in the tight reservoirs of adjacent source rocks; above the boundary, natural gas accumulates only in the upper part of the structure. Figure 55 shows the comparison of predicted buoyancy-driven oil accumulation thresholds with drilling results in the Putaohua reservoir in the Songliao Basin. As shown in Fig. 55, liquid oil also has a buoyancy-controlled threshold, and the force equilibrium boundary corresponds to the burial depth with a porosity between 10 % and 11 %. Above this boundary, liquid oil reservoirs have the features of high-point accumulation, high-porosity enrichment, high-point sealing, and high pressure; below the boundary, liquid oil reservoirs have the features of low-depression accumulation, low-stand inversion, low-porosity enrichment, and low-pressure stability. The consistency of the theoretic prediction results with the actual drilling results reflects the existence of a buoyancy-controlled threshold and the practicality and reliability of the prediction model for both natural gas and liquid oil. Their difference is small (Pang et al. 2013, 2014a).

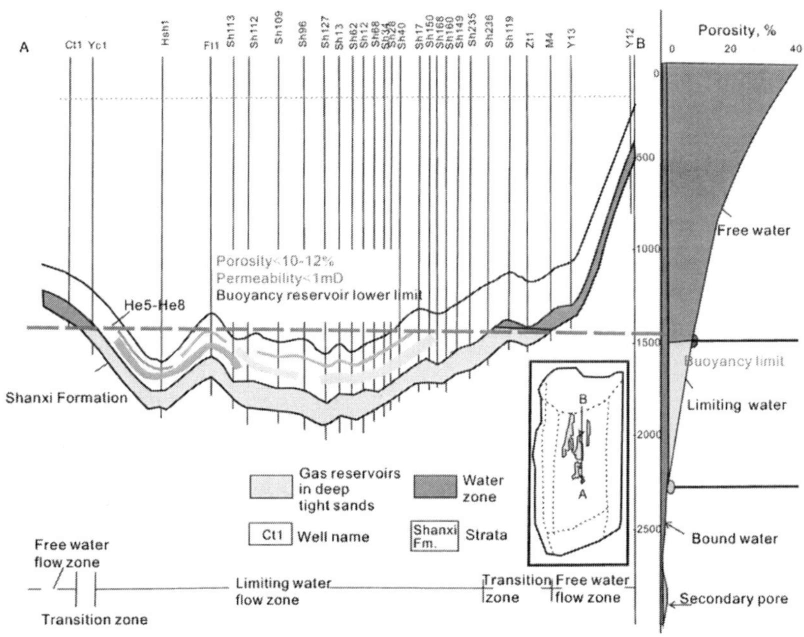

Figure. 54. Comparison of predicted buoyancy-controlled threshold of the Upper Paleozoic with drilling results in the Ordos Basin (Pang et al. 2014a)

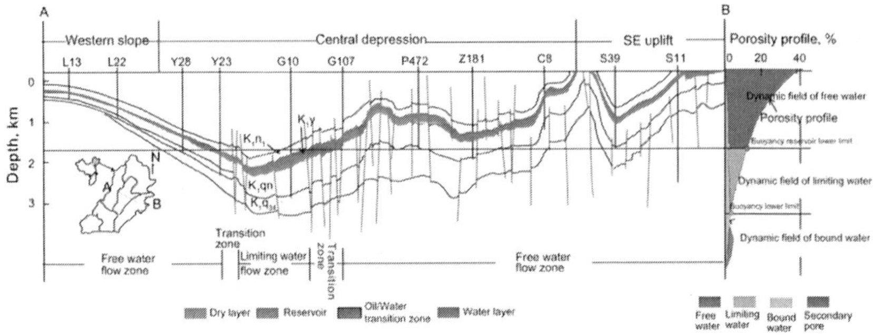

Figure. 55. Comparison of predicted buoyancy-controlled threshold with drilling results in the Putaohua reservoir in the Songliao Basin (Pang et al. 2014a)

The second dynamic boundary is the hydrocarbon accumulation limit. It is the maximum depth for hydrocarbon accumulation under geological conditions. The enrichment and accumulation of hydrocarbons require certain temperature and pressure conditions. Theoretically, each hydrocarbon basin may have a hydrocarbon accumulation threshold at a certain depth, below which hydrocarbon accumulations do not exist. The dynamic boundary corresponding to the lower limit of hydrocarbon accumulation is regarded as the hydrocarbon accumulation threshold. According to Pang et al. (2014a), the hydrocarbon accumulation limit is the maximum burial depth or corresponding critical geological condition for hydrocarbon accumulation in hydrocarbon basins, which can be characterized by the porosity, permeability, or pore throat radius of a reservoir (Fig. 56). Pang et al. (2014a) recognized through analysis that the corresponding critical values of the hydrocarbon accumulation limit in clastic petroliferous basins are generally as follows: porosity is less than 2 %–2.4 %, permeability is less than 0.01 mD, pore throat radius is less than 0.01 μm, and burial depth is 5,000–8,000 m.

Many scholars discovered through research using different methods that there is a physical threshold for sedimentary basin oil–gas-bearing reservoirs, below which hydrocarbon does not accumulate or has no exploration significance (Wan et al. 1999; Guo 2004; Shao et al. 2008). Through analysis of actual mass exploration results, four different methods have been used to determine hydrocarbon accumulation limits in the deep part of hydrocarbon basins by Pang et al. (2014a): (1) As the burial depth increases, the bound water saturation in the reservoir will reach 100 %, resulting in the termination of hydrocarbon accumulation. When the bound water saturation reaches 100 % in the Fuyang sandstone reservoir in the Songliao Basin, the hydrocarbon accumulation limit porosity is 2.4 %–4 %; when the bound water saturation reaches 100 % in the hydrocarbon-bearing sandstone in the Kuqa Depression in the Tarim Basin, the hydrocarbon accumulation limit porosity is about 2.4 %. Figure 56 shows an example of this. (2) The hydrocarbon accumulation limit can be determined when the porosity and permeability as a function of buried depth do not allow the migration of oil and gas normally. As the burial depth increases, the reservoir permeability under actual geological condition decreases as a result of compaction strengthening. Figure 57 shows an example. (3) The hydrocarbon accumulation limit can be determined when capillary pressure difference between inside and outside the reservoir or the potential difference tends to disappear as a function of buried depth. As the burial depth under actual geological conditions increases, the potential difference between inside and outside the reservoir will disappear, resulting in the termination of hydrocarbon accumulation. Figure 58 shows the hydrocarbon accumulation limits in terms of inner and outer potential difference of major target reservoirs in the Jiyang Depression in the Bohai Bay Basin in eastern China and the Kuqa Depression in the Tarim Basin in western China are at burial depths of 6,000 and 8,500 m, respectively (Fig. 58). (4) The hydrocarbon accumulation limit can be determined based on the exploration well data from a 100 % dry bed. When a 100 % dry bed was met in the exploration well during drilling for the purpose of understanding the hydrocarbon and water distribution in the Central Tarim Basin region, the reservoir porosity was less than 2 %, so it is taken as the hydrocarbon accumulation limit (Fig. 59).

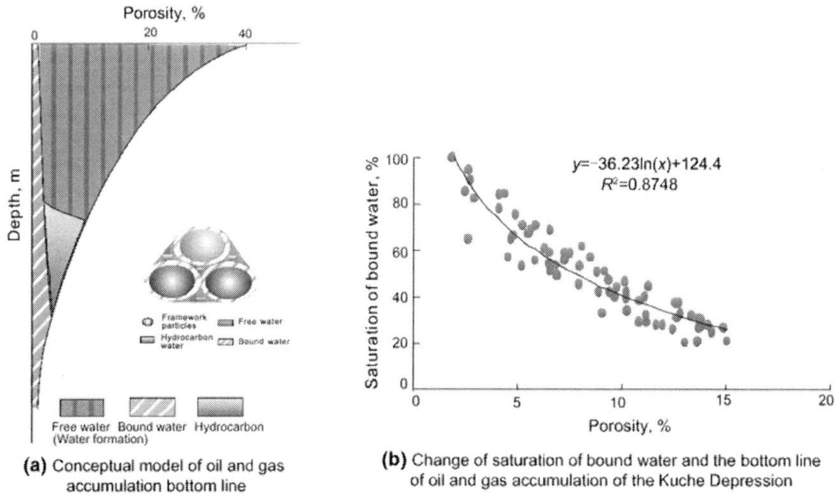

Figure. 56. Bound water saturation in an oil–gas-bearing target reservoir as a function of buried depth and hydrocarbon accumulation limits

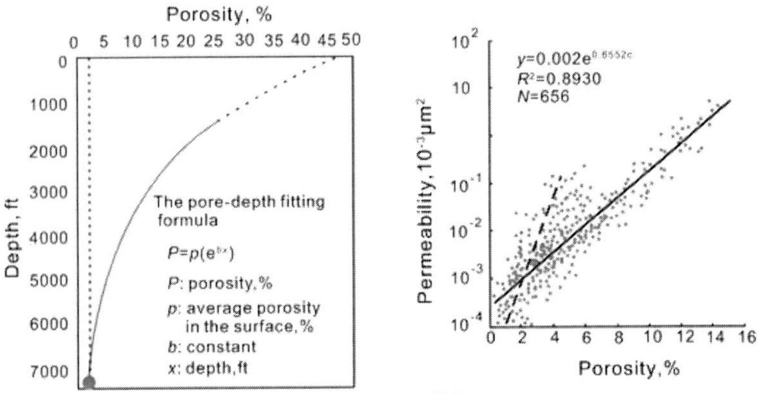

(a) Relationship between porosity and depth

(b) Relationship between permeability and effective porosity in Xu6,Chuanxi section

Figure. 57. Porosity and permeability changes and hydrocarbon accumulation limits in the burial process of a hydrocarbon-bearing target layer

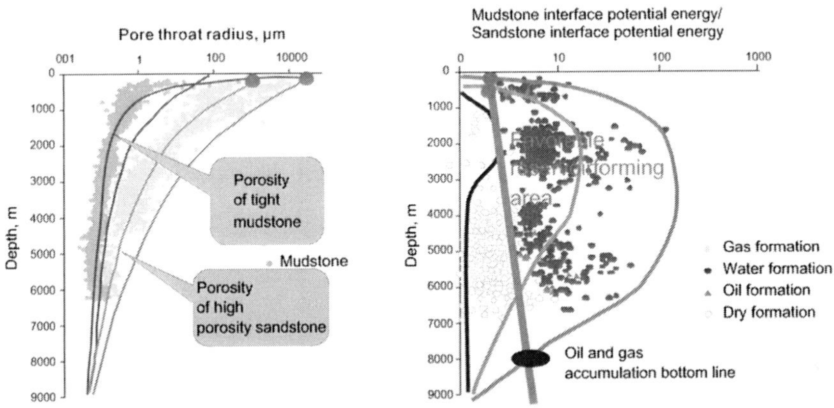

(a) Yeyun2 well pore throat radius changes with depth

(b) Changes of potential difference of inner-outer interface in Kuqa Depression, Tarim as a function of depth

Figure. 58. Changes of inner–outer surface potential and capillary pressure difference and hydrocarbon accumulation limits

We highlight here that the hydrocarbon accumulation limit is not a threshold indicating the existence of hydrocarbon. When hydrocarbon exists in reservoirs below this limit, it was probably accumulated before entering the threshold, below which there may be exploration risks because the strata porosity is low and accumulation has terminated. There may be hydrocarbon (liquid oil) reservoirs of industrial value in strata with a burial depth of 6,000 m, and even with a burial depth exceeding 8,000 m in some basins in the world. The threshold of deep hydrocarbon accumulation in hydrocarbon basins is in a wide range and changes according to certain rules. As the sand grain size increases or the sand grain sorting difficulty decreases, the hydrocarbon accumulation limit becomes deeper.

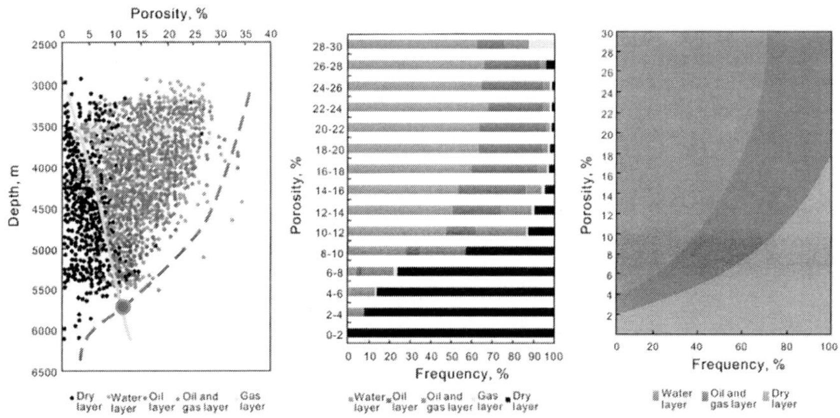

(a) Diagram showing the fluid porosity of the Tarim reservoirs

(b) Frequency histogram of changes of fluid in the Tazhong reservoirs as a function of porosity

(c) Change model of fluid in the Tazhong reservoirs as a function of porosity

Figurw. 59. Distribution of oil–gas, water bearing, and dry layers in the Central Tarim Basin region and accumulation thresholds

The third dynamic boundary is the lower limit of hydrocarbon generation of source rocks. It usually refers to the critical geological condition that hydrocarbon output of organic-rich source rocks is less than 1 % of the total. Methods used to determine the lower limit of hydrocarbon generation mainly include the amount of residual hydrocarbon, organic element variation, hydrocarbon generation potential, and efficiency of hydrocarbon expulsion. The residual hydrocarbon capacity of source rocks is expressed by the amount of residual hydrocarbon in organic carbon (S_1/TOC) or chloroform asphalt ("A"/TOC). The residual hydrocarbon S_1/TOC or "A"/TOC of the source rocks increases at first followed by a gradual decrease as the depth or R_o increases, generating a "belly-shaped" curve. When S_1/TOC or "A"/TOC reaches a minimal value which is so small that change is hardly visible, it means that hydrocarbon is no longer generated and expelled from the organic matter, and the R_o value at such minimal value is the lower limit of hydrocarbon generation

(Fig. 60a). The organic element variation method is described as follows: When H/C and O/C in the source bed reaches a minimal value which is so small that change is hardly visible, the source rock will no longer yield hydrogen-rich hydrocarbons. That is the lower limit of hydrocarbon generation. Theoretically, when hydrocarbon generation terminates, H/C will no longer change when reaching a minimal value. Huo et al. (2014b) determined the lower limit of hydrocarbon generation of carbonate rocks in the basin-platform region of the Tarim Basin, by supposing that the minimal value H/C = 0.1 and the H/C value at R_o = 0.5 % as the maximum value (Fig. 60b). In the diagram of hydrocarbon generation potential changes of source rocks, $(S_1 + S_2)$/TOC is the current hydrocarbon generation potential index of the source rocks. Due to hydrocarbon generation and expulsion from the source rocks, $(S_1 + S_2)$/TOC increases first followed by a gradual decrease as the depth or R_o increases, generating a "belly-shaped" curve. When the hydrocarbon generation potential reaches a certain minimal value, the hydrocarbon generation potential stops changing. This means that the source rock stops generating hydrocarbon and the lower limit of hydrocarbon generation of the source rock is reached (Fig. 60c). The amount of hydrocarbon expelled from source rock increases at first followed by a gradual decrease as the depth or R_o increases. When the amount of expelled hydrocarbon drops to zero or a minimal value, it means that hydrocarbon expulsion stops, and hydrocarbon generation may also have stopped. The R_o value when the amount of expelled hydrocarbon is zero or reaches a minimal value is the lower limit of hydrocarbon generation. When the amount of expelled hydrocarbon is zero, the hydrocarbon expulsion rate is also zero, but the hydrocarbon expulsion efficiency increases to the maximum of nearly 100 %. The lower limit of hydrocarbon generation can be determined according to the hydrocarbon expulsion rate and efficiency instead of the amount of expelled hydrocarbon. According to the lower limit of hydrocarbon generation concept, the R_o value when F_e = 99 % is the lower limit of hydrocarbon generation (Fig. 60d).

5.5 The formation and distribution of deep hydrocarbon reservoirs are controlled by three fluid dynamic fields

Research on fluid dynamic fields originated from fluid dynamics research in the domain of geodynamics. In 1953, Hubbert (1953) proposed the fluid potential concept in order to describe energy changes and migration rules of underground fluids. Ye et al. (1999) made a detailed description of the concept, pointing out that a fluid dynamic field is an integration of temperature, pressure, fluid potential, and structural stress fields in a sedimentary basin and their relations. Based on the above analysis that there is a buoyancy-controlled threshold and a hydrocarbon accumulation limit in the deep part of each hydrocarbon basin, Pang et al. (2014a) suggested that a hydrocarbon basin may be divided into three dynamic fields according to buoyancy-controlled threshold and hydrocarbon accumulation limit. Strata above the buoyancy-controlled threshold are called the free fluid dynamic field; strata between the buoyancy-controlled threshold and hydrocarbon accumulation limit are in the limited fluid dynamic field; and strata below the hydrocarbon accumulation limit are in the bound fluid dynamic field (Fig. 61). The fluid dynamic field is referred to the strata field where the hydrocarbon has identical or similar media, migration–accumulation force and reservoir-forming rule.

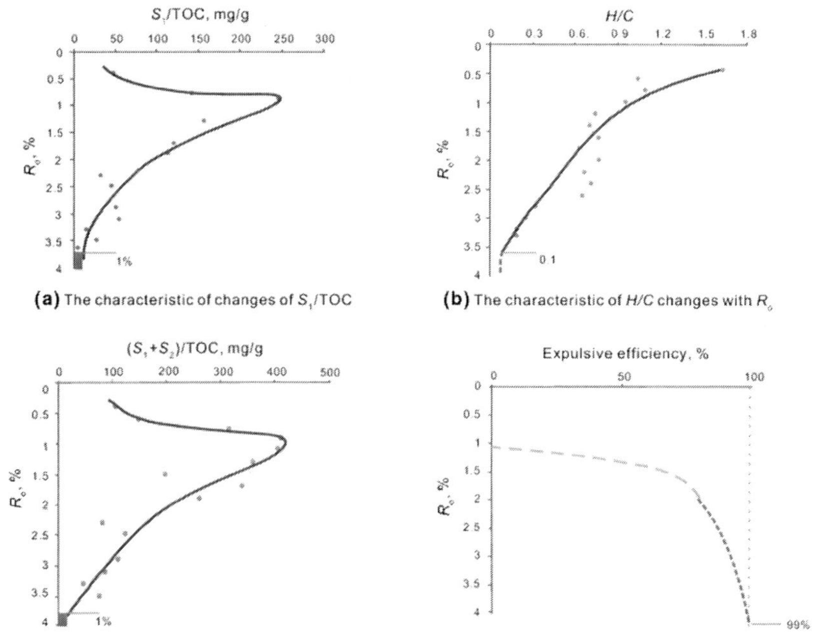

(a) The characteristic of changes of S_1/TOC **(b)** The characteristic of H/C changes with R_o

(c) The characteristic of the changes of (S_1+S_2)/TOC **(d)** The characteristic of changes of expulsive efficiency

Figure. 60. Changes of hydrocarbon generation and expulsion in the source rocks as a function of buried depth and the lower limit of hydrocarbon accumulation (*blue line*) (Huo et al. 2014b)

The research conducted by Pang et al. (2014a) shows that deep hydrocarbon reservoirs are controlled by free and limited fluid dynamic fields. The free fluid dynamic field is where conventional hydrocarbon reservoirs develop, and where buoyancy has a dominant effect on the migration and accumulation of hydrocarbons, with the hydrocarbon distribution features of high-point sealing and accumulation, high-porosity enrichment, and high-pressure stability. The limited fluid dynamic field is where tight hydrocarbon reservoirs develop, and where buoyancy has less effect on the migration and accumulation of hydrocarbons. The sandstone porosity in this field is generally 2.4 %–12 %, permeability is 0.01–1 mD, and pore throat radius is 0.01–2 μm. In this formation area, three major tight hydrocarbon reservoirs are formed. (1) the conventional tight hydrocarbon reservoirs are developed from pre-existing conventional reservoirs under compaction, i.e., first forming reservoirs and then being compacted, such reservoirs are characterized by the features of high-point accumulation, high-stand sealing, and high-pressure stability; (2) the tight deep basin gas reservoirs underwent the process of first compaction and then reservoir forming. They show the features of low-depression accumulation, low-stand inversion, low-porosity enrichment, and low-pressure stability. They are mainly formed by the expansion of the petroleum area caused by the volume expansion of hydrocarbon when migrating into the tight reservoirs; (3) the tight superimposed hydrocarbon reservoirs are formed by the combination of the two hydrocarbon reservoir types mentioned above, which are formed by compaction and molecule volume expansion. They underwent three stages, i.e., reservoir forming—compaction—reservoir forming. They generally show the features as follows: the coexistence of oil and gas of high and low points; the coexistence of oil and gas of high and low porosity; the coexistence of oil–gas-bearing zone of high and low pressure; the coexistence of oil–gas reservoirs of high and low production. A superimposed continuous hydrocarbon reservoir is formed by the superimposition of the tight hydrocarbon reservoirs mentioned above (Pang et al. 2014a). The bound fluid dynamic field is at the bottom of the basin, and the oil and gas in it were accumulated at an early stage and retained from that earlier time. The risk of exploration and development is huge in this field, as the target formation is characterized by deep-buried depth, low porosity and permeability, and lack of formation energy (Fig. 62). Later tectonism may damage and reconstruct the hydrocarbon reservoirs and develop fracture type, cave-type, or fracture-cave complex-type hydrocarbon reservoirs. They are tight reservoirs, but the porosity and permeability may be good in some parts of the area, which then show the geological features of conventional hydrocarbon reservoirs.

The distribution of fluid dynamic fields varies in different basins. It is affected by the following three factors in the deep strata of hydrocarbon basins: (1) The changing rate of porosity and permeability with the burial depth. The depth of a fluid dynamic field decreases as the changing rate increases, and vice versa; (2) The uplifting of deep strata as a result of erosion of overlying strata leads to the uplifting of tight strata as a whole and the limited and bound fluid dynamic fields in the basins uplift to a shallow formation or even to the surface; (3) Faults caused by structural changes may damage the fluid dynamic field boundary in hydrocarbon basins. Faults may change a limited fluid dynamic field into a free fluid dynamic field near the faults and damage some reservoirs. Overall faulting may change the deep part into a free fluid dynamic field to form conventional hydrocarbon reservoirs under the effect of buoyancy.

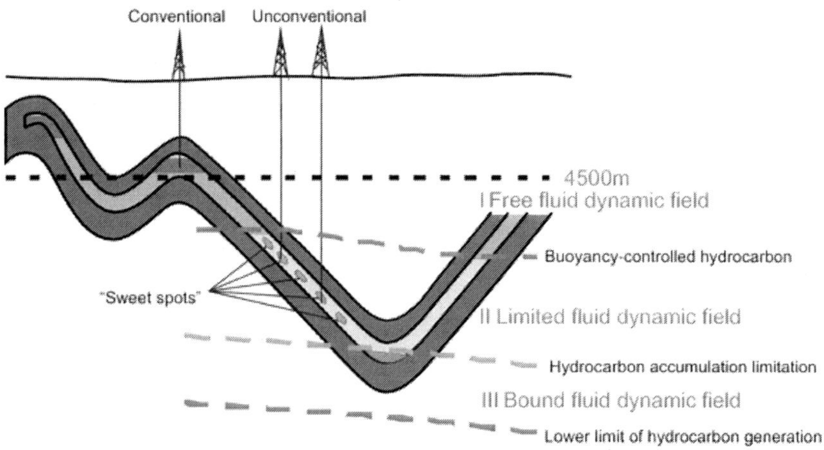

Figure. 61. Diagram showing the division of dynamic field boundary and fluid dynamic field of hydrocarbon migration and accumulation in deep petroliferous basins

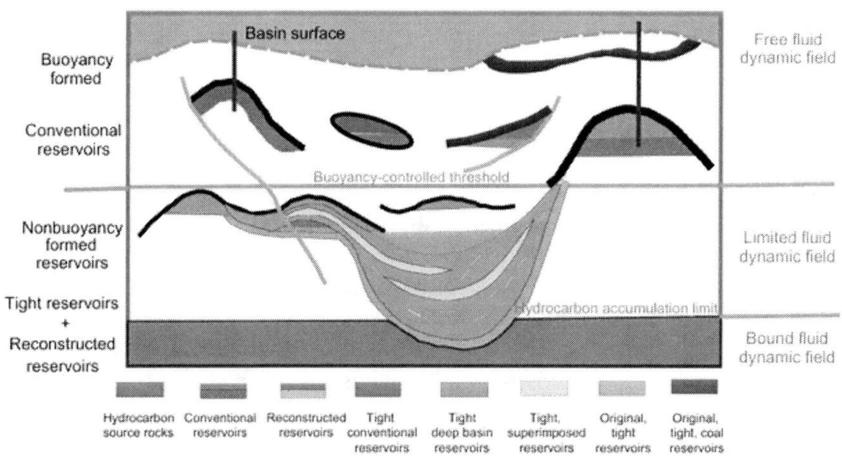

Figure. 62. Distribution of deep fluid dynamic field-controlled hydrocarbons in hydrocarbon basins (Pang et al. 2014a)

5.6 Tight hydrocarbon resources in limited deep fluid dynamic fields and exploration prospects

Tight hydrocarbon reservoirs in limited deep fluid dynamic fields can be divided into three types by the development mode: conventional reservoirs, deep basin reservoirs, and composite reservoirs, and each has its own unique formation process. Conventional tight reservoirs were deep-buried conventional reservoirs formed under the action of buoyancy after long years of compaction and diagenesis, characterized by an "accumulation followed by tightening" process; deep basin tight reservoirs were formed by hydrocarbons expelled from source rocks and accumulated in adjacent tight rocks without being controlled by buoyancy forces. These reservoirs are characterized by a "tightening followed by accumulation" process; composite tight reservoirs are a combination of conventional tight reservoirs and deep basin tight reservoirs, characterized by an "accumulation-tightening-accumulation" process. Figure 63shows the migration–accumulation force, source-reservoir matching, major controlling factors, and reservoir forming modes in the three tight hydrocarbon reservoirs.

Deep reservoirs in hydrocarbon basins have better forming and preserving conditions than shallow ones in the following aspects: (1) Reservoir tightness. Buoyancy has less effect on the migration and accumulation of hydrocarbons. This means that all hydrocarbons from source rocks during this period are not easily dispersed, and the preservation conditions are much better than those in the free fluid dynamic field. (2) Thermal maturity of source rocks. The geotemperature is high, between 100 and 200 °C, and R_o is 1.2 %–2.5 %. The hydrocarbon yield of per unit weight of parent matter is 1.0–2.2 t/tc, and hydrocarbon expulsion efficiency is 25 %–99 %, equivalent to 2 and 5 times those in the free fluid dynamic field, respectively. (3) Hydrocarbon migration and accumulation efficiency. Hydrocarbons from source rocks accumulate in adjacent rocks to form reservoir resources. The migration and accumulation efficiency is 3–10 times that in the free fluid dynamic field. (4) The hydrocarbon targets in limited deep fluid dynamic fields have experienced the evolution period of free fluid dynamic fields, and the conventional reservoirs formed in early periods accumulated in limited fluid dynamic fields, where they were compacted and changed into conventional tight reservoirs as part of the hydrocarbon resources of limited fluid dynamic fields. The oil and gas resource evaluating results of limited fluid dynamic fields in several China major basins evaluated by Pang et al. (2014a) indicate the oil and gas resource accumulated in the field accounts for more than 84 % of the total basin resource.

Classification		Mechanism of accumulation		Mode of accumulation			Example
Categories	Sub-categories	Power for migration	Matching of source rock and reservoir rock	Conceptual model	Factor combination	Regularities of distribution	At home and abroad
Tight oil and gas reservoirs in function of non-buoyancy	Conventional tight oil and gas oil reservoir	Static formation pressure	Accumulating first, and then becoming tight reservoir	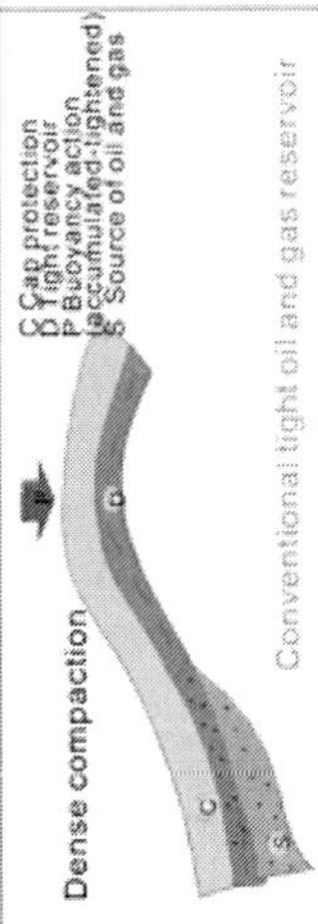	T-CDPS	Stable tectonic zone in restricted dynamic field of liquid	Dina 2 and tight conventional oil and gas reservoirs currently discovered below 4500m
	Deep-basin tight oil and gas reservoir	Expansion stress of hydrocarbon; the differences of capillary force and concentration of hydrocarbon between the source rock and reservoir rock	Becoming tight reservoir first, and then accumulating	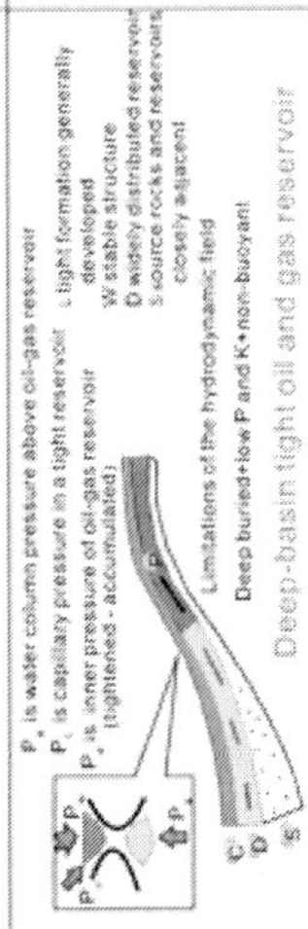	T-LWDS		Yinan 2 and deep-basin reservoirs in the Alberta Basin
	Compound tight oil and gas reservoir	Static formation pressure; expansion stress of hydrocarbon; the difference between inner and outer	Accumulating first, becoming tight reservoir later, accumulating again	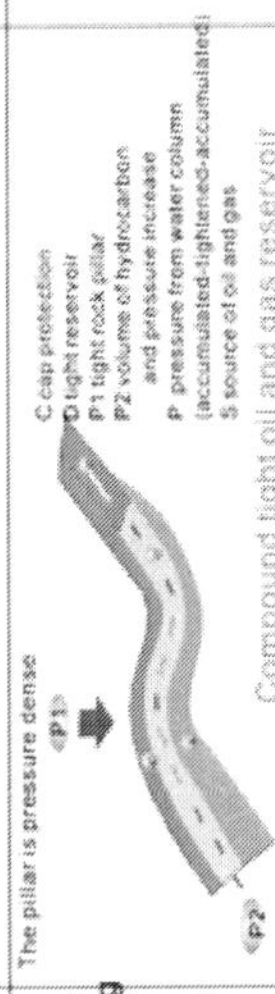	T-CDPS + T-LWDS		Dibei Oil field structural high gas reservoir in Paleozoic tight gas reservoir of the Erdos Basin

Figure. 63. Three unconventional non-buoyancy tight hydrocarbon reservoirs in limited fluid dynamic field and basic modes

The hydrocarbon resource potential is different in different types of tight reservoirs in deep petroliferous basins. Deep hydrocarbon reservoir distribution patterns are affected and controlled by single factor (Fig. 64). (1) under the control of terrestrial heat flow or geothermal gradient in the basin, "cold basins" characterized by low geothermal gradient and terrestrial heat flow values are favorable for deep oil and gas exploration. They can develop both free fluid dynamic fields and limited or bound fluid dynamic fields below a depth of 4,500 m with an extensive formation area for favorable exploration. Their resource potential is huge and a variety of hydrocarbon reservoir types are found in the basins with promising exploration prospects. "Hot basins" characterized by a high geothermal gradient and terrestrial heat flow values are unfavorable for deep oil and gas exploration, only developing a limited fluid dynamic field below the depth of 4,500 m with limited formation area favorable for exploration. Their resource potential and reservoir types are limited, with huge exploration risk. "Warm basins" whose geothermal gradient and terrestrial heat flow values are between "cold basins" and "hot basins" have an extensive formation area favorable for exploration and the resource potential and exploration prospects are between the two extremes. (2) When hydrocarbon source rocks are widely developed and hydrocarbon generation and expulsion amounts are large, the exploration potential is favorable. When a source-reservoir-cap rock combination is completed and the cap rock condition is good, the exploration potential is also favorable. (3) When intense tectonic movements occurred with fractures dominating, deep oil and gas reservoirs are susceptible to damage leading to low exploration potential. When a deep fluid dynamic field is uplifted to middle–shallow formation due to the denudation of overlying strata, the oil and gas exploration potential is weakened below a depth of 4,500 m. For example, the limited fluid dynamic field is currently above a depth of 4,500 m in the Chu-Saleisu Basin, Kazakhstan, resulting from the overall uplifting caused by the denudation of overlying strata. So the bound fluid dynamic field, which is unfavorable for oil and gas exploration, is currently developed below a depth of 4,500 m.

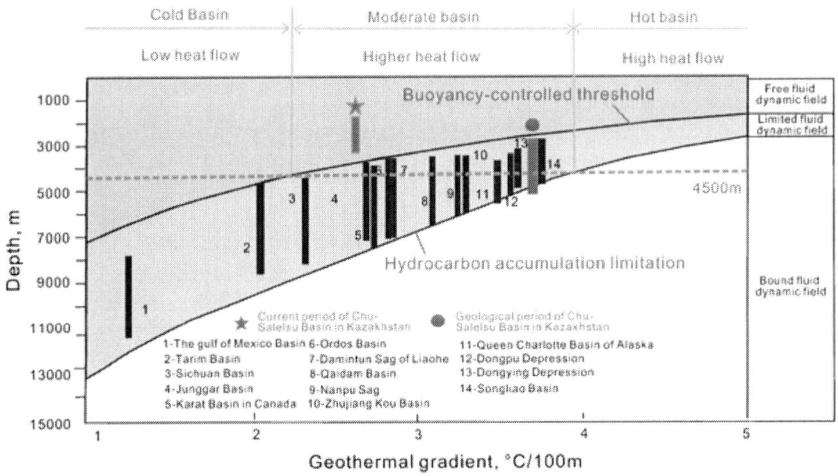

Figure. 64. Relations between the distribution of limited fluid dynamic fields and geothermal gradients in a deep petroliferous basin

6. GEOLOGICAL RESEARCH DIRECTIONS FOR HYDROCARBONS IN DEEP PETROLIFEROUS BASINS

6.1 Identification of deep hydrocarbon sources and relative contribution evaluation

The sources of deep hydrocarbons are complex. Some have single source and some have mixed sources. The hydrocarbon can originate from degradation of organic matter, from thermal cracking of asphalt sand or dispersed organic matter, or from the catalytic action of deep hydrothermal activity so it is very important to understand the origin of deep hydrocarbon and evaluate relative contributions. This will play a leading role in determining the hydrocarbon resource potential and favorable exploration directions. Scholars have done a lot of research into the identification of hydrocarbon sources and quantitative evaluation of hydrocarbon contributions using a variety of methods, laying a solid foundation for further work.

Methods for hydrocarbon source identification mainly include total hydrocarbon gas chromatographic fingerprints, *n*- and iso-alkane hydrocarbon ratios, steroids/terpenoids, aromatics, non-aromatics, as well as family composition and carbon isotope studies (Gormly et al.

1994; Telnæs and Cooper 1991; Stahl 1978; Seifert 1978; Hirner et al. 1981); index comparison using aromatics and thiophene compounds (Michael et al. 1989; Mukhopadhyay et al. 1995; Jing 2005) and comprehensive crude oil light hydrocarbon analysis for multi-period and multi-source reservoirs (Chen et al. 2006; Philippi 1981; Odden et al. 1998; Chen et al. 2003).

Methods for quantitative evaluation of the relative contribution of mixed hydrocarbon sources mainly include special compound absolute concentration quantification (Zhang et al. 2005a), biomarker parameters (Li et al. 2002), carbon isotope ratios (Song et al. 2004), and chart matching (Wang et al. 1999, 2004).

In deep strata, there are multiple sets and multiple varieties of hydrocarbon sources (including shale, carbonate, paleo-reservoir asphalt, and inorganic hydrocarbons) and multiple hydrocarbon-generating points, which experienced a number of periods of hydrocarbon generation and expulsion and multiple hydrocarbon migration pathways (faults, unconformities, carrier systems, and combinations). For source rocks that experienced many periods of evolution, the results of comparison using a single biomarker will be seriously affected by the degree of maturity. For multiple sets of source rocks, the results of analysis using the carbon isotope method will be affected by the sedimentary environment and climate conditions. Therefore, integrated analysis is required for the determination of deep hydrocarbon sources, including geological (altitude, log response, color, and mud content) and geochemical (biomarker, carbon isotope) analysis. The determination of deep hydrocarbon sources and evaluation of relative contributions are important to the evaluation of deep hydrocarbon resources and determination of hydrocarbon exploration fields.

6.2 Genetic mechanisms and preservation conditions of deep-buried high-quality reservoirs

Deep reservoirs include clastic, carbonate, volcanic, and metamorphic reservoirs. Research on the genesis and preservation of high-quality, high-porosity, and high-permeability reservoirs has drawn the attention of many geologists, but there is hardly any research done on the physical thresholds of deep-buried reservoirs. Although the limited fluid dynamic field mainly forms tight hydrocarbon and unconventional reservoirs, what concerns people is still "sweet point" formation in high-porosity and high-permeability reservoirs. The most important part of deep oil and gas exploration is to find these reservoirs under current conditions.

The genesis of deep-buried high-quality reservoirs mainly includes dissolution by organic acids (Surdam et al. 1984), effects of hydrocarbon charging and deep thermal fluids (Navon et al. 1988), clay mineral membranes (Ehrenberg 1993; Dolbier 2001), temperature and depth (Ezat 1997), faulting (Moretti et al. 2002), abnormal pressure (Wilkinson et al. 1997; Osborne and Swarbrick 1999), effects of fractures (Harris and Bustin 2002), sedimentary environment (Amthor and Okkerman 1998; Khidir and Catuneanu 2003; Pape et al. 2005; Rossi et al. 2001), and tectonics (Watkinson and Ward 2006). Preservation mechanisms mainly include early hydrocarbon charging (Gluyas et al. 1990; Robinson and Gluyas 1992; Rothwell et al. 1993), grain coating (Heald and Larese 1974; Ramm et al. 1997), and overpressure (Ramm et al. 1997; Osborne and Swarbrick 1999).

There is little research on physical thresholds of deep reservoirs. Deep clastic reservoirs are extensively distributed, ranging from 3,000 to 6,000 m, but most are distributed between 3,500 and 4,000 m (Wood and Hewett 1984; Surdam et al. 1984; Gaupp et al. 1993; Ehrenberg 1993; Gu 1996; Gu et al. 1998, 2001; Aase et al. 1996; Wilkinson et al. 1997). Deep carbonate reservoirs have developed secondary pores (such as leaching pores and dissolution pores), cracks, and fissures, which greatly improve the physical properties of the reservoirs even at considerable depths (Shi et al. 2005; Xie et al. 2009). Deep volcanic reservoirs are buried deep, with complicated geological conditions, which add to the particularity and uncertainty of the reservoirs (Zhang and Wu 1994; Liu et al. 2010a). The physical properties of deep metamorphic reservoirs are decided by the development of fractures and are affected by structural movements and faulting activities (Nelson 1985; Waples 1990; Walker and James 1992; Nelson 2000).

Due to the burial depth, long evolutional history, multiple formation mechanisms, and complicated distribution, geological and geophysical exploration for deep reservoirs is difficult, and limited by technical conditions. The accuracy of geophysical data is low, the number of deep wells and ultra-deep wells are limited, and the availability of original data cannot be guaranteed. Besides, there are intense diagenesis and structural volcanic activities and abnormal pressures. All of these add to the difficulty in understanding the genesis of deep-buried high-porosity and high-permeability reservoirs. Traditional petroleum and geological theories can neither explain why there are porosities of 20 % or more at such depths nor guide the exploration and development of deep reservoirs. Therefore, understanding the genesis and preservation mechanisms of deep-buried high-quality reservoirs and studying the physical thresholds of different reservoirs are of great significance to improving the success rate of deep effective reservoir exploration and reducing exploration risks.

6.3 Phase behavior and conversion mechanisms of deep hydrocarbon and its distribution prediction

Deep hydrocarbons are subject to complicated temperature and pressure conditions and are affected by multiple fluid compositions. Understanding the phase behavior of oil and gas accumulation in deep basins plays a leading role in revealing its genetic mechanism and distribution regularity.

The temperature and pressure environment of deep hydrocarbon reservoirs mainly include high temperature + high pressure, high temperature + low pressure, low temperature + high pressure, and low temperature + low pressure (Miao et al. 2000; Gu et al. 2001; Jiao et al. 2002; Ma et al. 2005; Meng et al. 2006; Zhang et al. 2008). Deep hydrocarbon reservoirs are primarily composed of gaseous hydrocarbons, oil gas mixtures, oil, water vapor, and water (Zhang 2006). Due to the unique temperature and pressure environment, deep hydrocarbons exist in three forms: free, dissolved, and adsorbed, and three phases: oil, gas, and mixed (Tuo 2002; Shi et al. 2005; Wu and Xian 2006; Huang et al. 2007).

The migration of deep hydrocarbons is affected by a variety of forces, such as buoyancy, (oil) gas molecular expansive force, capillary force, molecular adsorption, and binding forces, which jointly act on the accumulation and entrapment of hydrocarbons (Pang et al. 2007b). For deep conventional reservoirs, buoyancy is the main force on the migration of hydrocarbons; for deep unconventional reservoirs, gas molecular expansion is the main force (Gies 1984; Pang et al. 2003; Xiao et al. 2008; Ma et al. 2009; Jiang et al. 2010; Zhu et al. 2010).

Affected by the widely different temperature and pressure conditions and complicated fluid compositions, deep hydrocarbon reservoirs occur in different forms, which are difficult to predict. Besides, due to the unique porosity and permeability conditions, the fluid moving forces are so complicated that it is impossible to characterize the dynamic mechanism that controls the hydrocarbons. In addition, the hydrocarbon driving forces are related to the conditions in a complicated manner. Different hydrocarbon phases have different forces, and the forces are affected by the hydrocarbon phases. Therefore, the formation of deep hydrocarbon reservoirs is not only related to fluid composition and occurrence conditions, but also is jointly controlled by a range of forces including buoyancy. Prediction of the occurrence conditions of deep fluids and characterization of their dynamic mechanism are important to the scientific prediction of deep hydrocarbon resources and the determination of favorable exploration directions.

6.4 Economic feasibility evaluation for deep-buried tight hydrocarbon exploration and development

The exploration and development of deep hydrocarbons is difficult, and the investment return is low. It not only relies on the further understanding of related theories, but also is affected by technical (well drilling and completion techniques) and economic feasibilities.

Economic feasibility is a key to the development of deep hydrocarbon exploration and can decide the exploration direction. It is related to development costs and profits. Profits are affected by the international oil price. Oil exploration is a system, in which drilling costs account for about 50 %–70 % of the total production costs and drilling costs increase considerably as the drilling depth increases (Du and Yao 2001). According to research by some scholars, drilling costs increase exponentially, instead of linearly, as the depth increases (Guan et al. 2012). Profits are mainly affected by oil price and output. The higher the oil price is, the more output there will be, and the more profit it will bring. Output is controlled by the market as well as geological and engineering factors. The higher the international oil price is, the more output there will be. Rapid increase of output will in turn restrain the price rise.

According to the 2014 International Energy Agency, the United States will close 2 % of its shale wells if the international oil price drops below $80/bbl; it will close 18 % of its shale wells if the international oil price drops below $60/bbl. This suggests that production profits are the motive force of hydrocarbon exploration. Cost and breakeven analysis and economic feasibility research are important factors for the prediction of deep hydrocarbon exploration prospects.

ACKNOWLEDGMENTS

The paper is completed based on the National Basic Research Program of China (973 Program, 2011CB201100) “Complex hydrocarbon accumulation mechanism and enrichment regularities of deep superimposed basins in Western China” and National Natural Science Foundation of China (U1262205) under the guidance of related department heads and experts. Here, the authors express gratitude and appreciation for their contributions. In addition, we thank the postgraduates including Li-Ming Zhou, Hua Bai, Jing Bai, Lu-Ya Wu, Qian-Wen Li, and Rui Yu for their help in data gathering and sorting.

REFERENCES

1. Aase NE, Bjφrkum PA, Nadeau PH. The effect of grain coating micro quartz on preservation of reservoir porosity. AAPG Bull. 1996;80(10):1654–73.
2. Amthor JE, Okkerman J. Influence of early diagenesis on reservoir quality of Rotliegende sandstones, northern Netherlands. AAPG Bull. 1998;82(12):2246–65.
3. Bai GP, Cao BF. Characteristic and distribution patterns of deep petroleum accumulations in the world. Oil Gas Geol. 2014;01:7–19 (in Chinese).
4. Barker C, Takach NE. Prediction of natural gas composition in ultradeep sandstone reservoirs. AAPG Bull. 1992;76(12):1859–73.
5. Barker C. Calculated volume and pressure changes during the thermal cracking of oil to gas in reservoirs. AAPG Bull. 1990;74:1254–61.
6. Barker C. Primary migration: The importance of water organic mineral matter interaction in the source rock. Tulsa: AAPG Studies in Geology; 1980.
7. Baskin DK. Atomic ratio H/C of kerogen as an estimate of thermal maturity and organic matter conversion. AAPG Bull. 1997;81(9):1437–50.
8. Berkenpas PG. The Milk River shallow gas pool: role of the up dip water trap and connate water in gas production from the pool. SPE. 1991;229(22):371–80.
9. Bluokeny TB. Translated by Shi D. Development properties under deep abnormally high pressure. Nat Gas Geosci. 2001;12(4–5):61–4 (in Chinese).
10. Burruss RC. Stability and flux of methane in the deep crust a review. In: The future of energy gases. US Geological Survey Professional Paper; 1993. p. 21–29.
11. Cant DJ. Diagenetic traps in sandstones. AAPG Bull. 1986;70(2):155–60.
12. Chen DX, Pang XQ, Qiu NS, et al. Accumulation and filling mechanism of lenticular sand body reservoirs. Earth Sci. 2004;29(4):483–8 (in Chinese).
13. Chen JP, Liang DG, Wang XL, et al. Mixed oils derived from multiple source rocks in the Cainan oilfield, Junggar Basin, Northwest China. Part II: artificial mixing experiments on typical crude oils and quantitative oil-source correlation. Org Geochem. 2003;34(7):911–30.
14. Chen SJ, Zhang Y, Lu JG, et al. Limitation of biomarkers in the oil-source correlation. J Southwest Petrol Instit. 2006;28(5):11–4 (in Chinese).
15. Cheng KM, Wang ZY, Zhong NN, et al. Carbonate hydrocarbon generation theory and practice. Beijing: Petroleum Industry Press; 1996. p. 290–303 (in Chinese).
16. Cluff RM, Cluff SG. The origin of Jonah field, Northern Green River Basin, Wyoming. In: Robinson JW and Shanley KW (eds.) Jonah field: case study of a tight-gas fluvial reservoir. AAPG Stud Geol. 2004;52:215–41.
17. Dai JX, Qin SF, Tao SZ, et al. Developing trends of natural gas industry and the significant progress on natural gas geological theories in China. Nat Gas Geosci. 2005;16(2):127–42 (in Chinese).
18. Dai JX. Characteristic of abiogenic gas resource and resource perspective. Nat Gas Geosci. 2006;17(1):1–6 (in Chinese).
19. Dai JX. Enhance the studies on natural gas geology and find more large gas field in China. Nat Gas Geosci. 2003;14(1):3–14 (in Chinese).
20. Darouich T, Behar F, Largeau C. Thermal cracking of the light aromatic fraction of Safaniya crude oil—experimental study and compositional modelling of molecular classes. Org Geochem. 2006;37:1130–54.
21. Davies GR, Smith LB. Structurally controlled hydrothermal dolomite reservoir facies: an overview. AAPG Bull. 2006;90(11):1641–90.
22. Dickey PA. Possible primary migration of oil from source rock in oil phase. AAPG Bull. 1975;59(2):337–45.
23. Dolbier RA. Origin, distribution, and diagenesis of clay minerals in the Albian Pinda Formation, offshore Cabinda, Angola. Reno: University of Nevada; 2001.
24. Dominé F, Dessort D, Brevart O. Towards a new method of geochemical kinetic modelling: implications for the stability of crude oils. Org Geochem. 1998;28:576–612.

25. Du CG, Hao F, Zou HY, et al. Process and mechanism for oil and gas accumulation, adjustment and reconstruction in Puguang gas field, Northeast Sichuan Basin, China. Sci China. 2009;52(12):1721–31 (in Chinese).
26. Du JH, Wang ZM, Li QM, et al. Oil and gas exploration of cambrian-ordovician carbonate in Tarim Basin. Beijing: Petroleum Industry Press; 2010. p. 1–4 (in Chinese).
27. Du XD, Yao C. The necessary need of deep oil and gas exploration. Mar Oil and Gas Geol. 2001;6(1):1–5 (in Chinese).
28. Durand B. Understanding of HC migration in sedimentary basins (present state of knowledge). Org Geochem. 1988;13(1–3):445–59.
29. Dyman TS, Cook TA. Summary of deep oil and gas wells in the United States through 1998, Chapter B. In: Dyman TS, Kuuskraa BN, editors. Geologic Studies of Deep Natural Gas Resources. U.S. Geological Survey Digital Data Series DDS-67, Version 1.00; 2001. p. B1–B9.
30. Dyman TS, Crovelli RA, Bartberger CE, et al. Worldwide estimates of deep natural gas resources based on the US Geological Survey World Petroleum Assessment 2000. Nat Resour Res. 2002;11(6):207–18.
31. Dyman TS, Litinsky VA, Ulmishek GF. Geology and natural gas potential of deep sedimentary basins in the Former Soviet Union, Chapter C. In: Dyman TS, Kuuskraa VA, editors. U. S. Geological Survey studies of deep natural gas resources. U. S. Geological Survey Digital Data Series DDS-67, Version 1.00; 2001. p. C1–C29.
32. Dyman TS, Spencer CW, Baird JK, et al. Geologic and production characteristics of deep natural gas resources based on data from significant fields and reservoirs, Chapter C. In: Dyman TS, Rice DD, Westcott WA, editors. Geologic controls of deep natural gas resources in the United States. U.S. Geological Survey Bulletin 2146-C; 1997. p. 19–38.
33. Ehrenberg SN. Preservation of anomalously high porosity in deeply buried sandstones by grain-coating chlorite: examples from the Norwegian Continental Shelf. AAPG Bull. 1993;77(7):1260–86.
34. Ezat H. The role of burial diagenesis in hydrocarbon destruction and H2S accumulation, Upper Jurassic Smackover Formation, Black Creek Field, Mississippi. AAPG Bull. 1997;81(1):26–45.
35. Fan JS. Characteristics of carbonate reservoirs for oil and gas field in the world and essential controlling factors for their formation. Earth Sci Front. 2005;13(3):23–30 (in Chinese).
36. Fan SF, Zhou ZY, Jie QD. Characteristics and simulation experimental study of generation and conservation of deep carbonate oils and gases. Acta Sedimentol Sin. 1997;2:114–7 (in Chinese).
37. Feng ZQ. Exploration of large Qingshen gas field in the Songliao Basin. Nat Gas Ind. 2006;26(6):1–5 (in Chinese).
38. Fisher JB, Boles JR. Water rock interaction in tertiary sandstones, San Joaquin Basin, California, USA: diagenetic controls on water composition. Chem Geol. 1990;82:83–101.
39. Gaupp R, Matter A, Platt J, et al. Diagenesis and fluid evolution of deeply buried Permian (Rotliegende) gas reservoir, Northwest Germany. AAPG Bull. 1993;77(7):1111–28.
40. Gies RM. Case history for a major Alberta deep basin gas trap: the Cadomin Formation. AAPG Mem. 1984;38:115–40.
41. Gluyas JG, Leonard AJ, Oxtoby NH. Diagenesis and oil emplacement: the race for space Ultra Trend, North Sea. In: 13th International sedimentological congress. International Association of Sedimentologists; 1990. p. 193.
42. Gong S, Peng PA, Lu YH, et al. The model experiments for secondary pyrolysis of biodegraded bituminous sandstone. Chin Sci Bull. 2004;49:39–47 (in Chinese).
43. Gormly JR, Buck SP, Chung HM. Oil-source rock correlation in the North Viking Graben. Org Geochem. 1994;22(3–5):403–13.
44. Gu JY. Sedimentary environment and reservoir characters of the Carboniferous Donghe sandstone in the Tarim Basin. Acta Geol Sinica. 1996;70(2):153–61 (in Chinese).
45. Gu JY, Ning CQ, Jia JH. High-quality reservoir features and generate analysis of clastic rocks in the Tarim Basin. Geol Rev. 1998;44(1):83–9 (in Chinese).

46. Gu JY, Zhou XX, Liu WL. Buried-hill Karst and the distribution of hydrocarbon in Lunnan Area, Tarim Basin. Beijing: Petroleum Industry Press; 2001 (in Chinese).
47. Guan DS, Niu JY. Unconventional oil and gas geology in China. Beijing: Petroleum Industry Press; 1995. p. 60–5 (in Chinese).
48. Guan D, Luo Y, Zhang XD, et al. Research on prediction methods and analysis of cost influential factors for offshore drilling. Drill Prod Technol. 2012;35(4):41–9 (in Chinese).
49. Guo R. Supplement to determining method of cut-off value of net play. Petrol Explor Dev. 2004;31(5):140–4 (in Chinese).
50. Hao F, Zou HY, Huang BJ. Gas generation model and geo-fluid response, Yinggehai Basin. Science China Ser D. 2002;32(11):889–96 (in Chinese).
51. Hao SS, Gao G, Wang FY. Highly-over mature marine source rocks. Beijing: Petroleum Industry Press; 1996. p. 98–109 (in Chinese).
52. Hao SS, Liu GD, Huang ZL, et al. Dynamic equilibrium model of migration and accumulation for natural gas resource evaluation. Petrol Explor Dev. 1993;20(3):16–21 (in Chinese).
53. Harris RG, Bustin RM. Diagenesis, reservoir quality, and production trends of Doig Formation sand bodies in the Peace River area of Western Canada. Bull Can Pet Geol. 2002;48(4):339–59.
54. He DF, Jia CZ, Zhou XY, et al. Control principles of structures and tectonics over hydrocarbon accumulation in multi-stage super imposed basins. Acta Petrol Sin. 2005;26(3):1–9 (in Chinese).
55. He ZL, Wei XC, Qian YX, et al. Forming mechanism and distribution prediction of quality marine carbonate reservoirs. Oil Gas Geol. 2011;32(4):489–98 (in Chinese).
56. Heald MT, Larese RE. Influence of coating on quartz cementation. J Sediment Petrol. 1974;44(4):1269–74.
57. Hirner A, Graf W, Hahn-Weinheimer P. A contribution to geochemical correlation between crude oils and potential source rocks in the eastern Molasse Basin (Southern Germany). J Geochem Explor. 1981;15(1–3):663–70.
58. Hu GY, Xiao ZY, Luo X, et al. Light hydrocarbon composition difference between two kinds of cracked gases and its application. Nat Gas Ind. 2005;25(9):23–5 (in Chinese).
59. Huang JW, Gu Y, Chen QL, et al. Thermal simulation of Silurian bituminous sandstone of hydrocarbon supply in northern Tarim Basin. Petrol Geol Exp. 2012;34(4):445–50 (in Chinese).
60. Huang J, Zhu RK, Hou DJ, et al. The new advances of secondary porosity genesis mechanism in deep clastic reservoir. Geol Sci Technol Inf. 2007;26(6):76–82 (in Chinese).
61. Hubbert MK. Entrapment of petroleum under hydrodynamic conditions. AAPG Bull. 1953;37(8):1954–2026.
62. Hunt JM. Petroleum geochemistry and geology. 1st ed. New York: Freeman; 1979. p. 261–73.
63. Huo ZP, Pang XQ, Fan K, et al. Application of facies-potential coupling reservoir-controlling effect in typical lithologic reservoirs in Jiyang depression. Petrol Geol Exp. 2014a;36(5):574–83 (in Chinese).
64. Huo ZP, Pang XQ, Ouyang XC, et al. Upper limit of maturity for hydrocarbon generation in carbonate source rocks in the Tarim Basin, China. Arab J Geosci. 2014b;.
65. Ingersoll RV. Tectonics of sedimentary basins with revise nomenclature. In: Busby CJ, Perez AA, editors. Tectonics of sedimentary basins: recent advances. Cambridge, MA: Blackwell Science; 1995. p. 1–151.
66. Jemison RM. Geology and development of Mills Ranch complex—world's deepest field. AAPG Bulletin. 1979;63(5):804–9.
67. Jia CZ, Zhou XY, Wang ZM, et al. Petroleum geological features of Kela-2 gas field. Chin Sci Bull. 2002;47(S1):90–6 (in Chinese).
68. Jiang FJ, Pang XQ, Wu L. Geologic thresholds and its gas controlling function during forming process of tight sandstone gas reservoir. Acta Petrol Sin. 2010;31(1):49–54 (in Chinese).
69. Jiao HS, Fang CL, Niu JY, et al. Deep petroleum geology of Eastern China. Beijing: Petroleum Industry Press; 2002. p. 24–162 (in Chinese).

70. Jin ZJ, Zhang LP, Yang L, et al. Primary study of geochemical features of deep fluids and their effectiveness on oil-gas reservoir formation in sedimentary basins. Earth Sci. 2002;27(6):659–65 (in Chinese).
71. Jing GL. Aromatic compounds in crude oils and source rocks and their application to oil-source rock correlations in the Tarim Basin, NW China. J Asian Earth Sci. 2005;25(2):251–68.
72. Jones RW, Edison TA. Microscopic observations of kerogen related to geochemical parameters with emphasis on thermal maturation. In: Oltz DF, editor. Low temperature metamorphism of kerogen and clay minerals. Los Angeles: Pacific Section, SEPM; 1978. p. 1–12.
73. Kang YZ. Geological conditions of the formation of the large Tahe oilfield in the Tarim Basin and its prospects. Geol China. 2003;30(3):315–9 (in Chinese).
74. Khidir A, Catuneanu O. Sedimentology and diagenesis of the Scollard sandstones in the Red Deer Valley area, central Alberta. Bull Can Pet Geol. 2003;51(1):45–69.
75. Kutcherov V, Lundin A, Ross RG, et al. Theory of abyssal abiotic petroleum origin: challenge for petroleum industry. AAPG Eur Reg Newslett. 2008;3:2–4.
76. Land LS, MacPherson GL. Origin of saline formation waters, Cenozoic section, Gulf of Mexico sedimentary basin. AAPG Bull. 1992;76(9):1344–62.
77. Li GY, Lü MG. Atlas of petroliferous basins in China. Beijing: Petroleum Industry Press; 2002 (in Chinese).
78. Li GY. Atlas of world petroliferous basins. Beijing: Petroleum Industry Press; 2009 (in Chinese).
79. Li HX, Ren JH, Ma JY, et al. Abnormal high pore fluid pressure and clastic rock deep reservoir. Petrol Explor Dev. 2001;28(6):5–8 (in Chinese).
80. Li L, Ren ZW, Sun HB. An integrated evaluation on petroleum geology of the deep reservoirs in the west depression, Liaohe Basin, China. Acta Petrol Sin. 1999;20(6):9–15 (in Chinese).
81. Li MC, Shan XQ, Ma CH, et al. Dynamics of sand lens reservoir. Oil Gas Geol. 2007;28(2):209–15 (in Chinese).
82. Li SM, Pang XQ, Jin ZJ, et al. Geochemical characteristics of the mixed oil in Jinhu Sag of Subei Basin. J China Univ Petrol. 2002;26(1):11–5 (in Chinese).
83. Li Z, Li HS. An approach to genesis and evolution of secondary porosity in deeply buried sandstone reservoirs, Dongpu depression. Sci Geol Sin. 1994;29(3):267–75 (in Chinese).
84. Li Z, Chen JS, Guan P. Scientific problems and frontiers of sedimentary diagenesis research in oil-gas-bearing basins. Acta Petrol Sin. 2006;22(8):2113–22 (in Chinese).
85. Li Z, Huang SJ, Liu JQ, et al. Buried diagenesis, structurally controlled thermal-fluid process and their effect on Ordovician carbonate reservoirs in Tahe, Tarim Basin. Acta Sedimentol Sin. 2010;28(5):969–74 (in Chinese).
86. Lin CS, Li H, Liu JY. Major unconformities, tectonostratigraphic framework, and evolution of the superimposed Tarim Basin, northwest China. J Earth Sci. 2012;23(4):395–407.
87. Liu JQ, Meng FC, Cui Y, et al. Discussion on the formation mechanism of volcanic oil and gas reservoirs. Acta Petrol Sin. 2010a;26(1):1–13 (in Chinese).
88. Liu QY, Jin ZJ, Gao B, et al. Characterization of gas pyrolysates from different types of Permian source rocks in Sichuan Basin. Nat Gas Geosci. 2010b;21(5):700–4 (in Chinese).
89. Liu WH, Chen MJ, Guan P, et al. Ternary geochemical-tracing system in natural gas accumulation. Sci China Ser D. 2007a;37(7):908–15 (in Chinese).
90. Liu WH, Zhang JY, Fan M, et al. Gas generation character of dissipated soluble organic matter. Petrol Geol Exp. 2007b;29(1):1–6 (in Chinese).
91. Liu Z, Zhu WQ, Sun Q, et al. Characteristics of geotemperature-geopressure systems in petroliferous basins of China. Acta Petrol Sin. 2012;33(1):1–17 (in Chinese).
92. Luo P, Qiu YN, Jia AL, et al. The present challenges of Chinese petroleum reservoir geology and research direction. Acta Sedimentol Sin. 2003;21(1):142–7 (in Chinese).
93. Luo XR, Yu J, Zhang FQ, et al. Numerical migration model designing and its application in migration studies in the 8th segment in the Longdong area of Ordos Basin, China. Sci China Ser D. 2007;37(S1):73–82 (in Chinese).

94. Luo XR. Review of hydrocarbon migration and accumulation dynamics. Nat Gas Geosci. 2003;14(5):337–46 (in Chinese).
95. Ma WM, Wang XL, Ren LY, et al. Over-pressures and secondary pores in Dongpu Depression. J Northwest Univ. 2005;35(3):325–30 (in Chinese).
96. Ma XZ, Pang XQ, Meng QY, et al. Hydrocarbon expulsion characteristics and resource potential of deep source rocks in the Liaodong Bay. Oil Gas Geol. 2011;32(2):251–9 (in Chinese).
97. Ma YS, Chu ZH. Building-up process of carbonate platform and high-resolution sequence stratigraphy of reservoirs of reef and oolitic shoal facies in Puguang gas field. Oil Gas Geol. 2008;29(5):548–56 (in Chinese).
98. Ma YS, Guo TL, Zhao XF, et al. Formation mechanism of high porosity and permeability dolomites deeply buried in Puguang gas field. Sci China Ser D. 2007;37:43–52 (in Chinese).
99. Ma YS, Zhang SC, Guo TL, et al. Petroleum geology of the Puguang sour gas field in the Sichuan Basin, SW China. Mar Petrol Geol. 2008;25(4–5):357–70.
100. Ma ZZ, Wu HY, Dai GW. Progress of study on reservoir forming mechanism of deep basin gas. Petrol Geol Oilfield Dev Daqing. 2009;28(6):57–61 (in Chinese).
101. Magara K. Compaction and fluid migration: practical petroleum geology. London: Elsevier; 1978. p. 319.
102. Mango FD. The stability of hydrocarbon under the time/temperature conditions of petroleum genesis. Nature. 1991;352:146–8.
103. Manhadieph NR. Translated by Shi D. Geothermal conditions of deep hydrocarbon-bearing layers. Nat Gas Geosci. 2001;12(2):56–60 (in Chinese).
104. Masters JA. Deep basin gas trap, Western Canada. AAPG Bull. 1979;63(z2):152–81.
105. Melienvski BN. Translated by Shi D. Discussion on the deep zonation of the oil-gas formation. Nat Gas Geosci. 2001;12(4–5):52–5 (in Chinese).
106. Meng YL, Gao JJ, Liu DL, et al. Diagenetic facies analysis and anomalously high porosity zone prediction of the Yuanyanggou area in the Liaohe depression. J Jilin Univ. 2006;36(2):227–33 (in Chinese).
107. Miao JY, Zhu ZQ, Liu WR, et al. Relationship between temperature, pressure and secondary pores of deep reservoirs in eogene in the Jiyang depression. Acta Petrolei Sinica. 2000;21(3):36–40 (in Chinese).
108. Michael GE, Lin LH, Philp RP, et al. Biodegradation of tar-sand bitumens from the Ardmore/Anadarko Basins, Oklahoma—II correlation of oils, tar sands and source rocks. Org Geochem. 1989;14(6):619–33.
109. Ministry of Land and Resources. Resource evaluation of the new round national oil and gas resources. Beijing: China Land Press; 2005 (in Chinese).
110. Moretti I, Labaume P, Sheppard SMF, et al. Compartmentalization of fluid migration pathways in the sub-Andean zone, Bolivia. Tectonophysics. 2002;348(1–3):5–24.
111. Mukhopadhyay PK, Wade JA, Kruge MA. Organic facies and maturation of Jurassic/Cretaceous rocks, and possible oil-source rock correlation based on pyrolysis of asphaltenes, Scotian Basin, Canada. Org Geochem. 1995;22(1):85–104.
112. Navon O, Hutcheon ID, Rossman GR, et al. Mantle-derived fluids in diamond micro-inclusions. Nature. 1988;335:784–9.
113. Nelson RA. Production characteristics of the fractured reservoirs of the La Paz field, Maracaibo Basin, Venezuela. AAPG Bull. 2000;84(11):1791–809.
114. Nelson RA. Geologic analysis of naturally fractured reservoirs. Houston: Gulf Publishing Company; 1985. p. 304–51.
115. Odden W, Patience RL, Van Graas GW. Application of light hydrocarbons (C4-C13) to oil/source rock correlations: a study of the light hydrocarbon compositions of source rocks and test fluids from offshore Mid-Norway. Org Geochem. 1998;28(12):823–47.
116. Osborne MJ, Swarbrick RE. Diagenesis in North Sea HPHT clastic reservoirs—consequences for porosity and overpressure prediction. Mar Pet Geol. 1999;16:337–53.
117. Pan WQ, Liu YF, Dickson J, et al. The geological model of hydrothermal activity in outcrop and the characteristics of carbonate hydrothermal karst of Lower Paleozoic in Tarim Basin. Acta Sedimentol Sin. 2009;27(5):983–94 (in Chinese).

118. Pang XQ. Key challenges and research methods of petroleum exploration in the deep of superimposed basins in western China. Oil Gas Geol. 2010;31(5):517–41 (in Chinese).
119. Pang XQ, Chen DX, Zhang J, et al. Concept and categorize of subtle reservoir and problems in its application. Lithol Reserv. 2007a;19(1):1–8 (in Chinese).
120. Pang XQ, Chen DX, Zhang J, et al. Physical simulation experimental study on mechanism for hydrocarbon accumulation controlled by facies—potential—source coupling. Journal of Palaeogeography. 2013;5:575–92 (in Chinese).
121. Pang XQ, Gao JB, Meng QY. A discussion on the relationship between tectonization and hydrocarbon accumulation and dissipation in the platform-basin transitional area of Tarim Basin. Oil Gas Geol. 2006;27(5):594–603 (in Chinese).
122. Pang XQ, Gao JB, Lü XX, et al. Reservoir accumulation pattern of multi-factor recombination and procession superimposition and its application in Tarim Basin. Acta Petrol Sin. 2008;29(2):159–66 (in Chinese).
123. Pang XQ, Jiang ZX, Huang HD, et al. Genetic mechanism, development mode and distribution forecast of overlapping continuous oil and gas reservoir. Acta Petrol Sin. 2014a;35(5):1–34 (in Chinese).
124. Pang XQ, Jin ZJ, Zuo SJ. Dynamics models and classification of hydrocarbon accumulations. Earth Sci Front. 2000;7(4):507–14 (in Chinese).
125. Pang XQ, Jin ZJ, Jiang ZX, et al. Critical condition for gas accumulation in the deep basin trap and physical modeling. Nat Gas Geosci. 2003;14(2):207–14 (in Chinese).
126. Pang XQ, Li QW, Chen JF, et al. Recovery method of original TOC and its application in source rocks at high mature—over mature stage in deep petroliferous basin. J Palaeogeogr. 2014b;16(6):769–88 (in Chinese).
127. Pang XQ, Li SM, Jin ZJ, et al. Geochemical evidence of hydrocarbon expulsion threshold and its application. Earth Sci. 2004;29(4):384–9 (in Chinese).
128. Pang XQ, Luo XR, Jiang ZX, et al. Advancements and problems on hydrocarbon accumulation research of complicated superimposed basins in Western China. Adv Earth Sci. 2007b;22(9):879–87 (in Chinese).
129. Pape H, Clauser C, Iffland J, et al. Anhydrite cementation and compaction in geothermal reservoirs: interaction of pore-space structure with flow, transport, P-T conditions, and chemical reactions. Int J Rock Mech Min Sci. 2005;42(7–8):1056–69.
130. Perry W Jr. Structural settings of deep natural gas accumulations in the conterminous United States. In: Dyman TS, Rice DD and Westcott PA (eds.), geologic controls of deep natural gas resources in the United States. USGS Bull. 1981;2146-D:41–6.
131. Philippi GT. Correlation of crude oils with their oil source formation, using high resolution GLC C6-C7 component analyses. Geochim Cosmochim Acta. 1981;45(9):1495–513.
132. Qiao HS, Zhao CL, Tian KQ, et al. Petroleum geology of deep reservoirs in the eastern China. Beijing: Petroleum Industry Press; 2002. p. 254–8 (in Chinese).
133. Qin JZ, Jin JC, Liu BQ. Thermal evolution pattern of organic matter abundance in various marine source rocks. Oil Gas Geol. 2005;26(2):177–84 (in Chinese).
134. Ramm M, Forsberg AW, Jahren J S. Porosity-depth trends in deeply buried Upper Jurassic reservoirs in the Norwegian Central Graben: an example of porosity preservation beneath the normal economic basement by grain-coating microquartz. In: AAPG Memoir 69: reservoir quality prediction in sandstones and carbonates; 1997. p. 177–199.
135. Robinson A, Gluyas J. Duration of quartz cementation in sandstones, North Sea and Haltenbanken Basins. Mar Pet Geol. 1992;9:324–7.
136. Rodrenvskaya МИ. Translated by Shi D. Hydrocarbon potential in deep layers. Nat Gas Geosci. 2001;12(4–5):49–51 (in Chinese).
137. Rossi C, Marfil R, Ramseyer KR, et al. Facies related diagenesis and multiphase siderite cementation and dissolution in the reservoir sandstones of the Khatatba Formation, Egypt's Western Desert. J Sediment Res. 2001;71(3):459–72.
138. Rothwell NR, Sorensen A, Peak JL, et al. Gyda: recovery of difficult reserves by flexible development and conventional reservoir management. In: SPE26778, Proceedings of the offshore Europe conference; 1993. p. 271–280.

139. Samvelov PΓ. Translated by Guan F X Deep oil-gas reservoir forming characteristics and distribution. Northwest Oil Gas Explor. 1997;9(1):52–7 (in Chinese).
140. Seifert WK. Steranes and terpanes in kerogen pyrolysis for correlation of oils and source rocks. Geochim Cosmochim Acta. 1978;42(5):473–84.
141. Shao CX, Wang YZ, Cao YZ. Determine the effective reservoir lower limit—two new methods and application to Paleogene deep clastic reservoirs in Dongying Sag as an example. J Oil Gas Technol. 2008;30(2):414–6 (in Chinese).
142. Shi JA, Wang Q. A discussion on main controlling factors on the properties of clastic gas reservoirs. Acta Sedimentol Sin. 1995;13(2):128–39 (in Chinese).
143. Shi X, Dai JX, Zhao WZ. Analysis of deep oil and gas reservoir exploration prospect. China Petrol Explor. 2005;10(1):1–10 (in Chinese).
144. Simmons MR. The world's giant oilfields. Hubbert Center Newslett. 2002;1:1–62.
145. Song CC, Peng YM, Qiao YL, et al. Direction for future oil & gas exploration in SINOPEC's exploration area in the Junggar Basin. Oil Gas Geol. 2008;29(4):453–9 (in Chinese).
146. Song FQ, Zhang DJ, Wang PR, et al. A contribution evaluation method for biodegraded mixed oils. Petrol Explor Dev. 2004;31(2):67–70 (in Chinese).
147. Spencer CW. Review of characteristics of low-permeability gas reservoirs in western United States. AAPG Bull. 1989;73(5):613–29.
148. Stahl WJ. Source rock-crude oil correlation by isotopic type-curves. Geochim Cosmochim Acta. 1978;42(10):1573–7.
149. Sugisaki R. Deep-seated gas emission induced by the earth tide: a basic observation for geochemical earthquake prediction. Science. 1981;212(4500):1264–6.
150. Sun LD, Fang CL, Li F, et al. Petroleum exploration and development practices of sedimentary basins in China and research progress of sedimentology. Petrol Explor Dev. 2010;37(4):385–96 (in Chinese).
151. Sun LD, Zou CN, Zhu RK, et al. Formation, distribution and potential of deep hydrocarbon resources in China. Petrol Explor Dev. 2013;40(6):641–9 (in Chinese).
152. Surdam RC, Boese SW, Crossey LJ. The chemistry of secondary porosity. AAPG Mem. 1984;37:127–49.
153. Surdam RC, Crossey LJ, Hagen ES, et al. Organic-inorganic and sandstone diagenesis. AAPG Bull. 1989;73(1):1–23.
154. Svetlakova EA. Model of formation regularities of distribution of hydrocarbon pools in the North Caspian Basin. In: Krylov NA, Nekhrikova NA, editors. Petroleum potential of the North Caspian Basin and Adjacent Areas (Neftegazonosnost Prikaspiyskoy vpadinyi sopredelnykh rayonov). Moscow: Nauka; 1987. p. 151–4.
155. Telnæs N, Cooper BS. Oil-source rock correlation using biological markers, Norwegian continental shelf. Mar Pet Geol. 1991;8(3):302–10.
156. Tissot BP, Welte DH. Petroleum formation and occurrence: a new approach to oil and gas exploration. Berlin: Springer; 1978.
157. Tissot B, Durand B, Espitalé J, et al. Influence of the nature and diagenesis of organic matter in the formation of petroleum. AAPG Bull. 1974;58:499–506.
158. Tuo JC. Research status and advances in deep oil and gas exploration. Adv Earth Sci. 2002;17(4):565–71 (in Chinese).
159. Tuo JC. Secondary hydrocarbon generation in carbonate rocks. Nat Gas Geosci. 1994;3:9–13 (in Chinese).
160. Tuo JC, Huang XZ, Ma WY. The lagging phenomenon of the petroleum generation in carbonate rocks. Petrol Explor Dev. 1994;21(6):1–5 (in Chinese).
161. Tuo JC, Wang XB, Zhou SX. Prospect of oil-gas resources in the Bohai Bay Basin. Nat Gas Geosci. 1999a;06:27–31 (in Chinese).
162. Tuo JC, Wang XB, Zhou SX, et al. Exploration situation and research development of deep hydrocarbon. Nat Gas Geosci. 1999b;06:1–8 (in Chinese).
163. Walker RG, James NP. Facies models: response to sea level change. Geol Assoc Can. 1992;4(5):153–70.
164. Wan L, Sun Y, Wei GQ. A new method used to determine the lower limit of the petrophysical parameters for reservoir and its application: a case study on Zhongbu Gas Field in Ordos Basin. Acta Sedimentol Sin. 1999;17(3):454–7 (in Chinese).

165. Wang GQ, Chen YD, Zhou YH. Difficulty analysis and solution discussion for deep and ultra-deep exploration well. Oil Drill Prod Technol. 1998;20(1):1–17 (in Chinese).
166. Wang Q, Song YJ, Zhang SM. The study on the methods used to delineate the deep tight gas bearing sands in Daqing oil field. J Daqing Petrol Inst. 1994;18(2):22–6 (in Chinese).
167. Wang TG, Wang CJ, He FQ, et al. Determination of double filling ratio of mixed crude oils in the Ordovician oil reservoir, Tahe Oilfield. Petrol Geol Exp. 2004;26(1):74–9 (in Chinese).
168. Wang T. Deep Basin Gas in China. Beijing: Petroleum Industry Press; 2002 (in Chinese).
169. Wang WJ, Song N, Jiang NH, et al. The mixture laboratory experiment of immature oils and mature oils, making and using the mixture source theory plat and its application. Petrol Explor Devel. 1999;26(4):34–7 (in Chinese).
170. Wang XB, Tuo JC, Yan H, et al. The formation mechanism of deep oil and gas. The meeting abstract on Chinese sedimentology in 2001. Wuhan: China Univ Geosci; 2001. p. 163–4 (in Chinese).
171. Wang YM, Niu JY, Qiao HS, et al. An approach to the hydrocarbon resource potential of Bohai Bay Basin. Petrol Explor Dev. 2002;29(2):21–5 (in Chinese).
172. Wang YS, Pang XQ, Liu HM, et al. Low-potential reservoir-controlling characteristics and dynamics mechanism and its role in oil and gas exploration. J Earth Sci. 2013;38(1):165–72 (in Chinese).
173. Wang Y, Su J, Wang K, et al. Distribution and accumulation of global deep oil and gas. Nat Gas Geosci. 2012;23(3):526–34 (in Chinese).
174. Wang ZM, Wang QH, Wang Y. Poll formation condition and controlling factor for Hetian gas field in the Tarim Basin. Mar Oil Gas Geol. 2000;5(1):124–32 (in Chinese).
175. Waples DW. Time and temperature in petroleum formation: application and detachment free deformation. J Struct Geol. 1990;12(3):355–81.
176. Watkinson AJ, Ward EMG. Reactivation of pressure solution seams by a strike slip fault sequential, dilational jog formation and fluid flow. AAPG Bull. 2006;90(8):1187–200.
177. White IC. The geology of natural gas. Science. 1885;6(128):42–4.
178. Wilkinson M, Darby D, Haszeldine RS, et al. Secondary porosity generation during deep burial associated with overpressure leak-off: Fulmar Formation, United Kingdom Central Graben. AAPG Bull. 1997;81(5):803–13.
179. Williams LB, Hervig RL, Wieser ME, et al. The influence of organic matter on the boron isotope geochemistry of the gulf coast sedimentary basin, USA. Chem Geol. 2001;174:445–61.
180. Wood JR, Hewett TA. Reservoir diagenesis and convective fluid flow. AAPG Mem. 1984;37:99–110.
181. Wu FQ, Xian XF. Current state and countermeasure of deep reservoirs exploration. Sediment Geol Tethyan Geol. 2006;26(2):68–71 (in Chinese).
182. Wu YX, Wang PJ, Bian WH, et al. Reservoir properties of deep volcanic reservoirs in the Songliao Basin. Oil Gas Geol. 2012;33(4):236–47 (in Chinese).
183. Xiao ZH, Zhong NN, Huang ZL, et al. A study on hydrocarbon pooling conditions in tight sandstones through simulated experiments. Oil Gas Geol. 2008;29(6):721–5 (in Chinese).
184. Xie JL, Huang C, Wang XX. Distribution of proved reserves of carbonate oil and gas pools in China. Mar Orig Petrol Geol. 2009;14(2):24–30 (in Chinese).
185. Xie ZY, Jiang ZS, Zhang Y, et al. Novel method of whole rock thermal simulation and application to the evaluation of gas source rock. Acta Sedimentol Sin. 2002;20(3):510–4 (in Chinese).
186. Yang XH, Ye JR, Sun YC, et al. Diagenesis evolution and preservation of porosity in the member2, Shahejie Formation, Qinan fault step zone. China Offshore Oil Gas (Geol). 1998;12(4):242–8 (in Chinese).
187. Ye JR, Wang LJ, Shao R. Fluid dynamic fields in hydrocarbon accumulation dynamics. Oil Gas Geol. 1999;20(2):86–9 (in Chinese).
188. Zappaterra E. Source-rock distribution model of the Periadriatic region. AAPG Bull. 1994;78(3):333–54.
189. Zhang DY. Fluid characteristics and reservoir forming condition in deep oil and gas reservoirs. Petrol Geol Recovery Effic. 2006;13(5):45–9 (in Chinese).

190. Zhang HF, Fang CL, Gao XZ, et al. Petroleum geology. Beijing: Petroleum Industry Press; 1999 (in Chinese).
191. Zhang WC, Li H, Li HJ, et al. Genesis and distribution of secondary porosity in the deep horizon of Gaoliu area, Nanpu Sag. Petrol Explor Dev. 2008;35(3):308–12 (in Chinese).
192. Zhang XD, Tan XC, Chen JS. Reservoir and controlling factors of the secondary member of the Jialing River Formation in the hit and the south transition strip of Sichuan. Nat Gas Geosci. 2005a;16(3):338–42 (in Chinese).
193. Zhang YW, Jin ZJ, Liu GC, et al. Study on the formation of unconformities and the amount of eroded sedimentation in Tarim Basin. Earth Sci Front. 2000;7(4):449–57 (in Chinese).
194. Zhang ZS, Wu BH. Research on volcanic reservoir and exploration technology home and abroad. Nat Gas Explor Dev. 1994;1:1–26 (in Chinese).
195. Zhang ZY, Shao LY, Zhang SH, et al. Distribution of mixed crude in western arc structural belt of Taibei Depression in Turpan-Hami Basin. Acta Petrol Sin. 2005b;26(2):15–20 (in Chinese).
196. Zhao MJ, Zeng Q, Zhang BM. The petroleum geological condition and prospecting orientation in Tarim Basin. Xinjiang Petrol Geol. 2001;22(2):93–6 (in Chinese).
197. Zhao WZ, Wang ZY, Zhang SC, et al. Successive generation of natural gas from organic materials and its significance in future exploration. Petrol Explor Devel. 2005;32(2):1–7 (in Chinese).
198. Zhao WZ, Zou CN, Gu ZD, et al. Preliminary discussion on accumulation mechanism of sand lens reservoirs. Petrol Explor Dev. 2007;34(3):273–83 (in Chinese).
199. Zhong DK, Zhu XM. Pore evolution and genesis of secondary pores in Paleogene clastic reservoirs in Dongying Sag. Oil Gas Geol. 2003;24(3):281–5 (in Chinese).
200. Zhong DK, Zhu XM, Wang HJ. Characteristics and formation mechanism analysis of deep buried sandstone reservoirs in China. Sci China. 2008;38(S1):11–8 (in Chinese).
201. Zhou J, Pang XQ. A method for calculation the quantity of hydrocarbon generation and expulsion. Petrol Explor Dev. 2002;29(1):24–7 (in Chinese).
202. Zhou SX, Wang XB, Tuo JC, et al. Advances in geochemistry of deep hydrocarbon. Nat Gas Geosc. 1999;10(6):9–15 (in Chinese).
203. Zhu GY, Zhang SC. Hydrocarbon accumulation conditions and exploration potential of deep reservoirs in China. Acta Petrol Sin. 2009;30(6):793–802 (in Chinese).
204. Zhu GY, Zhang SC, Liang YB, et al. Dissolution and alteration of the deep carbonate reservoirs by TSR: an important type of deep-buried high-quality carbonate reservoirs in Sichuan Basin. Acta Petrol Sin. 2006;22(8):2182–94 (in Chinese).
205. Zhu ZQ, Zeng JH, Wang JJ. Microscopic experiment on oil flowing through porous media in low permeability sandstone. J Southwest Petrol Univ Sci Technol. 2010;32(1):16–20 (in Chinese).
206. Zou CN. Unconventional oil and gas geology. Beijing: Geological Publishing House; 2011. p. 50–2 (in Chinese).

CHAPTER 10

Late Pleistocene and Holocene Glacier Extent in the Georgian Caucasus

Levan G. Tielidze[1,2]

[1]*Department of Geomorphology, Vakhushti Bagrationi Institute of Geography, Ivane Javakhishvili Tbilisi State University, Tbilisi, Georgia*

[2]*Department of Earth Sciences, Georgian National Academy of Sciences, Tbilisi, Georgia*

ABSTRACT

This article presents the information of the Georgian Caucasus glaciation at the Late Pleistocene and Holocene period (~126,000-11,700 calendar years ago). Our primary aim was to numerically reproduce the ice extent deduced from geological and geomorphological mapping. We used the analog method with the 30 m resolution SRTM DEM (Shuttle Radar Topography Mission Digital Elevation Model). In addition, the rates of glaciation of those times are identified based on the stadial moraines and erratic boulders. The current investigation has revealed that in the Late Pleistocene, the central and western Caucasus characterized the highest glaciation, while the eastern Caucasus boasted the lowest glaciated area, and in the southern Georgia's highland glaciation has almost the same form as there is in the Eastern Caucasus today. The longest glaciers were located in the Enguri (Nenskra ~36 km, Mulkhura ~35.1 km, Dolra ~34.5 km), Kodori (Sakeni ~25 km, Klichi ~20 km, Marukhi ~17.3 km) Rioni (Buba ~23 km, Kirtisho ~20.5 km, Jejora ~17.5 km) and Tergi (Devdoraki ~38.5 km, Suatisi ~32 km) river basins. We found, that topography thresholds related to the elevation and hypsometry of individual catchments controlled the gradient of the rate of glacier expansion in the domain at that time.

Keywords: Late Pleistocene, Holocene, Würm, Glacier Reconstruction, Analog Method, Caucasus Mountains

1. INTRODUCTION

One of the problematic issues of the studies of modern geographical science is the study of the paleoglacial issues. The changes in the natural physical-geo- graphical conditions in the earth surface in the Upper Quaternary and Holocene are much depended on the glaciation dynamics. The glaciers in this period changed not only the structure of the geosphere, but they also established the new relationship between the geographical elements. It is known that in the Quaternary Age, the earth surface was subject to multiple glaciations and there are many views about the limits of their extension. Sometimes, these views complement each other, but sometimes they are mutually exclusive.

Basically, all research methods related to the reconstruction of the old glaciation are based on the analysis of the morpho-sculptural forms created by glaciers. The analysis of the glaciogenic forms does not always give the desirable result, particularly, when dealing with multiple glaciations. Moraines, as the forms of multiple glaciations, are not always maintained in the relief, or sometimes, a very strong glaciation removes the trace of the previous less strong glaciation from the earth surface. There are many such examples with the Caucasus and other mountainous massifs, where the most recent Würm glaciation removed the real traces of Mindel and Riss glaciations.

At the present stage, the glacial and fluvioglacial forms and deposits in the region of the Caucasus Mountain Range are quite common, indicating the quite large scales of glaciers extension. Particularly well preserved are the glacial forms in the basins of the major rivers of the Central and Western Caucasus and on the slopes of the Caucasus subranges [1] . Good distribution of the terminal moraines, existence of the associated trough forms and sometimes the allocation of the lateral moraines on the slopes of the river gorges, as well as the well preserved cirque forms allow identifying the borders of the glaciers extension quite accurately.

2 PREVIOUS STUDIES

After the famous scheme—the Paleo-Glacial Concept by A. Penk and E. Brukner (1901/1909) [2] almost all researchers (e.g. Reingard, [3] ; Vardanyants, [4] ; Tsereteli, [5] ; Kovalev, [6] ; Shcherbakova, [7]) tried to compare the Caucasus to the Alps, where the old glaciers were emerged in the plain from the foothill zone. Drawing such a picture for the Caucasus was acceptable for the researchers. They did not took into account that the Alps and the Caucasus are located at different latitudes, and what was common to the Alps at a spatial scale, could not be identical to the Caucasus. Surely, the glaciation both, in the Alps and Caucasus was of coincided character. Coincided was the fact as well, that the glaciation of that period was the strongest, but their expansion limits could not be identical. There were many opponents to this view, who considered that comparing the Alps with other mountainous regions was inappropriate. Solving this issue is problematic and urgent even today.

As far back as in 1956 L. Maruashvili, on assessing critically the study of old glaciers in the Caucasus in his monograph [8] , indicates the mistakes accompanied their works. L. Maruashvili criticizes the Alpine school followers—A. Reinhardt [3] and L. Vardaniantz [4] , who, based on the Alpine Scheme, compiled the paleoglacial map of the Caucasus (a paleographic scheme of the Caucasus in the glaciation epoch). These materials were the basis for all following researchers. This view was unanimously shared almost by everyone.

The Alpine Scheme was opposed by L. Maruashvili, who refused to adapt and disseminate it to the Caucasus and who stated that the glaciation in the Caucasus was weaker than it was stated by the representatives of this school. L. Maruashvili considered that in the Quaternary the glacier tongues flowed down to the height of ~1000 - 1500 m in the Western and Central Caucasus (Bzipi-Kodori basin) and to ~2000 - 2500 m in the Eastern Caucasus (Tergi, Asa). The snow line depression made ~600 - 800 m.

In the opinion of D. Tsereteli [5] , the glaciation of the Caucasus can be compared to that of the Alps and even to other mountainous regions. In his opinion, the course of the glaciation in the Caucasus is behind the Alps glaciation by a half cycle. He considered that the glaciers in the Caucasus descended quite below (if judging by Tsebelda moraine), and the snow line depression made ~1200 m. This opinion is close to all opinions about the Alps and to our mind, it is generally true. As for the descending of the glacier tongues in the basins of the rivers of Enguri and Rioni to ~800 - 1000 - 1100 m, this point of view is correct and is proved by the existence of the moraine complex there.

The Nenskra, Mestia, Dolra and Ushguli moraines are located at the different altitudes in the Enguri River basin. Due to this, D. Tsereteli [5] considered them to be of different ages (stages). To our mind, this opinion needs further specification.

R. Khazaradze [9] studied the questions of glaciation by using the petropraphic method. His works give the borders of the glacier extension in the basins of the rivers of Georgia with the locations of the erratic boulders.

The old Caucasus glaciation is considered in the works by G. Abich [10] , A. Reinhardt [11] [3] , I. Kuznetsov [12] , L. Vardaniantz [13] [4] , B. Klopotovskiy [14] , K. Paffenholz [15] , S. Nemanishvili [16] , G. Dondua [17] , N. Astakhov [18] , E. Milanovskiy [19] , N. Dumitrashko [20] , P. Kovalev [6] , D. Tabidze [21] , D. Tsereteli [5] , L. Serebruanny and A. Orlov [22] , R. Gobejishvili [23] and etc.

Unfortunately, the problems of Caucasus Pleistocene glaciation are still far from being resolved. There are very few absolute dates for the Late Pleistocene and Early Holocene, and even the Late-glacial and Post-glacial pollen spectrum is not dated properly. Nevertheless, there are two major schools established around the old Caucasus glaciation: 1) the followers of the Alpine school think that the glaciation during the Würm was of a sheet or semi-sheet with the snow line depression of ~1100 - 1300 m and the glacier tongues descending to a very low hypsometric level. The glaciers in the river gorges joined each other and emerged in the foothill zone as a single tongue (the rivers of Tergi, Enguri, Baksan, Cherek, Kuban, Kodori and etc.); 2) By the opinion of another group of scientists, the glaciation was weak during the Würm, with the snow line depression at ~600 - 800 m, and the glacier tongues descending to ~1500 - 2000 m (from Kodori to Tergi).

Both opinions are based on certain factual material and at one glance, it is difficult to assess which opinion is right, though, we think that there is a truth in both opinions and their comparative analysis will yield in the desirable result.

3 METHODS

For this purpose, we decided to study these questions using analog method following the Gobejishvili et al., [24] approach. The essence of the analog method is as follows: in case of similar natural conditions (factors), the same process or phenomenon can be studied in one region and apply the obtained information to another region. The analog method is widely used in glaciology and geomorphology. For instance, let us consider the glaciers located on the southern slope of the Caucasus. If considering a valley glacier and studying its thawing or accumulation, we will see that the numerical data typical to it can be applied to the similar glaciers found in the adjacent basins. The glaciologists call such a glacier the model glacier.

When reconstructing the morphology and sizes of the old glaciation, an objective difficulty is that the trace of old glaciation is slightly preserved in the relief and many sedimentary forms very much look like the glacial forms by their appearance. The trough form of the valley at many locations is modified due to the weathering, erosion or denudation processes. Under their impact, the lateral moraines on the slopes are absolutely removed, or are survived as fragments. The terminal moraines are preserved not so well, as they are washed off under the action of fluvioglacial or mudflow currents. Due to the active exogenous processes, the old glacial material is transported to the lower hypsometric levels and forms a false image of the real extension of the glaciers.

Despite the many-year detailed studies, the genesis of these formations is impossible to specify by using the traditional glacial-geomorphological approach. Because of this there are essential differences in the identification of the morphology and scale of the glaciation in many mountainous regions of the world.

It is important that the data on modern glaciers are not used in solving the paleo-glacial problems. In order to fill this gap, we studied the modern glaciers of Georgia.

We identified the link between the length of the modern glaciers and their feeding firn basins. The analysis of aerial images and cartographic materials showed that the tongues of the valley glaciers are fed from the firn within the vast cirque forms located at higher altitudes. The larger the firn basin is, the longer the glacier is. If the glacier of a certain length corresponds to the feeding area of the modern glacier, it is natural that the feeding area of the old glacier would have formed a glacier of a certain length [24] . This idea can be expressed by a directly proportional dependence formula:

$$\frac{Lc}{Sc} = \frac{Ld}{Sd}$$

where the Lc—is the length of the modern glacier, Sc—is the area of the firn basin of the modern glacier, Ld—is the length of the Pleistocene glacier and Sd—is the area of the firn of the Pleistocene glacier.

In this formula, the values of Lc and Sc are identified by using the topographic maps for the valley glaciers of some mountains in Georgia and Eurasia (Tian Shan, Spitsbergen, Tibet, Himalayas) [23] .

For comparison purposes, we classified the gained quantitative indexes by their values and grouped them. We singled out four groups of glaciers, which clearly differ from each other with the above mentioned dependence. We denoted this dependence by coefficient K:

$$\frac{Lc}{Sc} = K$$

The first group included the hanging-valley and simple-valley glaciers; for them K = 0.81 (with the correlation coefficient of 0.96); the second group included the simple-valley glaciers with a multi-chamber firn with K = 0.50 (0.95); the third group included the compound-valley glaciers with K = 0.33 (0.99); and the fourth group included the branched glaciers with K = 0.13 (0.86).

Today there are no glaciers of the fourth group found in the Caucasus. Classical examples of such glaciers are Inylchek, Koendi and Mushketov in Tian Shan, Rongbuk—in the Himalayas and etc. Further division of the glaciers of this group for the other mountainous regions (Alaska, Karakorum, Hindu Kush) will be necessary to improve the accuracy of the data (according to the number of branches composed the glacier). For the higher accuracy the coefficient is better to calculate for the individual types of the glaciers.

Based on the data gained for individual groups, a nomogram showing the relation between the area of the glacier firn basin and the glacier length was drafted. After defining coefficient (K) the proportional dependence formula will be as follows:

$$\frac{Ld}{Sd} = K, \text{ then } Ld = Sd \times K.$$

In this formula, the Sd—is the feeding area of the old glacier, i.e. it is the cirque of the old glacier, which was the source of feeding the glacier.

In addition, we used the geomorphological, cartographical, petrographical and aerial images decoding methods (in order to identify the areas of the distribution of moraines). The complex application of these methods led us to the different method of study, which will allow identifying the limits of the old glaciation according to the individual orographic units and the river basins, which finally gives the full picture of the Würm glaciation extension in the Georgian Caucasus. Based on these data we compiled the map of the Würm glaciation of the Georgian Caucasus (Figure 1).

4 RESULTS

The morphological and morphometric analysis of the relief and processing of modern aerial images showed that the old cirques developed in the axial zone of the Caucasus built with crystal rocks are well preserved in the relief. The old cirques are mostly originated during the glaciation of the Late Pleistocene. This glaciation is known as the Würm Glaciation in literature. Post-Würm stage glaciation led to the slight re-deepening of the existing cirque. We mapped almost all glacial cirques and measured their areas. In addition, we identified the height of the lower limits of the cirques. To our opinion, the height of the lower limits of the cirques is the indicator of the firn line location. Based on the gained materials, we calculated the lengths of the Late Pleistocene glaciers, the altitudes of their terminus and location of the firn lines (Figure 1, Table 1).

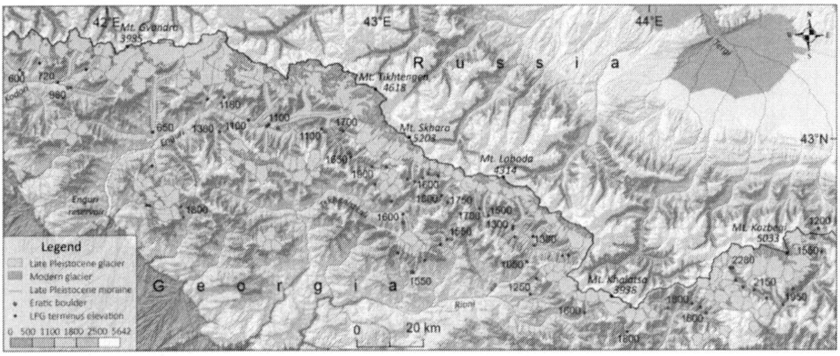

Figure 1. Modern and Late Pleistocene glaciers of the Kodori, Enguri, Rioni and Tergi River basins.

4.1. Late Pleistocene Glacier Extent

4.1.1. Western Caucasus

The Late Pleistocene glaciation in the Western Caucasus was of a mountain-valley type. The principal glaciation center was the Caucasus watershed, while the other centers were found in the Gagra, Bzipi, Chkhalta (Apkhazeti) and Kodori subranges [25] . Despite the fact that the Western Caucasus is not high hypsometrically, the signs of the old glaciation are well preserved in the form of cirques and moraines. The lower threshold of the cirque forms is located at the altitude of ~1900 - 2100 m. This altitude during the Würm was the lower limit of the nival zone. This evidences that in the Late Pleistocene the firn line depression in the river basins of the Western Caucasus was ~1000 - 1200 m. The largest glaciers on the southern slope of the Caucasus descended to ~600 - 700 m. The glaciers were of a compound-valley type and they reached ~16 - 17 km in length. The largest glacier was in the Sakeni valley (with the length of ~25 km).

Quite a strong glaciation occurred on the southern slopes of the Western Caucasus watershed during the Würm. Large glaciers descended to the bottom of the Kodori River and ended at different hypsometrical levels (Figure 1, Table 1).

4.1.2. Central Caucasus

The high hypsometrical location of the Central Caucasus, the morphological peculiarities of the relief and climatic conditions determine the development of intense glaciation in the Late Pleistocene and Holocene. The lower threshold of the cirques formed by the glaciers of the Late Pleistocene is located at different hypsometric heights. Despite this, it is well seen that the height of the lower threshold of the cirques in the Central Caucasus increases from ~2100 m (the Nenskra River basin) to ~2500 m (the Tergi River basin) from the west to the east. The main glaciation center in the Late Pleistocene was the watershed of the Caucasus range (Figure 1, Table 1). There were quite large glaciers in the Svaneti range, while the relatively small valley and cirque glaciers were formed in the other subranges of the Caucasus (Samegrelo, Lechkhumi, Shoda-Kedela, Racha, Germukhi and Kharuli). Due to the hypsometric conditions of the subranges, there was a glacial zone developed here as the individual areas. On the other hand, the entire crest zone was included in the nival zone. As the subranges are of general Caucasian or lateral direction, the glaciation degree here is determined by the relief hypsometry, morphology and distance from the Black Sea. The glaciation in the Svaneti range had a sharply asymmetrical nature in the Late Pleistocene and still has today, what in our opinion, is caused not only by exposition, but also by the morphological conditions of the relief.

Among the river basins located on the southern slope of the Central Caucasus, the Enguri River basin was distinguished by the intense glaciation. Our studies showed that in the Enguri River gorge, especially the glaciers descending from the basins of the right tributaries blocked the Enguri River gorge and sometimes followed it for 1 to 3 km (the rivers of Dolra, Mulkhura, Khalde and Lailchala). In other basins (the Nenskra, Nakra, Adishchala and Laila) massive valley glaciers were developed during the Würm, but they did not emerge in the Enguri River bed.

Table 1. Morphometric indicators of the Late Pleistocene glaciers on the Georgian Caucasus.

Name of the glacier	Old cirque area, ~km^2	Length, ~km	Terminus elevation, a.s.l. ~m
Western Caucasus			
Kodori River basin			
Marukhi	36.0	17.3	1450
Sopruju	7.7	6.7	1000
Adeba	8.2	7.2	900
Atsiashi	20.8	8.7	760
Ptishi	27.0	14.0	600
Khetskvara	12.7	7.0	1200
Klichi	59.0	20.0	720
Ghvandra	17.5	11.5	1400
Sakeni	89.0	25.0	980
Adange	8.0	7.5	1700
Central Caucasus			
Enguri River basin			
Nenskra	275.0	36.0	650
Nakra	63.0	20.0	1180
Dolra	105.0	34.5	1050
Mulkhura	270.0	35.1	1150
Adishi	24.0	19.4	1700
Khalde	31.0	15.5	1650
Enguri (Skhara)	35.0	17.5	1900
Lailchala	23.0	11.5	1100
Laila	17.0	13.5	1380
Tkheishi	7.0	5.6	1450
Kveishkhi	4.5	3.6	1600
Didgali	6.0	4.8	1350
Magana	7.5	6.0	1450
Khobistskali	8.0	6.4	1800
Rioni River basin			
Koruldashi	15.0	11.7	1600
Zeskho	9.0	7.5	1600
Shari	11.5	9.0	1800
Edena	16.5	14.1	1650
Zopkhito	21.5	17.4	1500
Kirtisho	42.0	20.5	1300
Notsara	14.8	12.6	1280

Continued

Boko	20.9	18.0	1100
Buba	40.5	23.0	1050
Chanchakhi	13.1	11.0	1750
Gharula	17.5	14.0	1250
Jejora	19.8	17.5	1600
Latashuri	11.0	8.5	1500
Sokhortuli	7.5	6.0	1600
Ghobishuri	8.5	7.0	1800
Shodura	5.4	4.0	1600
Liakhvi River basin			
Zekara	8.5	7.0	1900
Kveshelta	10.0	8.0	1800
Jomagi	11.2	9.1	1550
Sba	9.2	7.5	1800
Cheliata	8.0	6.5	1850
Kalasani	11.0	9.0	1800
Eastern Caucasus			
Tergi River basin			
Devdoraki	38.5	14.2	1220
Gergeti	21.0	17.0	1550
Mna	23.0	18.0	1950
Suatisi	32.0	15.0	2150
Tergi headwater	20.0	10.0	2270
Khdestskali	24.0	11.9	2140
Chaukhi	7.0	5.7	2200
Aragvi River basin			
Abudelauri	14.2	11.5	1600
Asa River basin			
Chkhotani	-	6.5	1700
Akhieli	-	7.3	1750
Bisna	-	5.5	2060
Arghuni River basin			
Shatili	10.5	8.5	1800
Makhi	9.5	7.0	1900
Kamkhi	5.5	4.5	2100
Pirikita Alazani River basin			
Tebulo	14.0	11.0	2400
Amugho	7.5	6.0	2500
Larovani	6.8	5.5	2560
Kachu	11.5	6.0	2200
Didkhevi	13.7	7.0	2220
Cheros Khevi	10.0	5.0	2140
Diklo (Khao)	14.5	7.0	2100

4.1.3. Eastern Caucasus

The relief of the Eastern Caucasus with its morphological and morphometric properties is behind the Central Caucasus, resulting in, together with the climatic factors, the small scales of glaciation [26] both, today and in the entire Upper Quaternary and Holocene. As we have seen above, the modern glaciers are associated with the individual high-elevated massifs. A similar picture was observed in the Würm, as by our calculations, the lower limit of the glacial zone in the Würm was at the height of ~3400 - 3600 m. The nival zone was presented at a vast area in the Eastern Caucasus, but it was not extended over the entire territory. It was spread continuously only along the Caucasus watershed range. As for the Würm glaciers, they were linked with the massifs with the heights of about 3400 - 3600 m.

The glaciation in the Würm in the Eastern Caucasus had the properties similar to those of glaciers on the southern slope of the Central Caucasus at present.

The Tergi River head and the basins of its tributaries are located on the northern slope of the Eastern Caucasus. Here, the northern exposition of the slope plays a certain role in the origination of both, modern and Würm glaciations. The old cirque forms here evidence that during the Würm there were quite strong glaciers in the basins of the right tributaries of the Tergi River with their tongues outcropping into the bottom of the Truso basin (the lower threshold of the glacial cirques is at the height of ~2500 - 2600 m). The longest glaciers were in the head of the Tergi River and the Desistskali gorge. Their length was ~10 - 12 km (Figure 1, Table 1).

4.1.4. Southern Georgia's Highland

Under the southern Georgia's highland, we consider all those ranges and plateaus, which are located to the south of the intermountain plain: the ranges of Adjara-Imereti, Shavsheti, Arsiani, Trialeti, Samsari and Javakheti and the Erusheti upland. In the elevated part of these orographic units there were small glaciers developed during the Late Pleistocene evidenced by the glacial forms remained there.

All previous researchers indicate the presence of old glacial forms there (B. Klopotovskiy, [14] ; L. Maruashvili, [8] ; D. Tabidze, [21] ; D. Tsereteli, [5] and others). The only difference among them is in the scales of glaciation.

Our studies showed that in the southern Georgia's highland the Würm glaciation has almost the same form as there is in the Eastern Caucasus today. In the Würm, the firn line was located in the Adjara-Imereti range at the lowest altitude, at ~2200 - 2300 m, and in the Samsari and Javakheti ranges—at the highest altitude, at ~2500 - 2700 m. Neither of the ranges reached the glacial zone. However, the nival zone was presented quite clearly, where small cirque and valley glaciers were formed under the influence of the atmospheric precipitations. These glaciers were concentrated at the small sections of the ranges with their heights more than ~2400 m, for example: Sakornia, Mepistskaro and Loboroti ? in the Adjara-Imereti range; Karakaya and Mount Sakvelo—in the Trialeti range; Kanli Dagi—in the Arsiani range, and Abuli, Samsari and others—in the Samsari range.

The degraded cirque forms remained there are of small sizes and accordingly, the glacier lengths were short—up to ~3.0 km. The largest glaciers were in the Samsari range. Valley glaciers were developed around the volcanic peaks of Abuli and Samsari; their maximum length did not exceed ~5 - 5.5 km [8] . The cirque forms start at the height of ~2600 - 2700 m. The genesis of the small lakes at this location is associated with the glacial and periglacial processes. The forms typical to the nival zone, both of Würm and post-Würm, are well preserved in the Samsari and Javakheri ranges. The sediments of old rock glaciers are a good evidence of this.

4.2. Holocene Glacier Extent

In contrast to a hypothesis of a stable Holocene climate, as indicated by oxygen isotope records from the Greenland Ice core [27] and northern Alpine lake sediments [28] , a growing number of studies (e.g. Mayewski et al., [29] and references therein) have demonstrated that distinct periods of climate change occurred repeatedly throughout the Holocene [30] . Mountain glaciers are potentially good climatic indicators, but most of the information on the dynamic responses of mountain glaciers to past climate changes comes from European glaciers which only form 3% of the world's glaciated area [31] . Knowledge of the Holocene glacier behaviour of remote and inaccessible mountain regions, often characterized by a high degree of glaciation, is rather limited. For better understanding of the global spatial-temporal glacier variability and throughout the Holocene, it is very important to study worldwide glacier dynamics, including the Caucasus Mountains.

There were several stadial glaciations in the Holocene and their paleoglacial study allows reconstructing the landscape zones and paleo-climate. With this purpose, the glacial deposits were studied on the southern slope of the Caucasus range and their correlation to the river terraces has been conducted. The absolute age (^{14}C) of organic deposits has been identified at several locations.

The trace of the stage glaciation in the Late Holocene (Neogene) is particularly well preserved in the Kodori River basin in the form of lateral and terminal moraines. The signs of an early times stadial glaciation are survived in the Sakeni River gorge, while in other gorges their traces are washed out or they are in the form of the separate fragments making it difficult to identify their number and limits of expansion.

The first stadial moraine in the Sakeni River gorge is at the height of ~1550 - 1600 m, in the environs of its right tributary of Olbaki [8] . At a distance of ~1.5 km above it, there is another moraine, which is seen quite well in the relief.

At the heights of ~1860 m and ~1930 m, there are two moraine complexes in the Sakeni River gorge distanced by ~500 - 600 m from each other. The upper (fourth) moraine of them is well seen in the relief.

After the mentioned stadial glaciation there was a lake; after it was filled up, a straight, slightly inclined and flat-bottomed valley was originated [8] . The origination of the fourth stadial moraine and the lake should have taken place at the beginning of the Late Holocene. Conditionally, we attributed the third moraine to the Upper Holocene. During this stadial glaciation, the Sakeni and Achapara glaciers were merged with each other.

There are two stadial moraines in the basins of almost all glaciers in the Sakeni River gorge. The distance between them is about 0.5 km. The young lateral and terminal moraines of them belong to the LIA maximum glaciation, and the intensely denuded moraine in the relief belongs to the historical stage (~2000 - 2500 years) [23] .

In the Sakeni River gorge we see the traces of six stadial glaciations. The 1st - 3rd moraines of them were formed in the Late Pleistocene and Early Holocene, and the 4th - 6th stadial moraines were originated during the Neo-glacial (Late Holocene) glaciation.

The stadial moraines of the Late Holocene age are remained in the heads of the rivers of Klichi and Marukhi at the height more than ~1900 m above sea level. The 4th stadial moraine is located at the height of ~1900 - 1950 m; the 5th historical stadial moraine is remained at the height of ~2150 - 2200 m and the 6th LIA maximum moraine is remained at the height of ~2250 - 2300 m and is clearly seen in the relief. It is distanced from the glacier by ~1.0 - 1.5 km [23] .

The trace of stadial glaciation in the Enguri River basin is best fixed in the gorges of the rivers of Mulkhura and Adishchala.

We distinguish the six stadial glaciations in the Mulkhura River gorge. A full spectrum of the stadial moraines is seen in the Mestiachala River gorge, on the western slope of Gvalda range, near the airport. The lateral stadial moraines are fixed as steps, while the Würm moraine is clearly seen as a hillock; the Würm moraine is found as a hillock on the right side of the gorge, as well. The trace of (1st - 2nd) two stadial glaciations is fixed as a step at the height of ~1750 m and ~1810 m above sea level. Below that, at the height of ~1650 and ~1620 m, there are two more steps built with glacial material. The 4th moraine of them is in the form of a hillock, which is stretched to the village of Spardishi. The moraine of the same stage is found at the village of Lanchvali, which is covered with delluvial cones. Lanchvali and Spardishi lateral moraines arch near the Lanchvali Bridge. The glacial boulders here evidence that during the 4th stage, the glacier ended at this location. However, the terminal moraine is washed due to the river action [32] [33] .

The moraine material in the environs of a former touristic base in Mestia is a continuation of the 3rdstadial moraine, while its morphological continuation is found even higher, in the Mulkhura-Mestiachala flat watershed.

The traces of the 1st and 2nd stadial glaciations are remained on the left slope of the Mulkhura River, in front of Mestia-Lenjeri (Figure 2). The 2nd stadial moraine is presented as a hillock descending to the gorge bottom above Lenjeri. The trace of the 1st stadial glacier is seen very slightly as individual steps and reaches the village of Kashveti. During these stages the Tviberi and Lekhziri glaciers were merged with each other.

The fragments of the 5th (historical stage) stadial moraine are remained in the environs of the village of Lavladashi and as fragments on the left side up along the gorge. There are a number of granite boulders on the both sides of the gorge, both, in the floodplain environs and beyond it (Figure 3). It should be noted that a morphologically well-seen debris cone in front of Lavladashi gorge is of three-step identified through the correlation to the glaciation in the Neo-Glacial (Late Holocene) period. During the LIA maximum the Lekhziri and Chalaati glaciers did not merge with each other. The Chalaati glacier descended to the Lekhziri River bed and created the plug. The 6thstadial moraines are clearly seen in the relief.

The trace of the stadial glaciation in the Mulkhura River basin is best remained in the environs of the Ughviri Pass and Lagvzalieri area. The relative height of the Ughviri Pass is 300 - 350 m from the Mulkhura gorge. During the stadial glaciation (the 1st and 2nd stages) the thickness of the Tviberi-Tsaneri glacier was more and one flow of the glacier (the Nageba glacier) moved to the Enguri River gorge. On the slope of the Enguri River, within the Würm moraines, there are two series of lateral moraines and a corresponding terminal moraine. Such state of moraines evidences that the glacier flow of these stages did not reach the bottom of the Enguri River. On the northern slope of Ughviri two lateral stadial moraines (3rd - 4th) are identified. One of them is located immediately near the watershed and another one is located in the environs of the Ughviri Lake (the lake itself is originated during this stage).

Figure 2. The trace of the 1st and 2nd stadial glaciations on the left slope of the Mulkhura River in front of Mestia-Lenjeri (photo by: L. G. Tielidze).

Figure 3. Erratic boulders in the Mestiachala River valley (photo by: L. G. Tielidze).

On the right slope of the Mulkhura River gorge, at Lagvzalieri location, at the height of ~1900 - 1950 m, there is a ~1.0 - 1.5 km long moraine, which is located at the same hypsometrical height as Ughviri 3rd and 4th moraines. Even higher (~2120 - 2150 m), there is another moraine, which in our opinion, is synchronous to the 1st and 2nd stages of Ughviri. The fragments of the terminal moraines of the 3rd and 4th stages are found on the territory of Mestia community (at the beginning of the Mulkhura gorge), at the height of ~1450 m.

The 5th stadial moraine is located at the confluence of the rivers of Tviberi and Tsaneri. This moraine is morphologically seen in the relief and without a doubt it was formed by the action of the Tviberi glacier. Corresponding to it the moraine is found in the Tsaneri gorge, in the environs of the mineral waters. These moraines belong to the historical stage. The LIA maximum moraine is completely presented in the basins of all glaciers.

Identification of the age of a stadial moraine became possible only by correlating them to the river terraces. [23] sampled an organic substance (wood) from ~8 - 10 m high terrace at the village of Jorkvali (the Enguri River gorge) with its age (~5000 ± 200 years) identified at the Tallinn radiocarbon laboratory. This terrace seems to be formed at the end of the Middle Holocene. This terrace of the given height is fixed at many locations in the Enguri River gorge (villages of Pari, Tskhumari, Lenjeri, etc.) and ends at the 4th stadial moraine near the settlement of Mestia.

There is a 2 - 3 m high terrace found at Mestia and at the village of Lenjeri, which stretches long the gorges of the rivers of Mestiachala and Mulkhura. In the Mestiachala River gorge an airport and a new settlement are built on this terrace, which is morphologically clearly seen and stretches to the village of Lavladashi, where it is transformed into the historical stadial (5th) moraine. In the Mulkhura River gorge this low terrace is seen everywhere and is transformed into the stadial moraine at the confluence of the rivers of Tviberi and Tsaneri.

The analysis of the materials allows making the following conclusion: the 4th - 5th and 6th stadial moraines were originated in the Late Holocene (Neo-glacial period).

5. CONCLUSION

In this paper, we have presented the Late Pleistocene and Holocene glacier simulation of the entire Georgian Caucasus, using analog method and guiding constraints from geological data. Our experimentation has explored that the Late Pleistocene glaciation of the Georgian Caucasus consisted of valley glaciers, with ice caps only on some peaks. The glacier tongues which terminated at the lowest altitudes were those of the Ptishi and Nenskra Glacier on the south slope at ~600 and ~650 m a.s.l. respectively. Tongues of other large glaciers descended down to 1600 - 1200 m a.s.l. The firn line in the Central Caucasus was found at 2000 - 2500 m in the Late Pleistocene, increasing in the altitude from west to east. This implies that the firn line in the Caucasus was depressed by 1200 - 1300 m during the Late Pleistocene (allowing for neotectonic activity), and the depression increased from west to east. This is equivalent to a decline in mean annual temperature of ca. 7°C - 8°C (with a gradient of 0.6°/100 m altitude). We expect that the approach used in this article is a useful technique for determination the Late Pleistocene glacier parameter, configuration and geometries.

CITE THIS PAPER

Tielidze, L.G. (2017) Late Pleistocene and Holocene Gla- cier Extent in the Georgian Caucasus. Open Journal of Geology, 7, 517-532.

REFERENCES

1. Tielidze, L. (2016) Glaciers Catalog of Georgia. Publishing House "Samshoblo", Tbilisi, 116.
2. Penck, A. and Bruckner, E. (1901/1909) Die Alpen im Eiszeitalter, 3 Vols. Tauchnitz, Leipzig, 1199 p.
3. Reingard, A.L. (1937) About the Problem of the Stratigraphy of the Glacial Time of the Caucasus. Transactions of International INQUA Congress Transactions, Section M 1, 9-31. (In Russian)
4. Vardaniants, L.A. (1937) The Ancient Glaciation of the Rivers Iraf (Urukh) and Tsey (Central Caucasus). Proc. State Geogr. Soc., 69, 537-562. (In Russian)
5. Tsereteli, D.V. (1966) Pleistocene Deposits of Georgia. Publ. House "Metsniereba", Tbilisi, 582 p. (In Russian)
6. Kovalev, P.V. (1961) Modern and Ancient Glaciation Enguri River Basin. Mat. Kavkaz. Eksped. T. 2. Xarkov. Izd XGU. (In Russian)
7. Shcherbakova, E.M. (1973) The Ancient Glaciation of the Great Caucasus. Moscow University Press, Moscow, 272 p. (In Russian)
8. Maruashvili, L.I. (1956) Advisability of Revising Existing Paleogeographic Conditions of the Glaciation Age in the Caucasus. Publ. Academy of Sciences of the Georgian SSR, Tbilisi. (In Russian)
9. Khazaradze, R.D. (1971) Relief, Continental Sediments and Pleistocene Glaciation Enguri River basin. Avtoreferat kand, Disertacii, Tbilisi. (In Russian)
10. Abich, H. (1865) The Study of Current and Ancient Glaciers of the Caucasus. Cb, cvedenii o Kavkaze, T. 1, Tiflis. (In Russian)
11. Reinhardt, A.L. (1936) Ice Age of the Caucasus and Its Relation to the Glaciation of the Alps and the Altai. tr. II. mejd. konf. po izuchenie chetvertichnogo perioda evropi. L-M, bip. 2. (In Russian)
12. Kuznetsov, I.G. (1931) Geological Structure Shamshovy Resort to the Chanchakhi River Basin. tr. vsesoiuznogo geolog, razvitia obedinenia. vipusk 151. (In Russian)
13. Vardaniants, L.A. (1930) On a New Method Estimate Depression Snow Line in Relation the Study Stage Glacier Retreat Mining Group Adai-Hoch in the Central Caucasus. Izd. RGO. T. 62. Vip. 2 (In Russian)
14. Klopotovskiy, B.A. (1949) Khorlakelskaya Moraine. Soobsh. AN. GSSR. T. 10, N.6. (In Russian)
15. Paffenholz, K.N. (1958) The New Stratografic Data of Kazbegi District and the Keli Volcanic Plateau (Central Caucasus), and Ancient Glaciation in This Area. Sovetskaia geologia, N. 12. (In Russian)
16. Nemanishvili, S. (1962) Geomorphology of the Northern Part of Lower Racha. Vakhushtis saxelobis geogr. Institutis shroebi, T. 12. (In Georgian)
17. Dondua, G. (1959) Geomorphology of Jejora nad Gharula River Basins. Vaxushtis saxelobis geografiis isntitutis shromebi, T. 12. (In Georgian)
18. Astakhov, N.E. (1973) Structural Geomorphology of Georgia. Pub. House "Metsniereba", Tbilisi. (In Russian)
19. Milanovskiy, E.E. (1960) Traces Late Pleistocene Glaciation in the High Part of the Central Caucasus. Dokladi an SSSR. T. 130, N. 1. (In Russian)

20. Dumitrashko, N.V., Lilienberg, L.A., Antonov, B.A. and Baljan, S.P. (1961) Ancient Glaciations of the Caucasus aud Their Correlation with Glaciations of the Russian Plain. Trudi komisii po izucheniju chetvertichnogo perioda, 19, 170-180. (In Russian)
21. Tabidze, D.D. (1965) Kodori River Basin Geomorphology, Avtoreferat kand. Disertacii, Tbilisi. (In Russian)
22. Serebruanny, L.P. and Orlov, A.V. (1989) Moraines as the Source of Glacial Information. Unpublished Manuscript, 235 p.
23. Gobejishvili, R.G. (1995) The Evolution of the Modern Ice Age Glaciers and Mountains of Eurasia in the Late Pleistocene and Holocene. The Thesis of Doctor of Science Degree in Geography. (In Georgian)
24. Gobejishvili, R., Lomidze, N. and Tielidze, L. (2011) Late Pleistocene (Wurmian) Glaciations of the Caucasus. In: Ehlers, J., Gibbard, P.L. and Hughes, P.D., Eds., Quaternary Glaciations: Extent and Chronology, Elsevier, Amsterdam, 141-147.
25. Tielidze, L. (2017) Glaciers of Georgia. Springer, Geography of the Physical Environment.
26. Tielidze, L.G. (2016) Glacier Change over the Last Century, Caucasus Mountains, Georgia, Observed from Old Topographical Maps, Landsat and ASTER Satellite Imagery. The Cryosphere, 10, 713-725.
27. Johnsen, S.J., Clausen, H.B., Dansgaard, W., Gundestrup, N.S., Hammer, C.U., Andersen, U., Andersen, K.K., Hvidberg, C.S., DahlJensen, D., Steffensen, J.P., Shoji, H., Sveinbjornsdottir, A.E., White, J., Jouzel, J. and Fisher, D. (1997) The δ18O Record along the Greenland Ice Core Project Deep Ice Core and the Problem of Possible Eemian Climatic Instability. Journal of Geophysical Research: Oceans, 102, 26397-26410.
28. Von Grafenstein, U., Erlenkeuser, H., Brauer, A., Jouzel, J. and Johnsen, S.J. (1999) A Mid-European Decadal Isotope-Climate Record from 15,500 to 5000 Years B.P. Science, 284, 1654-1657.
29. Mayewski, P.A., Rohling, E.E., Curt Stager, J., Karlen, W., Maasch, K.A., David Meeker, L., Meyerson, E.A., Gasse, F., van Kreveld, S. and Holmgren, K.U. (2004) Holocene Climate Variability. Quaternary Research, 62, 243-255.
30. Agatova, A.R., Nazarov, A.N., Nepop, R.K. and Rodnight, H. (2012) Holocene Glacier Fluctuations and Climate Changes in the Southeastern Part of the Russian Altai (South Siberia) Based on a Radiocarbon Chronology. Quaternary Science Reviews, 43, 74-93.
31. Dyurgerov, M.B. and Meier, M.F. (1997) Year-to-Year Fluctuations of Global Mass Balance of Small Glaciers and Their Contribution to Sea-Level Changes. Arctic and Alpine Research, 29, 392-402.
32. Gobejishvili, R.G., Tielidze, L., Lomidze, N. and Javakhishvili, A. (2012) Monitoring of the Glaciers against the Background of the Climate Change. Publ. House "Universali", Tbilisi. (In Georgian)
33. Tielidze, L. (2016) Modern and Old Glaciers of Georgia. Publishing House "Samshoblo", Tbilisi, 216.

CHAPTER 11

Early Cenozoic Multiple Thrust in the Tibetan Plateau

Zhenhan Wu,[1] Peisheng Ye,[2] Patrick J. Barosh,[3] Daogong Hu,[2] and Lu Lu[1]

[1]*Chinese Academy of Geological Sciences, Beijing 100037, China*

[2]*Institute of Geomechanics, Chinese Academy of Geological Sciences, Beijing 100081, China.*

[3]*P. J. Barosh and Associates, 103 Aaron Avenue, Bristol, RI 02809, USA*

ABSTRACT

Recently completed regional geological mapping at a scale of 1 : 250,000 or larger across all of the Tibetan Plateau coupled with deep seismic surveys reveals for the first time a comprehensive depiction of the major early Cenozoic thrust systems resulting from the northward subduction of the Indian Continental Plate. These systems define a series of overlapping north-dipping thrust sheets that thickened the Tibetan crust and lead to the rise of the plateau. The few south-dipping thrusts present apparently developed within a sheet when the back moved faster than the toe. Many of the thrusts are shown to extend to the middle-lower crustal depths by seismic data. The regional thrust systems are the Main Central, Renbu-Zedong, Gangdese, Central Gangdese, North Gangdese, Bangoin-Nujiang, Qiangtang, Hohxil, and South Kunlun Thrusts. The minimal southward displacements of the South Kunlun, Hohxil, South Qiangtang, and Central Gangdese Thrusts are estimated to be 30 km, 25 km, 150 km and 50 km, respectively. Deep thrusting began in the Himalaya-Tibetan region soon after India-Eurasia continental collision and led to crustal thickening and subsequent uplift of the Tibetan Plateau during Late Eocene-Early Miocene when the systems were mainly active. The major thrust systems ceased moving in Early Miocene and many were soon covered by lacustrine strata. This activity succeeded in the late Cenozoic to crustal extension and strike-slip movement in the central Tibetan Plateau. The revelation of the full array of the early Cenozoic thrust systems provides a much more complete understanding of the tectonic framework of the Tibetan Plateau.

1. INTRODUCTION

A great change in landforms and environments took place during the Cenozoic era as the Indian Continental Plate was subducted northward beneath central Asia to cause crustal thickening and uplift of the Tibetan Plateau. The Tethys Ocean had occupied the Himalayan and southern Tibetan areas at a time when volcanic rocks of the Linzizong Group formed in the Paleocene-Early Eocene magmatic belt along the Gangdese island arc, while the oceanic plate plunged northward along the Yarlung Zangbo suture [1, 2]. Continental red-beds of Fenghuoshan Group were being deposited farther north in central and northern Tibet in Late Cretaceous-Early Cenozoic, and a sea existed beyond Tibet in southwestern Tarim basin [3] (Figure 1). The subduction of the oceanic plate, volcanic eruptions, and intrusions, which ended before ~45 Ma, was followed by the subduction of the Indian Continental Plate [4] and the subsequent contractional deformation and crustal shortening across the Tibetan Plateau [5]. A variety of hypotheses have been proposed to explain the tectonic process [6–9], but many unanswered questions remained due to the lack of regional data on Cenozoic stratigraphy and structure and tectonic movements as well as a limited understanding of the deep structure of the Tibetan Plateau. There also has been in debate as to which direction various thrust zones dip and if different dips mean different episodes of thrusting. The newly completed regional geologic mapping and seismic surveys across the plateau now provide critical information relevant to these questions.

Much of what has been described in the past about the Tibetan Plateau was primarily based on road traverses, such as the 1 : 500,000 scale geological mapping along Golmud-Lhasa Highway during the Chinese-British collaboration in mid 1980s [25]. Now, however, regional mapping has been completed across the region by Chinese geologists after decades of working, and the understanding of the geology keeps improving as the field work progresses. These maps include the following: the 1 : 1,000,000 scale geological map of southern and central Tibetan Plateau [26], 1 : 200,000 scale geological maps of northeastern Tibetan Plateau [27], and 1 : 200,000 scale geological maps of eastern Tibetan Plateau [28, 29]. The China Geological Survey also carried out 1 : 250,000 scale geological mapping throughout the Tibetan Plateau since 2000 [30, 31], and released 1 : 250,000 scale geological maps of the southern, central, and northern plateau in 2002–2007 and revised the earlier 1 : 250,000 scale geological maps of eastern Tibet. The authors finished a 1 : 250,000 scale geological map of the Damxung quadrangle in the central Lhasa Block [32] and 1 : 100,000 scale structural maps of the central Qiangtang Block [19] and portions of the East Kunlun Mountains. This was augmented in complex areas by detailed 1 : 50,000 and 1 : 25,000 scale geological mapping in the southern East Kunlun Mountains [11, 33, 34]. All of this mapping, especially since 2000, has provided new information on the stratigraphy and structures that sheds considerable insight on the Cenozoic tectonic movements of the region [30].

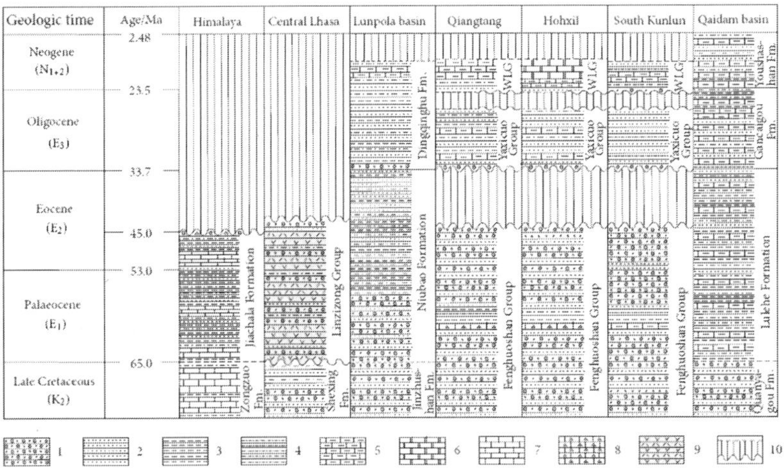

Figure 1: Tertiary stratigraphic systems in the Tibetan Plateau. Explanation: 1: conglomerate and sandstone; 2: sandstone; 3: shale; 4: mudstone and muddy siltstone; 5: marl; 6: dolomite and dolomitic limestone; 7: limestone; 8: breccia limestone; 9: Island-arc-type volcanic rocks; 10: unconformity; WLG: Wudaoliang Group [10]. The Zongzuo and Jiachala Formations are dominated by marine deposits [1], and the other Formations are terrestrial deposits. Geological time scale is that recommended by China National Committee of Stratigraphy (2002).

From the 1 : 250,000 scale geological maps, field studies of typical thrust fault from the Kunlun to Himalaya Mountains, and previous studies we compiled selected data on early Cenozoic strata, structures, plutons, and volcanic rocks to produce a Late Eocene-Oligocene paleotectonic map of the Tibetan Plateau (Figure 2). We also identified deep thrusts within the middle-lower crust from seismic reflection profiles and wide-angle seismic data to tie into the surface structures. The results reveal the principal structural systems and deep structures that yield a greatly improved comprehension of the geodynamic movements, which caused the crustal thickening and uplift of the Tibetan Plateau.

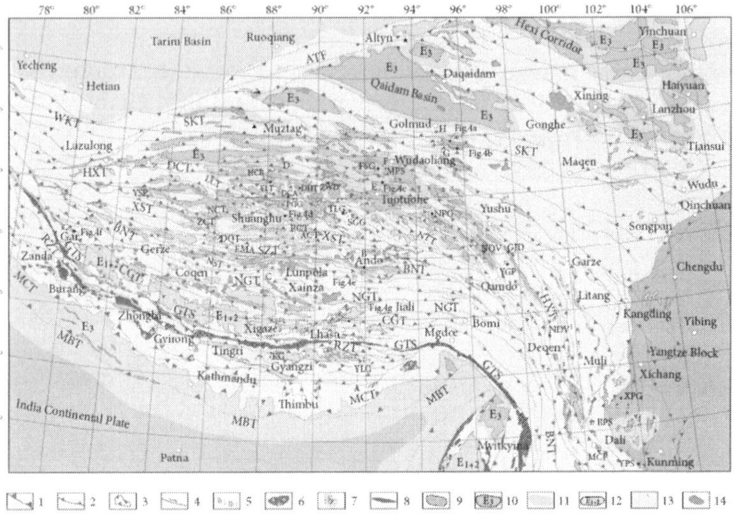

Figure 2: Paleotectonic map of the Tibetan Plateau for the Late Eocene-Oligocene. Explanation: 1: major thrust fault; 2: thrust fault; 3: nappe or outlier; 4: strike-slip fault; 5: ductile shear zone; 6: Late Eocene-Oligocene granite; 7: granitic dome; 8: ophiolite along the Yarlung Zangbo Suture (YZS); 9: Late Eocene-Oligocene volcanic rocks; 10: Oligocene (E_3) terrestrial strata; 11: Oligocene basin; 12: Paleocene-Eocene (E_{1+2}) strata; 13: erosion area; 14: present-day lake. Major fault systems: ATF: Altyn Tagh Fault [5, 8]; WKT: West Kunlun Thrust; SKT: South Kunlun Thrust [11]; HXT: Hohxil Thrust; NTT: North Tangula Thrust [12], DCT: Dogai Coren Thrust; LCT: Longwei Co Thrust; NCT: north central Thrust; XST: Xiaocaka-Shuanghu Thrust; DQT: Doma-Qixiang Co Thrust; SZT: Saibu Co-Zagya Thrust; NST: Nima-Silin Co Thrust; BNT: Bangoin-Nujiang Thrust; NGT: North Gangdese Thrust; CGT: Central Gangdese Thrust; GTS: Gangdese Thrust System [13]; RZT: Renbu-Zedong Thrust [13, 14]; MCT: Main Central Thrust [15, 16]; MBT: Main Boundary Thrust after Mid Miocene [5]. Black dots mark sampling location for isotopic ages and outcrops for photos, and dotted lines A-B, C-D, E-F, and G-H mark positions of regional cross sections.

This paper shows both previously known thrust systems and newly discovered ones to present for the first time a comprehensive depiction of thrust zones due to early Cenozoic compression. The areas that were theoretically predicted are now thoroughly understood through geologic and seismic data. This research also demonstrates that a single essentially north-dipping series of thrust sheets forms the Tibetan Plateau, although the related deep processes may be very complex. The purpose of this paper is to present a brief outline of the stratigraphy and descriptions of the major early Cenozoic thrust systems and the derived crustal structures across the entire Himalayan-Tibetan Plateau region.

2. STRATIGRAPHIC SYSTEMS

The stratigraphy is of particular importance in recognizing the dislocations and offsets across thrust systems. The early Cenozoic strata, which reflect the contrasting environments of deposition across the region, provide important information related to the structural control of sedimentation. Paleocene-Eocene marine strata were being deposited in the northern Himalaya belt [1], while Paleocene-Early Eocene volcanic-sedimentary rocks of the Linzizong Group accumulated in the Gangdese Magmatic Arc and early Cenozoic continental stratigraphic units were laid down in the northern Lhasa, Qiangtang, Hohxil, and Kunlun-Qaidam Blocks [30]. Farther north marine strata were accumulating in the southwestern Tarim basin about the same time [35]. The unconformities that reflect uplift also are strikingly different between these areas.

The Paleocene to early Eocene strata in the Himalaya Block consists of marine dolomite, dolomitic limestone, and shale of the Jiachala Formation that sits conformably on Late Cretaceous limestone [1, 2]. This is in sharp contrast with the volcanic rocks in the northern Gangdese Magmatic Arc in the central Lhasa Block to the north. There the section includes Early Eocene rhyolite, trachyte, conglomerate and mudstone of the Pana Formation, Late Paleocene rhyolite, andesite, conglomerate and sandstone of the Nianbo Formation, Early Paleocene andesite, dacite, basalt-andesite breccia and conglomerate of the Dianzhong Formation, and Paleocene volcanic rocks and conglomerate of the Meisu Formation [30]. These units lie with an angular unconformity over Late Cretaceous shale, sandstone, and conglomerate of the Shexing Formation (Figure 1).

The Paleocene-Eocene lacustrine, fluvial and alluvial deposits formed to the north in continental basins of the central and northern Tibetan areas [30, 36]. These constitute the Paleocene-Eocene marl, shale, sandstone, and mudstone of the Lulehe Formation in the Qaidam Basin, Late Cretaceous-Eocene reddish conglomerate and sandstone of the Fenghuoshan Group in the Qiangtang and Hohxil Blocks, and Paleocene-Eocene conglomerate, sandstone, marl, and mudstone of Niubao Formation unconformably overlying Late Cretaceous red-beds of the Jingzhushan Formation, which is dominated by conglomerate and sandstone in Lunpola Basin (Figure 1).

Oligocene lacustrine mudstone, marl, and sandstone that intercalated with gypsum of the Dingqinghu Formation, Yaxicuo Group, and Gancaigou Formation covered the Niubao Formation, Fenghuoshan Group, and Lulehe Formation in the Lunpola, Hohxil, and Qaidam Basins, respectively (Figure 1). The ages of the early Tertiary continental strata are constrained by sporopollenin assemblages [10] and magnetostratigraphic data [36, 37].

Farther north in the southwest Tarim basin a marine sequence of Paleocene to basal Oligocene marine shale, siltstone, limestone, marl, and intercalated thick gypsum of the Kashi Group is unconformably covered by Late Oligocene-Miocene lacustrine sandstone, siltstone, marl, and minor shale of the Wuqia Group [3]. This sequence rests on the Late Cretaceous radiolarian limestone and shale of the Yingjisha Group [35].

3. LARGE THRUST SYSTEMS IN THE UPPER CRUST

The early Cenozoic stratigraphic sequences are now disrupted and telescoped by intensive thrusting in the Himalayan Mountains and Tibetan Plateau as a result of the India-Eurasia continental collision. The thrusts form several major systems, each of which contains several thrusts that extend over the entire region, and thrust faults constitute the early Cenozoic structural framework of Tibet (Figure 2). These thrust systems, from south to north are the Main Central Thrust (MCT) [6, 15], Renbu-Zedong Thrust (RZT) and Gangdese Thrust system (GTS) in southern Tibet [13], the Central Gangdese Thrust (CGT), North Gangdese Thrust (NGT) and Bangoin-Nujiang Thrust (BNT) in central Tibet, and the Qiangtang Thrust system (QTS), Hohxil Thrust (HXT), and South Kunlun Thrust (SKT) [11] in northern Tibet. Most thrust faults of the CGT, NGT, BNT, QTS, HXT, and SKT are the ones newly discovered by mapping since 2000. The West Kunlun Thrust (WKT) [38] in northwestern Tibetan Plateau and the North Qaidam, South Qilian, Central Qilian, and North Qilian Thrust systems in northeastern Tibetan Plateau [5, 39] may have been activated in early Cenozoic, but their main movement occurred in late Cenozoic [8]. The thrust systems provide boundaries of the regional structural blocks as well as internal breaks. The chief systems and the strata displaced are briefly summarized below from south to north.

3.1. Main Central Thrust

The Main Central Thrust (MCT) and associated thrusts in combination with the Main Boundary Thrust (MBT) form a complex zone whose movement lifted the Himalayan Mountains [5, 6, 40]. Proterozoic and Early Paleozoic metamorphic rocks are thrust southward over Late Paleozoic strata to form large-scale nappes of gneiss, ductile shear zones and tight folds along MCT (Figure 2). These thrust faults converged into the Main Himalaya Thrust (MHT), which forms the boundary with the subducted Indian Continental Plate, to cause tectonic emplacement of the Greater Himalaya (GH), Lower Lesser Himalaya (LLH) and Upper Lesser Himalaya (ULT) along with partial melting and magmatic emplacement in the deep crust of the Himalaya fold-thrust belt [15]. The MCT was initiated in Early Eocene, ~45–42 Ma [40, 41], and developed southward during the Oligocene and Early Miocene, followed by the Main Boundary Thrust (MBT) after Late Miocene, ~10 Ma, south of the Himalaya Mountains, according to chronological data in Nepal [16]. An Oligocene-Early Miocene foreland basin formed south of the MCT (Figure 2). However, late Cenozoic thrusting (e.g., MBT) and uplift of the Himalaya Mountains lead to destruction of the Oligocene basin and erosion of its deposits and then formed the Siwalik foreland basin along the MBT after Late Miocene [5, 6, 31].

3.2. Gangdese and Renbu-Zedong Thrust Systems

The Gangdese Thrust System (GTS) and Renbu-Zedong Thrust (RZT) both formed along the Yarlung Zangbo Suture, which forms the boundary between the Himalaya and Lhasa Blocks (Figure 2). The GTS in the southern Lhasa Block is characterized by southward thrusting along north-dipping faults of the Gangdese batholith and Linzizong volcanic rocks over the Late Cretaceous marine forearc sequence of the Xigaze Group and the Eocene-Oligocene Dazhuka Group [13]. The RZT in the northern Himalaya Block is characterized by northward thrusting of Mesozoic marine strata over the Gangdese batholith and Xigaze and Dazhuka Groups along south-dipping faults (Figure 3(a)). Major GTS faults were truncated and overthrusted by RZT along the suture in south of Lhasa to west of Xigaze (Figure 2). Analysis of ^{39}Ar-^{40}Ar chronology suggests that GTS formed in 27–18.3 Ma with minimal displacement and slip rate of ~ km and mm/a, respectively, [13] and that RZT formed in 19–11 Ma with minimal displacement and slip rate of 12 km and ~2 mm/a, respectively [14].

3.3. Central Gangdese and North Gangdese Thrusts

The Central Gangdese Thrust (CGT) in central Lhasa Block (Figure 2) displaced tectonic slices of Carboniferous-Permian metamorphic rocks and Triassic-Jurassic limestone southward over Cretaceous volcanic-sedimentary strata (Figure 3(a)). Late Triassic limestone is thrust southward over Late Cretaceous lacustrine deposits of the Shexing Formation along the CGT north of Lhasa (Figure 4(g)), and Jurassic limestone of the Duodigou Formation is thrust southward over Cretaceous volcanic-sedimentary strata to form nappes along outer faults of the CGT south of Mozhugongka (Figure 3(a)). The CGT thrust faults north of Lhasa also are named the South Damxung Thrust (SDT) [32]. Carboniferous-Lower Permian slates are thrust over late Cretaceous-early Cenozoic red-beds of the Jingzhushan Formation along back faults of SDT east of Damxung [42]. Volcanic rocks of the Linzizong Group are offset by CGT thrust faults east of Mozhugongka. The minimal southward displacement of CGT is estimated to be 50 km according to the displacement of Jurassic limestone of the Dodigou Formation south of Mozhugongka (Figure 3(a)). Emplacement of Miocene porphyry granite produced local skarn in the Jurassic limestone nappes along the outer faults of CGT and formed metallic deposits at ~15 Ma [43]. These data indicate that CGT mainly formed in the Oligocene-Early Miocene.

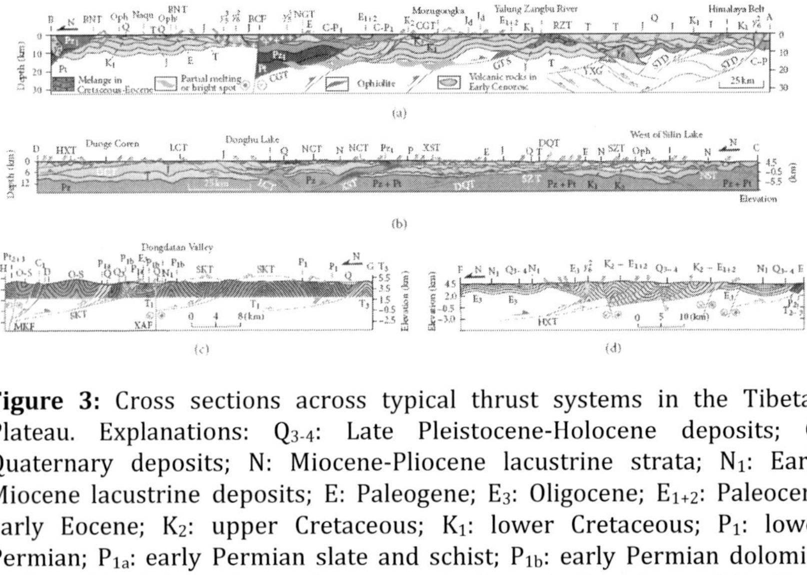

Figure 3: Cross sections across typical thrust systems in the Tibetan Plateau. Explanations: $Q_{3\text{-}4}$: Late Pleistocene-Holocene deposits; Q: Quaternary deposits; N: Miocene-Pliocene lacustrine strata; N_1: Early Miocene lacustrine deposits; E: Paleogene; E_3: Oligocene; E_{1+2}: Paleocene early Eocene; K_2: upper Cretaceous; K_1: lower Cretaceous; P_1: lower Permian; P_{1a}: early Permian slate and schist; P_{1b}: early Permian dolomite and marble; J: Jurassic marine strata; J_d: Jurassic limestone of Dodigou Formation; T: Triassic shale and sandstone; T_3, upper Triassic slate and conglomerate; $T_{2\text{-}3}$, middle-lower Triassic slate; T_1: lower Triassic slate and limestone; C_1: Carboniferous slate; D: Devonian metamorphic conglomerate and sandstone; O-S: Ordovician-Silurian schist and marble; O_3: Upper Ordovician limestone; Pz_1: Early Paleozoic schist and gneiss; Pt_{2+3}: middle-upper Proterozoic dolomite, marble and slate; Pt: Proterozoic gneiss; Oph: ophiolite; YXG: Yalaxianbo granite, deep extension of Yadoi leucogranite (YLG). BCF: Beng Co dextral-slip fault; MKF: Middle Kunlun fault; XAF: Xidatan active fault; other symbols are the same as in Figure 2. Locations of cross section A-B, C-D, E-F, and G-H are marked in Figure 2.

Figure 4: Photos of Late Eocene-Oligocene southward thrust in the Tibetan Plateau. Explanation: (a): view northward at Triassic granodiorite (Tg) thrust over Paleocene-Eocene red-beds (E_{1+2}) along SKT in southern Kunlun Mountains; (b): view northward at Early Triassic slate (T_1) thrust over Oligocene red-beds (E_3) along SKT in southern Kunlun Mts.; (c): view

northward at Upper Cretaceous-Eocene red-beds (K_2-E_{1+2}) thrust over Oligocene lacustrine deposits (E_3) along HXT in north of Tuotuohe; (d): view northwestward at Jurassic limestone (J) thrust over Paleocene-Eocene red-beds (E_{1+2}) along NCT north of Shuanghu; (e): view northwestward at Jurassic limestone (J) thrust over Paleocene-Eocene red-beds (E_{1+2}) and breccia formed by thrusting along BNT in south of Ando; (f): view northwestward at Cretaceous porphyry granite (Kg) thrust over Eocene-Oligocene red-beds ($E_{2\text{-}3}$) along BNT in east of Gar; (g): view northward at upper Triassic limestone (T_3) thrust over upper Cretaceous-Paleocene lacustrine strata (K_2) along CGT in northeast of Lhasa. Locations for photos are marked in Figure 2.

The North Gangdese Thrust (NGT), which formed in the northern Lhasa Block, merges with CGT southeast of Jiali and west of Coqen (Figure 2). The NGT displaces varied units relatively southward. East of Xainza Proterozoic metamorphic rocks and Ordovician-Devonian limestone move over Carboniferous-Permian strata and Jurassic ophiolite. Paleozoic strata are thrust southward over Early Cretaceous marine strata and Late Cretaceous-Early Cenozoic red-beds west of Namco along a large thrust system, which has been termed the West Namco Thrust (WNT) [32]. Jurassic and upper Permian limestone are thrust southward over Late Cretaceous-Early Cenozoic red-beds and Carboniferous-Early Permian slate along its outer faults west of Jiali and north of Damxung (Figures 2 and 3).

3.4. Qiangtang Thrust Systems

Intense compression affecting the Qiangtang Block created several thrust systems that raised a central uplift, and this uplift divides the block into southern and northern parts (Figures 2 and 3(b)). The Saibu Co-Zagya Thrust (SZT), which forms the southern base of the block, has an associated reverse Nima-Silin Thrust (NST) in the adjacent northern Lhasa Block. To the north the higher Doma-Qixiang Co (DQT) and Xiaocaka-Shuanghu Thrusts (XST) cut the southern part of the Qiangtang Block, and the central uplift is raised above the latter. The northern Qiangtang Block is broken by the Longwei Co (LCT) and Northern Central (NCT) Thrusts. The Dogai Coren Thrust (DCT) lies at the upper contact of the Qiangtang Block with the Hohxil Block to the north.

Permian limestone and dolomite, Triassic sandstone and shale, and Jurassic limestone are thrust relatively southward over Paleocene-Eocene continental red-beds along faults of DCT, NCT, XST, and DQT (Figure 4(d)). Permian marble and slate, Jurassic limestone and ophiolite, and Cretaceous andesite are transported southward over Paleogene red-beds along the SZT (Figure 3(b)). However, northward directed thrust or reverse movement occurs along the south dipping North Tangula Thrust (NTT) [12], LCT, and NST (Figure 2). Jurassic limestone and sandstone are thrust northward over Paleocene-Eocene red-beds along LCT, and Early Cretaceous limestone, mudstone, and sandstone are thrust northward over Late Cretaceous conglomerate and Paleocene-Eocene red-beds along NST in the adjacent northern Lhasa Block (Figure 3(b)). These south-dipping thrusts are considered to be adjustments within the relatively southward moving, north-dipping Qiangtang Block and not separate movements. The early Cenozoic thrusts also formed a variety of tectonic slices, outliers, and nappes of Permian-Jurassic rocks and structural windows of Paleogene red-beds within the Qiangtang Block (Figure 2).

The Northern Central (NCT), Xiaocaka-Shuanghu (XST), Doma-Qixiang Co (DQT), and Saibu Co-Zagya Thrusts (SZT) formed a mega thrust system during the Early Cenozoic. This large system merges with the Bangoin-Nujiang Thrust (BNT) at Bangoin-Nujiang Suture (BNS), which marks the boundary between the Lhasa and Qiangtang Blocks. Jurassic limestone is thrust southward over Paleocene-Eocene red-beds along BNT south of Ando; ophiolite slices are thrust southward along BNS over the northern Lhasa block (Figure 4(e)), and Cretaceous porphyry granite is thrust southeastward over Eocene-Oligocene red-beds, accompanied by tight folding along BNT west of Gar (Figure 4(f)). Southward displacement of this great Early Cenozoic thrust system is estimated to be ~150 km as measured by the displacement of tectonic slices of Triassic-Jurassic marine strata overriding late Cretaceous-Eocene red-beds in the southern Qiangtang terrain (Figure 3(b)). The average southward slip rate ranges from 5.6 mm/a to 7.4 mm/a in Late Eocene-Oligocene based on ages of volcanic rock overlying the thrust faults [19]. Northward displacement of the reverse LCT is 16 km according to the offset of Jurassic limestone thrust over Paleocene-Eocene red-beds (Figure 3(b)).

3.5. Hohxil Thrust System

The Hohxil Thrust System (HXT) consists of north-dipping thrust faults with some minor reverse faults that cut the Hohxil Block above the Dogai Coren Thrust (DCT), which marks the border with the Qiangtang block (Figures 2 and 3). Jurassic limestone is moved southward over late Cretaceous-Early Cenozoic red-beds of the Fenghuoshan Group along the DCT with a minimal southward displacement of 25 km north of Dogai Coren Lake. Triassic shale and sandstone are thrust southward over red-beds of Fenghuoshan Group along HXT in Hohxil Block, and Late Cretaceous-Eocene red-beds of Fenghuoshan Group are thrusted southward over Oligocene brownish conglomerate and sandstone along gently-dipping faults of HXT north of Tuotuohe (Figures 3(d) and 4(c)). Oligocene basins and granites formed along the major thrust faults (Figure 2), which are covered unconformably by Early Miocene lacustrine deposits of Wudaoliang Group [10], demonstrating that the major thrusting occurred in the Oligocene.

3.6. South Kunlun Thrust

Southward thrusting occurred during Late Oligocene-Early Miocene in the southern East Kunlun Mountains to form the South Kunlun Thrust (SKT). Permian strata and Triassic rocks are thrust southward over the Paleocene-Eocene red-beds of Fenghuoshan Group (Figure 4(a)) and Oligocene brownish red conglomerate and sandstone of the Yaxicuo Group (Figure 4(b)). This movement along the SKT formed tectonic slices, low-angle thrust faults, multiscaled outliers, and nappe structures on both sides of the Dongdatan valley. Farther north to the Middle Kunlun Fault (MKF) a series of more steeply dipping thrusts repeat lower Paleozoic strata [33] (Figure 3(c)). Many of these steep thrusts are reactivated Silurian faults [34].

The SKT mainly developed during the Oligocene-Early Miocene and major thrust faults of SKT are covered unconformably by Miocene lacustrine deposits. Possible earlier movement is suggested by fault slivers of schist with ^{39}Ar-^{40}Ar isotopic age of 26.5 Ma and one of foliated granite with a fission track age of apatite of 26.0 Ma that lie in thrusts of the SKT along the Dongdatan Valley [11]. The MKF continued to move with sinistral-slip thrusting during the Late Miocene-Pliocene [11]. The southward displacement of Lower Permian dolomite over early Cenozoic red-beds along SKT is estimated to be 30–35 km with a minimal slip rate of 2.3~2.6 mm/a [11]. Some minor reverse faults may occur within the northern Kunlun Mountains, southwest of the Qaidam Basin (Figure 2).

4. DEEP STRUCTURES IN MIDDLE-LOWER CRUST

The thrust systems mapped at the surface have been extended downward with some confidence with the aid of seismic surveys. Several deep seismic profiles have revealed structures in the middle-lower crustal depths of the Tibetan Plateau, and these deep structures are compatible with the faults observed on the surface. The International Deep Profiling of Tibet and the Himalaya Mountains known as INDEPTH has been conducted in four stages that now extend from the Himalaya Mountains to the Qaidam Basin. INDEPTH-I made seismic reflections of the Main Himalaya Thrust and Moho beneath Himalaya Mountains [17], INDEPTH-II discovered bright-spots and partial melting of the middle crust in southern Tibet [18, 44], and INDEPTH-III and INDEPTH-IV provided broad-band seismic data and receive function profiles of deep structures in northern Tibetan Plateau [22, 45]. Additional seismic reflections across Qiangtang Block have shown deep structures of the central Tibetan Plateau [20, 21].

The Main Himalaya Thrust (MHT) at the upper boundary of the subducted Indian Continental Plate is seen only in southern and central Himalaya and disappears in the northern Himalaya belt [17], although north-dipping faults do occur in the middle-lower crust of northern Himalaya and southern Lhasa Blocks in the INDEPTH-I and II data [17, 18, 44] (Figure 5). These deep, generally parallel faults that show the northward subduction of the Indian continental plate caused dextral shearing of the crust into tectonic slices whose relatively movement southward has stacked them to cause crustal thickening in the Himalaya terrain and southern Lhasa Block.

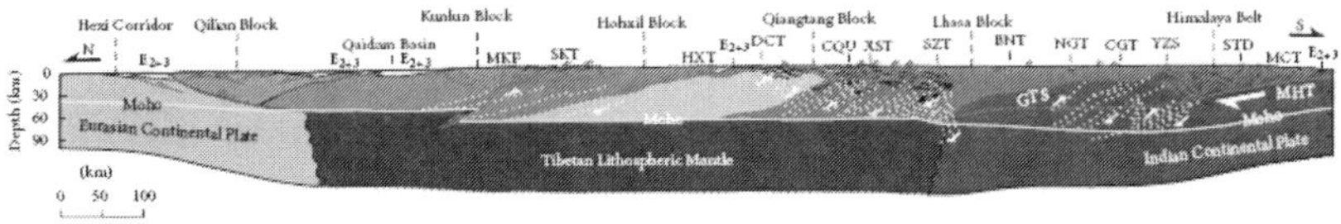

Figure 5: Generalized crustal profile across the Tibetan Plateau. Explanation: red line: major thrust system; dashed red line: upper boundary of Indian continental crust shearing thrusted away from subducted Indian Continental Plate; dotted white lines: deep shearing thrust in middle-lower crust; black fine lines: strata or foliation in middle-upper crust. Deep structures of Himalaya and southern Tibet are inferred from seismic reflection profiles of INDEPTH-I~II [17, 18]; deep structures of central Tibetan Plateau are inferred [19] from seismic reflection profiles across Qiangtang Block [20, 21], and the receiver function profile and wide angle seismic data of INDEPTH-IV [22, 23] in the northern Tibetan Plateau were used to interpret middle-lower crustal structures shown herein [24]. Deep structures marked by the black and white dotted lines are new interpretations of the seismic profiles. Letter symbols are the same as on Figures 2 and 3.

A similar deep shearing of the middle-lower crust also is found in the central and northern Tibetan Plateau (Figure 5). Most faults dip northward ~30°, although some faults curve and dip southward along more gently-dipping thrusts in central Tibet. The structural pattern of deep faults and associated folds indicate a thrusting and southward movement of the crust beneath the Qiangtang Block to cause thickening of middle-lower crust in central plateau. Curved faults exist locally in the lower crust, and some occur in the lithospheric mantle beneath Silin Co, where the Moho increases in depth to the south. Such structures near the Moho fit well with the hypothesis that Indian Continental Plate subducted northward to the

Bangoin-Nujiang Suture (BNS). However, the Indian crust was sheared away from subducted continental plate, thrust southward along north-dipping faults and became accreted to the Himalaya terrain in the middle-lower crust. Deep extensions of gently northward-dipping thrust faults also are present in the middle-lower crust in the northern Tibetan Plateau; some bound the Kunlun, Hohxil, and Qiangtang Blocks, and the Middle Kunlun fault offsets the Moho.

5. AGE OF CENOZOIC THRUSTS

Thrusts in the upper crust such as GTS in south Tibet and BNT in central Tibet (Figure 5) formed in the brittle-ductile transition zone at depths of ~15–20 km, and seismic bright spots suggesting partial melting occurs along GTS in the southern Lhasa Block (Figure 3(a)). Partial melting along GTS may have resulted in the emplacement of Early Miocene granites in central Lhasa block [46]. The isotopic ages of the granites and metallogeny of porphyry granites provide time constraints for movement along the GTS detachment. The foliated syntectonic Nyainqentanglha granite formed in 18.3–11.1 Ma [47], and copper-bearing porphyry granites in southern Gangdese Mountains formed in 14.5–17.7 Ma [43], indicating that thrusting was underway along the GTS in Early Miocene (18–11 Ma) beneath the central Lhasa Block. This is ~10 Ma later than the indicated age of the southern outer thrusts of GTS [13]. GTS apparently began earlier in south and developed northward.

The geochemistry of the Eocene-Oligocene igneous rocks [48] in the Qiangtang and Hohxil Blocks of the central and northern plateau is similar to Miocene adakitic porphyry granites in the Gangdese tectonic belt [49]. Because the Miocene porphyry granites in the Gangdese tectonic belt are geodynamically related to partial melting along GTS detachment [46], these Eocene-Oligocene rocks also are inferred to be the result of this melting thus, their isotopic ages provide time constraints for the deep thrusts. The U-Pb, ^{39}Ar-^{40}Ar, and K-Ar isotopic ages of trachyte, dacite, granite, and syenite formed in Eocene-Oligocene in Qiangtang and Hohxil Blocks range from 44.3–44.0 Ma in northern Qiangtang Block to 38.3–26.5 Ma in southern Qiangtang Block (Table 1). This suggests that the deep thrusts of the Qiangtang Block may have been initiated soon after the India-Eurasia continental collision in early Cenozoic and developed southward with associated partial melting.

Table 1: Isotopic ages of early Cenozoic volcanic rocks and Miocene thermal events in the Tibetan Plateau.

Magmatic rock and sampling location	Dating method/material	Isotopic ages	References
Dongyuehu andesite (DHT)	SHRIMP U-Pb/zircon	44.3 ± 1.8 Ma	Dong et al., 2008 [50]
Gongjue dacite (GJD)	K-Ar linear isochron	44.2 ± 0.9 Ma	Li et al., 2004 [51]
Yuejinla trachyandesite (YLT)	K-Ar/whole rock	43.0–32.6 Ma	Zhu et al., 2005 [52]
Purogangri granite (PGG)	K-Ar/biobite	41.1 ± 0.5 Ma	Zhu et al., 2005 [52]
Zurkanwula dacite (ZWD)	^{39}Ar-^{40}Ar/whole rock	41.1 ± 0.8 Ma	Lin et al., 2003 [53]
Dogecuoren andesite (DCA)	K-Ar/whole rock	41.6–32.6 Ma	Zhu et al., 2005 [52]
Saiduopu granite (SGG)	U-Pb/zircon	40.6 ± 3.1 Ma	Duan et al., 2005 [54]
Tangula granite at Geladandong (SGG)	U-Pb/zircon	40.0 ± 3.0 Ma	Li et al., 2006 [12]
Bilog Co trachite (BCT)	K-Ar/whole rock	38.3 ± 3.3 Ma	Lai et al., 2006 [48]
Nuangqian volcanic rocks (NQV)	K-Ar/whole rock	38.0–34.0 Ma	Lai et al., 2006 [48]
Yulong Porphyry Copper Deposit (YCP)	Re-Os/Molybdenite	35.6 ± 0.2 Ma	Li et al., 2004 [1, 51]
Yaoan porphyry syenite (YPS)	K-Ar/whole rock	34.0–47.0 Ma	Hou et al., 2004 [55]
Beiya porphyry syenite (BPS)	K-Ar/whole rock	45.0 ± 2.0 Ma	Hou et al., 2004 [55]
Machangqing Copper porphyrite (MCP)	Rb-Sr linear isochron	36.0 ± 1.2 Ma	Hou et al., 2004 [55]
Ejimaima andesite (EMA)	K-Ar/whole rock	34.8 ± 1.2 Ma	Zeng et al., 2002 [56]
North Deqen volcanic rocks (NDV)	K-Ar/whole rock	33.0 ± 1.5 Ma	Li et al., 2004 [1, 51]
Andar Co trachite (ACT)	K-Ar/whole rock	32.6 ± 0.6 Ma	Wang et al., 2005 [57]
Zougoyouchaco trachyandesite (ZCT)	^{39}Ar-^{40}Ar/whole rock	30.5 ± 0.6 Ma	Li et al., 2006 [58]
Yulinshan leucite phonolite (YSP)	^{39}Ar-^{40}Ar/leucite	27.8 ± 0.8 Ma	Cai et al., 2002 [59]
Fenghuoshan granite (FSG)	SHRIMP U-Pb/zircon	27.6 ± 0.5 Ma	Wu et al., 2008 [10]
Huochetou Shan basanite (HCB)	^{39}Ar-^{40}Ar/biobite	26.7 ± 0.7 Ma	Liu et al., 2004b [60]
Manshan porphyry syenite (MPS)	SHRIMP U-Pb/zircon	26.5 ± 0.8 Ma	Wei et al., 2007 [61]
Yadoi leucogranite (YLG)	SHRIMP U-Pb/zircon	35.3 ± 1.1 Ma	Gao et al., 2009 [62]
Kude granite (KG)	U-Pb/zircon	27.5 ± 0.5 Ma	Zhang et al., 2004 [63]

Explanation: sampling locations of Magmatic rocks are Marked in Figure 2.

Major faults moved over Late Cretaceous-Eocene red-beds and Oligocene brownish red-beds in the northern Lhasa, Qiangtang, Hohxil, and Kunlun Blocks, and most thrust faults were covered unconformably by only weakly deformed Miocene lacustrine deposits in the central and northern Tibetan Plateau [10]. This demonstrates that the intense thrusting occurred in Late Eocene-Early Miocene in the Tibetan Plateau. However, the Himalaya Thrust developed progressively southward from the Oligocene-Early Miocene MCT through Late Miocene MBT to Quaternary MFT along MHT. Far to the north the intensive thrusting in the Qilian Mountains was mainly initiated since Miocene [5, 39].

6. CONCLUSIONS AND DISCUSSION

A comprehensive depiction of the major thrust systems forming the early Cenozoic structural framework across the Himalayan Mountains, and the Tibetan Plateau now is revealed by the regional mapping and deep seismic surveys. The seismic data show many thrusts to mid-lower crustal depth. Some thrust systems bound the Himalaya, Lhasa, Qiangtang, Hohxil, and Kunlun Blocks, and a few mark suture zones.

Early Cenozoic compression due to the northward subduction of the Indian continental plate squeezed and telescoped the Tibetan region to create a series of north-dipping thrust systems in the upper and middle-lower crust of the present narrower Himalayan Mountains and Tibetan Plateau. Continued movement stacked up the overlapping thrust sheets to form the thickened crust of the region. The majority of the thrust faults in the upper crust dip northward, leading to relative southward displacement, although some reverse, south-dipping thrusts exist in the Himalaya (RZT), northern Lhasa (NST), northern Qiangtang (NTS), and northern Kunlun Blocks, but southward movement is dominate in the middle-lower crust of the Tibetan Plateau. The south-dipping thrusts appear to be due to local adjustments within thrust sheets when the back of a sheet moves faster than the toe. The displacement by thrusting first caused shortening and thickening of middle-lower crust and then to shortening and southward thrusting in the upper crust to form the multiple thrust systems.

The thrusting took place between the Eocene and Early Miocene. The MCT chiefly formed in Late Eocene-Early Miocene; GTS formed in Late Oligocene-Early Miocene, 27–18.3 Ma, in southern Gangdese and extended into the deep crust in 18.3–11.1 Ma in the central Lhasa block; SKT formed in Oligocene-Early Miocene; HXT formed in Late Eocene-Oligocene. Intensive thrusting was initiated in the northern Qiangtang block soon after the India-Eurasia continental collision in early Cenozoic and then developed southward to form the Southern Qiangtang Thrust in Late Eocene-Early

Oligocene. Displacement and rate of movement during the early Cenozoic are difficult to determine due to erosion of the upper thrust plates and the few available precise dates. Normally only minimal offsets and rates can be determined, such as the more accurate ones for the SKT, HXT, QTS, CGT, and GTS. The minimal southward displacements of SKT, HXT, South Qiangtang Thrust, and CGT are estimated at 30 km, 25 km, 150 km, and 50 km, respectively, with average slip rates of SKT, South Qiangtang Thrust, and GTS being 2.3~2.6 mm/a, 5.6–7.4 mm/a, and mm/a, respectively, during the Oligocene-Early Miocene.

This timing, which is derived from the field data, is in general agreement with inferences from paleomagnetic studies. Such studies provide estimates of the subduction velocity of the Indian continental plate that relates to the thrusting and crustal thickening process in the Himalayan-Tibetan region. The suggested northward plate velocity changed from ~90 mm/a in the Paleocene-Early Eocene to ~55 mm/a in Late Eocene-Oligocene and has remained ~50 mm/a since Early Miocene with the rapid decrease of northward velocity of the Indian plate occurring ~45–43 Ma [64]. This corresponds to a change from the Tethys oceanic plate subduction to India-Eurasia continental collision, which is marked by the initial northward subduction of Indian Continental Plate in ~54–40 Ma in eastern Himalaya [65], ~50–42 Ma in central Himalaya [6, 40, 41, 66, 67], and ~57 Ma in western Himalaya [68]. This timing is consistent with earlier paleomagnetic data which suggested that the India-Eurasia continent collision began ~50 Ma [69, 70]. It is further inferred that the northward subduction of the Indian continental plate is ~1,183 km with northward slip rate ~55 mm/a during Middle Eocene-Oligocene, ~45–23.5 Ma [64]. This would increase the thickness of the Tibetan crust to ~60 km, which is about the present Moho depth in northern Tibet, before 23.5 Ma, if all the crust sheared away from the subducted Indian continental plate were accreted to the crust of the Himalaya terrain and Tibetan Blocks along deep thrusts (Figure 5). Or the crust might double in thickness in the Tibetan Plateau several million years after 23.5 Ma if part of the Indian Continental Crust were subducted to the mantle.

Geologic field evidence demonstrates that the thickened Tibet crust was buoyed upwards in Early-Middle Miocene. Southern Tibet was uplifted to between m– m before ~15 Ma [71] and the Lunpola Basin rose to its present elevation before Early Miocene [72]. Large lakes formed in Early Miocene [10], and their lacustrine deposits, which only show weak deformation, are unconformably over major thrust systems in northern Lhasa, Qiangtang, Hohxil, and Kunlun Blocks. The intense thrusting is indicated to have stopped before Early Miocene in the central and northern Tibetan Plateau. This agrees with the crustal extension beginning in Middle and Late Miocene in the Qiangtang and Lhasa Blocks, respectively [73, 74]. However, intensive thrusting and tectonic contraction appear to have continued active in the southern Himalaya [5], Namche Barwa Syntaxis [65], West Kunlun [38], Qilian [39], Longmenshan Mountains [7] and the Jiuquan [75], Guide [76, 77], and Linxa Basins [78] in late Cenozoic.

ACKNOWLEDGMENTS

This research was supported by Ministry of Land and Resources of China under Grant Sinoprobe-02, China Geological Survey under Grant nos. 1212011221111 and 1212011120185, and Ministry of Science and Technology of China under Grant 2006DFB21330. The authors thank Professor Keiko Hattori and Dr. Josep Pares for giving their detail suggestions to improve the paper. They are also grateful to Professor Gao Rui and Dr. Lu Zhanwu who provided their seismic reflection data for interpretation of deep structures in the central Tibetan Plateau.

REFERENCES

1. G. Li, X. Wan, W. Liu, D. Liang, and H. Yun, "Discovery of Palaeogene marine strata along Yaluzangbu suture and its tectonic significance," Science in China D, vol. 34, no. 3, pp. 228–240, 2004.
2. X. Wan, J. Li, S. Zhang, and L. Wu, "The late cretaceous-paleocene planktonic foraminifera from Zanda, western Tibet and their chronolostratigraphic implications," ACTA Micropalaeontologica Sinica, vol. 22, no. 1, pp. 10–18, 2005.
3. B. Sun, Z. Liu, and Z. Wang, "New knowledge on geology of Karshi depression in southwest Tarim," Xinjiang Geology, vol. 21, no. 1, pp. 78–84, 2003.
4. K. J. Matthew and P. D. Christopher, "Petrologic case for Eocene slab breakoff during the Indo-Asian collision," Geology, vol. 30, no. 7, pp. 591–594, 2002.
5. A. Yin and T. M. Harrison, "Geologic evolution of the Himalayan-Tibetan orogen," Annual Review of Earth and Planetary Sciences, vol. 28, pp. 211–280, 2000.
6. J. F. Dewey, R. M. Shackleton, C. Chengfa, and Y. Sun, "The tectonic evolution of the Tibetan plateau," Philosophical Transactions of the Royal Society A, vol. 327, no. 1594, pp. 379–413, 1988. View at Publisher•
7. M. K. Clark and L. H. Royden, "Topographic ooze: building the eastern margin of Tibet by lower crustal flow," Geology, vol. 28, no. 8, pp. 703–706, 2000.
8. P. Tapponnier, Z. Xu, F. Roger et al., "Oblique stepwise rise and growth of the tibet plateau," Science, vol. 294, no. 5547, pp. 1671–1677, 2001.
9. C. Beaumont, R. A. Jamieson, M. H. Nguyen, and S. Medvedev, "Crustal channel flows: 1. Numerical models with applications to the tectonics of the Himalayan-Tibetan orogen," Journal of Geophysical Research B, vol. 109, no. 6, Article ID B06406, 29 pages, 2004. • View at Scopus
10. Z. Wu, P. J. Barosh, Zh. Wu, D. Hu, X. Zhao, and P. Ye, "Vast early Miocene lakes of the central Tibetan plateau," Bulletin of the Geological Society of America, vol. 120, no. 9-10, pp. 1326–1337, 2008.
11. Z. Wu, P. Ye, P. J. Barosh, D. Hu, W. Zhao, and Z. Wu, "Late oligocene-early miocene thrusting in southern East Kunlun Mountains, northern Tibetan plateau," Journal of Earth Science, vol. 20, no. 2, pp. 381–390, 2009.

12. Y. Li, C. Wang, H. Yi, Z. Liu, and Y. Li, "Cenozoic thrust system and uplifting of the Tanggula Mountain, northern Tibet," Acta Geologica Sinica, vol. 80, no. 8, pp. 1118–1130, 2006.
13. A. Yin, T. M. Harrison, F. J. Reyerson, W. Chen, W. S. F. Kidd, and P. Copeland, "Tertiary structural evolution of the Gangdese thrust system in southern Tibet," Journal of Geophysical Research, vol. 99, pp. 18175–18201, 1994.
14. Q. Xavier, M. Grove, O. M. Lovera, M. Harrison, and A. Yin, "Thermal evolution and slip history of the Renbu Zedong Thrust, southeastern Tibet," Journal of Geophysical Research B, vol. 102, no. 2, pp. 2659–2679, 1997.
15. P. G. Decelles, D. M. Robinson, J. Quade et al., "Stratigraphy, structure, and tectonic evolution of the Himalayan fold-thrust belt in Western Nepal," Tectonics, vol. 20, no. 4, pp. 487–509, 2001.
16. D. M. Robinson, P. G. Decelles, C. N. Garzione, O. N. Pearson, T. M. Harrison, and E. J. Catlos, "Kinematic model for the Main Central Thrust in Nepal," Geology, vol. 31, no. 4, pp. 356–362, 2003.
17. K. Zhang, G. Wang, J. Ji et al., "Paleogene-Neogene stratigraphic realm and sedimentary sequence of the Qinghai-Tibet Plateau and their response to uplift of the Plateau," Science China Earth Sciences, vol. 53, no. 9, pp. 1271–1294.
18. L. D. Brown, W. Zhao, K. D. Nelson et al., "Bright spots, structure, and magmatism in southern Tibet from INDEPTH seismic reflection profiling," Science, vol. 274, no. 5293, pp. 1688–1690, 1996.
19. Z. Wu, P. Ye, D. Hu, and L. Lu, "Paleogene thrust system in Southern Qiangtang Basin, Central Tibetan Plateau," Geological Bulletin of China, vol. 30, no. 7, pp. 1009–1016, 2011.
20. J. Wang, J. Ding, C. Wang, and F. Tan, Survey and Evaluation on Target Areas for Oil-Gas Exploration in the Tibetan Plateau, Geological Publishing House, Beijing, China, 2009.
21. R. Gao, Z. Lu, X. Xiong et al., "SINOPROBE deep seismic reflection profiling across the Bangong-Nujiang suture, central Tibet," in Proceedings for the 25th Himalaya-Karakoram-Tibet Workshop, M. L. Leech, Ed., 2010.
22. M. S. Karplus, W. Zhao, S. L. Klemperer et al., "Injection of Tibetan crust beneath the south Qaidam Basin: evidence from INDEPTH IV wide-angle seismic data," Journal of Geophysical Research B, vol. 116, no. 7, Article ID B07301, 2011.
23. D. Shi, W. Zhao, M. Feng et al., "Primary results of receive function profile of INDEPTH-IV across northern Tibetan Plateau," in Proceedings of the Workshop on International Deep Profiling of Tibet and Himalaya, 2011.
24. P. J. Barosh, Z. Wu, D. Hu, P. Ye, W. Zhao, and Zh. Wu, "Crustal structure of the northern Tibetan Plateau along the INDEPTH-IV seismic traverse," American Geophysical Union, Fall Meeting, Abstracts and Proceedings, 2012.
25. W. S. F. Kidd, Y. Pan, C. Chang et al., "Geological mapping of the 1985 Chinese-British Tibetan (Xizang-Qinghai) Plateau geotraverse route," Philosophical Transactions of the Royal Society A, vol. 327, pp. 287–305, 1988.
26. Tibetan Bureau of Geology and Mineral Resources, Regional Geology of Tibetan Autonomous Region, Geological Publishing House, Beijing, China.
27. Qinghai Bureau of Geology and Mineral Resources, Regional Geology of Qinghai Province, Geological Publishing House, Beijing, China, 1991.
28. Yunan Bureau of Geology and Mineral Resources, Regional Geology of Yunan Province, Geological Publishing House, Beijing, China, 1990.
29. Sichuan Bureau of Geology and Mineral Resources, Regional Geology of Sichuan Province, Geological Publishing House, Beijing, China, 1991.
30. G. Pan, J. Ding, D. Yao, and L. Wang, Geological map of Qinghai-Xizang (Tibet) Plateau and adjacent areas, Chengdu Cartographic Publishing House, Chengdu, China, 1–140, 2004.
31. G. Pan, L. Wang, R. Li et al., "Tectonic evolution of the Tibetan Plateau," Journal of Asian Earth Sciences, vol. 53, pp. 3–14, 2012.
32. Z. Wu, D. Hu, P. Ye, X. Zhao, and Q. Liu, "Thrusting of the North Lhasa Block in the Tibetan Plateau," ACTA Geologica Sinica, vol. 78, no. 1, pp. 246–259, 2004.
33. P. J. Barosh, D. Hu, Z. Wu, and Y. Zhang, "Map at scale 1:25, 000 and geology of the Shuinichang (Cement Factory) quadrangle, northern Kunlun Mountains," Published by Institute of Geomechanics, Chinese Academy of Geological Sciences, Beijing, China, 2009.

34. P. J. Barosh, Z. Wu, D. Hu et al., "Deep, late? Silurian and late oligocene-early miocene thrusts crossing the INDEPTH-IV survey traverse in the Middle Kunlun Mountains, Northern Tibetan Plateau," in Proceedings of the International Symposium on Deep Exploration into the Lithosphere, Sinoprobe, pp. 195–196, Chinese Academy of Geological Sciences, Beijing, China, 2011.
35. K. Zhang, G. Wang, J. Ji et al., "Paleogene-Neogene stratigraphic realm and sedimentary sequence of the Qinghai-Tibet Plateau and their response to uplift of the plateau," Science China Earth Sciences, vol. 53, no. 9, pp. 1271–1294, 2010.
36. C. Wang, X. Zhao, Z. Liu et al., "Constraints on the early uplift history of the Tibetan Plateau," Proceedings of the National Academy of Sciences of the United States of America, vol. 105, no. 13, pp. 4987–4992, 2008.
37. Z. Liu, X. Zhao, C. Wang, S. Liu, and H. Yi, "Magnetostratigraphy of tertiary sediments from the Hoh Xil Basin: implications for the Cenozoic tectonic history of the Tibetan Plateau," Geophysical Journal International, vol. 154, no. 2, pp. 233–252, 2003.
38. S. Liu, B. Qiu, H. Yin, J. L. Li, S. M. Zhai, and W. H. Li, "Structural characteristics of Wuboer thrust belts in the foreland of West Kunlun Mountain," Acta Petrolei Sinica, vol. 26, no. 6, pp. 16–19, 2005.
39. A. Yin, Y. Dang, L. Wang et al., "Cenozoic tectonic evolution of Qaidam basin and its surrounding regions (Part 1): The southern Qilian Shan-Nan Shan thrust belt and northern Qaidam basin," Bulletin of the Geological Society of America, vol. 120, no. 7-8, pp. 813–846, 2008.
40. K. V. Hodges, R. R. Parrish, and M. P. Searle, "Tectonic evolution of the central Annapurna Range, Nepalese Himalayas," Tectonics, vol. 15, no. 6, pp. 1264–1291, 1996.
41. L. Zeng, L. Gao, K. Xie, and J. Zeng, "Mid-Eocene high Sr/Y granites in the Northern Himalayan Gneiss Domes: melting thickened lower continental crust," Earth and Planetary Science Letters, vol. 303, no. 3-4, pp. 251–266, 2011.
42. P. Kapp, P. G. DeCelles, A. L. Leier et al., "The Gangdese retroarc thrust belt revealed," GSA Today, vol. 17, no. 7, pp. 4–9, 2007.
43. Z. Rui, Z. Hou, G. Li, L. Zhang, L. Wang, and S. Tang, "Subduction, collision, deep fault, adakite and porphyry copper deposits," Geology and Exploration, vol. 42, no. 1, pp. 1–6, 2006.
44. K. D. Nelson, W. Zhao, L. D. Brown et al., "Partially molten middle crust beneath southern Tibet: synthesis of project INDEPTH results," Science, vol. 274, no. 5293, pp. 1684–1688, 1996.
45. W. Zhao, K. D. Nelson, J. Che et al., "Deep seismic reflection evidence for continental underthrusting beneath southern Tibet," Nature, vol. 366, no. 6455, pp. 557–559, 1993.
46. Z. Wu, P. J. Barosh, X. Zhao, Zh. Wu, D. Hu, and Q. Liu, "Miocene tectonic evolution from dextral-slip thrusting to extension in the Nyainqentanglha region of the Tibetan Plateau," ACTA Geologica Sinica, vol. 81, no. 3, pp. 365–384, 2007.
47. Q. Liu, Z. Wu, D. Hu et al., "SHRIMP U-Pb zircon dating on Nyainqentanglha granite in central Lhasa block," Chinese Science Bulletin, vol. 49, no. 1, pp. 76–82, 2004.
48. S. Lai, J. Qing, Y. Li, and X. Zheng, "Cenozoic volcanic rocks in the Bêlog Co area, Qiangtang, northern Tibet, China: petrochemical evidence for partial melting of the mantle-crust transition zone," Geological Bulletin of China, vol. 25, no. 1-2, pp. 64–69, 2006.
49. Z. Hou, X. Meng, X. Qu, and Y. Gao, "Copper ore potential of adatic intrusives in the Gangdise copper belt: constraints from rock phase and deep melting process," Mineral Deposits, vol. 24, no. 2, pp. 108–121, 2005.
50. Y. Dong, Q. Wang, J. Xu, and F. Zi, "Dongyue Lake adakitic volcanic rocks with high Mg in north Qiangtang block: petrogenesis and its tectonic implication," Acta Petrologica Sinica, vol. 24, no. 2, pp. 291–302, 2008.
51. Z. Li, Z. Chen, X. Li, C. Gizbert, and B. C. Burchfiel, "K-Ar ages of Cenozoic volcanic rocks from Gongjue Basin in eastern Tibet," Journal of China University of Geosciences, vol. 29, no. 3, pp. 278–282, 2004.
52. W. Zhao, P. Kumar, J. Mechie et al., "Tibetan plate overriding the Asian plate in central and northern Tibet," Nature Geoscience, vol. 4, pp. 870–873, 2011.

53. J. Lin, H. Yi, B. Zhao, B. H. Li, Z. Q. Shi, and J. J. Huang, "39Ar-40Ar isotopic dating and its implication of Cenozoic volcanic rocks from Zurkenwula Mountain area, northern Tibetan," Mineral and Petrology, vol. 23, no. 3, pp. 31–34, 2003.
54. Z. Duan, Y. Li, Y. Zhang, Y. Li, and M. Wang, "Zircon U-Pb age, continent dynamics significance and geochemical characteristics of the Mesozoic and Cenozoic granites from the Tanggula range in the Qinghai-Tibet Plateau," Acta Geologica Sinica, vol. 79, no. 1, pp. 88–97, 2005.
55. Z. Hou, D. Zhong, and W. Deng, "A tectonic model for porphyry copper-molybdenum-gold metallogenic belts on eastern margin of the Qianghai-Tibet Plateau," Geology in China, vol. 31, no. 1, pp. 1–14, 2004.
56. Q. Zeng, G. Mao, G. Cheng et al., "Geological survey report and geological map at scale 1:250, 000 of Ripeigan Co Quadrangle," Published by Tibet Geological Survey, Lhasa, China, 2002.
57. Q. Wang, W. Yang, Z. Zhang, Y. Yang, J. Wu, and A. Dong, "Geological characteristics of Neogene volcanic rocks in the Heishi North Lake area, northwestern Tibet and their implications for Neogene tectonic evolution," Geological Bulletin of China, vol. 24, no. 1, pp. 80–86, 2005.
58. C. Li, X. Huang, S. Mou, and X. Chi, "Age dating of the Zougouyouchacuo volcanic rocks and age determination of the Kangtog Formation in southern Qiangtang, northern Tibet, China," Geological Bulletin of China, vol. 25, no. 1-2, pp. 226–228, 2006.
59. C. Li, Z. Zhu, and X. Chi, "Isotope chronology of volcanic rocks in the Yulinshan Formation in the Gaize area, northern Tibet," Geological Bulletin of China, vol. 21, no. 11, pp. 732–734, 2002.
60. H. Liu, B. Xia, W. Deng, and Y. Zhang, "Study of 39Ar-40Ar and K-Ar dating on the high-K volcanic rock from Bamaoqiongzong to Qiangbaqian in the northern Tibet," Mineral and Petrology, vol. 24, no. 1, pp. 71–75, 2004.
61. Q. Wei, D. Li, G. Wang, and J. Zheng, "Zircon SHRIMP U-Pb dating and geochemical characteristics of Chabaoma Formation volcanic rocks in northern Tibetan plateau and its petrogenesis," Acta Petrologica Sinica, vol. 23, no. 11, pp. 2727–2736, 2007.
62. L. Gao, L. Zeng, J. Liu, and K. Xie, "Early Oligocene Na-rich peraluminous leucogranites in Yadoi gneiss dome, southern Tibet: formation mechanism and tectonic implications," ACTA Petrologica Sinica, vol. 25, no. 9, pp. 2289–2302, 2009.
63. H. F. Zhang, N. Harris, R. Parrish, L. Zhang, and Z. Zhao, "U-Pb ages of Kude and Sajia leucogranites in Sajia dome from North Himalaya and their geological implications," Chinese Science Bulletin, vol. 49, no. 19, pp. 2087–2092, 2004.
64. T. Y. Lee and L. A. Lawver, "Cenozoic plate reconstruction of Southeast Asia," Tectonophysics, vol. 251, no. 1–4, pp. 85–138, 1995.
65. Z. Xu, S. Ji, Z. Cai, L. Zeng, Q. Geng, and H. Cao, "Kinematics and dynamics of the Namche Barwa Syntaxis, eastern Himalaya: constraints from deformation, fabrics and geochronology," Gondwana Research, vol. 21, no. 1, pp. 19–36, 2012.
66. D. B. Rowley, "Age of initiation of collision between India and Asia: a review of stratigraphic data," Earth and Planetary Science Letters, vol. 145, no. 1–4, pp. 1–13, 1996.
67. T. Zhu, Y. Yu, C. Jin, B. Zhu, Q. Zhang, and M. Zhou, "Geological survey report and geological map at scale 1:250, 000 of Dogai Corin Quadrangle," Published by Chengdu Institute of Geology and Mineral Resources, China Geological Survey, 2005.
68. S. Guillot, G. Mahéo, J. de Sigoyer, K. H. Hattori, and A. Pêcher, "Tethyan and Indian subduction viewed from the Himalayan high- to ultrahigh-pressure metamorphic rocks," Tectonophysics, vol. 451, no. 1–4, pp. 225–241, 2008.
69. J. Besse, V. Courtillot, J. P. Pozzi, M. Westphal, and Y. X. Zhou, "Palaeomagnetic estimates of crustal shortening in the Himalayan thrusts and Zangbo suture," Nature, vol. 311, no. 5987, pp. 621–626, 1984.
70. C. T. Klootwijk, P. J. Conaghan, and C. M. Powell, "The Himalayan Arc: large-scale continental subduction, oroclinal bending and back-arc spreading," Earth and Planetary Science Letters, vol. 75, no. 2-3, pp. 167–183, 1985.
71. R. A. Spicer, N. B. W. Harris, M. Widdowson et al., "Constant elevation of southern Tibet over the past 15 million years," Nature, vol. 421, no. 6923, pp. 622–624, 2003.

72. D. B. Rowley and B. S. Currie, "Palaeo-altimetry of the late Eocene to Miocene Lunpola basin, central Tibet," Nature, vol. 439, no. 7077, pp. 677–681, 2006.
73. P. M. Blisniuk, B. R. Hacker, J. Glodny et al., "Normal faulting in central Tibet since at least 13.5 Myr ago," Nature, vol. 412, no. 6847, pp. 628–632, 2001.
74. T. M. Harrison, P. Copeland, W. S. F. Kidd, and O. M. Lovera, "Activation of the Nyainqentanghla shear zone: implications for uplift of the southern Tibetan Plateau," Tectonics, vol. 14, no. 3, pp. 658–676, 1995.
75. X. Fang, Z. Zhao, J. Li et al., "Magnetostratigraphy of the late Cenozoic Laojunmiao anticline in the northern Qilian Mountains and its implications for the northern Tibetan Plateau uplift," Science in China D, vol. 48, no. 7, pp. 1040–1051, 2005.
76. J. M. Parés, R. Van der Voo, W. R. Downs, M. Yan, and X. Fang, "Northeastward growth and uplift of the Tibetan Plateau: magnetostratigraphic insights from the Guide Basin," Journal of Geophysical Research-Solid Earth, vol. 108, no. 2017, 11 pages, 2003.
77. X. Fang, M. Yan, R. Van der Voo et al., "Late Cenozoic deformation and uplift of the NE Tibetan Plateau: evidence from high-resolution magnetostratigraphy of the Guide Basin, Qinghai Province, China," Geological Society of America Bulletin, vol. 117, no. 9-10, pp. 1208–1225, 2005.
78. X. Fang, C. Garzione, R. Van der Voo, J. Li, and M. Fan, "Flexural subsidence by 29 Ma on the NE edge of Tibet from the magnetostratigraphy of Linxia Basin, China," Earth and Planetary Science Letters, vol. 210, no. 3-4, pp. 545–560, 2003.

CHAPTER 12

Erosion Vulnerability Assessment of Sperchios River Basin, in East Central Greece—A GIS Based Analysis

Stathopoulos Nikolaos[1*], Lykoudi Evdoxia[1], Vasileiou Eleni[1], Rozos Dimitrios[1], Dimitrakopoulos Dimitrios[2]

[1]*Department of Geological Sciences, School of Mining Engineering and Metallurgy, National Technical University of Athens, Athens, Greece*

[2]*Department of Hydrogeological Studies, Direction of Studies and Development of Mines, Public Power*

ABSTRACT

One of the most important natural processes, with significant environmental impacts, is soil erosion. The aim of this paper is to evaluate and assess the erosion vulnerability of the geological formations constituting Sperchios River basin, in Eastern Central Greece, as well as to distinguish and locate the areas of such risk and estimate it. The developed methodology in this paper is based on Multi-Criteria Evaluation (MCE), with the implementation of Weighted Cartographic Overlay (WCO) technique in Geographic Information Systems (GIS) environment. By applying this method, an attempt was made to combine a set of factors, such as geological-hydrogeological characteristics, morphological slopes and hydrographic texture (causative factors of the phenomenon) along with rainfall (triggering factor) and land cover/use (competitive factor). The weighted combination of these factors defines the vulnerability of the formations in the area of interest. The main results of the methodology are two thematic maps of erosion vulnerability of Sperchios basin's formations, one that does not include land use/cover factor and one with current existing conditions of the basin (co-evaluating land use/cover). In the first case, 36% of the basin surface undergoes high erosion vulnerability, while 53% present medium vulnerability. These percentages reduced significantly in the second case (3% is the high erosion vulnerability category and 38% the medium one), fact that proves the importance of vegetation in erosion restriction and protection of the geological formations consisting the Sperchios River basin.

Keywords: Erosion Vulnerability, Sperchios River Basin, Geographical Information Systems, Multi-Criteria Evaluation

1. INTRODUCTION

Soil erosion is a natural process, with significant environmental impacts. This natural process constitutes one of the main environmental problems affecting Sperchios River basin in Eastern Central Greece, causing soil degradation and augmentation of sediment yield in the river. The high rate of sediment transport in the river system causes unique changes of the deltaic part and shoreline throughout Greece. Vulnerability assessment and mapping of erosion are of high importance for managing a hydrological basin and constitute useful tools for environmental planning.

Many methods and models are used to assess and analyze this process, both quantitatively and qualitatively. The RUSLE (Revised Universal Soil Loss Equation) is considered as a simple model for predicting soil erosion [1] . It is a model that is an evolution of the Universal Soil Loss Equation (USLE) model, as it considers upstream areas contributing to downstream surface runoff. Thus it is a model with better predictability than the USLE. In this regard, the topographic parameter gains a more grounded scientific connotation, physically improving the previous approach [2] .

In general, the previous methods are used in arable lands with low terrain slopes and mostly in small basins.

From a more qualitative aspect, analysis of spatial information is becoming an emerging approach which is capable of acquiring, managing and analyzing complex problems of river basins and lake watersheds. In recent years, GIS have proven to be a good alternative to serve as a better decision support tool in planning, management and sustainable development of soil and water resources. Consequently, the integration of multi-criteria evaluation (MCE) within a GIS context could help the relevant authorities to improve decision making pro- cesses.

The main purpose of the MCE technique is to investigate a number of alternatives in the light of multiple criteria and conflicting objectives [3] . In the last decade, MCE has received renewed attention in the context of GIS-based decision making [4] [5] [6] . Different studies have been conducted using MCE technique in the area of the natural resources management [7] - [12] .

In this paper, an attempt is made on spatial estimation of soil erosion in large mountainous catchments of Mediterranean type like Sperchios River, using the MCE method. For this purpose, an evaluation based on GIS was adopted. The applied methodology comprises qualitative analysis. The application of MCE in Sperchios River catchment, aiming at spatial vulnerability assessment of the geological formations, offers in addition valuable assessment of the expected spatial distribution of soil erosion.

Thus, in order to evaluate and assess the erosion vulnerability of the geological formations in a region like Sperchios River basin as well as to distinguish the areas of high erosion vulnerability is of high importance. The main factors that control erosion are the lithological composition of the geological formations, along with morphology and hydrographic network texture of the study site. The combination of these factors defines the vulnerability of the formations.

This paper is based on the methodology presented by [13] and later by [14], with specific modifications that were considered to be appropriate in this implementation. In particular, the rainfall factor was co-evaluated in the erosion model and weights were assigned in each factor during the phase of synthesis, according to bibliography and experts' opinions. The development of this method offers a more effective and integrated approach in erosion vulnerability research. Conclusively, the methodology developed in this paper is based on multi-criteria evaluation (MCE), with the implementation of Weighted Cartographic Overlay (WCO) technique in Geographic Information System environment. This method combines a set of factors, such as geological characteristics, morphological slopes, hydrographic network, along with rainfall and land cover/use which are the causative, triggering and competitive factors respectively (Figure 1). The weighted combination of these factors defines the vulnerability of the formations in the study area and is indicated generally as a primary estimation method for their vulnerability, as it can easily be adjusted based on the updated data of every new simulation.

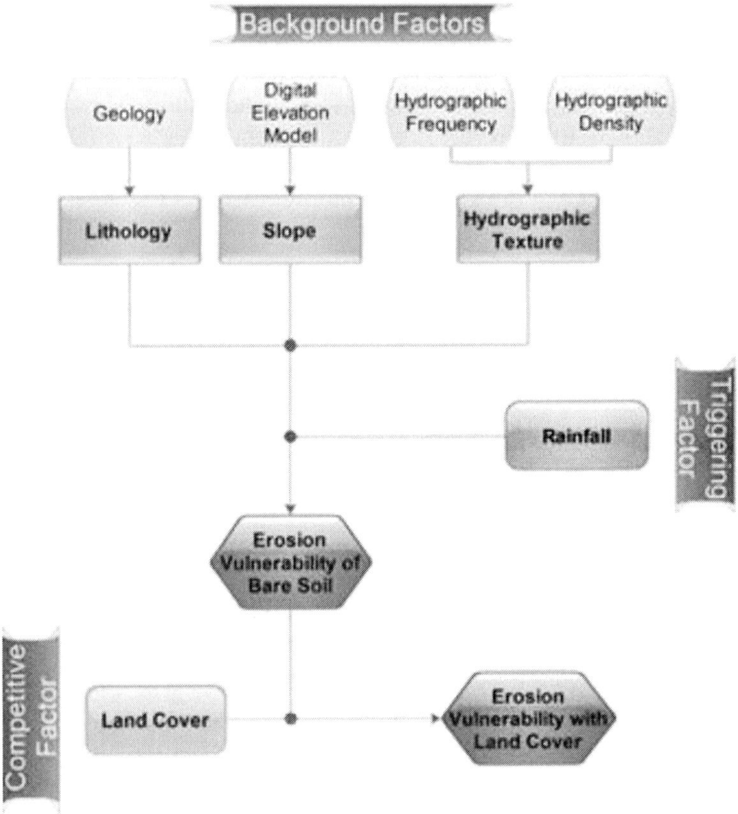

Figure 1. Flowchart of the methodology.

The results derived from this attempt provide important and very useful information for the behavior of geological formations under the current conditions of Sperchios basin (with the existing vegetation setting), as well as for the behavior of geological formations under extreme events and unfavorable circumstances such as fire, deforestation, and other natural disasters. In order to accomplish this purpose, two thematic maps were produced, one combining all the factors of the analysis along with land cover/use depicting thus the current condition of the basin and one excluding the land cover/use factor i.e. simulating the worst case scenario for the basin. Therefore, MCE seems to be applicable to GIS-based spatial delineation of erosion vulnerability, which helps to assess the most erosion prone areas in Sperchios River Basin, especially on mountainous and semi-mountainous zones.

2. MATERIAL AND METHODS

2.1. Geomorphological Conditions of the Study Area

Sperchios river basin (Figure 2) extends in the northern part of Eastern Central Greece Water Compartment and is surrounded by Orthris Mountain (1727 m) from North, by Vardousia (2437 m), Oiti (2152 m) and Kalidromo (1372 m) mountains from South, and by Timfristos Mountain (2316 m) from the west. The runoff of Sperchios River ends up in Maliakos Gulf, the coastline of which is the eastern limit of Sperchios basin. This basin occupies an area of 2031 km^2 with an average altitude of 605 m. Sperchios riverbed is recharged by many streams of permanent and periodic flow. The high gradients, which are present along with almost the 2/3 of the total length of the river course, form a rather mountainous—streamy topography, with crucial flooding peaks and very intense sediment yield. On the contrary, within the last downstream part of its course, the river is transformed gradually into a lowland relief, where cases of severe flooding have been recorded. The deltaic alluvial part of the valley covers about 200 km^2 and is continuously forming with a unique rate throughout Greece. This rate, according to [15] , was estimated approximately in 130 acres annually, with augmentation trends in the last 150 - 200 years.

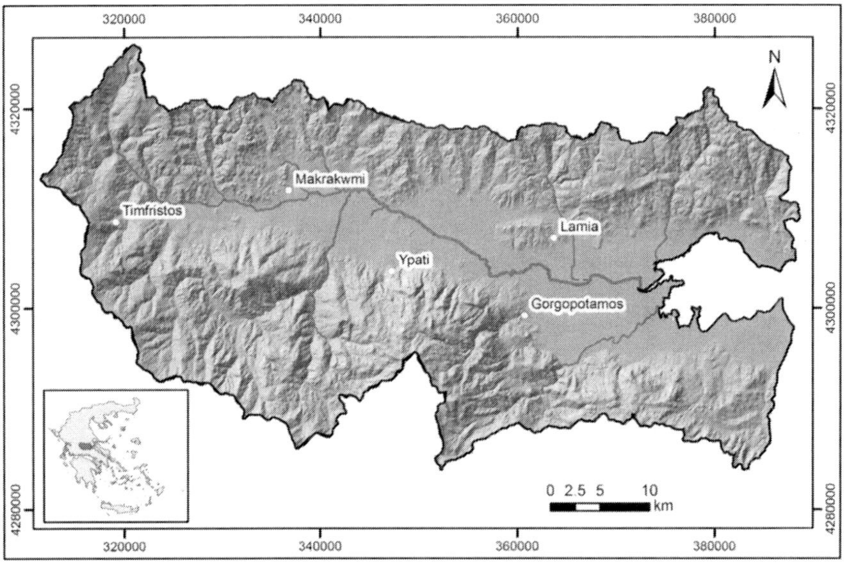

Figure 2. Hydrological basin of Sperchios River.

The morphology of the area is particularly steep in the south part of the basin, while in the north part is gentler and with lower elevations. The largest part of the area (767 km^2) is characterized by medium to high morphological slopes i.e. from 12% to 35%. Slopes lower than 12% as well as higher than 35%, are observed in equally large areas covering 690 km^2 and 539 km^2 respectively. Furthermore it should be stressed that the southwestern area of the basin exhibits the higher morphological slope inclination (>35%), while almost the whole riverbed flows through a 42 km^2lowland part. The Valley of Sperchios River varies in width from 2 up to 13 km, throughout a distance of approximately 50 km up to Maliakos Gulf.

The type of the hydrographic network is characterized in general as elongated dendritic and Sperchios as a 7th class river. Furthermore, asymmetry is observed in the basin, almost exclusively due to its neotectonic as well as lithological structure.

The river course has shifted many times, but its riverbed and delta still remains in the south part of the valley depositing high volumes of sediments. The overflow channel created in 1957, north of today's riverbed, has created (due to sediment deposit) a wide shallow area in the northwest part of Maliakos Gulf [16] .

Regarding the geological conditions (Figure 3), the main rocks that constitute Sperchios valley are igneous and sedimentary [17] - [24] . Analytically, the main geological formations in the west part of the basin are limestones, flysch and clastic sediments. In the north, northeast and southeast parts of the basin, and specifically in Kallidromo Mountain, the main rocks are limestone and schist-crests with ophiolites. In the south part (Oiti mountain), prevailing geological formations are limestones and, in a limited extent, flysch formations.

Finally, the lower parts of Sperchios River valley are covered by quaternary deposits, which constitute the 20% of the basin's extent. These sediments consist of conglomerates, lacustrine deposits, quaternary alluvial deposits, scree, talus cones, as well as Holocene alluvial deposits.

Sperchios River basin, as well as Maliakos Gulf, constitutes an asymmetric tectonic rift.

2.2. Data—Spatial Database

The main objective of this study, as mentioned before, is the application of MCE method in Sperchios River basin aiming at the spatial vulnerability assessment of the geological formations and consequently the expected spatial distribution of soil erosion. The application of MCE technique produces as a final result an "Erosion vulnerability map" This map shows the differentiation of the factors (lithology, geomorphology, hydrology and vegetation) in the basin, namely the variation of the sub-basins based on the erosion susceptibility of their formations.

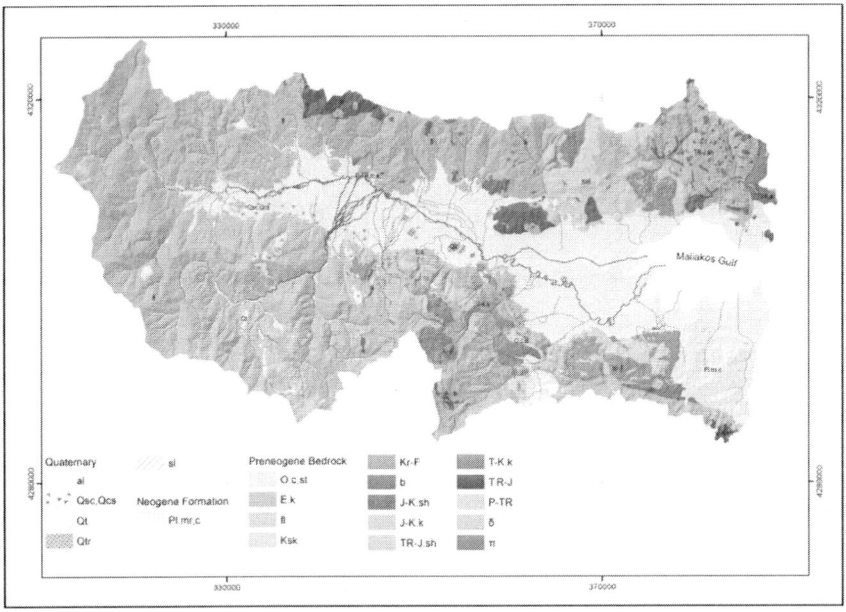

al: Alluvial deposits	**fl:** Flysch (sandstones, graywackes, conglomerates, marls)	**T-K.k:** Dolomite and limestone (Triassic-Jurassic)
Qsc, Qcs: Talus cones and scree	**Kr-F:** Upper-cretaceous Flysch	**T.R-J:** Radiolarite (Triassic-Jurassic)
Qt: Fluvial terraces	**Ksk:** Upper-cretaceous transgressive limestones	**P-TR:** Schists, Sandstones, Limestones (Upper Permio Lower Triassic
Qtr: Terra Rossa	**b:** Bauxite	**Ol:** Ophiolite:
Pl.mr,c: Neogene Formation (marls, conglomerates etc)	**J-K.k:** Limestones (Jurassic-Lower Cretaceous)	**δ:** Diabas-Dolerite. Tuffs of igneous basic rocks
O.c,st: Molassik sediments	**J-K.sh:** Shale-Sandstone-cherty formation (Jurassic)	**π:** Peridotite
E.k: Thinly bedded limestones (Eocene)	**TR-J.sh:** Shales-cherts group with ophoplites (Up Triassic-Jurassic)	**Sl:** Landslides

Figure 3. Geology of Sperchios River basin.

For implementing the methodology of this study, in order to research the erosion vulnerability of the area, various data were collected, evaluated and also created, developing thus a spatial database, described below:

- Digital Elevation Model (DEM) with pixel size 25 × 25 m
- Digital point shapefiles, concerning the names of the towns
- Digital polyline shapefiles, concerning the hydrographic network

- Digital polygon shapefiles, concerning the basin's extent and limits, geological formations and land cover/use
- Tables with descriptive attributes of the above features
- Coordinates of the meteorological stations inside the basin
- Time series of rainfall values (measured by the meteorological stations).
- The data were derived from the following sources:
- Topographical maps 1:50000, from G.Y.S (Hellenic Military Geographical Service)
- Geological maps 1:50000, from I.G.M.E. (Institute of Geology & Mineral Exploration)
- Aster SRTM project (DEM)
- Corine 2000 project (Land Use/Cover)
- E.M.Y (Hellenic National Meteorological Service).

The new features that were created after the spatial data analysis, for the study area, are layers concerning lithology, slope, hydrographic density, frequency and texture, rainfall distribution and land cover:

By co-evaluating the above layers the following were produced:

- Erosion Vulnerability without Land Cover
- Erosion Vulnerability with Land Cover

The Coordinate Reference System used for data processing and model analysis is EGSA '87 (Greek Grid).

2.3. Model Analysis

2.3.1. Geology Factor (L)

In order to assess the effect of geology in the model an "Erosion vulnerability map of lithological formations" was compiled (Figure 4). For the compilation of this map, a number of geological characteristics were co-evaluated, such as lithology, faults, infiltration and permeability of the geological formations in the study area. Formations were classified in three erosion vulnerability classes according to lithological and hydro-lithological characteristics. The geological and hydrogeological data were evaluated based on bibliography.

Formations with high infiltration, such as carbonate rocks which are fractured and so highly karstified and with discontinuities and cracks, as well as formations with low infiltration, such as schists, are of low erosion vulnerability. On the other hand, formations with medium infiltration rates, such as Neogene sediments and the weathering mantle, can become saturated by long-term rainfalls and thus loosen and erode easily [25] .

Accordingly, formations of high permeability, such as karstic formations, exhibit high infiltration rates about 30% - 40% and low surficial runoff resulting in low erosion susceptibility. On the contrary, formations with increased surficial runoff, e.g. schists or Neogene sediments, exhibit higher erosion vulnerability.

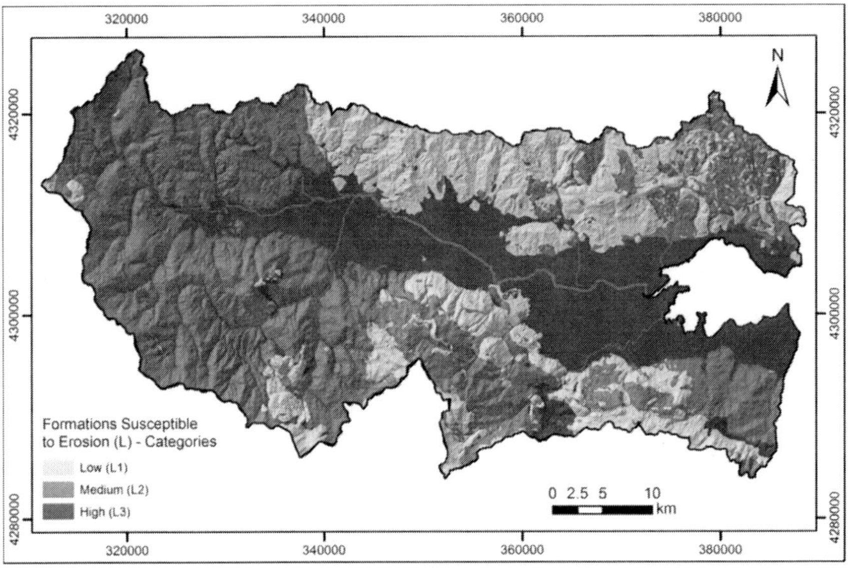

Figure 4. Classified map of the formations based on their erosion susceptibility.

The map was synthesized by lithological, hydrolithological and hydrogeological information for the geological formations of the study area and constitutes a "Map of areas with lithological composition susceptible to erosion (L)".

In Figure 4, three geological formation categories are distinguished. Category L1 includes relatively parent formations of low mass or without weathered mantle, as well as formations with high infiltration rates and of high or medium permeability, e.g. limestone, dolomite, etc. L2 includes erodible formations with high infiltration rates and low permeability, as well as those covered by weathered mantle of considerable thickness, e.g. flysch, schist, marl, sandstone, etc. Finally, L3 includes loose and more prone to erosion formations, e.g. silts, sales, alluvial deposits, scree, etc.

2.3.2. Slope Factor (S)

In the second map (Figure 5) the morphological slopes (S) of the study area were classified. The "Map of morphological slopes (S)" was produced by analyzing the Digital Elevation Model (pixel size 25 m). Three categories were created based on the slope values (Figure 5). Category S1 includes sites with slope inclination percentages lower than 12%, S2 sites with slope inclination percentages between 12% - 35% and S3 values higher than 35%.

The surface extent of the three categories can be seen in Figure 6. S1 category occupies 689.4 km^2, namely the 34% of the whole surface of Sperchios basin, S2 corresponds to 767.1 km^2, i.e. 37.9% of the basin's surface and finally S3 covers 568.8 km^2, i.e. 28.1% of the basin's surface.

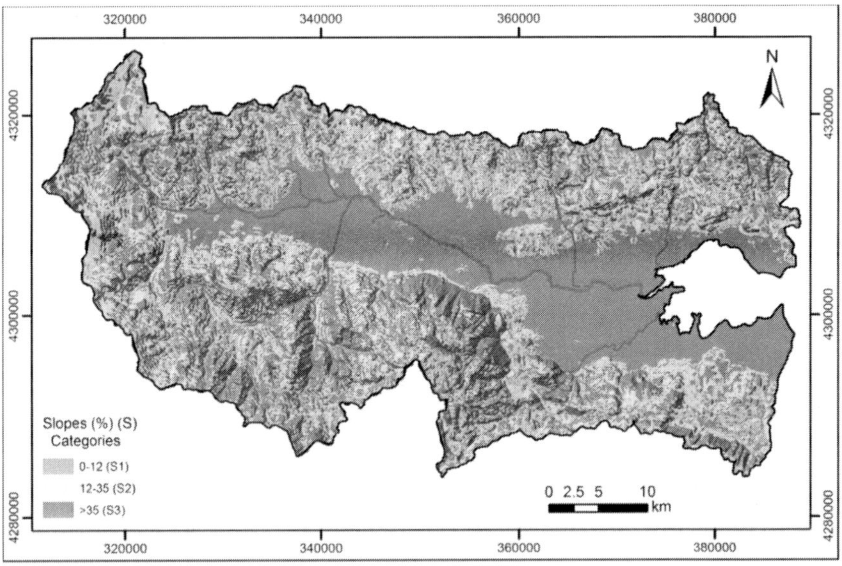

Figure 5. Classified map of the area based on morphological slopes.

The inclination with percentage from 0 to 12 corresponds to slopes consisting of scree, which are created in high and continuous flow conditions, while the upper limit (12%) is considered to be the slope limit of stable deposition. Consequently, the erosion processes are milder in sites with lower inclination values and higher in sites with increased inclination i.e. more than 12% [25] .

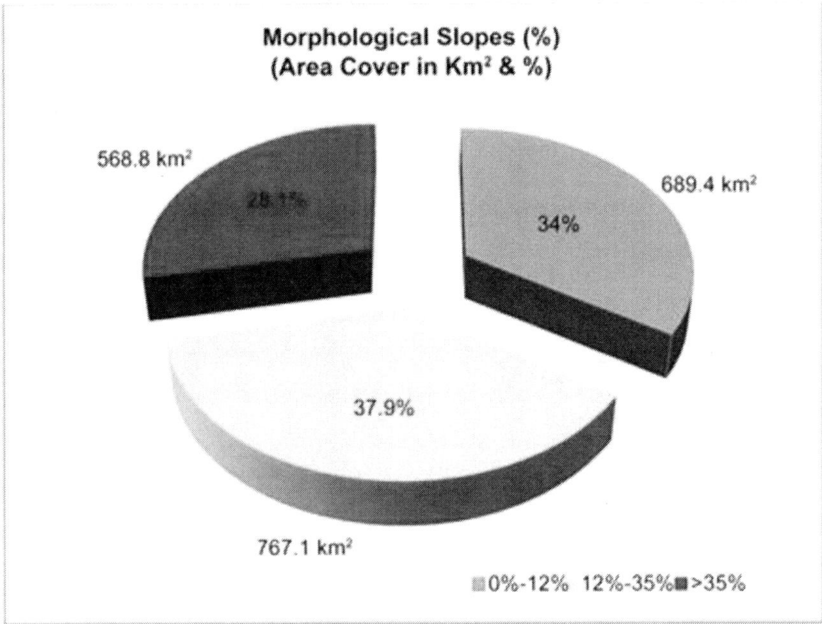

Figure 6. Morphological slopes of the area in km² and percentages (%).

2.3.3. Hydrographic Texture Factor (T)

The third group of maps (Figures 7-9) refers to hydrographic texture analysis of the basin (T), which is expressed by the morphometric parameters of hydrographic density (Dd) and hydrographic frequency (Fs). In order to compile these maps, the classes of the river's branches (channel networks) and of sub basins were calculated, as well as the number and the total length of the river's branches for every sub-basin class.

The drainage density and stream frequency are given in Table 1, according to [26] [27] . Drainage density (Dd) is the ratio of total channel segment lengths, cumulated for all orders within a basin, to the basin area, which is expressed in terms of km/sq · km [26] . Stream frequency or channel frequency (Fs) is the total number of stream segments of all orders per unit area [26] .

On the next step, the sub-basins were classified in 3 categories according to hydrographic density and frequency, namely high, medium, low, with designations D1, D2, D3 (Figure 7) and F1, F2, F3 (Figure 8) accordingly.

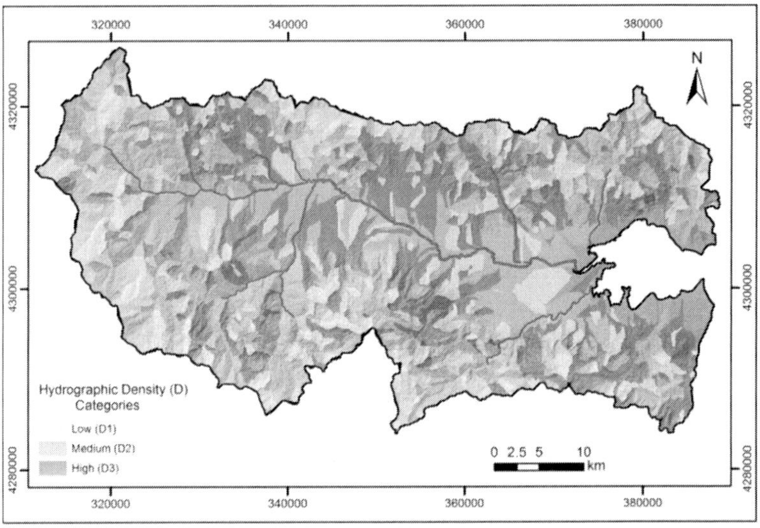

Figure 7. Classification map of the study area based on hydrographic density.

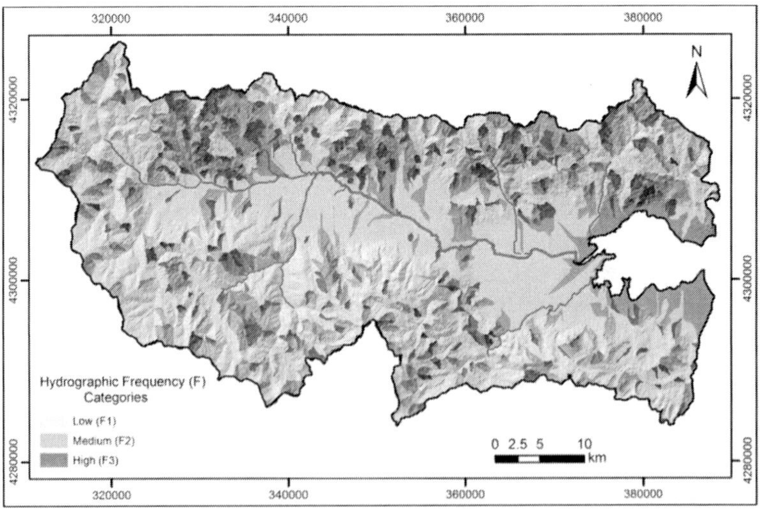

Figure 8. Classification map of the study area, based on hydrographic frequency.

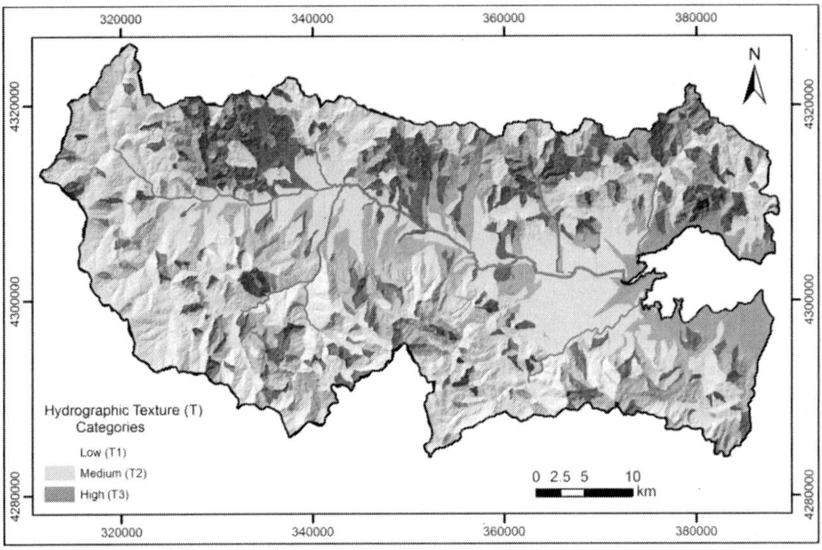

Figure 9. Classification map of the study area based on hydrographic texture.

The combination of the two thematic maps of hydrographic density and frequency resulted in the "Map of hydrographic texture" (T). In this map the study area is classified in three categories of low (T1), medium (T2) and high (T3) hydrographic texture (Figure 9) according to Table 2 presented in paragraph 2.3.6.

2.3.4. Rainfall Factor

Rainfall data in Sperchios River basin were evaluated from the records of eight meteorological stations, which cover efficiently the surface of the study area. From the spatial and elevation distribution of the stations, the correlation between rainfall and altitude of the stations was estimated and it was found 0.6547, which is considered good (Figure 10).

Table 1. Drainage density and stream frequency description.

Drainage Density (Dd)	Dd = Lu/A	Horton (1932, 1945)	Dd is measured as the length of stream channels per unit area of the drainage basin.
Stream Frequency (Fs)	Fs = ΣNu/A	Horton (1932, 1945)	Fs is computed as the ratio between the total number of streams and area of the basin.

where, u = Stream order [28]
Nu = Total no. of streams of all orders,
Lu = Total stream length of all orders (km)
A = Area of basin (km²)

Table 2. Categorization of the erosion vulnerability parameters.

	Model Factors (Vulnerability Parameters)	Erosion Vulnerability		
		Low	Medium	High
	L	L1	L2	L3
1)	Geological formations prone to erosion (Lithology)	Limestones, dolomites, phyllite, etc.	Flysch, schist, marl, sandstone, conglomerate, grauwacke, etc	Alluvial, clastic sediments, etc.
	S	S1	S2	S3
2)	Relief gradient (Slopes)	$S \leq 12\%$	$12\% < S \leq 35\%$	$S > 35\%$
	Fs	F1	F2	F3
3a)	Hydrographic Frequency	$F \leq 0.62$ km^{-2}	$0.62 < F \leq 1.09$ km^{-2}	$F > 1.09$ km^{-2}
	Dd	D1	D2	D3
3b)	Hydrographic Density	$D \leq 0.63$ km/km^2	$0.63 < D \leq 0.95$ km/km^2	$D > 0.95$ km/km^2
	T	T1	T2	T3
3)	Hydrographic Texture	Combinations of: F1 and D1, D2 F2 and D1	Combinations of: F1 and D3, F2 and D2 F3 and D1	Combinations of: F2 and D3 F3 and D2 F3 and D3
Triggering Factor		Erosion Influence		
		Low	Medium	High
	R	R1	R2	R3
4)	Rainfall	$R \leq 700$ mm	$700 < R \leq 900$ mm	$R > 900$ mm
Competitive Factor		Erosion Containment		
		Low	Medium	High
	C	C1	C2	C3
5)	Vegetation cover (Land cover)	Sparse bushes, pasture lands and areas without vegetation	Various cultivations	Forests

The annual rainfall was found to be between 575 mm and 1426 mm, and so the total water volume of the basin was estimated in 1.687 × 109 m^3.

Most of the higher rainfall events are observed from October until December (35% of total) and the lower between June and September (Figure 11). The average infiltration rate is about 18% and the surface runoff 12% of the total precipitation. These percentages resulted from the estimation of the hydrological equilibrium of the area (estimation of evapotranspiration based on Thornwaite method, of infiltration rate based on the area's lithology and the active porosity of the geological formations, of rainfall volume from the rainfall data).

The fourth map refers to rainfall distribution in Sperchios River basin (Figure 12). The map was created by the implementation of the linear regression equation (Figure 10), using DEM as the elevation factor ('x' factor in the equation). Rainfall is the dynamic factor of erosion and acts as an operative cause of the phenomenon on the specific soil and geomorphological characteristics (lithology, slope, hydrographic network, land cover, etc.).

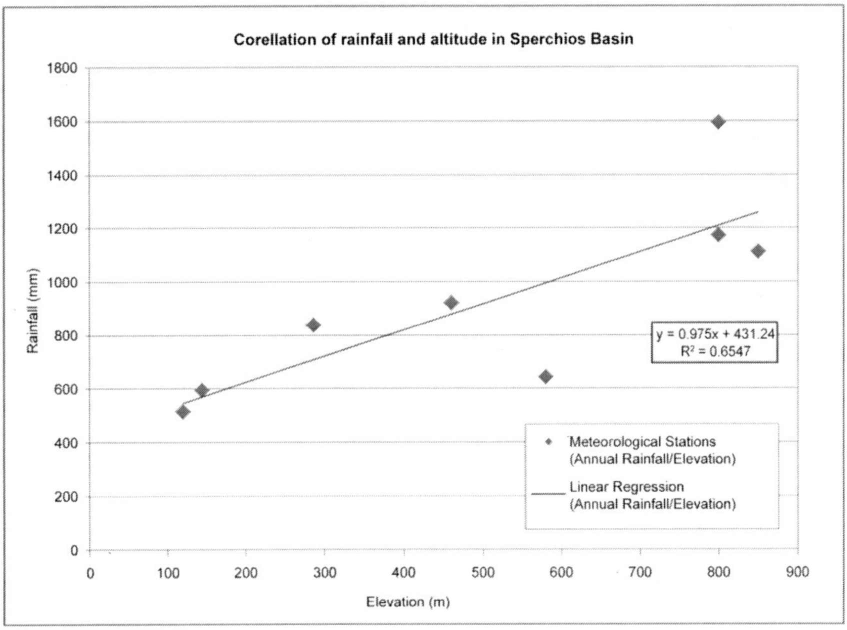

Figure 10. Correlation of annual rainfall and altitude in Sperchios River basin.

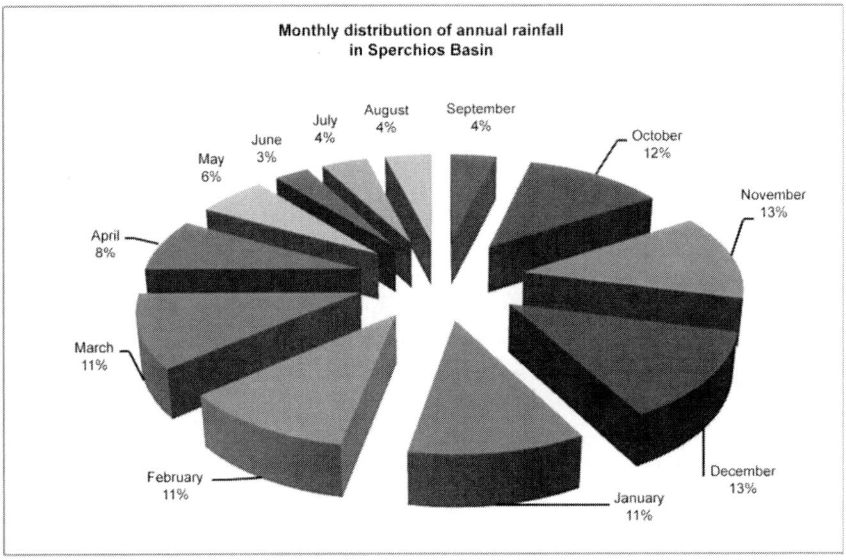

Figure 11. Monthly average rainfall in Sperchios River basin.

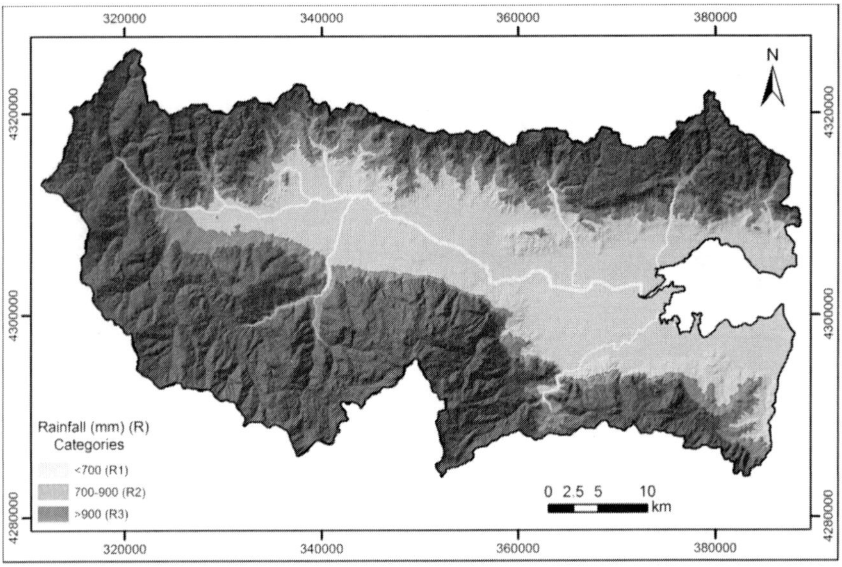

Figure 12. Classification map of the study area based on rainfall distribution.

2.3.5. Land Cover/Use Factor

The next step, which refers to assessing the erosion vulnerability of the basin's formations, concerns the creation of the "Classified map of land use/cover". Land cover is a competitive and protective factor against erosion. Land use/cover data were obtained from Corine 2000, and were classified into three categories (Figure 13). The first category (C1) includes moorlands and scrublands, natural pastures, areas with sparse vegetation, urban construction, etc. Because overgrazing is associated with desertification in Mediterranean Europe, particularly in the areas prone to fire [29] , it was considered that pastures (due to overgrazing) belong to the vegetation class of insufficient protection against erosion, in order to estimate the worst case scenario of soil erosion in the study area. The second category (C2) includes farmland, hard leaf vegetation, olive groves, rice paddies, vineyards, non-irrigated arable land, etc. Finally, the third category (C3) includes coniferous forests, broadleaved forests, mixed woodland, transitional woodland-shrubland, etc. As it is obvious from the above, C1 category provides low protection in terms of erosion, C2 medium and C3 high protection.

As shown in Figure 14, the highest percentage of land cover/use is hardleaf vegetation (22%) and non-arable land (17%), while the percentage of bare rocks (0.67%) and vineyards (0.07%) is low.

2.3.6. Classification of Factors

Summarizing the previous steps, Table 2 presents the categorization of the factors used in order to synthesize the erosion vulnerability model of Sperchios River basin. The categorization was based on the erosion vulnerability of each parameter's attributes (by grouping them in 3 categories as described in previous paragraphs), or its effect in the phenomenon (rainfall, land cover).

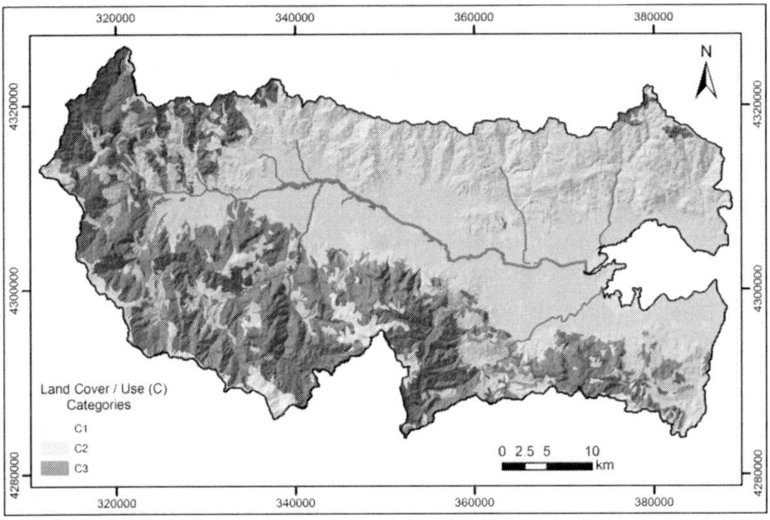

Figure 13. Classification map of the study area based on land use/cover.

2.3.7. Synthesis of Erosion Vulnerability Factors

For the creation of the "Erosion vulnerability map without land cover factor", weights were assigned for each factor that was previous described, such as:

a) geological formations vulnerability factor (L)

b) morphological slope factor (S)

c) hydrographic texture factor (T)

d) rainfall distribution factor (R)

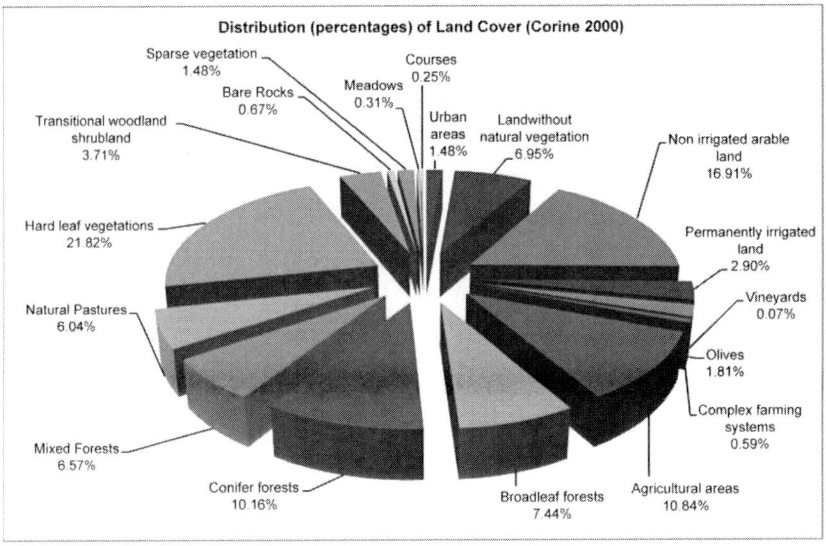

Figure 14. Land Cover distribution of Sperchios River basin.

Lithology was evaluated as the main/key factor that contributes to erosion. Geological formations are the source that offer the erosion material, thus this factor was considered to have the highest importance and contribution in the model (35%). The morphological and hydrographic characteristics of the basin are also of high importance, as these factors constitute the background base for erosion to occur and in this study were considered to contribute equally in the model (20% each). In order for erosion to take place, a triggering factor is needed and rainfall volume is the main one, especially when a hydrological basin is studied. Thus, it was considered as the second most important factor of the model (25%). It is highly important to mention that this study is targeted in erosion caused by rainfall and not by other triggering factors (e.g. wind, snowfall and snow coverage, etc.). Furthermore, this study doesn't focus on erosion caused by extreme events (e.g. storms, floods, etc.) but mainly on the volume of rainfall an area receives throughout a year. Finally, it must be underlined, that the contributing significance of each factor in the model was evaluated by bibliographical research and the scientific opinion and experience of the authors.

The map of the erosion vulnerability of Sperchios River basin is a combination of categorized thematic maps of "bare" soil vulnerability (Eb) and land cover/use. Three categories of vulnerability were created, low (Ec1), medium (Ec2) and high (Ec3), according to Table 3 below.

The categorization of the factors (lithology, geomorphology, hydrology and vegetation) in the basin, namely the differentiation of the sub-basins based on the vulnerability of their formations, in this paper consists of three classes low, medium and high. The last category (high) estimates high soil erosion risk areas.

Taking under consideration that land cover (mainly vegetation cover) is the main competitive factor of erosion, Table 3 and the map of soil vulnerability with land cover (the corresponding figureis presented and discussed in the "Results & discussion" chapter) were created. This new map of the erosion vulnerability with land cover factor depicts the real condition of Sperchios River basin. It must be noted that Table 3 was created taking under consideration that as land cover and vegetation cover category rises the soil protection against erosion rises, thus improving its final categorization, in comparison with the one evaluated by the model (erosion vulnerability of bare soil).

2.3.8. Sensitivity Analysis

In order to evaluate the model's stability and reliability, 4 different scenarios of sensitivity analysis were implemented. As shown in Table 4, in the 1st scenario the contribution of lithology factor in the model was increased by 3%, while the rainfall factor was reduced by 3%. In the 2nd scenario lithology factor was increased by 5% and the rainfall factor reduced by 5%. Scenarios 3 and 4 are the reverse ones of 1 and 2 correspondingly.

Table 3. Correlation of "bare" soil vulnerability with land cover/use factor.

Soil Erosion Vulnerability with Vegetation Cover (Ec)		Erosion Vulnerability of Bare Soil (Eb)		
		Eb1	Eb2	Eb3
Land Cover (C)	C1	Ec1	Ec2	Ec3
	C2	Ec1	Ec1	Ec2
	C3	Ec1	Ec1	Ec2

Based on the results of the 4 scenarios, area cover and percentage for each new erosion susceptibility category was calculated and compared with the model's corresponding values (Table 5).

The previous table presents the area extent (in km^2) that each category, of the model and the four scenarios, covers. The same extent is shown also in percentage. Finally, the difference between the extents of each category of the four scenarios with those of the model is presented.

Table 4. Sensitivity analysis scenarios.

SCENARIOS	FACTORS	
	LITHOLOGY	RAINFALL
1	+3%	−3%
2	+5%	−5%
3	−3%	+3%
4	−5%	+5%

3. RESULTS AND DISCUSSION

Many models are based on assigning weights on the specified factors. The proposed model of this paper is relatively simple and easily applicable and in addition is capable of altering the weights of the factors, in each sub-basin, according to which ones (factors) exceed the rest and enhance the vulnerability of the formations (lithological, geomorphological), which ones enhance the soil material transportation (slopes) and which ones the protection against erosion (vegetation). Additionally, this model can be further developed by including more factors related to soil erosion, in a simple way as this paper indicates where the hydrographic texture was taken under consideration, which is a result of neotectonic activity as well as of precipitation.

The applied methodology, both by overestimating or underestimating, produces acceptable and significant results and allows the qualitative detection of the most sensitive areas to soil erosion, making it an important tool for territorial and environmental management.

As it was discussed above, the "Map of Sperchios River Basin Erosion Vulnerability without Land Cover Factor" that was created by the model of this study is presented in Figure 15.

As it is revealed from this map, sites, with the highest erosion vulnerability, are found, in the parts of the basin, which are covered by highly erodible formations, have high morphological gradients, are characterized by high hydrographic texture and receive the highest rainfall volumes. The extents of the sites with low, medium and high vulnerability are presented in Figure 16. According to this figure, the largest extent of the basin (1077.7 km^2) is covered by formations with medium erosion vulnerability (53.2%). Sites of higher erosion vulnerability occupy the 35.6% of the whole surface of the basin (722.1 km^2).

Table 5. Results of sensitivity analysis scenarios.

EROSION VULNERABILITY CATEGORY	AREA (km^2)	PERCENTAGE (%)	DIFFERENCE (%)
	MODEL		
1	226.5	11.2	-
2	1077.7	53.2	-
3	722.1	35.6	-
	SCENARIO 1		
1	163.8	8.1	3.1
2	1205.5	59.5	−6.3
3	657.0	32.4	3.2
	SCENARIO 2		
1	163.8	8.1	3.1
2	1205.5	59.5	−6.3
3	657.0	32.4	3.2
	SCENARIO 3		
1	208.8	10.3	0.9
2	912.0	45.0	8.2
3	905.5	44.7	−9.1
	SCENARIO 4		
1	240.0	11.8	−0.7
2	880.8	43.5	9.7
3	905.5	44.7	−9.1

The final "Map of Sperchios River Basin Erosion Vulnerability" that was created by the combination of categorized thematic maps of "bare" soil vulnerability (Eb) and land cover/use is presented in Figure 17.

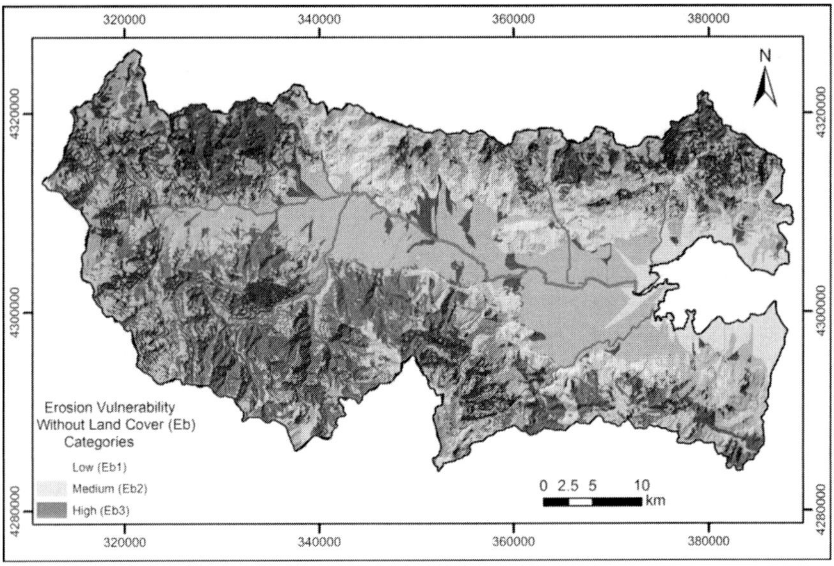

Figure 15. Classification map of the study area according to erosion vulnerability (without land cover factor).

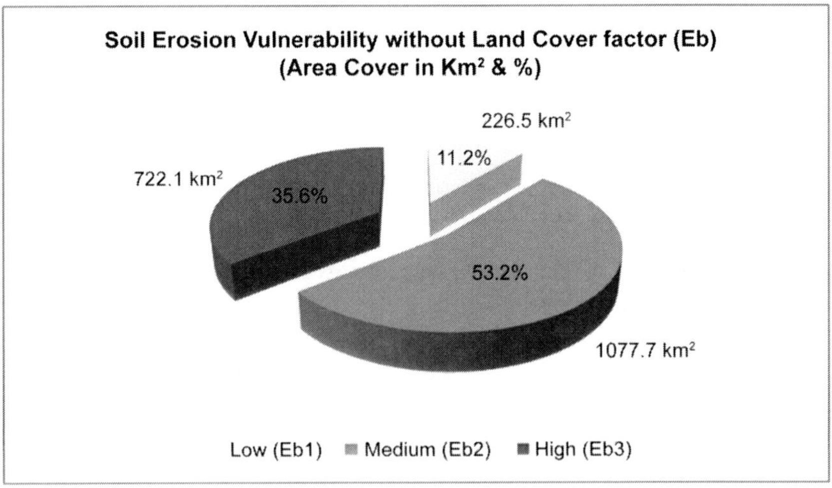

Figure 16. Area's extents according to their erosion vulnerability classification (without land cover) in km^2 and %.

Figure 18 presents the extents of the three categories derived from the map of Figure 17. According to this figure the erosion vulnerability for the low category (Ec1) the values in surface extension and percent are 1201.3 km^2 and 59.1% respectively, while for the medium one (Ec2) these values are 764.9 km^2 and 37.7%. Finally, for the high category (Ec1), the surface extension and percent are 65 km^2 and 3.2% correspondingly.

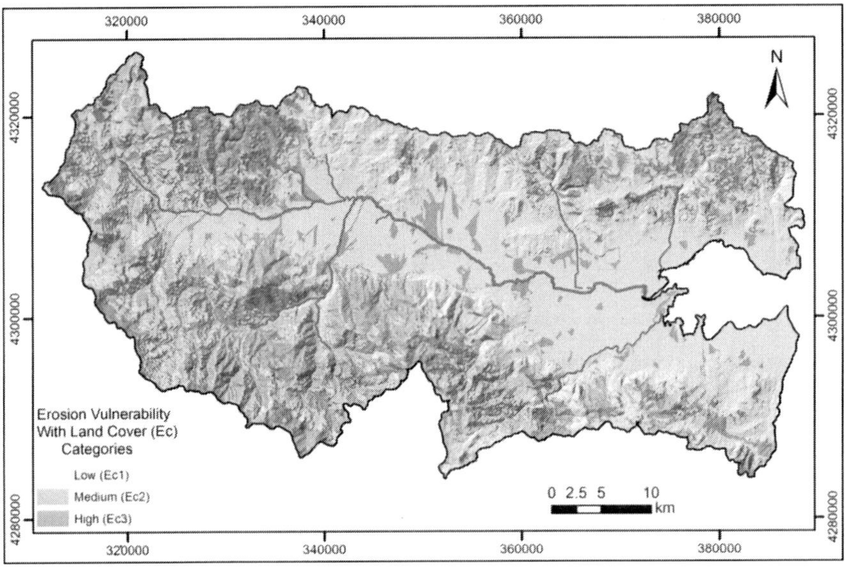

Figure 17. Classification map of the study area, according to their erosion vulnerability (with land cover factor).

Taking into account the land cover/use factor, a drastic reduction of the Ec2 and Ec3 categories is obvious.

Therefore, formations, whose vulnerability characterization was changed by co-evaluating the land cover factor, are mainly in medium and high vulnerability category (Figure 19).

Specifically, the area's extents that changed category (Figure 20), as to their erosion vulnerability classification by co-evaluation of the land cover factor, are from medium to low category 969.9 km^2 (47.7%) and from high to medium 657.1 km^2 (32.4%), while the area that has not changed category is 404.2 km^2 (19.9%).

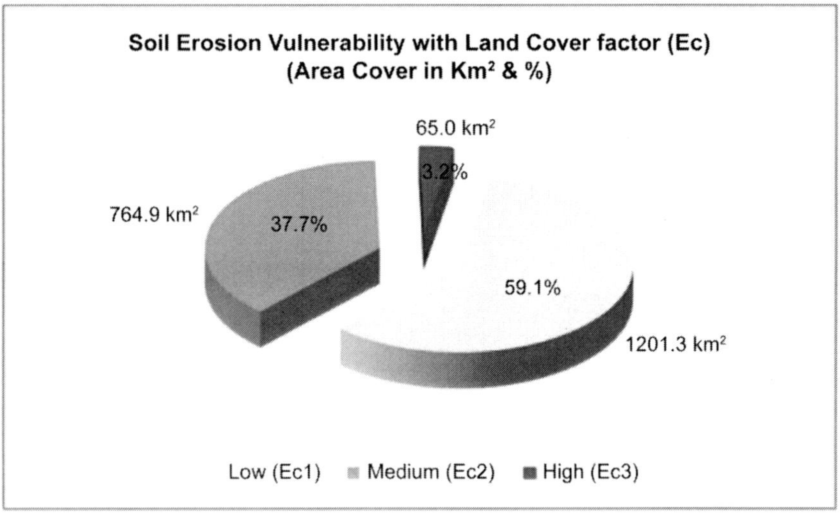

Figure 18. Area's extents, according to their erosion vulnerability classification (with land cover), in km^2 and %.

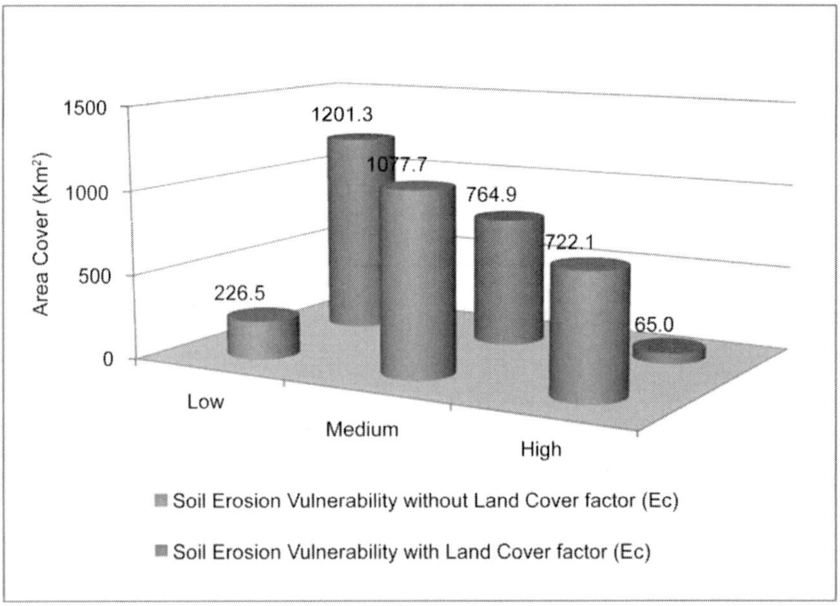

Figure 19. Area's extents, according to their erosion vulnerability classification, with and without land cover in km^2.

Finally, based on the results of the sensitivity analysis scenarios, the model is characterized as highly stable and reliable, because the highest divergence (area cover difference) that was estimated is below 10% (Figure 21, Figure 22). The highest area cover difference (compared with the model) is found in the 2nd erosion vulnerability category of the 4th scenario (9.7%) and the second highest in the 3rd category of both scenarios 3 & 4 (−9.1%), while scenarios 1 & 2 show the same results (area cover differences compared with the model) in all 3 categories (Figure 21, Figure 22). Figure 22 depicts the difference percentages per class, compared to the model, for every sensitivity analysis scenario, stating clearly that none exceeds the limit of 10% and verifying thus the model's stability.

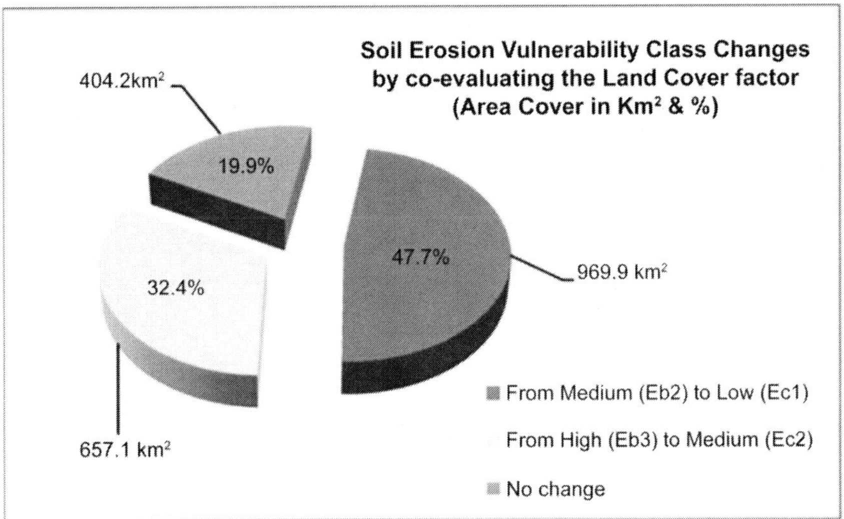

Figure 20. Area's extents that changed category, as to their erosion vulnerability classification, by co-evaluation of the land cover factor.

To sum up, from the methodology applied, in processing and analyzing the erosion vulnerability that affects the formations of Sperchios River basin, the following conclusions emerged:

- Areas of low vulnerability include mainly formations resistant to erosion (e.g. limestones) with low gradients (slopes) and low annual rainfall volumes. In this category belong formations with high infiltration rates, high permeability and thus low surficial runoff, such as carbonate rocks, which are fractured and highly karstified and with discontinuities and cracks, as well as formations with low infiltration, such as schists.

- Areas characterized by high vulnerability are mostly found in higher altitudes with high slope values and increased annual rainfall. The geological formations that are more susceptible to erosion in these sites belong to hard soil and soft rocks formations, like flysch sediments and Neogene deposits. In this category belong formations with medium infiltration rates, such as Neogene sediments and the weathering mantle, which can become saturated by long-term rainfalls and thus loosen and erode easily.

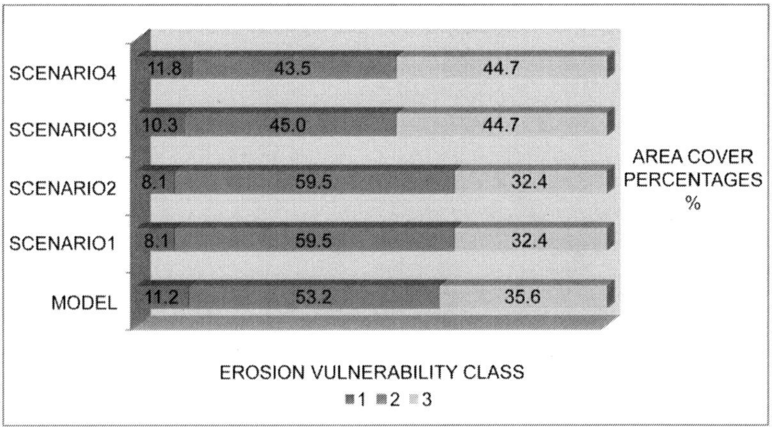

Figure 21. Percentages of area coverage, per erosion vulnerability category, for each sensitivity analysis scenario.

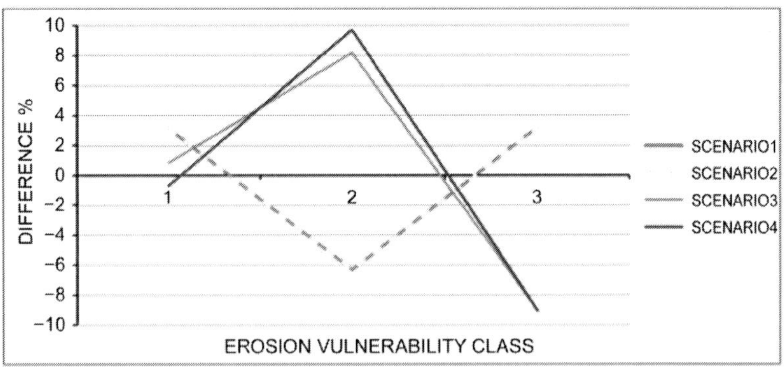

Figure 22. Differences (%) of area coverage, per erosion vulnerability class for each sensitivity analysis scenario.

The applied vulnerability analysis covers the first stage, which does not take into account land cover or other land uses in the area. It relies solely on co- evaluation of lithological-hydrogeological, morphological, hydrological and rainfall characteristics of the area.

The estimations and conclusions derived from this part of the survey provide information on unfavorable conditions, in extreme events, such as forest fires, destruction of crops etc. Furthermore, in the second stage of the method, in order to better access the actual and physical conditions of the area, the land cover and uses of the basin were taken into account, as land cover/use is a competitive factor to erosion.

From the final map it was concluded that the protective effect of land cover leads to demotion of vulnerability categories in terms of erosion respectively, compared with those that were assessed in the first stage of the model (without taking into account the factor of land cover/use). Thus, the majority of the areas belonging to high category in the vulnerability map of "bare" soil (Eb), changed to medium category in the erosion vulnerability map with land cover (Ec), i.e. 657.1 km^2(32.4%). Respectively, medium vulnerability category that changed into low, cover 969.9 km^2(47.7%).

In summary, in the final vulnerability map with the factor of land cover, the largest area belongs to low vulnerability category (1201.3 km^2, 59.1%). Medium class vulnerability is second in extent (764.9 km^2, 37.7%). In high vulnerability category there are very small scattered areas throughout the basin.

Finally, taking under consideration the applied sensitivity analysis the model can be characterized as sufficiently stable and reliable. From the 4 scenarios implemented the highest differentiation observed was under 10%, which is considered to be inside the acceptable limits.

4. CONCLUSIONS

In this paper, as mentioned before, an attempt is made on spatial estimation of soil erosion in large mountainous catchments of Mediterranean type like Sperchios River, using the MCE method. For this purpose, an evaluation based on GIS was adopted. The applied methodology comprises qualitative analysis. The application of MCE in Sperchios River catchment, aiming at spatial vulnerability assessment of the geological formations, offers in addition valuable assessment of the expected spatial distribution of soil erosion.

This method approximates sufficiently the real conditions of erodibility in Sperchios basin, while the results correspond highly to the ones of other similar methods applied to the area [30] .

The proposed method, described in this paper, is simple to use and can be implemented in all basins regardless of their extent. The application of this methodology, for spatial assessment of the formations' erosion vulnerability, with co-evaluation of selected factors, can constitute the basic research for exploring places in order to carry out technical projects. It can also help proper environmental planning and reduction of economic and social consequences caused by soil erosion in the study region.

Finally, it must be mentioned that this methodology is based on a local model that can be used (as it is or with suitable adjustments) in Mediterranean hydrological basins with same or similar geological formations.

CITE THIS PAPER

Stathopoulos, N., Lykoudi, E., Vasileiou, E., Rozos, D. and Di- mitrakopoulos, D. (2017) Erosion Vulnerability Assessment of Sperchios River Basin, in East Central Greece—A GIS Based Analysis. Open Journal of Geology, 7, 621-646. https://doi.org/10.4236/ojg.2017.75043

REFERENCES

1. Beskow, S., Mello, C.R., Norton, L.D., Curi, N., Viola, M.R. and Avanzi, J.C. (2009) Soil Erosion Prediction in the Grande River Basin, Brazil, Using Distributed Modeling. Catena, 79, 49-59.
2. Zhang, Z.-H., Gang, H.U. and Jian, N.I. (2013) Effects of Topographical and Edaphic Factors on the Distribution of Plant Communities in Two Subtropical Karst Forests, Southwestern China. Journal of Mountain Science, 10, 95-104.
3. Voogd, H. (1983) Multicriteria Evaluation for Urban and Regional Planning. Pion, London.
4. Pereira, J.M.C. and Duckstein, L. (1993) A Multiple Criteria Decision-Making Approach to GIS-Based Land Suitability Evaluation. International Journal of Geographical Information Systems, 7, 407-424.
5. Heywood, I., Oliver, J. and Tomlinson, S. (1995) Building an Exploratory Multi-Criteria Modeling Environment for Spatial Decision Support. In: Fisher, P., Ed., Innovations in GIS 2, Taylor & Francis, Bristol, 127-136.
6. Malczewski, J. (1996) A GIS-Based Approach to Multiple Criteria Group Decision-Making. International Journal of Geographical Information Systems, 10, 955-971.
7. Tecle, A. and Yitayew, M. (1990) Preference Ranking of Alternative Irrigation Technologies Via a Multicriterion Decision Making Procedure. Transactions of the ASAE, 33, 1509-1517.

8. Ceballos-Silva, A. and López-Blanco, J. (2003) Evaluating Biophysical Variables to Identify Suitable Areas for Oat in Central Mexico: A Multi-Criteria and GIS Approach. Agriculture Ecosystem and Environment Journal, 95, 371-377.
9. Ceballos-Silva, A. and Lopez-Blanco, J. (2003) Delineation of Suitable Areas for Crops Using a Multi-Criteria Evaluation Approach and Land Use/Cover Mapping: A Case Study in Central Mexico. Agricultural Systems, 77, 117-136.
10. Leskinena, P. and Kangas, J. (2005) Multi-Criteria Natural Resource Management with Preferentially Dependent Decision Criteria. Journal of Environmental Management, 77, 244-251.
11. Bello-Pineda, J., Ponce-Hernández, R. and Liceaga-Correa, M.A. (2006) Incorporating GIS and MCE for Suitability Assessment Modelling of Coral Reef Resources. Environmental Monitoring and Assessment, 114, 225-256.
12. Hajkowicz, S. and Higgins, A. (2006) A Comparison of Multiple Criteria Analysis Techniques for Water Resource Management. European Journal of Operational Research, 184, 255-265.
13. Marinos, P., Plessas, S. and Valadaki-Plessa, K. (1998) Thematic Maps Concerning Erosion Vulnerability Estimation and Sediment Production in Attica Prefecture. Proceedings of 4th Greek Geographical Conference, Athens, 584-616.
14. Alexouli Livaditi, A. (1999) Quantitative Geomorphological Study of Eurotas River Drainage Network (Greece). 5th Hellenic Geographical Conference, Athens.
15. Maroukian, H., Gaki-Papanastassiou, K., Pavlopoulos, K. and Zamani, A. (1995) Comparative Geomorphological Observations in the Kalamas Delta in Western Greece and the Sperkhios Delta in Eastern Greece. Rapport Commission International Mer Méditerranée, 34, 110.
16. Psomiadis, E. (2010) Research of Geomorphological and Environmental Changes in Sperchios River Hydrological Basin Using New Technologies. PhD Thesis, Agricultural University of Athens, Athens.
17. Kallergis, G., Koch, K.E. and Nicolaus, H.J. (1970) Geological Map of Greece, Sheet "Karpenision", Scale 1:50.000. IGME Publications.
18. Kallergis, G., Koch, K.E. and Nicolaus, H.J. (1970) Geological Maps of Greece, Sheet "Sperchias", Scale 1:50.000. IGME Publications.
19. Marinos, G., Anastopoulos, I., Maratos, G., Melidonis, N. and Andronopoulos, V. (1957) Geological Maps of Greece, Sheet "Anavra", Scale 1:50.000. IGME Publications.
20. Marinos, G., Anastopoulos, I., Maratos, G., Melidonis, N. and Andronopoulos, V. (1963) Geological Maps of Greece, Sheet "Stylis", Scale 1:50.000. IGME Publications.
21. Marinos, G., Anastopoulos, I., Maratos, G., Melidonis, N. and Andronopoulos, V. (1962) Geological Maps of Greece, Sheet "Leontarion", Scale 1:50.000. IGME Publications.
22. Marinos, G., Anastopoulos, I., Maratos, G., Melidonis, N., Andronopoulos, V. and Papastamatiou, I. (1967) Geological Maps of Greece, Sheet "Lamia", Scale 1:50.000. IGME Publications.
23. Papastamatiou, I., Tataris, A., Vetoulis, D., Katsikatsos, G., Lalexos, N. and Eleutheriou, A. (1962) Geological Maps of Greece, Sheet "Amfikleia", Scale 1:50.000. IGME Publications.
24. Papastamatiou, I., Tataris, A., Vetoulis, D., Mpornovas, I. and Xristodoulou, G. (1960) Geological Maps of Greece, Sheet "Amfissa", Scale 1:50.000. IGME Publications.
25. Livaditi-Alexouli, A., Livaditis, G. and Likouki, E. (2002) Evaluation of Erosion Vulnerability of Geological Formations and Sediment Production in Lesvos Island, Greece. 6th Greek Geographical Conference, Thessaloniki.
26. Horton, R.E. (1932) Drainage Basin Characteristics. American Geophysical Union Transactions, 13, 348-352.
27. Horton, R.E. (1945) Erosional Development of Streams and Their Drainage Basins; Hydrophysical Approach to Quantitative Morphology. Geological Society of America Bulletin, 56, 275-370.
28. Strahler, A.N. (1964) Quantitative Geomorphology of Drainage Basins and Channel Networks. In: Handbook of Applied Hydrology, Section 4 II, McGraw Hill Book Company, New York.

29. Papanastasis, V.P. (2000) Land Degradation Caused by Overgrazing and Wildfires and Management Strategies to Prevent and Mitigate Their Effects. In: Enne, G., Zanolla, Ch. and Peter, D., Eds., Desertification in Europe: Mitigations Strategies, Land-Use Planning, European Commission, EUR 19390, Luxembourg, 187-198.
30. Sigalos, G., Loukaidi, V., Dasaklis, S. and Alexouli-Livaditi, A. (2010) Assessment of the Quantity of the Material Transported Downstream of Sperchios River, Central Greece. Bulletin of the Geological Society of Greece, 42, 737-745.

CHAPTER 13

Non-Diapiric Salt Domes in the West Zanjan, Central Iran

Sina Alizadeh

Department of Geology, Islamic Azad University, Science and Research Branch, Tehran, Iran

ABSTRACT

The salt domes of the west Zanjan (Central Iran) are the most important structures in the study area. They have been formed by the uplifting and erosion together under low humidity and dry and warm climate condition. The salt rocks with near to 200 meters thickness are related to lower member of the Upper Red Formation (Early Miocene) that deposited in the inverted back arc basin. They have been formed in the Central Iran basin after the Arabian- Eurasian convergence. Based on filed works and preparation of geologic map, salt domes have been cropped out during regional uplifting and erosion along hinge zone of a longitudinal anticline. Also, there is no evidence for salt diapirism and so, they are different from some salt diapirs in the southwestern margin of Zanjan that is investigated by other researchers.

Keywords: Non-Diapiric, Salt Dome, Zanjan, Central Iran

1. INTRODUCTION

The Miocene gypsiferous rocks are named Upper Red Formation by the National Iranian Oil Company geologists [1] . Its thickness in the type section, near Qom ci- ty, is about 3000 m and comprises alternations of red fine-grained sandstones, compact gypsum, conglomerate and marly sandstones. The Upper Red Formation overlies the Qom Formation but, in some parts of NW and Central Iran, the con- tact comprises thick salt and gypsum [1] . The Miocene Upper Red Formation in the study area was deposited in a Marine regression condition (firstly) and fluvial system under dry and warm climatic condition (later) [1] . Salt domes or positive topographical features which have been formed by evaporate minerals (mainly halite) are the index structures in Upper Red Formation [2] .

There are many salt domes most of which have been cropped out and the aim of this study is to determine diapiric or non-diapiric condition of salt domes in the Zanjan area, Iran.

2. GEOLOGIC SETTING

Iran country has composed from different plates: Arabian plate in south and west (Zagros and Persian Gulf), Cimmerian miniplate in north and east, Eurasian plate in northeast margin [3] [4] . There are three physiographic-tectonic zo- nes that have salt rocks: 1) Zagros Mountain, 2) Persian Gulf basin and 3) Cental Iran basin on which the study area is situated. Dominant structural trends in Central Iran are NW-SE in northwestern part and N-S in eastern part. The base- ment of the plateau consists of Cimmerian miniplate. Volcanics of late Cretace- ous—early Miocene age in Central Iran represent Urmieh-Dokhtar magmatic arc. Volcanic rocks, evaporates and Turbidites successions up to 3 km thick re- present Eocene back-arc extension across Central Iran [5] .

This succession is commonly overlain in Central Iran by terrestrial clastics, evaporates and volcanics of the Lower Red Formation of Oligocene age. Marine deposition resumed across much of Central Iran with the carbonates of the largely lower Miocene Qom Formation. The Qom Formation is overlain by mi- ddle Miocene terrestrial clastics and evaporates of the Upper Red Formation. The salt deposits belonging to Cenozoic of Central Iran basin have two separate origins: The old type related to Upper Eocene deposits and Lower Red formation; and younger type related to Upper Red formation [5] .

Based on previous research [2] , the study area is situated in Urmieh-Dokhtar physiographic province (Figure 1). Dominant structural trend in

Urmieh-Do- khtar province is NW-SE. From tectonics view, it contains a magmatic arc that is result of subduction to beneath of southern active continental margin of Cimmerian plate. Also, Sahand and other cones (about 9) in lesser Caucasus hinterland are result of quaternary magmatism in this zone that it has marked by wide- spread Eocene volcanism. Urmieh-Dokhtar Province has continued to south of Black Sea and its width has increased from Naien city that many parts have covered by quaternary deposits of Dagh Sorkh Kavir, Southern Urmieh Lake, Namak and Hoz-e Soltan Lakes have formed on it. So, there are a few backland basins with Playa type sedimentation because of dip decreasing in Ben off zone (in NW part of magmatic arc). SE part of magmatic arc has formed on southwest margin of East-Central Iran microcontinent [3] .

3. MATERIAL AND METHODS

Salt domes of Iran have been presented by the large variety of shapes and elevation from high relief to entirely eroded structures. These different morphologies are controlled by various factors, for example, position related to other structures, the time of emergence, the rate of salt dissolution, which is controlled by the rate of erosion, the bulk rate of salt emission from the source layer, and provinces have distinguished in Iran and the study area has situated on Tasoj-Zanjan eventually by the tectonic activity [6] . Based on recent research [7] , nine diapiric Province.

The implemented studies indicate that the salts of this province are new type that cropped out along hinge zone of a burial longitudinal anticline on footwall of the Halab fault [8] . The salt domes of this province have occurred in an area covering about 10 km^2. They have been covered by thin marl and separated from others by Quaternary deposits. The salt structures have not got diapiric cores, thus, they are not salt diapirs (Figure 2 and Figure 3).

Also, there is a saline brine (Figure 4) that flows to the plain (Figure 5) for traditional mineral processing by evaporation.

4. RESULTS AND DISCUSSION

Salt domes in the study area have been caused by evaporation (Figure 6) and weathering of salt rocks under low humidity (about 15%) and dry climate condition. The salt rocks of study area are related to the marine regression and evaporate conditions in inverted back arc basin (Central Iran). They are related to lower member (M_1) of Upper Red formation that has been under weathering now. Thus, they are different from other salt domes and diapirism in Iran that investigated by many researchers [9] - [20] .

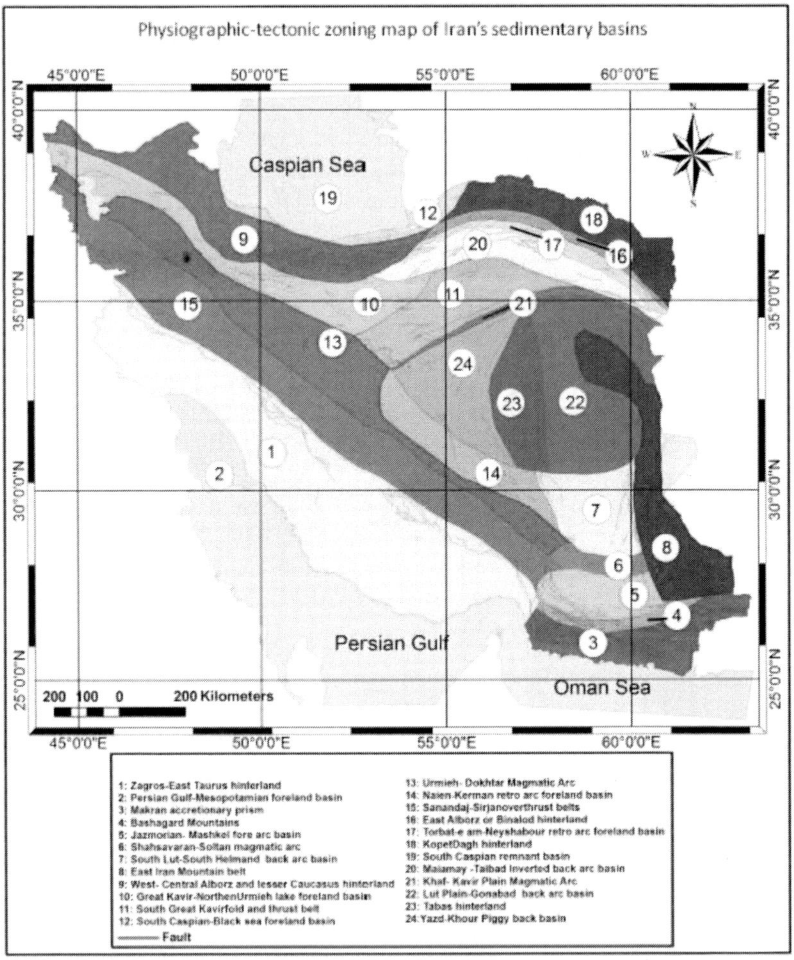

Figure 1. Physiographic-tectonic zoning map of Iran's sedimentary basins, modified from [2] . The study area is shown in the black rectangle.

Based on previous work on the mud diapirism [21] [22] and neotectonic regime in Iran [23] - [28] , Zagros in south Iran is the most active zone [29] - [69] . Then, Alborz [70] - [116] and Central Iran [117] - [145] have been situated in the next orders. Thus, salt domes in the study area have not been affected by regional shortening. The main characteristics of salt domes in the study area have sum- marized in Table 1.

Figure 2. A far view (to NE) from some salt domes of the study area.

Figure 3. A near view from the horizontal salt rock of the study area.

Figure 4. A saline brine in the southern part of study area.

Figure 5. A view from evaporative pools in the study area.

Figure 6. Salt deposition by evaporation of brine.

Table 1. Main characteristics of salt domes in the study area.

Age of salt	Early Miocene
Name of Formation	Upper Red
Thickness of salt rocks	Near to 200 m
Salt deposition Setting	Inverted back arc basin
Setting of Salt structures Evolution	Retro arc Basin
Folding Type	Open to Gentle
Humidity of the weather	Low
Salt dome Position	Hinge zone of anticline

5. CONCLUSION

The geologic map has prepared and based on field works on the salt domes of the west Zanjan (north Iran), salt rocks in the study area are caused by evaporation in the last marine regression condition. Then, their weathering under low humidity and dry climate condition has been formed individual salt domes. In the other words, the salt domes of study area are related to Miocene lower member of the Upper Red Formation that deposited in the inverted back arc basin. But they have been formed in the Central Iran retro arc basin. Therefore, they have been formed by the uplifting and regional shortening under dry and warm climate. There are some structures such as Halab fault and folds that cause the uplifting and cropping out of salt rocks. Thus, there is no evidence for salt diapirism and they are results of the uplifting and erosion together.

ACKNOWLEDGEMENTS

This work has been funded by the research project in Islamic Azad University, Science and Research Branch, Tehran, Iran. Also, special thanks to vice-president for research in Science and Research Branch, Tehran.

CITE THIS PAPER

Alizadeh, S. (2017) Non-Diapiric Salt Domes in the West Zanjan, Central Iran. Open Journal of Geology, 7, 132-146. https://doi.org/10.4236/ ojg. 2017.72009

REFERENCES

1. Rafiei, B., Ahmadi Ghomi, F. and Shahkaram, M. (2011) Depositional Environments and Sequence Stratigraphy of the Fluvial Upper Red Formation (Miocene) in the Avaj Area, Qazvin Province, Iran. Neues Jahrbuch für Geologie und Palaontologie—Abhandlungen, 259, 257-270.
2. Qorashi, M. and Arian, M. (2011) Tectonics of Iran. Geologic Survey of Iran, Tehran, 336 p.
3. Arian, M. (2013) Physiographic-Tectonic Zoning of Iran's Sedimentary Basins. Open Journal of Geology, 3, 169-177.
4. Arian, M. (2011) Basement Tectonics and Geology of Iran. Asar Nafis Press, Qum, 300 p.
5. Jackson, M.P.A., Cornelius, R.R., Craig, C.H., Gansser, A., Stocklin, J. and Talbot, J.C. (1990) Salt Diapirs of the Great Kavir, Central Iran. Geological Society of America Memoirs, 177, 1-150.
6. https://doi.org/10.1130/mem177-p1
7. Jahani, S., Callot, J.P., de Lamotte, F.D., Letouzey, J. and Leturmy, P. (2007) The Salt Diapirs of the Eastern Fars Province (Zagros, Iran): A Brief Outline of Their Past and Present. In: Lacombe, O., Roure, F., Lavé, J. and Vergés, J., Eds., Thrust Belts and Foreland Basins, Frontiers in Earth Sciences, 289-308.
8. Arian, M. and Noroozpour, H. (2015) Tectonic Geomorphology of Iran's Salt Structures. Open Journal of Geology, 5, 61-72.
9. Pourkermani, M. and Arian, M. (1999) Structural Analysis of Halab Fault. Proceeding of the 3rd Symposium of Geological Society of Iran, Shiraz, 31 August-2 September 1999, 130-132.
10. Arian, M. and Noroozpour, H. (2015) The Biggest Salt-Tongue Canopy of Central Iran. Open Journal of Geology, 5, 55-60.
11. Asadian, F., Pourkermani, M. and Arian, M. (2007) Tectonic Geomorphology of Salt Structures in the Garmsar—Lasjerd Area. Geographical Research, 39, 75-84.
12. Pourkermani, M. and Arian, M. (1997) Salt Domes of Central Iran. Journal of Humanities, 3, 29-41.
13. Arian, M. (2011) A Preface on Salt Diapirism of Iran. Asar Nafis Press, Qum, 309 p.
14. Asadian, F. and Arian, M. (2009) Identification of Diapiric Provinces of Central Iran through Geological and Geographical Analysis. International Journal of Agriculture Environment & Biotechnology, 2, 3443-3451.
15. Arian, M. (2012) Clustering of Diapiric Provinces in the Central Iran Basin. Carbonates and Evaporites, 27, 9-18.
16. Khodabakhshnezhad, A. and Arian, M. (2016) Salt Tectonics in the Southern Iran. International Journal of Geosciences, 7, 367-377.
17. Razaghian, G. and Arian, M. (2015) The Emergent Salt Diapirs in the East Zagros, Iran. Open Journal of Geology, 5, 718-726.
18. Arian, M. (2010) Tectonics and Sedimentation. Farazamin Press, Tehran, 307 p.
19. Arian, M. and Maleki, Z. (2010) Principals of Experimental Tectonics. Asar Nafis Publication, Qum, 224 p.
20. Pourkermani, M. and Arian, M. (1998) Tectonic Geomorphology of Salt Domes in West of Zanjan Province, Iran. Geographical Research, 47, 44-53.

21. Arian, M. and Feizi, F. (2010) The Significance of Faulting on the Surficial Spreading of Evaporitic Deposits in the Varamin-Semnan Area. Journal of Earth and Resources, 3, 1-20.
22. Feizi, F., Arian, M. and Arian, A. (2015) Mud Diapirism on the Makran, Iran: Case Study on the Napag Mud Volcano. Open Journal of Geology, 5, 300-308.
23. Arian, M. and Sistanipour, A. (2015) Mud Diapirism on the Gorgan, North Iran. Open Journal of Geology, 5, 442-450.
24. Arian, M. and Khodabakhshnezhad, A. (2015) Sedimentary Environments Can Be Changed by Geotechnology (Case Study: A Morphotectonic Idea for Design of Extensive Artificial Bay on the Iranian Plateau). International Journal of Geosciences, 6, 487-496.
25. Arian, M. (2011) Middle East Tectonics. Asar Nafis Press, Qum, 236 p.
26. Arian, M. (2010) Applied Seismotectonics. Farazamin Press, Tehran, 304 p.
27. Arian, M. and Maleki, R. (2008) Neotectonics. Farazamin Research Center, Tehran, 150.
28. Pourkermani, M. and Arian, M. (1998) Seismicity of Iran. Shahid Beheshti University Press, Tehran, 212.
29. Pourkermani, M. and Arian, M. (1997) Seismotectonics. DezAb Consulting Engineers Company Press, Tehran, 270.
30. Arian, M. and Aram, Z. (2014) Relative Tectonic Activity Classification in the Kermanshah Area, Western Iran. Solid Earth, 5, 1277-1291.
31. Mashal, M., PourKermani, M., Charchi, A., Almasian, M. and Arian, M. (2013) Pattern of Structural Geology Underground in Eastern of North Dezfol Embayment. Advances in Environmental Biology, 7, 260-268.
32. Pazhoohan, M., Arian, M., Ghorashi, M. and Khosrotehrani, K. (2014) A Study of Drainage Pattern Responses to Active Tectonics in Tadvan Region, SW Iran. Geodynamics, 1, 36-41.
33. Rahimi, N. and Arian, M. (2014) Tectonic Geomorphplogy of Kangavar-Sosangerd Region, West Iran. Advances in Environmental Biology, 8, 119-124.
34. Arian, M. and Hashemi, A. (2008) Seismotectonic Zoning in the Zagros. Journal of Sciences, 18, 63-76.
35. Arian, M., Ahmadnia, A., Qorashi, M. and Pourkermani, M. (2002) Structural Analysis of Mengharak Transcurrent Fault System in Zagros, Iran. Special GEO 2002 Conference Issue Geoarabia, 7, 209-210.
36. Arian, M., Qorashi, M., Pourkermani, M. and Ahmadnia, A. (2003) Fractal Analysis of Mengharak Transcurrent Fault System in Zagros, Iran. Proceedings of the 4th International Conference on Seismology and Earthquake Engineering, Tehran, 12-14 May 2003, 23.
37. Baharvand, S., Pourkermani, M., Ajalloian, R., Arian, M. and Nouryazdan, A.R. (2010) Seymareh Landslide and Its Role in Environmental and Geomorphologic Changes of the Pole-Dokhtar Area. Journal of the Earth, 4, 13-24.
38. Abdideh, M., Qorashi, M., Rangzan, K. and Arian, M. (2011) Assessment of Relative Active Tectonics Using Morphometric Analysis, Case Study of Dez River (Southwestern, Iran). Geosciences, 20, 33-46.
39. Arian, M., Qorashi, M., Pourkermani, M. and Ahmadnia, A. (2006) The Structural Significance Kareh Bas Transcurrent Fault System in the Zagros Fold and Thrust Belt. Geosciences, 15, 126-133.
40. Arian, M. and Noroozpour, H. (2015) Seismic Activity and Fractal Geometry of Kareh Bas Fault System in Zagros, South of Iran. Open Journal of Geology, 5, 291-299.
41. Ehsani, J. and Arian, M. (2015) Quantitative Analysis of Relative Tectonic Activity in the Jarahi-Hendijan Basin Area, Zagros Iran. Geosciences Journal, 19, 1-15.
42. Arian, M. and Mohammadian, R. (2011) Analysis of Fractures in the Asmari Reservoir of Marun Oil Field (Zagros, Iran). Geosciences, 20, 87-96.
43. Tajali, S. and Arian, M. (2016) Petrophysical Evaluation and the Effect of Shale Layers on Net Pay Zone Thickness in the Marun Oil Field, Iran. Open Journal of Geology, 6, 763-773.
44. Arian, M., Qorashi, M. and Ahmadnia, A. (2003) Analysis of Behbahan Shear Zone. Iranian Journal of Geology, 1, 1-4.
45. Omidali, M., Arian, M. and Sorbi, A. (2015) Neotectonics of Boroujerd Area, SW Iran by Index of Active Tectonics. Open Journal of Geology, 5, 309-324.

46. Chegini, A., Sorbi, A. and Arian, M. (2015) Active Tectonics of Hamedan Area, SW Iran by Index of Active Tectonics. International Journal of Geology, 4, 108-118.
47. Maleki, Z., Arian, M., Solgi, A. and Ganjavian, M.A. (2014) The Elements of Fold Style Analysis in the Khaftar Anticline, Zagros, Iran. Open Journal of Geology, 4, 79-92.
48. Maleki, Z., Arian, M. and Solgi, A. (2014) Structural Style and Hydrocarbon Trap of Karbasi Anticline, in the Interior Fars Region, Zagros, Iran. Solid Earth Discussions, 6, 2143-2167.
49. Ehsani, J., Arian, M. and Ghorashi, M. (2015) Geomorphic Signatures of Active Tectonics in the Jarahi-Hendijan Drainage Basin in the South West Iran. Geosciences, 24, 211-218.
50. Khodabakhshnezhad, A., Pourkermani, M., Arian, M., Matkan, A.A. and Charchi, A. (2015) Active Tectonics of Great Karounriver Basin. Geosciences, 24, 13-28.
51. Maleki, Z., Arian, M., Solgi, A. and Ganjavian, M.A. (2015) Elements of Fold Style Analysis in the Karbasi Anticline, Interior Fars Region, Zagros. Geosciences, 24, 293-302.
52. Baratpour, F., Arian, M. and Solgi, A. (2015) Geometric Analysis of Tukak and Kamarun Anticlines on Izeh Zone, Zagros. Geosciences, 24, 191-200.
53. Gholamhosein Fard, N., Sorbi, A. and Arian, M. (2015) Active Tectonics of Kangavar Area, West Iran. Open Journal of Geology, 5, 422-441.
54. Maleki, Z., Arian, M. and Solgi, A. (2015) Folding Pattern in the Fars Province, Zagros Folded Belt: Case Study on the Karbasi and Khaftar Anticlines, Interior Fars, Iran. Solid Earth Discussions, 7, 2347-2379.
55. Aram, Z. and Arian, M. (2016) Active Tectonics of the Gharasu River Basin in Zagros, Iran, Investigated by Calculation of Geomorphic Indices and Group Decision Using Analytic Hierarchy Process (AHP) Software. Episodes, 39, 39-44.
56. Abdolizadeh, S., Maleki, Z. and Arian, M. (2016) Earthquake Hazard Zonation and Seismotectonics of the Bandar Abbas Area, Zagros, Iran. Open Journal of Geology, 6, 210-224.
57. Hadizadeh, A., Arian, M. and Ganjouyan, M.A. (2011) Analysis Geometry Dashtak Detachment Zone in the Dashtak, Salamati and Sepidar Anticlines from Interior and Sub Coastal Fars Area. Journal of the Earth, 6, 33-44.
58. Khodabakhshnezhad, A., Arian, M. and Pourkermani, M. (2008) The Elements of Fold Style Analysis in the Asmari Anticline (Zagros). Journal of Sciences, 18, 129-138.
59. Khodabakhshnezhad, A., Arian, M. and Pourkermani, M. (2015) Folding Mechanism in the Asmari Anticline, Zagros, Iran. Open Journal of Geology, 5, 197-208.
60. Arian, M., Pourkermani, M., Khodabakhshnezhad, A. and Noroozpour, H. (2011) Investigation of Oil Trap in the Asmari Anticline (Zagros, Iran). Indian Journal of Science and Technology, 4, 1696-1699.
61. Dehbozorgi, M., Pourkermani, M., Arian, M., Matkan, A.A., Motamedi, H. and Hosseiniasl, A. (2010) Quantitative Analysis of Relative Tectonic Activity in the Sarvestan Area, Central Zagros, Iran. Geomorphology, 121, 329-341.
62. Rafiee, P., Baghbani, D., Aghanabati, A. and Arian, M. (2014) Micro Biostratigraphy and Lithostratigraphy of the Upper Permian Dalan Formation (Zagros Basin, Southwest Iran). International Journal of Geography and Geology, 3, 45-55.
63. Arian, M. (2012) Salt Diapirism and Tectonics. 2nd Edition, Asar Nafis Press, Qum, 319 p.
64. Maleki, Z., Arian, M., Solgi, A. and Ganjavian, M.A. (2013) Sediment Deformations on Strike Slip Fault Blocks and Analogue Modeling: A Case Study of the Nezamabad Fault, Interior Fars, Zagros. Journal of Sciences, 89, 39-51.
65. Vaseghi, H., Maleki, Z. and Arian, M. (2016) Structural Style in the Zagros Fold-Thrust Belt: The Gavbast Anticline, Coastal Fars. Open Journal of Geology, 6, 109-116.
66. Arian, M., Ahmadipour, M.R. and Khodaei, K. (2003) The Elements of Fold Style in Soltananticline, Northeast of Pole Dokhtar. Iranian Journal of Geology, 1, 1-8.
67. Arian, M., Pourkermani, M., Qoreshi, M. and Ahmadnia, A. (2002) Salt Diapirism along Mengharak Fault System. Proceedings of the 6th Symposium of Geological Society of Iran, Kerman, 2-4 September 2002, 45-49.
68. Manafi, M. and Arian, M. (2008) Basement Tectonics Investigation of Dezful Embayment, between Kazeron and Hendijan Faults. Journal of the Earth, 3, 47-61.

69. Asghari, M.R., Mokhtari, M., Arian, M. and Vanaki, M.R. (2015) Pseudo 3D Seismic Data Interpretation and Inversion in Ferdowsi Oil Field, Persian Gulf. Geosciences, 24, 341-348.
70. Badrkhani, M., Maleki, Z. and Arian, M. (2016) Folding Style Analysis of the Tabnak Sweet Gas Field, Zagros, Iran. Open Journal of Geology, 6, 971-981.
71. Alladin, Y., Talebian, M., Arian, M. and Ahmadi, M.M. (2015) Geotechnical Investigation and Seismic Zonation of Alluvial Deposits in Western Tehran. Geosciences, 24, 333-342.
72. Taherkhani, B., Nazari, H., Pourkermani, M. and Arian, M. (2015) Geometry and Recent Kinematics of the North Qazvin Fault: Morphotectonic Approach. Geosciences, 24, 29-38.
73. Manuchehri, H., Arian, M., Ghorashi, M., Solgi, M. and Sorbi, A. (2015) Geomorphic Signatures of Active Tectonics in the Chalus Drainage Basin in the Alborz, Iran. Geosciences, 24, 273-280.
74. Noroozpour, H., Arian, M. and Sorbi, A. (2015) Fault Movement Potentials in the Tehran-Semnan Region (North Iran). Open Journal of Geology, 5, 281-290.
75. Arian, M., Maleki, Z. and Noroozpour, H. (2011) Cenozoic Diastrophism and Deformational Events in the East Central Alborz. Journal of Basic and Applied Scientific Research, 1, 2394-2400.
76. Feizi, F., Arian, A. and Rahmani, R. (2007) Seismotectonic Zoning in the Eastern Part of the Central Alborz. Journal of Sciences, 17, 151-164.
77. Khavari, R., Arian, M. and Ghorashi, M. (2009) Neotectonics of the South Central Alborz Drainage Basin, in NW Tehran, N Iran. Journal of Applied Sciences, 9, 4115-4126.
78. Arian, M. and Bagha, N. (2012) Active Tectonics of Tehran Area, Iran. Journal of Basic and Applied Scientific Research, 2, 3805-3819.
79. Bagha, N., Arian, M., Ghorashi, M., Pourkermani, M., El Hamdouni, R. and Solgi, A. (2014) Evaluation of Relative Tectonic Activity in the Tehran Basin, Central Alborz, Northern Iran. Geomorphology, 213, 66-87.
80. Arian, M. and Feyzi, F. (2005) Application of Geomorphic Indices to the Assessment of Relative Tectonic Activity Levels in the Alborz-Central Iran Border Zone. Journal of Sciences, 15, 378-403.
81. Arian, M., Bagha, N., Khavari, R. and Noroozpour, H. (2012) Seismic Sources and Neo-Tectonics of Tehran Area (North Iran). Indian Journal of Science and Technology, 5, 2379-2383.
82. Moghimi, H., Arian, M. and Sorbi, A. (2015) Fault Movement Potential of Marzanabad Area, North Alborz, Iran. Open Journal of Geology, 5, 126-135.
83. Arian, M. and Pourkermani, M. (2004) Tectonic Elements of South Flank in the East-Central Alborz Mountain. Journal of Sciences, 4, 359-368.
84. Arian, M. and Qorashi, M. (2006) The Movement Potential Evaluation of the Major Quaternary Faults in Alborz-Central Iran Border Zone, from the East of Tehran to the East of Semnan. Journal of Geosciences, 15, 184-188.
85. Poroohan, N., Pourkermani, M. and Arian, M. (2013) An Assessment of Relationship in F-Parameter and Paleostress Fields in Heterogeneous Lithologies: Roudbar Area (Northwest of Iran). Australian Journal of Basic & Applied Sciences, 7, 933-942.
86. Poroohan, N., Poukermani, M. and Arian, M. (2009) An Assessment on Correlations of Seismotectonic Parameters Preceding and Following Roudbar-Manjil Earthquake (Gilan, North of Iran). Australian Journal of Basic & Applied Sciences, 3, 2643-2652.
87. Farrokhnia, A.R., Pirasteh, S., Pourkermani, M. and Arian, M. (2011) Geo-Information Technology for Mass Wasting Hazard Zonation: Central-West Alborz-Iran. Disaster Advances, 4, 24-33.
88. Khavari, R., Ghorashi, M. and Arian, M. (2009) Assessment of Relative Active Tectonics, South Central Alborz (North Iran). EGU General Assembly Conference Abstracts, 11, 1137.
89. Sorbi, A., Arian, M. and Pourkermani, M. (2009) The Movement Potential Evaluation of the Major Quaternary Faults in Tehran Quadrangle. Journal of the Earth, 19, 176-182.
90. Feizi, F. and Arian, M. (2006) The Classification of Thrust Fronts in the Alborz-Central Iran Border Zone from the East of Varamin to the East of Semnan. Journal of Sciences, 16, 75-87.

91. Arian, M. and Pourkermani, M. (2004) Structural Significance of North Semnan and Attary Faults in Alborz-Central Iran Border Zone. Journal of Science, 14, 4551-4569.
92. Arian, M. and Pourkermani, M. (2005) Cenozoic Diastrophism and Deformational Events in the Southern Flank of Central-East Alborz. Journal of Faculty Earth Sciences, 10, 43-51.
93. Sadeghi, R., Saeedi, A., Arian, M., Ghorashi, M. and Solgi, A. (2015) Comparison of Strain Ellipsoid Shape in the South of Ardabil Range (NW), Based on the Results of the Magnetic Susceptibility Anisotropy and Paleostress Methods. Open Journal of Geology, 5, 611-622.
94. Arian, M., Pourkermani, M., Qorashi, M. and Ghasemi, M.R. (2003) North Semnan Fault System and Its Role on Basin Division. Proceedings of the 8th Symposium of Geological Society of Iran, Shahrood, 4-6 September 2003, 11-17.
95. Pourkermani, M. and Arian, M. (2001) Structural Geomorphology of Northeastern Kurdistan. Journal of Humanities, 7, 37-48.
96. Mardani, Z., Ghorashi, M. and Arian, M. (2011) Geomorphic Signatures of Active Tectonics in the Talaghanrud, Shahrudand Sefidrud Drainage Basins in Central Alborz, N Iran. Geosciences, 20, 159-166.
97. Sorbi, A., Arian, M. and Pourkermani, M. (2011) The Application of Geomorphic Indices to the Assessment of Relative Tectonic Activity Levels in Tehran Quadrangle. Journal of the Earth, 6, 1-9.
98. Khavari, R., Ghorashi, M., Arian, M. and Khosrotehrani, K. (2010) Geomorphic Signatures of Active Tectonics in the Karaj Drainage Basin in South Central Alborz, N Iran. Geosciences, 19, 67-74.
99. Mousavi, E.J. and Arian, M. (2015) Tectonic Geomorphology of Atrak River, NE Iran. Open Journal of Geology, 5,106-114.
100. Nouri, R., Jafari, M.R., Arian, M., Feizi, F. and Afzal, P. (2013) Correlation between Cu Mineralization and Major Faults Using Multifractal Modelling in the Tarom Area (NW Iran). Geologica Carpathica, 64, 409-416.
101. Nouri, R., Jafari, M.R., Arian, M., Feizi, F. and Afzal, P. (2013) Prospection for Copper Mineralization with Contribution of Remote Sensing, Geochemical and Mineralographical Data in Abhar 1:100,000 Sheet, NW Iran. Archives of Mining Sciences, 58, 1071-1084.
102. Nouri, R., Afzal, P., Arian, M., Jafari, M. and Feizi, F. (2013) Reconnaissance of Copper and Gold Mineralization Using Analytical Hierarchy Process in the Rudbar 1:100,000 Map Sheet, Northwest Iran. Journal of Mining and Metallurgy, 49, 9-19.
103. Farrokhnia, A.R., Pirasteh, S., Pradhan, B., Pourkermani, M. and Arian, M. (2011) A Recent Scenario of Mass Wasting and Its Impact on the Transportation in Alborz Mountains, Iran Using Geo-Information Technology. Arabian Journal of Geosciences, 4, 1337-1349.
104. Arian, M. and Nouri, R. (2015) Lineament Tectonics and Mineralization in Tarom Area, North Iran. Open Journal of Geology, 5, 115-124.
105. Feizi, F. and Arian, M. (2011) The Role of Structural Controllers in Geneses of Copper Deposits in 1:50000 Map of Saiin Qaleh. Journal of Sciences, 21, 1-10.
106. Bahiraee, S., Arian, M., Qorashi, M. and Solgi, M. (2015) The Movement Potential Evaluation of the Mosha Fault (the West of Firoozkuh to the Shahrestanak). Geosciences, 24, 123-126.
107. Bagha, N., Ghorashi, M., Arian, M., Pourkermani, M. and Solgi, A. (2015) Neotectonic Analysis of Mosha-North Tehran Fault Zone, Based on Morphotectonic Features, Central Alborz, Northern Iran. Geosciences, 24, 41-52.
108. Mosavi, E. and Arian, M. (2015) Neotectonics of Kashaf Rud River, NE Iran by Modified Index of Active Tectonics (MIAT). International Journal of Geosciences, 6, 776-794.
109. Nouri, R. and Arian, M. (2015) Structural Control on the Distribution of Hydrothermal Alteration Zones and Mineralization in Dastjerdeh Area Based on Remote Sensing Data, NW Iran. Bulletin of the Georgian National Academy of Sciences, 9, 79-86.
110. Khosroshahizadeh, S., Pourkermani, M., Almasiyan, M., Arian, M. and Khakzad, A. (2015) Evaluation of Structural Patterns and Related Alteration and Mineralization Zones by Using ASAR-ASTER Imagery in Siyahrood Area (East Azarbaijan—NW Iran). Open Journal of Geology, 5, 589-610.
111. Pahlavani, A., Pourkermani, M., Arian, M. and Vahabzadeh, G. (2010) Structural Analysis of Emaft Fluorite Mine Region (SavaadKuh Mazandaran). Journal of the Earth, 5, 75-81.

112. Bazarchi, S., Jafari, M.R. and Arian, M. (2012) Geochemical Studies and Introduction of Anomaly Zones in Kandelus Sheet (Southwest of Baladeh). Journal of the Earth, 6, 67-75.
113. Bagheri, M., Feiznia, S., Arian, M., Shabanian, R. and Mahari, R. (2013) Continental Trace Fossils in the Semnan Area (Northern Iran). Open Journal of Geology, 3, 54-61.
114. Manafi, M., Arian, M., Raeesi, S. and Solgi, A. (2013) Tethys Subduction History in Caucasus Region. Open Journal of Geology, 3, 222-232.
115. Khosroshahizadeh, S., Pourkermani, M., Almasian, M., Arian, M. and Khakzad, A. (2016) Lineament Patterns and Mineralization Related to Alteration Zone by Using ASAR-ASTER Imagery in Hize Jan-Sharaf Abad Au-Ag Epithermal Mineralized Zone (East Azarbaijan—NW Iran). Open Journal of Geology, 6, 232-250.
116. Mousavi, E., Arian, M. and Ghorashi, M. (2017) Investigation of T Factor Variations along Atrak River to the Caspian Sea. Open Journal of Marine Science, 7, 35-50.
117. Yahyaee, H.N., Arian, M., Pourkermani, M. and Sorbi, A. (2010) Investigating and Analyzing the Present Lineaments in Tehran Quadrangle by Remote Sensing. Journal of Science, 5, 145-161.
118. Sistanipour, A. and Arian, M. (2015) Geometric Analysis of Davaran Fault System, Central Iran. Open Journal of Geology, 5, 458-469.
119. Nazemi, M., Ghorashi, M., Ghassemi, M.R. and Arian, M. (2015) Morphotectonics Features of Alluvial Fans Associated with Active Tectonics (Shotori Mountains, East of Tabas-Central Iran). Geosciences, 24, 91-100.
120. Alizadeh, H. and Arian, M. (2015) Rule of Structural Factors in Formation of Porphyry Copper Deposits in South Western Part of Kerman Area, Iran. Open Journal of Geology, 5, 489-498.
121. Mosavi, E.J. and Arian, M. (2015) Neotectonics of Tabas Area, Central Iran by Index of Active Tectonics (IAT). Open Journal of Geology, 5, 209-223.
122. Daryani, N.J., Arian, M. and Omran, N.R. (2015) Tectonics and Mineralization of Copper in the Ardestan-Kahang Area, Central Iran by Remote Sensing. Open Journal of Geology, 5, 188-196.
123. Arian, M. and Pourkermani, M. (2001) Rivers Morphology and Active Tectonic (Reviewing the Current Status of Ghezel Ozon River in the Province of Zanjan). Proceedings of the 5th Conference of Geological Society of Iran, Tehran, 28-30 August 2001, 556.
124. Eshghi, Z., Arian, M. and Pourkermani, M. (2012) Structural Investigation on the Lak Mining Area (Bueen Zahra) Based on Remote Sensing, Used for Its Mineralization. Journal of the Earth, 6, 145-155.
125. Arian, M., Toudeshki, V.H. and Noroozpour, H. (2011) Active Tectonics of Qezel Ozan River Basin, NW Iran. Journal of Applied Environmental and Biological Sciences, 1, 291-295.
126. Alizadeh, H., Arian, M., Lotfi, M., Ghorashi, M. and Ghorbani, M. (2015) Determination of Porphyry Copper Deposit Locations Using Photo Lineament Factor in Northern Parts of the Dehaj-Sardoiyeh Belt. Geosciences, 24, 247-252.
127. Toudeshki, V.H., Pourkermani, M., Arian, M. and Khosrotehrani, K.H. (2011) Influence of Structures on the Ghezel Ozan River. Geosciences, 21, 55-60.
128. Toudeshki, V.H. and Arian, M. (2011) Morphotectonic Analysis in the Ghezel Ozan River Basin, NW Iran. Journal of Geography and Geology, 3, 258-260.
129. Arian, M. (2015) Seismotectonic-Geologic Hazards Zoning of Iran. Earth Sciences Research Journal, 19, 7-13.
130. Arian, M., Pourkermani, M., Sistanipour, A. and Noroozpour, H. (2011) Kinematic Significance of Fold- and Fault-Related Fracture Systems in the Rafsanjan's Northeast Highlands (Central Iran). Journal of Basic and Applied Scientific Research, 1, 3398-3406.
131. Arian, M., Pourkermani, M., Sistanipour, A. and Noroozpour, H. (2011) Seismicity and Fault Segmentation of Bafq-Baghin Fault System (Central Iran). Journal of Applied Environmental and Biological Sciences, 1, 382-396.
132. Mosavi, E.J., Arian, M., Ghorashi, M. and Nazemi, M. (2012) Measurements of Geomorphic Indices in Tabas Area. Journal of the Earth, 7, 213-225.
133. Arian, M. (2010) Earthquake-Fault Hazard Investigations in the Kerman Quadrangle. Journal of Sciences, 19, 176-182.

134. Arian, M. and Sorbi, A. (2011) Remote Sensing Investigation of the Faults and Lineaments in the Karaj-Qazvin Region. Journal of the Earth, 5, 21-34.
135. Arian, M., Alizadeh, H. and Noroozpour, H. (2011) Satellite Geometry of Faults and Fractures and Its Relationship with Porphyry Deposits in Northern Parts of Dahaj-Sardoiyeh Belt, South of Iran. Indian Journal of Science and Technology, 4, 1303-1306.
136. Pourkermani, M., Mohajjel, M., Solgi, A., Arian, M. and Nadri, R. (2013) Active Tectonics of South Qom. Applied Geology, 8, 203-213.
137. . Noroozpour, H., Rad, Y.M., Arian, M., Solgi, A. and Vafa, N.N. (2013) Permo-Triassic Deposits of Shorjestan Area, Central Iran: The Palynological Report of the Greatest Phanerozoic Disaster in Iran. Disaster Advances, 6, 48-53.
138. Mirzeinaliyazdi, S.H., Pourkermani, M., Arian, M. and Almasian, M. (2012) Geometrical Analysis of Fractures in Sabzevar Ophiolite Zone. Journal of Geotechnical Geology (Applied Geology), 7, 359-371.
139. Pourkermani, M., Mohajjel, M., Solgi, A., Arian, M. and Nadri, R. (2012) Active Tectonics of South Qom, a Hazard for New Residential Areas. Journal of Geotechnical Gology (Applied Geology), 8, 203-213.
140. Pourkermani, M. and Arian, M. (2016) Analysis of Structural Specifications in Shah Neshin Area (the Eastern Border of Kordestan Province). Journal of Sciences, 11, 2859-2869.
141. Yazdi, S.H.M., Pourkermani, M., Arian, M. and Karkheiran, A. (2012) Determination of Stress Orientation in Sabzevar Ophiolite Zone in (Khorasan Razavi Province, Iran). Indian Journal of Science and Technology, 5, 3500-3505.
142. Keynezhad, A., Pourkermani, M., Arian, M., Saeedi, A. and Lotfi, M. (2010) Research on Fractures in North of the Torud-Moalleman Area (Central Iran, Southeastern of Damghan) and Their Relation with Mineralization. Journal of the Earth, 5, 81-97.
143. . Mahmoodi, M., Bahroudi, A., Ghorbani, M. and Arian, M. (2013) Preliminary Exploration of Copper Minerals in Jebal Barez Mountains, Iran. Open Journal of Geology, 3, 201-208.
144. Keynezhad, A., Pourkermani, M., Arian, M. and Saeedi, A. (2011) Dynamic Analysis of Fractures in North of Torud-Moalleman Area (Central Iran, East South of Damghan). Geosciences, 20, 3-16.
145. Pourkermani, M., Qorashi, M., Arian, M., Sorbi, A., Farokhnia, A. and Solgi, A. (2003) Structural Features in Western Part of the Lut Region. Iranian Journal of Geology, 1, 29-44.
146. Arian, M. (2003) Seismotectonic and Earthquake-Fault Hazard Investigations in the Kerman Region. Research Institute of Applied Sciences.

CHAPTER 14

Evaluate the Accuracy of Fargas and BLM Models for Identification of Erosion Intensity

Naser Abdi

Department of Geology, Ahar Branch, Islamic Azad University, Ahar, Iran

ABSTRACT

Erosion process not only changes the land use to badlands, but also produces sediments that are dumped in the dam reservoirs, reduces the reservoir volume and finally makes it useless. So, it is necessary to do study on erosion intensity and the sediment production evaluation for which there are some methods and models. The experimental Fargas and BLM models are largely used for this issue, in Iran as well as many other countries, separately or together based on the data availability. These studies results are as data for sediment supply estimation in different watershed management studies. So, the result accuracy is important for determination of sediment occurrence. This study evaluates these models' results accuracy, in order to find the limitations and any solutions. Therefore, these two models were used and run in the same area, Aghbolagh drainage basin, Iran; the results were compared and evaluated. These models are based on some factors like rock type, drainage density, surface erosion and litter cover. The study includes field and laboratory analysis. The data were combined in GIS software and processed. The results reveal that, Fargas model predicts 3.67%, 14.26% and 81.06% of the area susceptible for high, severe and very severe erosion respectively; whilst, referring to BLM model outcome, 42.96% area has high sensitivity; 42.96% and 24.94% of the area have high and severe sensitivity for erosion, respectively. Furthermore, both models show same severity for around 18% of the study area. So, for these two models results very low similarities are concluded, which could be an indication of low reliability of the results, especially when they are used separately without any combination or comparison. Finally, it is recommended to use both of them together, or use another method beside each mentioned models.

Keywords: Erosion, Fargas, GIS, BLM, Sediment

1. INTRODUCTION

Erosion is a type of soil and rock weathering in which superficial layer is worn away through the action of glaciers, wind and water [1] . Although both soil and rock erosion are important, cultivation and life system soil erosion is more important, and makes social and economic problem and ecosystem changes [2] [3] [4] .

Erosion monitoring is more effective than control of sediment using different methods. By detecting different erosion problems and classifying the area into different levels of susceptibility to erosion, it will be possible to use cost effective erosion control measurements [5] .

Different erosion types like rill, gully, stream and channel are used for erosion severity and sediment production intensity from which gully erosion is more important, as it is occurred in the last stages of erosion progress, indicating high sediment production, but its study has been limited recently [6] . Whilst, study on gully erosion based on historical evidence, photology or remotely sensed images, is very practical. The result is also very applicable, as there is a strong correlation between the length and volume of the gully [7] .

Drainage area of any basin is assumed as its sediment source, so the area is mapped and classified into erosion susceptive aprons to make remedial programs in order to reduce the erosion intensity, decrease soil erosion and overcome to sediment production and transform to dam reservoirs. This study uses different factors such as lithology, drainage pattern, steep slopes, erosion types and land use, to map and classify the drainage area using different models, and make comparison for the results.

2. STUDY AREA

The 78.2 Km^2 study area is located in the northern Hashtrood City, between 37°28'51"N - 37°36'39'N and 46°59'59"E - 47°05'36"E (Figure 1) with 2110 and 1668 m above sea level maximum and minimum elevations respectively. Average precipitation is 325 mm, and could be assumed as cold semiarid. The main land use, 82 percent of the study area, is dry farm.

Geological studies illustrate the area has rarely been affected by tectonic forces, and sedimentary rocks are the predominant lithology. Unconsolidated conglomerates, sandy marls and alluvial sediments cover 81% of study area, which cause high potential for erosion.

3. METHODOLOGY

Two empirical methods, Fargas and BLM were used in this study to identify the critical sediment resources. The first one as a qualitative method which was developed by Fargas et al. [8] was initially used in 1500 Km^2 area watershed of the Joaquin Costa reservoir (NE Spain). This method uses reliable but limited parameters of lithology and drainage density which is rated according to erosion class. It is supposed that too less factors are easy to obtain, but could be the weak point.

The second one, BLM methodology is also applicable method for sediment risks evaluation. This technique uses seven factors, and the results are classified according to five major erosion classes. If these factors study carefully, the result will be reliable.

4. DISCUSSION

Two main models, Fargas and BLM, which are generally used for erosion assessment and soil conservation planning, were used to categorize the erosion intensity, erosion risk assessment, sediment production and critical sediment resources detection. Finally the results of above mentioned methods were appraised in comparison with each other and the similarities were recognized.

4.1. Fargas Model

The Fargas model is based on two main factors, lithology and drainage density. So, these factors are prepared as geological maps, and used in GIS software (Figure 2) and each unit area was calculated as percentage of total area (Table 1). This study uses geological map which was prepared based on detail field study as well as laboratory analysis, considering its importance. According to the map and Table 1, Q^{t2} geological unit, composing of terrace alluviums which have high resistance index (erosion susceptible) covers about 46.5% of the area, whilst the hard rocks of gr, O_r, E_{va} and E_a have frequency of 14.26, 0.68, 0.68 and 3.67 percent, respectively. Qpl^c, $Qpl^m$and Q^{al} are other soft lithological units with 4.7, 27.56 and 2.21 percentage, correspondingly.

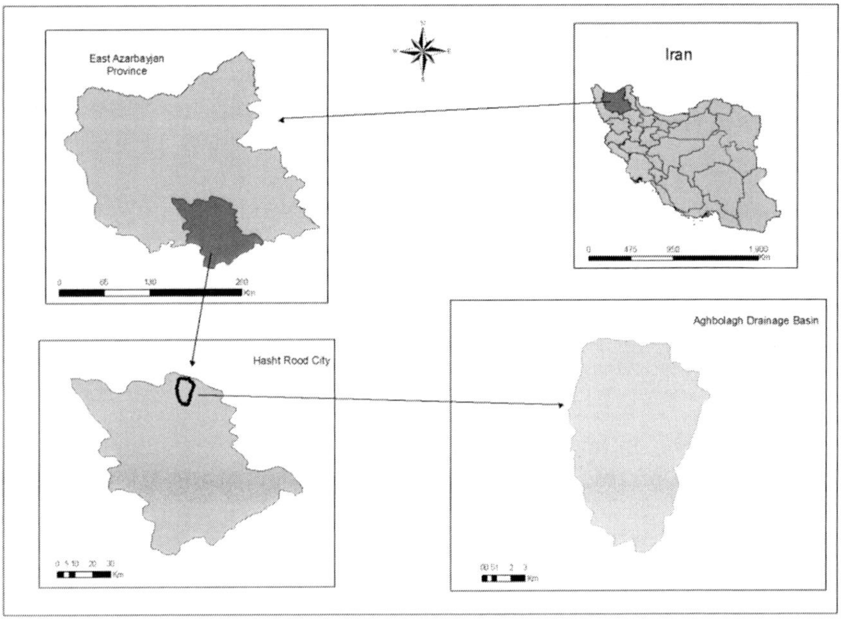

Figure 1. Location of study area.

The drainage density, the second main factor of this model, was prepared based on the detailed topography maps (Figure 3); then, erosion class which is attached to an elementary catchment depending on its drainage density [10] , were prepared (Table 2).

The coefficient value for each lithological unit from which the emission risk is obtained, is determined by multiplying the erosion class according the drainage densityand the weighting factor [6] . Table 3 shows the calculated values and the erosion risk for different geological units of study area. The critical sediment source could be calculated based on the risk data. This table shows the erosion risk of slight, moderate, high, severe and very severe as 0.68%, 0.33%, 3.67%, 14.26% and 81.06% of total area, respectively. The data is presented as map of erosion risk and critical sediment source (Figure 4).

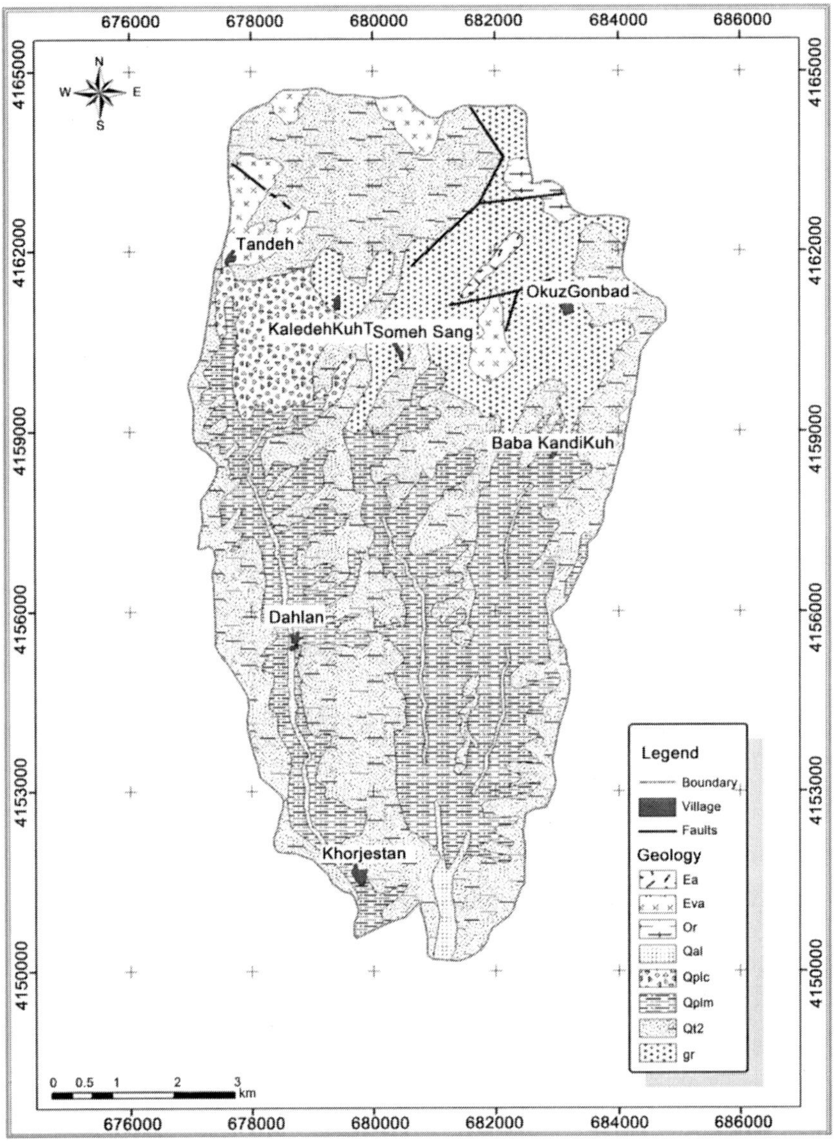

Figure 2. Geological map of the study area.

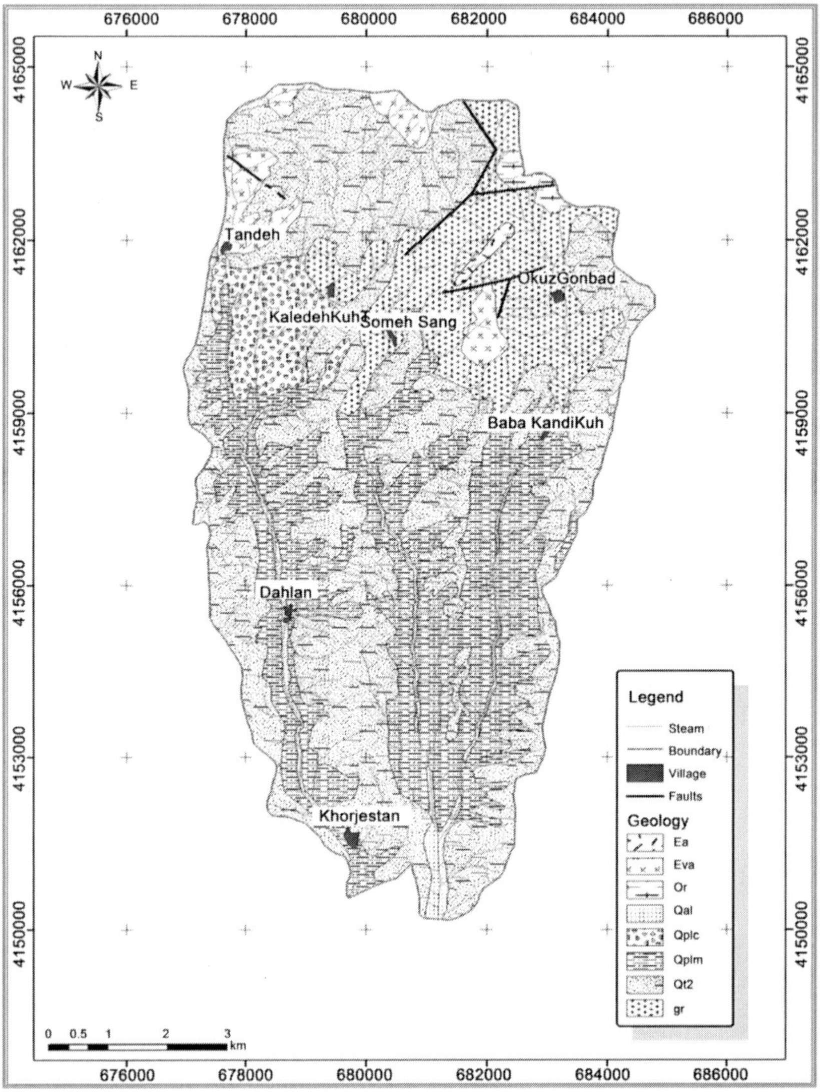

Figure 3. Drainage density on geological map.

4.2. BLM Model

BLM methodology, initiated by Bureau of Land Management (BLM) of the United State of America, is mainly base on the erosion types. This model is integrated seven factors of surface erosion variable from (0 - 14), Litter covering the surface (0 - 14), rock fragment (0 - 14), superficial degradation (0 - 14), rill erosion (0 - 14), runoff sediments precipitation (0 - 15) and gully erosion (0 - 15); to calculate the Soil Surface Factor (SSF), which is the base for assessment of erosion intensity and erosion factor determination (Table 4). Every factor of this model was studied and mapped in GIS software to get the results as Table 5, Figure 5 and Figure 6. According to Table 5, the study area could be dissection in 3 major erosion risk, including moderate, high and severe with area of 32.1%, 42.96% and 24.94% respectively.

Table 1. Geological units resistance index [9] .

Rocks and sediments	Geological units	Resistance index	Area (% of total area)
Hard Rocks	gr	4	14.26
	O_r	1	0.68
	E_{va}	3	0.68
	E_a	2	3.67
Soft Rocks (Unconsolidated Rocks)	Qpl^c	6	4.7
	Qpl^m	9	27.56
Old alluvial deposits	Q^{t2}	8	46.5
Recent alluvial deposits	Q^{al}	7	2.21

5. CONCLUSION

Fargas and BIM models are mostly used to determine erosion risk, erosion intensity and critical sediment source. Fargas model generally is more common, as it uses less factors, which are easy to produce the data. BLM model covers more components and needs data to be studied carefully and integrated to one aspect as soil surface factor. This study uses both model to evaluate erosion severity in the same area, in order to find the similarities and differences. The results show, generally there are similar classes between both model results, but deep investigation on the calculated classes area, re- veals that there are no enough coverage between the final outcomes of these models. Fargas model predicts 3.67%, 14.26% and 81.06% of the area susceptible for high, severe and very severe erosion respectively; whilst, referring to BLM model outcome, 42.96% area has high sensitivity; 42.96% and 24.94% of the area have high and severe sensitivity for erosion, respectively. Furthermore, both models show the same severity for around 18% of the study area. This means, by applying two models in the same study and making comparison between the products, we find advantages and disadvantages of the methods, and understand the shortage of any model, or any faults during the study. Finally it is recommended to never rely on one of these models, or maybe any others, even very common methods.

Table 2. Erosion class based on drainage density.

Class	Geological Unit	Erosion degree	Drainage Density (Km/Km^2)	Rating value
1	E_{va}	Very severe	4.34	10
2	E_a	Severe	1.76	8
3	O_r	Very severe	2.72	10
4	gr	Very severe	4.44	10
5	Qpl^c	Very severe	5.37	10
6	Qpl^m	Very severe	5.49	10
7	Q^{t2}	Very severe	3.04	10
8	Q^{al}	Very severe	13.85	10

Table 3. Classification of sediment emission risk.

Geological unit	Lithology	Resistance index	Area (% of total)	Drainage length	Drainage Density (Km/Km^2)	Rating value	Coefficient value	Risk
E^{va}	Andesitic lava, Tuff	3	3.67	12.49	4.34	10	30	High
E^a	Andesite, Andesitic basalt	2	0.33	0.45	1.76	8	16	Moderate
O_r	Rhyolite	1	0.68	1.45	2.72	10	10	Slight
gr	Granite	4	14.26	49.6	4.44	10	40	Severe
Qpl^c	Unconsolidated Conglomerate	6	4.71	19.77	5.37	10	60	Very Severe
Qpl^m	Sandy marl	9	27.57	118.5	5.49	10	90	Very Severe
Q^{t2}	Young terraces	8	46.46	11	3.04	10	80	Very Severe
Q^{al}	Alluvium	7	2.32	25.1	13.85	10	70	Very Severe

Table 4. The erosion risk based on the 7 factors (bureau of land management).

Sum 7 factors score	Erosion Risk
0 - 20	Low
21 - 40	Moderate
41 - 60	High
61 - 80	Severe
81 - 100	Very Severe

Table 5. Erosion risk in the study area based on SSF calculation.

Class	7 Factors of BLM							Total score	Risk	Area (% of total Area)
	F1	F2	F3	F4	F5	F6	F7			
E1	6	6	6	6	7	7	4	42	High	18.59
E2	6	6	6	6	8	8	0	40	Moderate	13.12
E3	4	4	4	4	5	5	5	31	Moderate	18.98
E4	6	6	6	6	8	8	8	48	High	24.37
E5	8	8	8	8	11	11	8	62	Severe	7.72
E6	8	8	8	8	11	11	11	65	Severe	17.22

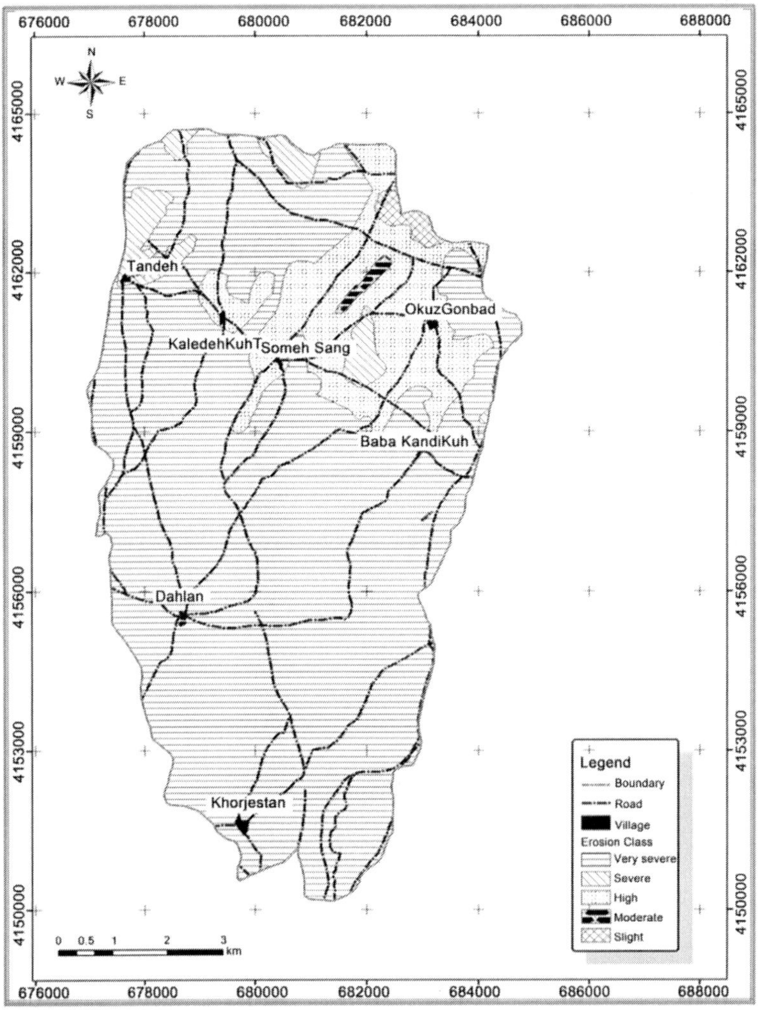

Figure 4. Risk of erosion based on Fargas method.

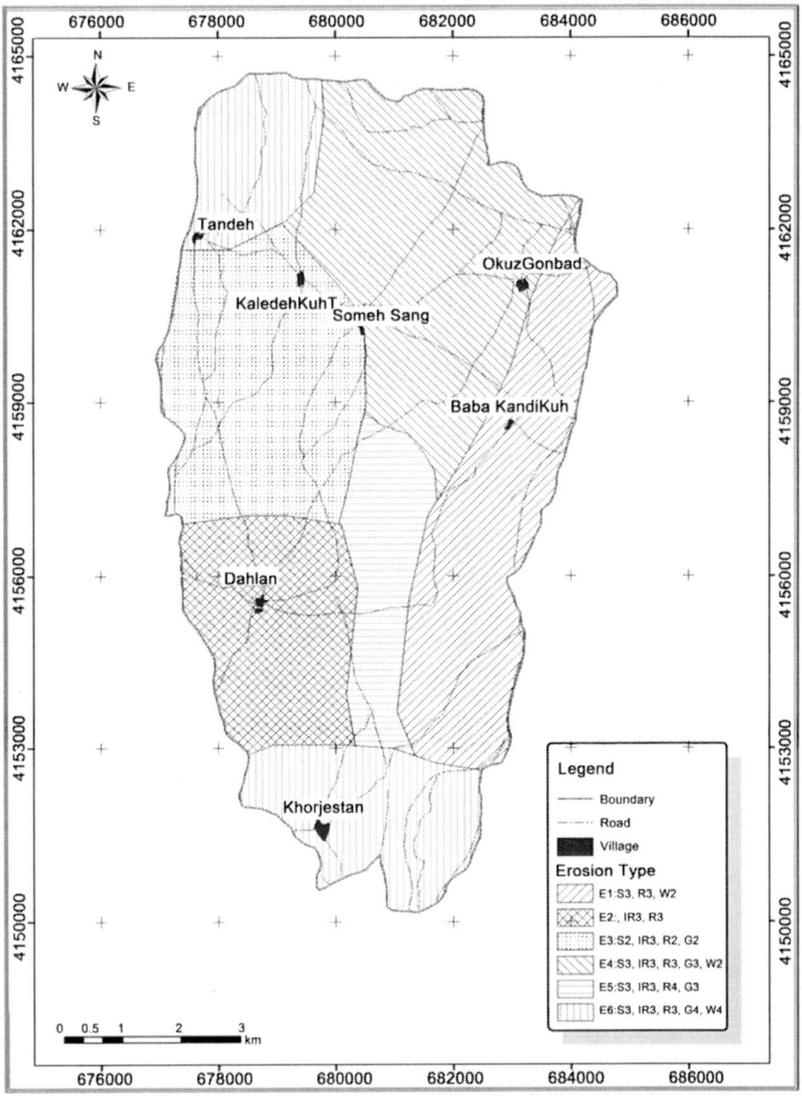

Figure 5. Erosion types in the study area.

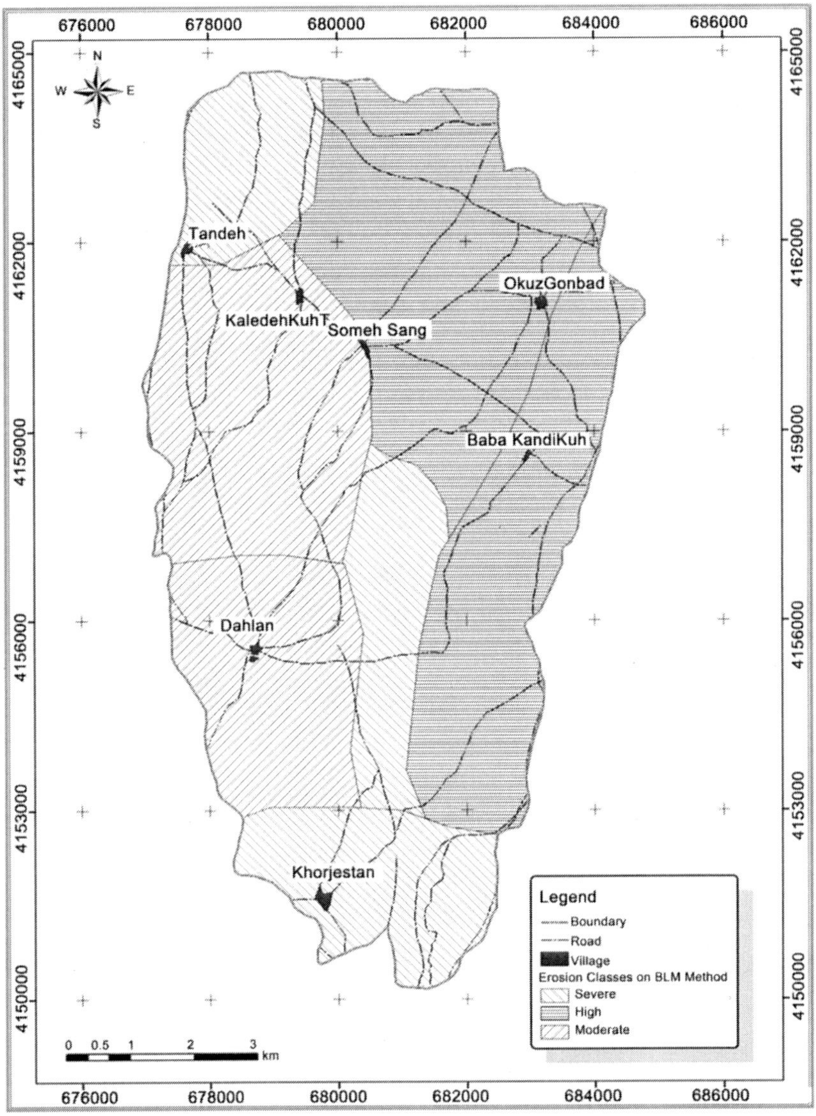

Figure 6. Erosion risk based on BLM method.

ACKNOWLEDGEMENTS

This study is based on the research plan of "Evaluation of Sediment Production intensity in Different phases of basic studies (Case Study, North of Hashtrood City)" which is supported by Islamic Azad University (IAU), Ahar Branch. We are thankful to the chancellor, research deputy and all of the staffs of the university which gave support us in this issue.

The author acknowledges the immense help received from the scholars whose articles are cited and included in references of this manuscript. The author is also grateful to authors/editors/publishers of all those articles, journals and books from where the literature for this article has been reviewed. The anonymous editor(s) of this paper is also respected for their comments.

CITE THIS PAPER

Abdi, N. (2016) Evaluate the Accuracy of Fargas and BLM Models for Identification of Erosion Intensity. Open Journal of Geology, 6, 1457-1468.

REFERENCES

1. Soil Science Society of America (2001) Glossary of Soil Science Terms. Soil Science Society of America, Inc., Madison.
2. Moran Lu, H. and Prosser, C.J. (2005) Modeling Sediment Delivery Ratio over the Murray Darling Basin. Environmental Modelling & Software, 21, 1297-1308.
3. Santhi, C., Srinivasan, R., Arnold, J.G. and Williams, J.R. (2006) A Modeling Approach to Evaluate the Impacts of Water Quality Management Plans Implemented in a Watershed in Texas. Environmental Modelling & Software, 21, 1141-1157.
4. Miller, S.N., Semmens, D.J., Goodrich, D.C., Hernandez, M., Miller, R.C., Kepner, W.G. and Guertin, D.P. (2007) The Automated Geospatial Watershed Assessment Tool. Environmental Modelling & Software, 22, 365-377.
5. Clean Water Services (2008) Erosion Prevention and Sediment Control Planning and Design Manual.
6. Poesen, J., Nachtergaele, J., Verstraeten, G. and Valentin, C. (2003) Gully Erosion and Environmental Change: Importance and Research Needs. Catena, 50, 91-133.
7. Cheng, H., Zou, X., Wu, C., Zhang, C., Zheng, Q. and Jiang, Z. (2007) Morphology Parameters of Ephemeral Gully in Characteristics Hillslopes on the Loess Plateau of China. Soil and Tillage Research, 94, 4-14.
8. Fargas, D., Martínez-Casanovas, J.A. and Poch, R. (1997) Identification of Critical Sediment Source Areas at Regional Level. Journal of Physics & Chemistry of the Earth, 22, 355-359.
9. MOPT (1992) Propuesta del proyecto de directrices. Cuenca del Ebro. Confederación Hidrográfica del Ebro-MOPT.
10. Stroosnijder, L. and Eppink, L.A. (1993) Principles of Soil and Water Conservation. Lecture Notes of Course K200-500/510. WAU, Wageningen.

CHAPTER 15

Soil Moisture Estimation in South-Eastern New Mexico Using High Resolution Synthetic Aperture Radar (SAR) Data

A.K.M. Azad Hossain [1,*] and Greg Easson [2]

[1]*Department of Geology and Geological Engineering, The University of Mississippi, 120A Carrier Hall, University, MS 38677, USA*

[2]*Mississippi Mineral Resources Institute, The University of Mississippi, University, MS 38677, USA*

ABSTRACT

Soil moisture monitoring and characterization of the spatial and temporal variability of this hydrologic parameter at scales from small catchments to large river basins continues to receive much attention, reflecting its critical role in subsurface-land surface-atmospheric interactions and its importance to drought analysis, irrigation planning, crop yield forecasting, flood protection, and forest fire prevention. Synthetic Aperture Radar (SAR) data acquired at different spatial resolutions have been successfully used to estimate soil moisture in different semi-arid areas of the world for many years. This research investigated the potential of linear multiple regressions and Artificial Neural Networks (ANN) based models that incorporate different geophysical variables with Radarsat 1 SAR fine imagery and concurrently measured soil moisture measurements to estimate surface soil moisture in Nash Draw, NM. An artificial neural network based model with vegetation density, soil type, and elevation data as input in addition to radar backscatter values was found suitable to estimate surface soil moisture in this area with reasonable accuracy. This model was applied to a time series of SAR data acquired in 2006 to produce soil moisture data covering a normal wet season in the study site.

Keywords: SAR backscatter; soil moisture estimation; semi-arid environment; multiple regressions; artificial neural networks; south eastern New Mexico

1. INTRODUCTION

Soil moisture, is an important hydrologic variable that controls the interactions (and feedbacks) between land surface and atmospheric processes [1]. It plays a very important role in the distribution of precipitation between runoff and infiltration. Soil moisture monitoring and characterization of the spatial and temporal variability of this hydrologic parameter at scales from small catchments to large river basins continues to receive much attention, reflecting its critical role in subsurface-land surface-atmospheric interactions and its importance to drought analysis, crop yield forecasting, irrigation planning, flood protection, and forest fire prevention [2,3,4].

In semi-arid environments ground water recharge is one of the most difficult parameters to quantify, where a number of recharge mechanisms, including soil moisture change, on variable temporal and spatial scales. Several studies showed that temporal analysis of soil moisture can be used to understand ground water recharge [5,6]. Remote sensing technology has been used successfully to estimate soil moisture [3,7,8,9,10] and map its spatio-temporal distribution in semi-arid environments and could potentially contribute to ground water recharge studies.

Synthetic Aperture Radar (SAR) data are particularly well suited for estimating soil moisture due to the relationship between the dielectric constant and soil moisture [11,12]. The microwave measurements are strongly dependent on the dielectric properties of the target. For a soil, dielectric properties are a function of the amount of water present. The real part of the complex dielectric constant of water in this spectral region (microwave) is approximately 80 comparing to the value for dry soil is about three. This large contrast provides a basis for estimating the moisture for dielectric values between these two extremes [13,14]. It is not always possible to take advantage of the dielectric constant and soil moisture relationship since microwave measurements are also influenced by surface roughness, which varies significantly from place to place due to diverse land use and land covers. That's why no global SAR based operational algorithm exists for estimating soil moisture [15].

Studies, particularly in the past two decades, have resulted in a multitude of methods, algorithms, and models relating satellite-based radar backscatter imagery to estimate surface soil moisture content [3,10,11,16,17,18,19,20]. The most commonly used algorithms are developed by a semi-empirical approach [21,22]. In a given bare soil condition, radar backscatter is linearly dependent on volumetric soil moisture content (θ_w) in the upper 2 to 5 cm of soil with a correlation (R^2)~ 0.8 to 0.9 [21,22]. However, the presence of vegetation cover complicates soil moisture estimation due to the interaction of the microwaves with the vegetation and soil [12]. The radar backscatter from a surface with vegetation consists of three components: (1) product of the backscatter contribution of bare soil surface ($\sigma_s°$) and the two way attenuation of the vegetation layer (τ^2), (2) the direct backscatter contribution of the vegetation layer ($\sigma°_{dv}$), and (3) multiple scattering involving the vegetation elements and the ground surface ($\sigma°_{int}$) [12].

$$\sigma^{\circ} = \tau^2 \sigma_s^{\circ} + \sigma_{dv}^{\circ} + \sigma_{\text{int}}^{\circ} \quad (1)$$

In soil moisture estimation for semi-arid environments using different SAR data the influence of sparse vegetation were found negligible by several studies [11,16,17,18,19,22]. That means, for a given soil with uniform surface roughness (*R*), θ_w can be estimated using the following simple linear regression expression since in this case $\sigma° \approx \sigma°_s$.

$$\theta_w = a + b\sigma^{\circ} \quad (2)$$

where *a* and *b* are regression coefficients, usually determined from field experiments encompassing the target-invariant, scene-invariant SAR wavelength, incidence angle, polarization, and calibration.

Using high resolution SAR data, however, it is not always possible to obtain strong linear relationship between measured soil moisture and radar backscatter even in semi-arid environments. Depending on the amount of vegetation present, its dielectric properties, height and geometry, the sensitivity of microwave backscatter to variations in volumetric soil moisture may be significantly reduced [23]. As shown by

[23], in a semi-arid environment of south-eastern New Mexico vegetation coverage can significantly reduce the accuracy of soil moisture estimation using numerical models based on simple linear (R^2 = 0.05 to 0.24) and non-linear (R^2 = 0.24) relationships between radar backscatter from high resolution SAR imagery and near real time *in situ* soil moisture measurements. In addition to the vegetation density, the SAR backscattering can also be influenced by the variation in soil types and soil salinity since they may affect the surface dielectric properties. Topographic variations can also influence the distribution of soil moisture in the field.

This paper investigated the potential of multiple linear regressions and Artificial Neural Networks (ANN) based models to improve soil moisture estimation in south-eastern New Mexico using high resolution Radarsat 1 SAR imagery. The models used SAR backscatters and near real time soil moisture measurements along with vegetation density, soil type, elevation, and soil salinity measurements. A time series of SAR based soil moisture estimation data were generated for an entire wet season in the study site using the developed numerical models.

2. STUDY SITE

A small area in south-eastern New Mexico called Nash Draw was selected as the study site for this study (Figure 1). It is located about 30 km east of Carlsbad, NM. It covers an area of about 400 km^2 and the study site occupies 225 km^2 of it. The extent of the site is limited between 103.78°W–103.92°W longitude and 32.23°N–32.36°N latitude. It is a part of the north-eastern Chihuahua desert, which is characterized by semi-arid environments and sparsely vegetated rangeland. The topographic relief in the region is not significant. The maximum relief across the area is approximately 200 m.

Nash Draw is a karst valley that developed in response to subsurface dissolution of evaporites (including halite and sulfate rocks) and subsidence of the overlying strata [24]. It is a complex example of the localized effects of evaporite karst on surface topography, near-surface geology, and hydrology [25]. Different areas of Nash Draw display small karst features, including caves, sinkholes, dolines, and larger integrated forms such as valleys or elongated depressions [25].

The hydrologic system within Nash Draw is poorly understood. Much of Nash Draw exhibits no significant integrated surface drainage [25], while much of the area is covered by a thick blanket of dune sand. That is why it is very difficult to identify the potential locations for ground water recharge in this area. Therefore, to assist in understanding the existing hydrologic processes modifying Nash Draw it is very important to determine the spatio-temporal distribution of surface soil moisture.

Since the study site covers an area of 225 km^2 high resolution SAR data would be very suitable to obtain remote sensing derived soil moisture estimation for this area. This is the reason Nash Draw was found suitable as a test site for the conducted study.

Rainfall in Nash Draw is unreliable and erratic, with August as the wettest part of the year and the rainy season ending in October. Therefore, it was considered that imagery acquired from August to November should record the maximum variation of soil moisture in the study site.

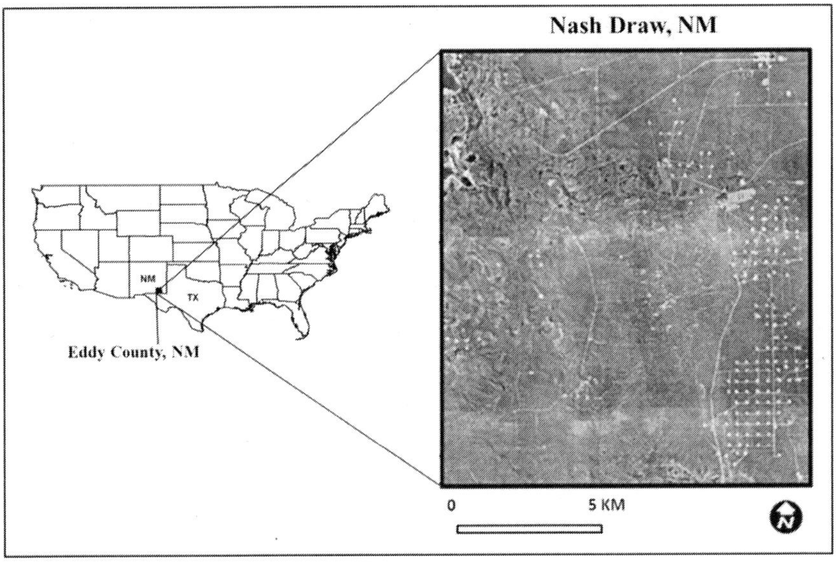

Figure 1. Location of study site.

3. DATA ACQUISITION AND PROCESSING

3.1. Synthetic Aperture Radar (SAR)

Due to its high spatial resolution, Radarsat 1 SAR Fine imagery was selected for estimating soil moisture in the study site. Five scenes of Radarsat 1 SAR Fine imagery, with 10 m spatial resolution and 50 km swath, covering the Nash Draw area were acquired from the Alaska Satellite Facility (ASF). The scenes were acquired in descending mode at a 37° incidence angle on 2 and 26 August, 19 September, 13 October, and 6 November of 2006. Figure 2 shows the Time series of the acquired SAR imagery.

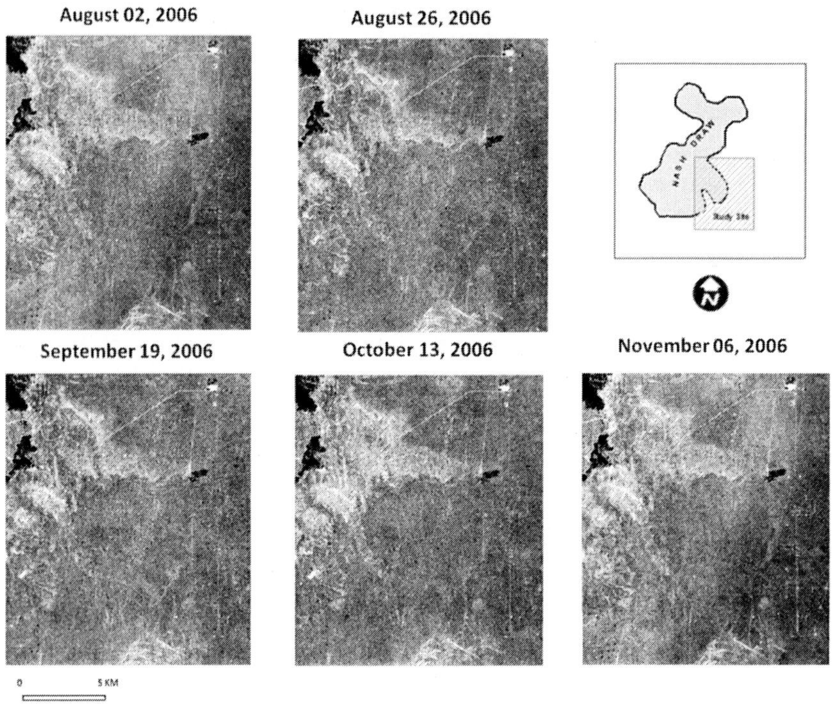

Figure 2. Time series of acquired SAR imagery.

3.1.2. SAR Pre-Processing

All acquired SAR imagery were received as Level 0 products and then converted to level 1 products. These Level 1 SAR data were terrain corrected, calibrated, and filtered in preparation for soil moisture estimation.

Data Calibration

Sigma-naught ($\sigma°$) describes the average reflectivity or scattering coefficient of a radar scene. Beta-naught ($\beta°$) interprets the brightness estimates of mean reflectivity and separates the radiometric response and reflectivity dependent on the terrain properties, such as local slope. According to [26]:

$$\sigma^o = \beta^o + 10\log_{10}(\sin I) \quad (3)$$

$$\beta^o = 10\log_{10}(DN^2 + A_3/A_2) \quad (4)$$

where, *DN*—Digital number, A_3—fixed offset from the radiometric data record, A_2—Look-Up Table (LUT) value, and *I*—local incidence angle.

Calibrating the radar scene in sigma naught (σ°) requires normalization using the knowledge of local incidence angle as we can see in Equation (3). However, this normalization is not required for beta naught (β°) and for an area with low relief (that is, no significant influence with the variation of local incidence angle). Since the topographic relief is not significant in the study site the acquired SAR data were calibrated as beta naught for this research. An experiment was conducted using one radar scene to determine the suitability of σ° or β° for soil moisture estimation in the study site. It was found that backscatter values as β° show better correlation with soil moisture than backscatter values as σ° in the study site. It is believed that β° produces a better estimation of soil moisture due to the low relief in the study site [27,28].

Removing Speckles

In radar remote sensing, coherent imaging systems produce images with a granular appearance. Bright and dark spots caused by random constructive and destructive interference of the wavelets returning from the various scatterers within the resolution cell of the system [29]. The effect of this interference process is regarded as a multiplicative noise, called speckle. Removing speckles is an essential component of SAR image pre-processing techniques. Speckles should be reduced, preserving edges and image texture to extract information from SAR imagery [30].

The best approach for removing speckles in SAR imagery is to use speckle removing filters. These speckle removing filters are developed on the basis of the multiplicative model with an exponential probability density function [31]. To remove the speckles in the acquired radar scenes different types of filtering techniques available in the Erdas Imagine image processing software were tested with different window sizes. The "Lee" filtering technique with 5 × 5 window size was found efficient in removing the speckles and was used to de-speckle the acquired radar scenes.

Geometric Correction

All SAR data were obtained as georectified using the Universal Transverse Mercator (UTM) projection system from the Alaska Satellite

Facility (ASF). The GIS coverage of the sample locations and all other GIS data were georectified to the same projection system. A recent aerial photograph was used to aid the georectification process.

3.2. Digital Elevation Model (DEM)

The knowledge of the local incident angle is essential for the quantitative estimation of soil moisture and roughness from SAR data. In the absence of topographic relief the local incident angle equals the radar look angle. In areas with large topographic relief, the incident angle becomes a function of the radar look angle and the local terrain slope. This makes the straightforward surface parameter estimation difficult. It is then necessary to correct the SAR data for terrain to allow geometric overlays of remotely sensed data from different sensors and/or geometries. The elevation of the study site varies from 900 m to 1100 m. Despite the minimal topographic influence, a terrain correction was performed using USGS 30 m Digital Elevation Model (DEM) (Figure 3) to improve geometric accuracy of the data. Since topographic variations influence the distribution of soil moisture in the field the acquired DEM was also used in the SAR soil moisture estimation models.

3.3. In Situ Soil Moisture Measurements

Near-real time soil moisture data is needed for quantitative estimation of soil moisture using SAR data. Soil samples were collected in the selected parts of Nash Draw and analyzed them to calculate volumetric soil moisture for calibrating the SAR imagery acquired on 2 August 2006 and 6 November 2006. Eighty soil samples were collected in the study site for each date.

3.3.1. Sampling Technique

A stratified soil sampling technique [32] was used to acquire the soil samples. The study site was divided into four equal parts and random sample points were selected in each part using 500 m grid spacing, with 20 samples collected in each part. Accessibility and variation in soil types was used to select sampling sites. Figure 4 shows the distribution of the soil sample locations in the study site.

RADARSAT 1 acquires imagery in C-band with HH polarization at 5.7 cm wavelength and 5.3 GHz frequency. According to [33] the RADARSAT 1 beam should be able to penetrate ground up to top few cm in dry bare soil condition. Therefore, samples were collected at a depth of approximately 3 cm below the land surface.

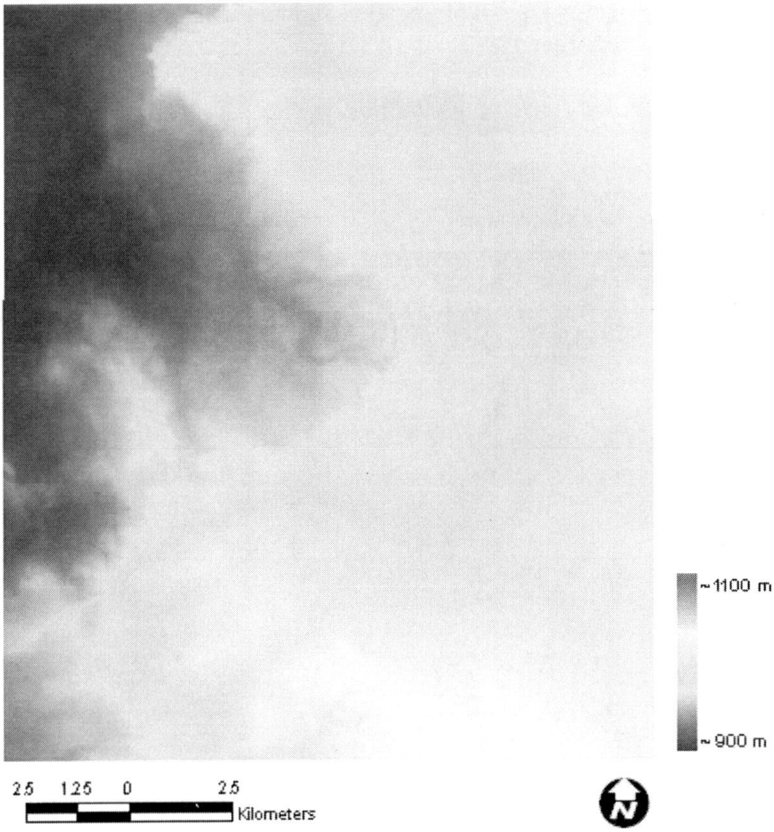

Figure 3. Digital Elevation Model (DEM) for the study site.

3.3.2. Volumetric Soil Moisture Measurement

Gravimetric soil moisture values were calculated from the obtained soil samples using the ASTM standard test method [34]. It was then converted to volumetric soil moisture using sample volume (Equations (5) and (6)). Table 1 shows the statistics of the soil moisture measurements for both August and November data sets.

$$V_w = \frac{(M_w - M_d)}{\rho_w}(cm^3) \tag{5}$$

where, V_w—the volume of water content in the obtained soil sample (cm^3), M_s—the mass of moist soil (g), M_d—the mass of dry soil (g), ρ_w—

water density (g/cm³), θ_v—volumetric soil moisture (%), and the volume of the sample container V_s = 76.55 (cm³).

$$\theta_v = \frac{V_w}{V_s} \times 100(\%) \tag{6}$$

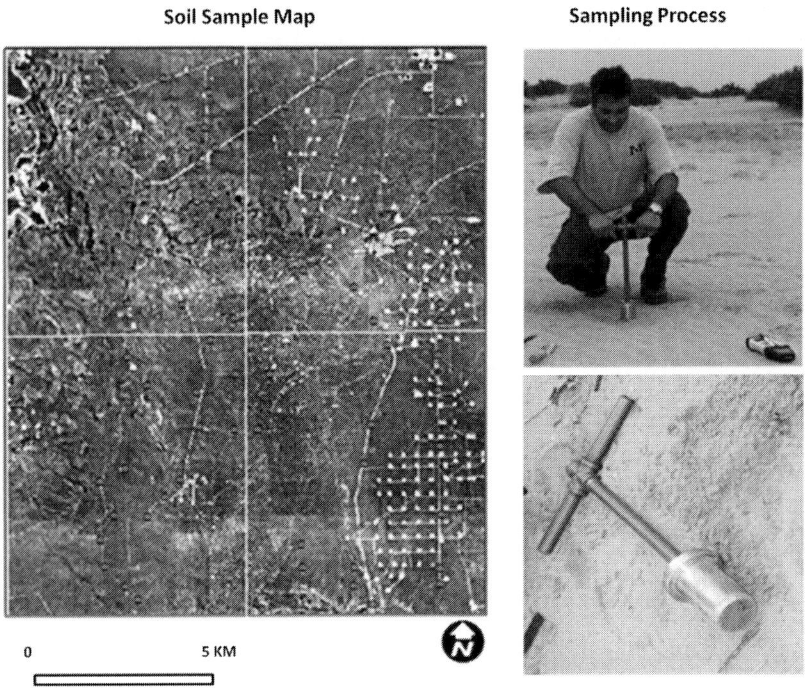

Figure 4. Distribution of soil samples collected for *in situ* soil moisture measurements.

Table 1. Statistics of *in situ* soil moisture measurements.

Statistical Parameters	Measurements (in %)	
	1–3 August 2006	**4–6 November 2006**
Mean	6.48	2.80
Minimum	2.55	0.26
Maximum	14.53	10.16
Standard Deviation	2.56	2.13

3.4. Vegetation Data

Vegetation cover strongly influences soil moisture estimation due to the interaction of the microwaves with the vegetation and soil. The amount of vegetation, its dielectric properties, and distribution pattern can significantly affect the sensitivity of microwave backscatter to soil moisture. The SAR imagery time series data were used to produce vegetation distribution maps for the study site and a vegetation density map for each SAR image acquisition date.

The ISODATA clustering algorithm was used to classify each SAR image to produce the vegetation density map for the study site. The image was classified into 25 classes and then recoded the data into four classes: thin or non-vegetated areas, sparsely vegetated areas, densely vegetated areas, and water. Field observations of vegetation were used to recode the data and produced vegetation density maps (Figure 5).

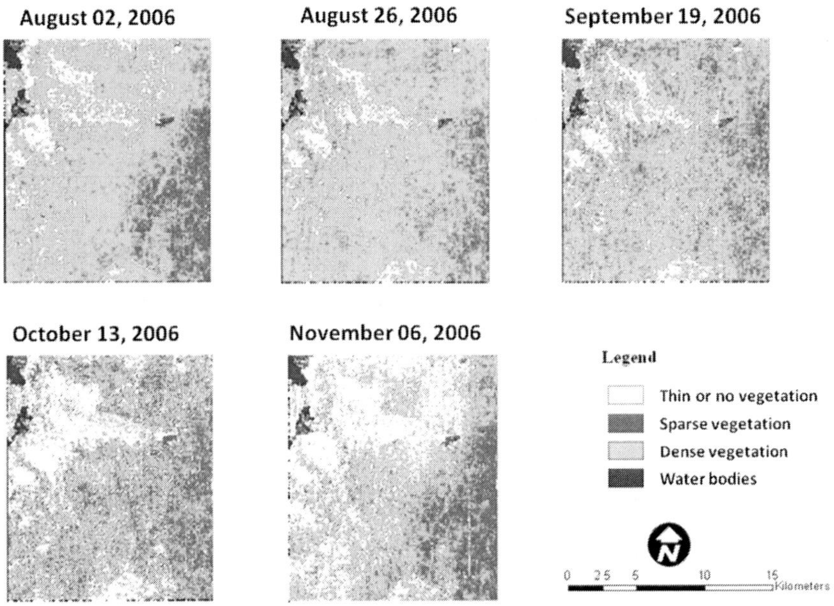

Figure 5. Vegetation density map for each SAR image acquisition date.

3.5. Soil Type and Soil Salinity

The soil data were acquired from the U. S. Department of Agriculture Natural Resource Conservation Service (USDA-NRCS) and converted into a 10 m gridded soil type dataset. This dataset was incorporated into the soil moisture estimation model. Figure 6 shows the soil type map for the study site. The soil type data was used in both multiple linear regressions and artificial neural networks based soil moisture estimation models.

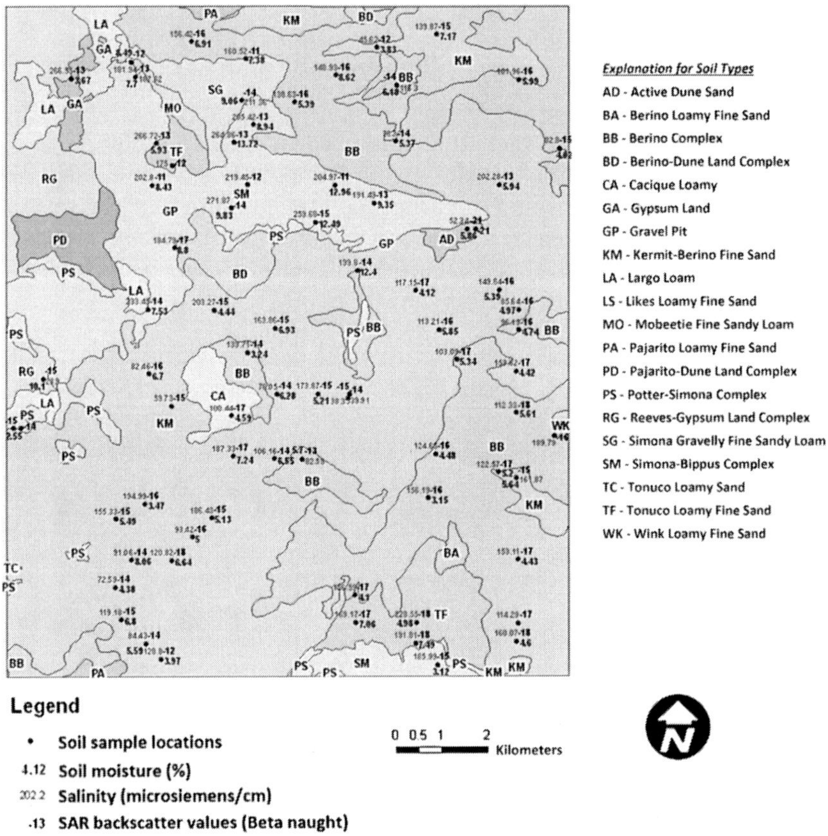

Figure 6. Measurements of soil moisture, soil salinity, and radar backscatter values shown on different soil types in Nash Draw (August 2006 data set).

The salinity of the 80 soil samples (collected for *in situ* soil moisture measurements in August 2006) were measured using USDA Soil Water Extract 1:5 methods [35]. The measurement is expressed by Electrical Conductivity (EC) in μS/cm. The soil salinity measurements were also used in both multiple linear regressions and artificial neural networks based soil moisture estimation models. Figure 6 shows the distribution of measured soil moisture, soil salinity, and radar backscatter values of different soil types in Nash Draw, NM.

4. METHODS

In this study, multiple linear regressions and Artificial Neural Networks (ANN) based models were used to investigate the influence of vegetation, elevation, soil type, and salinity on soil moisture estimation using microwave imagery. Coefficient of determination (R^2) values were used to evaluate the suitability of the different numerical models to estimate and map soil moisture distributions. Soil moisture maps were prepared for the five SAR imagery acquisition dates using the most suitable numerical model.

4.1. Multiple Linear Regressions

In semi-arid environments, the influence of sparse vegetation was found negligible in several soil moisture estimation studies using SAR data, e.g., [11,16,17,18,19]. However, as shown by [23] the vegetation distribution pattern in Nash Draw could significantly influence the soil moisture estimation using high resolution SAR data. The obtained vegetation density and distribution maps of Nash Draw (Figure 5) also support this observation.

In estimating soil moisture from low frequency radar backscattering (L band radar), the di-electric constant was found to be weakly sensitive to soil types [12]. The sensitivity of higher frequency radar backscattering (C band radar) to soil types, however, has not been fully analyzed in the published literature. Several studies reported that soil moisture estimation using L-band radar should consider the impact of soil salinity in the model [36,37,38]. Since in this research the soil moisture estimation model used C band radar, the impact of both soil type and soil salinity on soil moisture estimation was investigated.

Multiple linear regressions were conducted incorporating vegetation density information, elevation, soil type, and soil salinity in addition to radar backscattering (measured as β^{o}) to estimate soil moisture in the study site. The regression was done in a step fashion where independent variables (vegetation density information, elevation, soil type, and soil salinity) were sequentially added to the regression analysis to evaluate the effects of each independent variable in soil moisture estimation.

Equations (7)–(11) are the numerical models developed by multiple linear regressions with the observed *in situ* soil moisture measurements and different combination of independent variables to estimate soil moisture in Nash Draw, NM. The observed soil moisture data were acquired from the 2 August data set. Table 2 shows the coefficient of determination (R^2) values of the corresponding models. From the results of the simple linear regressions it was found that both the 2 August and the 6 November data sets produced similar results. Therefore, it was decided to use only one data set to evaluate the model performance.

$$\theta_w = 16.01 + 0.50\beta^o - 0.86V \quad (7)$$

$$\theta_w = 13.90 + 0.42\beta^o - 0.79V + 0.22S \quad (8)$$

$$\theta_w = 26.85 + 0.40\beta^o - 0.81V + 0.01E \quad (9)$$

$$\theta_w = 18.17 + 0.38\beta^o - 0.77V + 0.21S - 0.01E \quad (10)$$

$$\theta_w = 24.91 + 0.27\beta^o - 0.97V + 0.14S - 0.02E - 0.01SL \quad (11)$$

where, $\beta°$—Backscatter value, V—Vegetation, S—Soil type, E—Elevation, and SL—Soil salinity.

Table 2. Coefficient of determination (R^2) values of multiple linear regression based models.

Independent Variables	Coefficient of Determination (R^2)	Insignificant Variables
$\beta°$, V	0.50	--
$\beta°$, V, S	0.60	--
$\beta°$, V, E	0.51	E
$\beta°$, V, S, E	0.61	E
$\beta°$, V, S, E and SL	0.66	E and SL

Note: $\beta°$—Backscatter value; V—Vegetation; S—Soil type; E—Elevation; and SL—Soil salinity.

4.2. Artificial Neural Networks (ANN)

Most SAR based soil moisture estimation models are based on the assumption that soil moisture distribution is linearly related to the radar backscatter of the soil moisture surface, e.g., [11,12,21,39]. There are a few studies that explore the non-linear relationships between soil surface moisture and radar backscatters. In this study, Artificial Neural

Networks (ANN) based numerical models were developed to estimate soil surface moisture from SAR data and to explore the non-linear relationship between soil moisture and SAR backscatters. The significance of vegetation coverage, elevation, soil type, and soil salinity in soil moisture estimation using SAR data and artificial neural networks based models was also investigated.

Artificial neural networks, are a branch of artificial intelligence [40] in which the solution to a problem is learned from a set of examples [41]. A neural network can be regarded as a nonlinear mathematical function, which transforms a set of input variables into a set of output variables. The use of neural networks has been shown to be effective alternatives to more traditional statistical techniques [42]. Neural networks can be trained to approximate any smooth, measurable function [42], can model highly non-linear functions, and can be trained to be accurately generalized when presented with unseen data [26]. In a typical neural network model, a single neuron forms a weighted sum of the inputs $x_1, x_2, \ldots, x_d$ given by $a = \Sigma_i w_i x_i$, and then transforms this sum using a non-linear activation function $g(\ )$ to give a final output $z = g(a)$ (Figure 7).

A feed forward neural network can be regarded as a nonlinear mathematical function, which transforms a set of input variables into a set of output variables. The multilayer perceptron is the most widely used feed forward neural network. Figure 7 shows a single processing unit of neural networks. If we consider a set of *m* such units, all with common inputs, then we arrive at a neural network having a single layer of adaptive parameters (weights) as illustrated in Figure 8. The output variables are denoted by z_j and are given by Equation (12).

$$z_j = g\left(\sum_{i=0}^{d} w_{ji} x_i\right) \quad (12)$$

where w_{ji}—the weight from input *i* to unit *j*, and $g(\ )$—an activation function as discussed previously.

Artificial neural networks, based non-linear numerical models were developed for soil moisture estimation for the entire study site using the 2 August data set. Five different neural networks based models were developed, through addition of different variables, to estimate soil moisture in Nash Draw. The first model had only one input, the backscatter values. This model uses the non-linear relationship between radar backscatter and soil moisture content. The other four models used additional inputs (e.g., vegetation coverage, elevation, soil type, and soil salinity) in different combinations with radar backscatter values. JMP statistical software was used to perform the neural networks based analysis. The model coefficient of determination (R^2) and cross validation coefficient of determination (CV R^2) values were used to evaluate the model performance for soil moisture prediction. The impact of soil salinity was investigated and inclusion of this variable did not significantly improve model performance. Figure 9 shows the simplified schematic of the models that were developed. Table 3 shows the R^2 and CV R^2 values of the corresponding models.

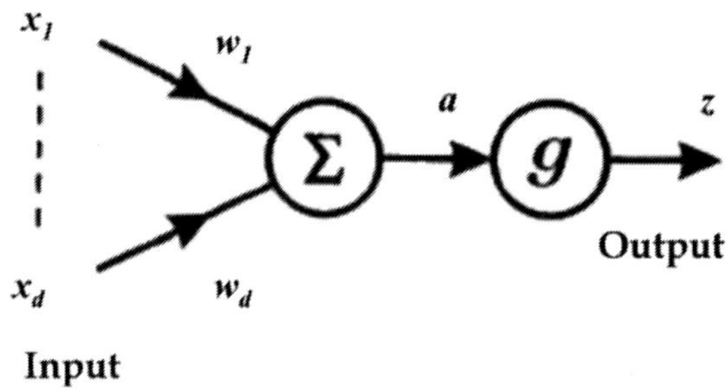

Figure 7. A single processing unit in neural networks.

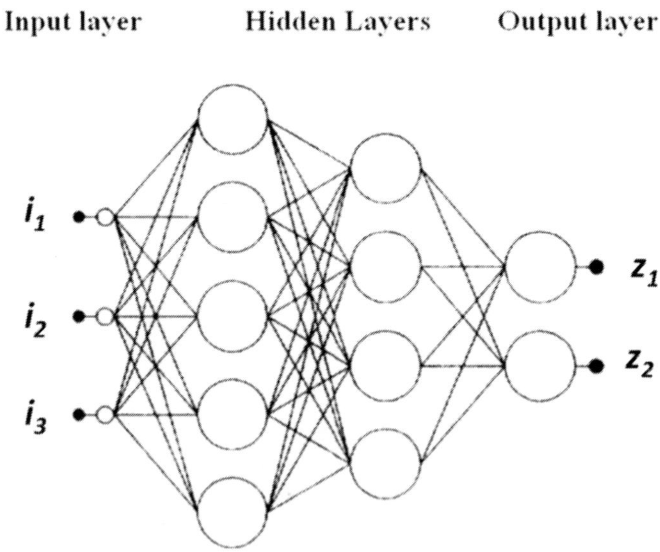

Figure 8. A multilayer perceptron with two hidden layers.

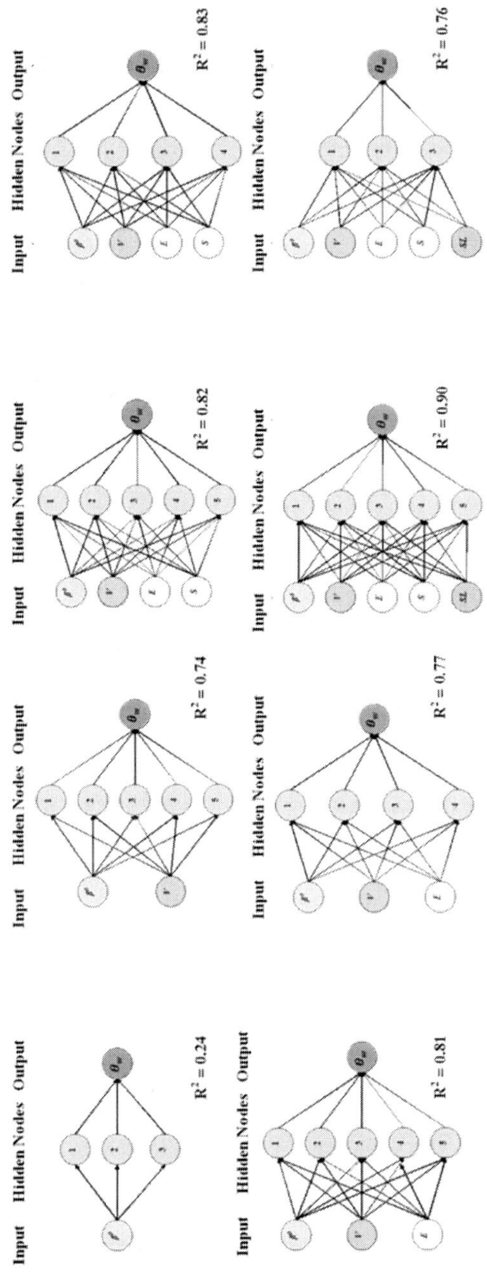

Figure 9. Neural networks based numerical models for soil moisture estimation in Nash Draw, NM, USA. Note: *β°*—Backscatter value; *V*—Vegetation; *S*—Soil type; *E*—Elevation; *SL*—Soil salinity.

Table 3. Coefficient of determination (R^2) and cross validation coefficient of determination (CV R^2) of the neural networks based numerical models for soil moisture estimation.

Input Variables	Hidden Nodes	Coefficient of Determination (R^2)	Cross Validation (CV) R^2
β^o	3	0.24	0.11
β^o, *V*	5	0.74	0.49
β^o, *V*, *E*	4	0.77	0.54
β^o, *V*, *E*	5	0.81	0.44
β^o, *V*, *E*, *S*	4	0.83	0.56
β^o, *V*, *E*, *S*	5	0.82	0.47
β^o, *V*, *E*, *S* and *SL*	3	0.76	0.45
β^o, *V*, *E*, *S* and *SL*	4	0.82	0.54
β^o, *V*, *E*, *S* and *SL*	5	0.90	0.58

Note: β^o—Backscatter value; *V*—Vegetation; *S*—Soil type; *E*—Elevation; *SL*—Soil salinity.

5. RESULTS

5.1. Soil Moisture Estimation

Near-real time field observations, acquired at the beginning and end of the time series, in conjunction with SAR data were used to estimate soil moisture from the SAR imagery acquired on five different dates in 2006. Model coefficient of determination (R^2) and model cross validation coefficient of determination (CV R^2) (for non-linear models) values were compared and used to evaluate their suitability for soil moisture estimation in Nash Draw. The model with the highest R^2 and CV R^2 values was considered as the most appropriate model for soil moisture estimation. The accuracy of the selected models was evaluated by Kappa statistics. The selected soil moisture estimation numerical models were then used to convert *ßº* SAR data into soil moisture data. The soil moisture data was divided into several categories to aid in the interpretation of the spatial distribution of the soil moisture in the study site.

The following observations were made after the evaluation of the R^2 and CV R^2 values of the models developed for soil moisture estimation in Nash Draw, NM.

1) Simple linear regression between radar backscatter values and *in situ* soil moisture measurements can be used to develop SAR-based soil moisture estimation models with model R^2 values of 0.51 to 0.61, but application of the model should be restricted to non-vegetated to thinly vegetated areas [23].

2) Multiple linear regressions using radar backscatter values, vegetation density, soil type, and elevation as independent variables

can be used to develop soil moisture estimation models for the entire study site, including areas with thicker vegetation.

3) Neural networks based models, using radar backscatter values, vegetation density, soil type, and elevation can also be used to estimate soil moisture for the entire study site, including areas with thicker vegetation. Neural networks based models achieved higher R^2 values and performed better than multiple linear regressions based models.

A neural network based numerical model using radar backscatter values, vegetation density, soil type and elevation was used to estimate soil moisture in the study site. This model was developed for both the 2 August and the 6 November data sets. The R^2 and CV R^2 values for the August model were 0.83 and 0.56 respectively, and 0.81 and 0.55 for the November model, respectively.

The model developed for the 2 August data set was also used to map soil moisture for the 26 August data set. The model developed for the 6 November data set was also used to map soil moisture for the 13 October data set. Since 19 September is approximately temporally equal from both 2 August and 6 November, we applied both the 2 August and the 6 November models to the 19 September data set and estimated soil moisture by taking the average of the two estimations.

Two sets of 50 m resolution soil moisture data were produced for each of the five dates of SAR data for the study site. The first set includes the unclassified soil moisture data, where the value of each pixel of the dataset is the volumetric soil moisture estimation (Figure 10). In the second dataset each pixel is classified into six categories of soil moisture to map the spatial variations in soil moisture at 2.5% intervals (Figure 11).

5.2. Accuracy Assessment

Kappa statistics [43,44] were used to evaluate the accuracy of the soil moisture estimation data produced from the SAR imagery. Kappa statistics have been used successfully for accuracy assessment in different remote sensing based studies, e.g., [13,45,46,47]. It is a discrete multivariate technique of accuracy assessment [28]. Kappa coefficients express the proportionate reduction in error generated by a classification process compared with the error of a completely random classification. For example, a value of 0.82 implies that the classification process is avoiding 82% of the errors that a completely random classification generates [43]. Kappa can be thought of as the chance-corrected proportional agreement, and possible values range from +1 (perfect agreement) to −1 (complete disagreement). A value of 0 indicates no agreement above that expected by chance. The calculation of the Kappa coefficients is explained below using an example of a 2 × 2 matrix (Table 4).

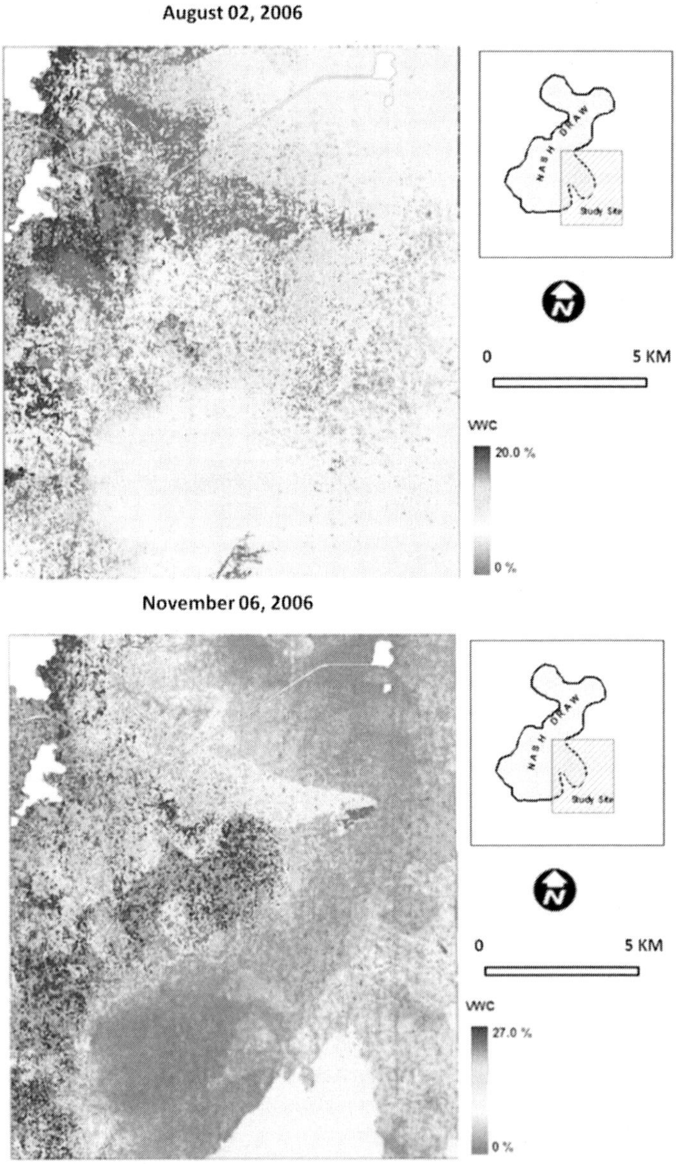

Figure 10. Unclassified soil moisture data generated for the study site.

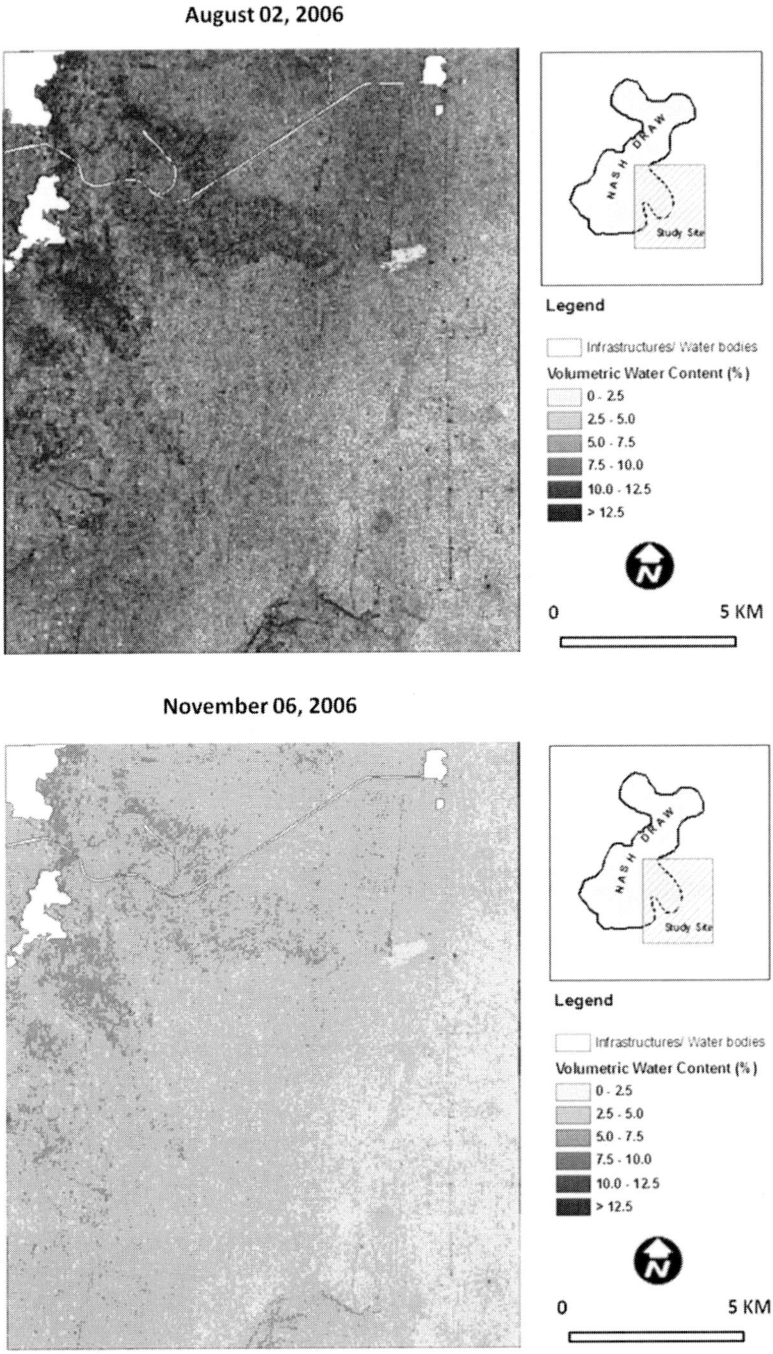

Figure 11. Classified soil moisture data generated for the study site.

Table 4. Computation of Kappa coefficients.

–		SAR Soil Moisture		Total
		Class 1	Class 2	
Reference Soil Moisture	Class 1	P_{11}	P_{12}	$P_{11} + P_{12} = \alpha$
	Class 2	P_{21}	P_{22}	$P_{21} + P_{22} = \beta$
Total		$P_{11} + P_{21} = \gamma$	$P_{12} + P_{22} = \delta$	$P_{11} + P_{22} = \chi$
Total number of data points: $Q = \alpha + \beta = \gamma + \delta$				

Observed agreement:

$$P_o = \frac{\chi}{Q} \tag{13}$$

Chance agreement:

$$P_e = \frac{\alpha}{Q} \times \frac{\gamma}{Q} + \frac{\delta}{Q} \times \frac{\beta}{Q} \tag{14}$$

Kappa coefficient:

$$K = \frac{P_o - P_e}{1 - P_e} \tag{15}$$

Kappa analysis requires continuous ground truth data so that a sufficient amount of random reference data can be obtained. Therefore, a continuous soil moisture surface was created from the *in situ* soil moisture data, using kriging [48]. Jackknife resampling techniques was used to correct for bias [49]. The measured *in situ* soil moisture data were randomly divided into a training set (90%) and testing set (10%) and were analyzed using the Arc GIS Geostatistical Analyst software. The training data set was used to create the krigged soil moisture surface and the testing data was used to evaluate the kriging results. Ten different krigged soil moisture data sets were generated using 10 different training data sets obtained from the same *in situ* soil moisture measurements. The RMS error, average standard error, standardized mean of error, and standardized RMS error (obtained from the evaluation of kriging results with the testing data) were used to select the appropriate kriging surface. Figure 12 shows the krigged surface generated for the 2 August and 6 November data sets.

Both SAR derived soil moisture data and the krigged soil moisture prediction surfaces were categorized into three class intervals. For the 2 August data set, the classes were 0.0%–5.0%, 5.0%–10.0%, and >10.0%. Since the soil moisture values were much lower in the November data set, these ranges were 0.0%–2.5%, 2.5%–5.0%, and >5.0%. Three hundred randomly generated points were used to calculate the Kappa coefficients and perform an accuracy assessment. Erdas Imagine image processing software was used to perform this accuracy assessment and Kappa analysis. Kappa coefficients were calculated for both the individual classes and the whole data sets.

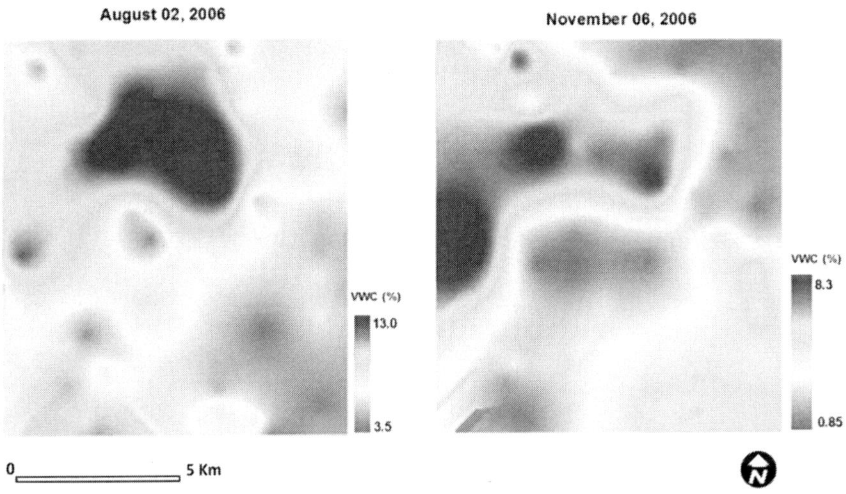

Figure 12. Krigged soil moisture surface generation for 2 August and 6 November 2006 field data field data.

The overall Kappa coefficients obtained for the 2 August and the 6 November data sets were 0.43 and 0.61, respectively. The overall accuracy was 75.67% and 77.67% for the 2 August and the 6 November data sets, respectively. Figure 13 and Figure 14 show the Kappa coefficients and classification accuracy for the individual classes. The evaluation of the individual classes indicate that the developed model performed well at low soil moisture regimes compared to the high soil moisture regimes in both 2 August and 6 November data sets.

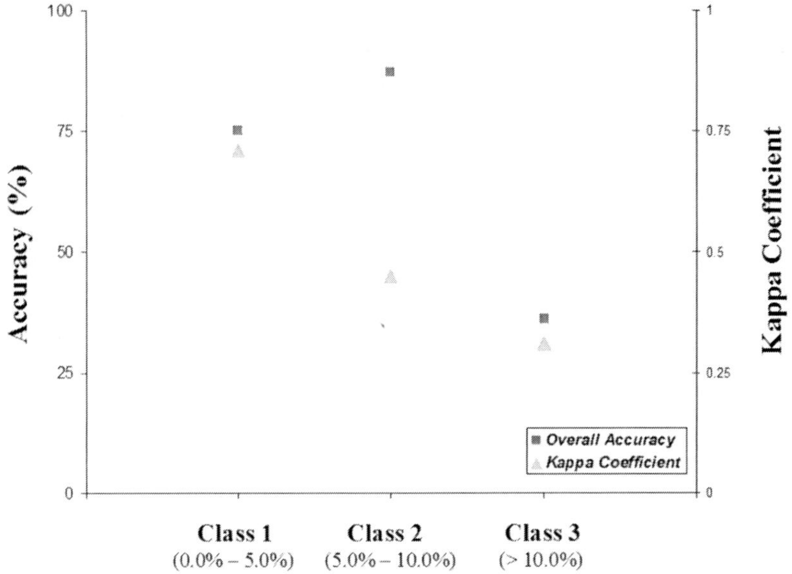

Figure 13. Kappa statistics for 2 August 2006 data set.

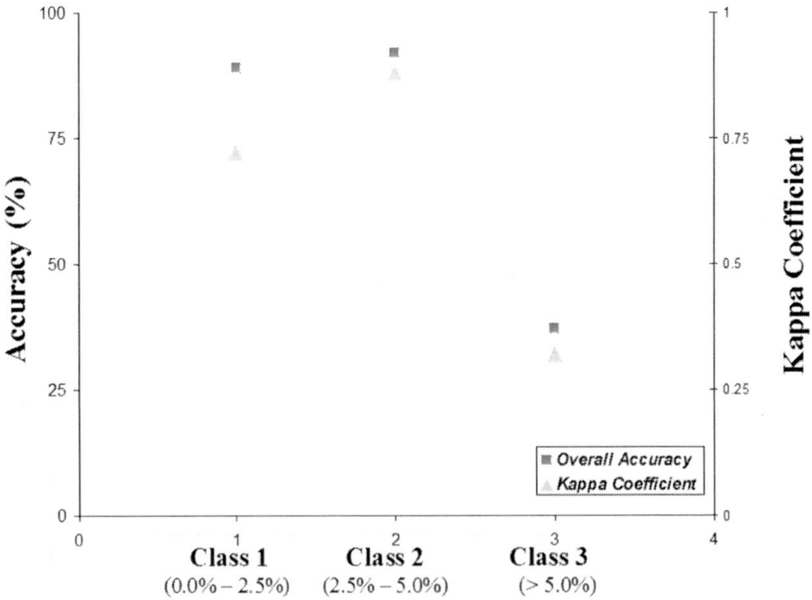

Figure 14. Kappa statistics for 6 November 2006 data set.

6. DISCUSSIONS

Research conducted by [23] has shown that in semi-arid environments even sparse vegetation can adversely influence the accuracy of soil moisture values estimated from radar backscatter (obtained from high resolution SAR data) using a linear relationship. This observation is supported by lower R^2 values (0.24 and 0.05 for the August and the November datasets respectively) for the linear numerical models developed for the entire study site and higher R^2 values (0.61 and 0.52 for the August and the November datasets respectively) for linear numerical models developed for portions of the study site with little or no vegetation. The lowest R^2 values were from linear numerical models developed for the more densely vegetated portions of the study site (0.01 to 0.04).

This research shows that soil moisture estimation using high resolution SAR data in a semi-arid environment can be improved by developing numerical models that use multiple linear regressions incorporating additional variables such as, vegetation density, soil type, and elevation in addition to radar backscatter values. Regression coefficients as high as 0.61 were obtained for a numerical model covering the entire study site using multiple linear regressions.

The non-linear relationship between radar backscatter values and soil moisture was also investigated by [23] using artificial neural networks. It showed that a neural network developed using only radar backscatter values and soil moisture had an R^2 value of 0.24, similar to the R^2 value obtained from numerical models using simple linear regressions. This study obtained significant improvement in soil moisture estimation when additional variables such as, vegetation density, soil type, and elevation were added as inputs to the artificial neural networks based models. The use of these additional inputs results in coefficients of determination of 0.83 and 0.81 for the entire study site for the 2 August and the 6 November datasets, respectively. The cross validation coefficients of determination (CV R^2) of the same models were 0.56 and 0.55, respectively.

Soil moisture data were produced using artificial neural networks based non-linear models that incorporated inputs of vegetation density, soil type, elevation, and radar backscatter values. The accuracy of the modeled soil moisture data was also evaluated by the comparison of a soil moisture distribution surface from the models with a soil moisture surface obtained by kriging the *in situ* measurements. The comparison was done by Kappa statistics. The overall accuracy and the Kappa coefficient for the soil moisture data obtained for 2 August and 6

November were 75.67% and 77.67%, and 0.43 and 0.61, respectively. Although the obtained kappa coefficient values indicate overall good agreement between the measured and estimated soil moisture values, however, it is worth looking into the possibility of having the influence of autocorrelation between these two soil moisture data sets. As suggested by [50] since most thematic maps have some degrees of spatial autocorrelation, the calculated expected agreement (by Kappa statistics) could be usually higher than the true expected agreement.

7. CONCLUSIONS AND FUTURE WORK

This research demonstrates the proof of concept of the application of Artificial Neural Networks (ANN) based models for estimating soil surface moisture in semi-arid environments of south-eastern New Mexico using high resolution SAR data (Radarsat 1 SAR Fine data). It is strongly believed that the methods developed in this research can be used to produce soil moisture data from high resolution SAR imagery in semi-arid environments, such as Nash Draw in New Mexico. The hydrology of Nash Draw in New Mexico is characterized by evaporite karst and is not well understood. The data produced in this research should be very useful to identify the pattern of soil moisture distribution in space and time and contribute to a better understanding of the groundwater recharge in the study area.

In the future, to develop an operational soil moisture estimation tool (for this particular area) using high resolution SAR data, the proposed modeling effort can be further enhanced/improved by considering: (1) inclusion of surface roughness parameters in the artificial neural networks based models in addition to vegetation density, soil type, and elevation; (2) incorporation of near real-time high resolution optical satellite data for mapping vegetation density; and (3) involving other statistical tests such as Root Mean Squared Error (RMSE), Mean Absolute Percent Error (MAPE) *etc.* for model accuracy assessments. It is also recommended to use high resolution L-band SAR data with the capability of multi-polarization. L band data will provide the opportunity to obtain moisture estimation in greater depth.

ACKNOWLEDGMENTS

Authors thankfully acknowledge the NASA Applied Sciences Program and the University of Mississippi Geoinformatics Center (UMGC) for funding this study through a Rapid Prototyping Capability (RPC) for Earth-Sun Systems Sciences project at The University of Mississippi. Thanks are also due to the Alaska Satellite Facility (ASF) for providing the SAR imagery; Robert Holt for proving technical support for regression and artificial neural networks analysis; Dennis Powers, Glen Garrett, and Dirk O'Daniel for their assistance in the field to acquire soil samples for soil moisture analysis; and Patrick Yamnik for assisting in image processing and GIS analysis. Authors are also grateful to Chung R. Song for his continuous support throughout the study.

AUTHOR CONTRIBUTIONS

Azad Hossain is responsible for data acquisition, processing and analysis (soil moisture estimation and accuracy assessments. Greg Easson is responsible for digital image processing aspect. Both authors prepare this manuscript together.

CONFLICTS OF INTEREST

The authors declare no conflict of interest.

REFERENCES

1. Hossain, F.; Anagnostou, E.N. Numerical investigation of the impact of uncertainties in satellite rainfall and land surface parameters on simulation of soil moisture. Adv. Water Resour. 2005, 28, 1336–1350.
2. Georgakakos, K.P.; Baumer, O.W. Measurement and utilization of on-site soil moisture data. J. Hydrol. 1996, 184, 131–152.
3. Lakshmi, V. Remote sensing of soil moisture. ISRN Soil Sci. 2013, 2013, 1–33.
4. Robock, A.; Luo, L.; Wood, E.F.; Wen, F.; Mitchell, K.E. Evaluation of the North American Land Data Assimilation System over the southern Great Plains during the warm season. J. Geophys. Res. 2003, 108.
5. Zhang, L.; Dawes, W.R.; Hatton, T.J.; Reece, P.H.; Beale, G.T.H.; Packer, I. Estimation of soil moisture and groundwater recharge using the TOPOG_IRM model. Water Resour. Res. 1999, 35, 149–161.
6. Silva, C.S.D.; Rushton, K.R. Groundwater recharge estimation using improved soil moisture balance methodology for a tropical climate with distinct dry seasons. Hydrol. Sci. J. 2007, 52, 1051–1067.
7. Hossain, A.; Easson, G. Mapping spatial variation in surface moisture using reflective and thermal ASTER imagery for Southern Africa. In Proceedings of 2006 ASPRS Annual Conference, Reno, NV, USA, 1–5 May 2006.

8. Narasimhan, B.; Srinivasan, R.; Arnold, J.G.; di Luzio, M. Estimation of long-term soil moisture using a distributed parameter hydrologic model and verification using remotely sensed data. Trans. ASAE 2005, 48, 1101–1113.
9. Albergel, C.; de Rosnay, P.; Gruhier, C.; Muñoz-Sabater, J.; Hasenauer, S.; Isaksen, L.; Kerr, Y.; Wagner, W. Evaluation of remotely sensed and modeled soil moisture products using global ground-based in-situ observations. Remote Sens. Environ.2012, 118, 215–226.
10. Klemas, V.; Finkl, C.W.; Kabbara, N. Remote sensing of soil moisture: An overview in relation to coastal soils. J. Coast. Res.2014, 30, 685–696.
11. Dubois, P.C.; van Zyl, J.; Engman, E.T. Measuring soil moisture with imaging radars. IEEE Trans. Geosci. Remote Sens. 1995, 33, 915–926.
12. Ulaby, F.T.; Dubois, P.C.; van Zyl, J. Radar mapping of surface soil moisture. J. Hydrol. 1996, 184, 57–84.
13. Jackson, T.J. Measuring surface soil moisture using passive microwave remote sensing. Hydrol. Process. 1993, 7, 139–152.
14. Jackson, T.J.; Schmugge, J.; Engman, E.T. Remote sensing applications to hydrology: Soil moisture. Hydrol. Sci. J. 1996, 41, 517–530.
15. Borgeaud, M.; Saich, P. Status of the retrieval of bio- and geophysical parameters from SAR data for land applications. In Proceedings of the 1999 IEEE International Geoscience and Remote Sensing Symposium, Hamburg, Germany, 28 June–2 July 1999.
16. Chanzy, A.; Kerr, Y.; Wigneron, J.P.; Calvet, J.C. Soil moisture estimation under sparse vegetation using microwave radiometry at C-band. In Proceedings of the 1997 International Geoscience and Remote Sensing Symposium, Singapore, 3–8 August 1997.
17. Demircan, A.; Rombach, M.; Mauser, W. Extraction of soil moisture from multitemporal ERS-1 SLC data of the Freiburg test site. In Proceedings of the 1993 IEEE International Geoscience and Remote Sensing Symposium, Tokyo, Japan, 18–21 August 1993.
18. Lin, D.S.; Wood, E.F. Behavior of AirSAR signals during MAC-Europe'91. In Proceedings of the 1993 IEEE International Geoscience and Remote Sensing Symposium, Tokyo, Japan, 18–21 August 1993.
19. Thoma, D.P.; Moran, M.S.; Bryant, R.; Rahman, M.; Holifield-Collins, C.D.; Skirvin, S.; Sano, E.E.; Slocum, K. Comparison of four models to determine surface soil moisture from C-band radar imagery in a sparsely vegetated semiarid landscape. Water Resour. Res. 2006, 42, 1–12.
20. Mladenovaa, I.E.; Jacksona, T.J.; Njokub, E.; Bindlishc, R.; Chanb, S.; Cosha, M.H.; Holmesa, T.R.H.; de Jeud, R.A.M.; Jonese, L.; Kimballe, J.; et al. Remote monitoring of soil moisture using passive microwave-based techniques—Theoretical basis and overview of selected algorithms for AMSR-E. Remote Sens. Environ. 2014, 144, 197–213.
21. Moran, M.S.; Hymer, D.C.; Qi, J.; Sano, E.E. Soil moisture evaluation using multi-temporal Synthetic Aperture Radar (SAR) in semiarid rangeland. Agric. For. Meteorol. 2000, 105, 69–80.
22. Moran, M.S.; Peters-Lidard, C.D.; Watts, J.M.; McElroy, S. Estimating soil moisture at the watershed scale with satellite-based radar and land surface models. Can. J. Remote Sens. 2004, 30, 805–826.
23. Hossain, A.; Easson, G. Microwave remote sensing of soil moisture. In Geoscience and Remote Sensing; Ho, P.P., Ed.; IN-TECH Press: Vukovar, Croatia, 2009; pp. 529–554.
24. Holt, R.M.; Beauheim, R.L.; Powers, D.W. Predicting fractured zones in the Culebra Dolomite. In Dynamics of Fluids and Transport in Fractured Rock: AGU Geophysical Monograph Series; Faybishenko, B., Witherspoon, P.A., Gale, J., Eds.; American Geophysical Union: Washington, DC, USA, 2005; pp. 103–116.
25. Powers, D.W.; Beauheim, R.L.; Holt, R.M.; Hughes, D.L. Evaporite karst features and processes at Nash Draw, Eddy County, New Mexico. In Caves & Karst of Southeastern New Mexico, NM Geological Society Fifty-seventh Annual Field Conference Guidebook; Land, L., Lueth, V.W., Raatz, W., Boston, P., Love, D.L., Eds.; New Mexico Geological Society: Socorro, NM, USA, 2006; pp. 253–266.
26. Glenn, N.F.; Carr, J.R. Establishing a relationship between soil moisture and RADARSAT-1 SAR data obtained over the Great Basin, Nevada, USA. Can. J. Remote Sens. 2004, 30, 176–181.

27. Hossain, A.; Easson, G.; Powers, D.W.; Holt, R.M. 2007 Impact of Variation in Reflectivity of Microwave Data for Soil Moisture Estimation in Semi-arid Environment; Sigma Xi Poster Presentation; The University of Mississippi: University, MS, USA, 2007.
28. Hossain, A.; Easson, G.; Holt, R.M.; Powers, D.W. Developing methods for mapping soil moisture in Nash Draw, NM using Radarsat 1 SAR fine imagery. In Proceedings of the 2006 AGU Fall Meeting, San Francisco, CA, USA, 11–15 December 2006.
29. Goodman, J.W. Statistical properties of laser speckle patterns. In Laser Speckle and related Phenomena; Springe: Berlin, Germany, 1975; pp. 9–75.
30. D'Elia, C.; Ferraiuolo, G.; Pascazio, V.; Schirnzi, G. An MRF based technique for speckle reduction in SAR images. In Proceedings of the 2004 International Geoscience and Remote Sensing Symposium, Anchorage, AK, USA, 20–24 September 2004.
31. Touzi, R. A review of speckle filtering in the context of estimation, theory. IEEE Trans. Geosci. Remote Sens. 2002, 40, 2392–2404.
32. Dane, J.H.; Topp, G.C. Methods of Soil Analysis, Part 4, Physical Methods; Soil Science Society of America: Madison, WI, USA, 2002.
33. Boisvert, J.B.; Gwyn, Q.H.J.; Brisco, B.; Major, D.; Brown, R.J. Evaluation of soil moisture techniques and microwave penetration depth for radar applications. Can. J. Remote Sens. 1995, 21, 110–123.
34. ASTM D2216. Standard Test Method for Laboratory Determination of Water (Moisture) Content of Soil and Rock by Mass; American Society Testing and Materials: Philadelphia, PA, USA, 1999.
35. Tanji, K.K. Agricultural Salinity Assessment and Management; ASCE Manuals and Reports on Engineering Practice No. 71; American Society of Civil Engineers: Reston, VA, USA, 1990.
36. Shao, Y.; Guo, H.; Hu, Q.; Lu, Y.; Dong, Q.; Han, C. Effect of dielectric properties of moist salinized soils on backscattering coefficients extracted from RADARSAT image. IEEE Trans. Geosci. Remote Sens. 2003, 41, 1879–1888.
37. Sreenivas, K.; Venkataratnam, L.; Narasimha Rao, P.V. Dielectric properties of salt-affected soils. Int. J. Remote Sens. 1995, 16, 641–649.
38. Taylor, G.R.; Mah, A.H.; Kruse, F.A.; Kierein-Yong, K.S.; Hewson, R.D.; Bennette, B.A. Characterization of saline soils using airborne radar imagery. Remote Sens. Environ. 1996, 57, 127–142.
39. Dobson, M.C.; Pierce, L.; Arabandi, K.; Ulaby, F.T.; Sharik, T. Preliminary analysis of ERS-1 SAR for forest ecosystem studies. IEEE Trans. Geosci. Remote Sens. 1992, 30, 203–221.
40. Gardner, M.W.; Dorling, S.R. Artificial neural networks (The multilayer perceptron)—A review of applications in the atmospheric sciences. Atmos. Environ. 1998, 32, 2627–2636.
41. Bishop, C.M. Neural networks and their applications. Rev. Sci. Instrum. 1994, 65, 1803–1833.
42. Hornik, K.; Stinchcombe, M.; White, H. Multilayer feed forward networks are universal approximators. Neural Netw. 1989, 2, 259–366.
43. Cohen, J. A coefficient of agreement for nominal scales. Educ. Psychol. Meas. 1960, 20, 37–46.
44. Congalton, R. A review of assessing the accuracy of classifications of remotely sensed data. Remote Sens. Environ. 1991, 37, 35–46.
45. Congalton, R.G.; Green, K. Assessing the Accuracy of Remotely Sensed Data: Principles and Practices; Lewis Publisher: Boca Raton, FL, USA, 1999.
46. Naesset, E. Use of the weighted Kappa coefficient in classification error assessment of thematic maps. Int. J. Geogr. Inf. Syst.1996, 10, 591–603.
47. Monserud, R.A.; Leemans, R. Comparing global vegetation maps with the Kappa statistic. Ecol. Model. 1992, 62, 275–293.
48. Isaaks, E.H.; Srivastava, R.M. An Introduction to Applied Geostatistics; Oxford University Press: Oxford, UK, 1989.
49. Efron, B.; Tibshirani, R. Bootstrap methods for standard errors, confidence intervals, and other measures of statistical accuracy. Stat. Sci. 1986, 62, 275–293.
50. Hagen-Zanker, A. An improved Fuzzy Kappa statistic that accounts for spatial Autocorrelation. Int. J. Geogr. Inf. Sci. 2009, 23, 61–73.

CHAPTER 16

Isotope Dating of the Potash and Rock Salt Deposit at Bamnet Narong, NE-Thailand

Bent T. Hansen[1*], Klaus Wemmer[1], Monique Eckhardt[1], Prinya Putthapiban[2], San Assavapatchara[3]

[1]*Geoscience Center Göttingen, University of Göttingen, Göttingen, Germany*

[2]*Geoscience Programme, Kanchanaburi Campus, Mahidol University, Kanchanaburi, Thailand*

[3]*Department of Mineral Resources, Bureau of Geological Survey, Bangkok, Thailand*

ABSTRACT

Bamnet Narong is located in northeastern Thailand (Chaiyaphum Province). It is the largest salt mine in the country and has been mined for decades. The landscape in this part of Thailand is characterised by a low plateau, which is called the Khorat Plateau. The plateau is divided into two basins by the Phu Phan Range, the Sakhon Nakhon Basin in the north and the Khorat Basin in the south. The analysed potashes and rock salts are deposited in the Maha Sarakham Formation, which represents the salt-bearing strata of the Khorat Basin. The stratigraphic age of this deposit has been debated since the late 1960's. The assigned ages range from Mid-Cretaceous to Late Cretaceous and up to the Eocene. In this study different isotopic dating systems (Rb-Sr, Sr-Sr, K-Ar and K-Ca) were applied. The stratigraphic age for the time of deposition was confirmed to be Mid Cretaceous (Cenomanian). Furthermore, the homogeneity of the carnallites was investigated in order to trace a possible redistribution of rubidium.

Keywords: Isotope Dating, Salt Deposits, Mid-Cretaceous, Asian Potash Mine, Thailand

1. INTRODUCTION

Bamnet Narong is located in northeastern Thailand (Chaiyaphum Province) and hosts the largest salt mine (Asian Potash Mine) in Thailand that was mined for decades. Geographically, it is considered part of the western edge of the Khorat Basin (Figure 1). Stratigraphically this salt deposit is part of the so-called Maha Sarakham Formation and represents one of the world's largest salt deposits. The stratigraphic and palaeontological study of the Maha Sarakham Formation, as well as the whole Khorat Basin began in the early 1960's, whereas the salt resources and related issues were studied during the 1970's. The stratigraphic age of this formation has been highly debated since the late 1970's. The Maha Sarakham Formation was first named by [1] who assigned an age from the Late Cretaceous to the Eocene. Based on palynological evidence [2] a Mid-Cretaceous age was proposed for the deposition. In 1990 doubt was cast by [3] who compared it with the depth of core samples and then specified an age of Mid-Late Albian to Cenomanian. Geochemical studies of [4] on the sulphur and oxygen isotopes of the Maha Sarakham anhydrites yielded an age of Neocomian (lower Cretaceous) to Cenomanian. Multiple isotope analyses on various halite, anhydrite and carnallite samples as well as one clay sample from the shaft at Bamnet Narong pointed to a deposition age of about 95 Ma (Cenomanian) [5] . This agrees with the assumed Aptian-Albian age from [6] , which was obtained from vertebrate fossils of the underlying Khok Kruat Formation.

The research objectives of this study are to date the Maha Sarakham Formation by using the K-Ca and Rb-Sr isotope systems in comparison with the existing data. K-Ca decay is used for the carnallite samples, which were collected from a 1200 m long profile in the shaft of the Bamnet Narong (Asian Potash Mine). Rb-Sr decay is used for the anhydrites which were in parts already analysed [5] . These data were subsequently compared with the seawater evolution curve, LOWESS curve [7] . Anhydrites are used because they are very stable in regards to the redistribution of rubidium, as their content of rubidium is negligible. Another aim of the present study is the geochemical characterisation of the carnallites in terms of the variation of specific elements in a small area concerning their potential heterogeneity.

2. GEOLOGICAL SETTING

2.1 Tectonic and Geologic Evolution of the Khorat Plateau

The Khorat Plateau is framed by an escarpment of mostly steeply dipping sediments, which form slopes on its western and southern margins rising from 600 - 1000 m above sea level. The plateau is situated between 14°N and 19°N and 101°E and 106°E. It is divided by the Phu Pan Range into two depositional basins, the Khorat Basin in the south and the smaller Sakhon Nakhon Basin in the north [8] . In a broader sense it is part of the Indochina Terrane, which originated by rifting of the margin of Gondwana. Furthermore, it represents a broad synclinorium bounded to the west by the Shan Thai Terrane and to the north by the South China Plate [9] [10] . The Khorat Plateau has, in connection with investigations of its hydrocarbon potential, been described as a "sauce- pan morphology" [11] .

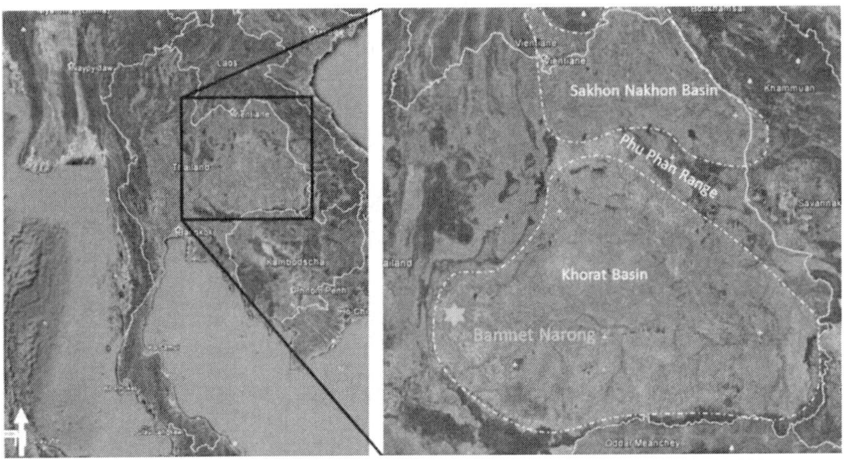

Figure 1. Geographical location of the Khorat Basin, the Phu Phan Range and the northern Sakhon Nakhon Basin. Bamnet Narong is highlighted by a pink star in the western part of the Khorat Basin. Source: Google Maps (query 10.07.2013).

The basement of the Khorat Plateau consists of igneous, metamorphic and sedimentary rocks which underlie the Huai Hin Lat Formation. Sedimentation of the Mesozoic rocks began in the Early Triassic, after the Khorat Basin was formed due to extension during the Indosinian Orogeny [13] . The deposition took place in half-gra- ben rift basins, which were developed during this orogeny, and filled with mainly lacustrine and fluviatile clastic sediments. This Triassic sequence is called the Huai Hin Lat Formation. The depositional period of this formation ended in the late Upper Triassic (Norian-Rhaetian) during the first episodes of the Indosinian Orogeny [14] . The following fluvial red-beds sequence was mapped as the Nam Phuong Formation, which represents the beginning of the Khorat Group. Afterwards uplift and erosion events as well as subsidence took place. The following red-bed formations of Cretaceous age are (from bottom to top): The Phu Kradung, the Phra Wihan, the Sao Khua, the Phu Pan and the Khok Kruat Formation. This Cretaceous succession is unconformably overlain by the continental evaporitic Maha Sarakham Formation [15] , which was deposited during the Mid-Cretaceous (Albian-Cenomanian) regional folding. This process initiated the uplift of a large scale basin, which was later filled with evaporitic deposits, caused by sea water inflow from the west into the western edge of the basin (Figure 2)): the Phu Kradung, the Phra Wihan, the Sao Khua, the The Maha Sarakham Formation is not part of the Khorat Group. Three transgressions and two regressions with evaporation can be reconstructed during the accumulation of this formation. The youngest Mesozoic formation which overlies the Maha Sarakham Formation is the Phu Tok Formation, which was formed when the sea water regressed entirely. Finally, the Khorat deposits were intruded by granites of Campanian and Cenomanian age. During the Early Paleocene, compression from the northeast, due to continental collision of the Indochina Terrane with the Southeast China Plate (Himalayan Orogeny), and also from back arc compression to the west, resulted in uplift and erosion of a part (~3000 m) of the Khorat sediments as well as the formation of the NW-SE trending Phu Phan Anticline. The Cenozoic sediments consist of alluvial, eolian and fluvial deposits, as described in [15] - [17] .

3. STRATIGRAPHIC AND SUPPOSED AGES

The stratigraphy of the entire Mesozoic succession of the plateau is shown in Figure 3. The continental red-beds of the Mesozoic Khorat Group, which built up most of the Khorat Plateau, unconformably overly the Triassic rocks. The definition of the Khorat Group differs from author to author and is commonly considered to comprise the sediments above the basal Nam Phuong Formation unconformity [17] . Also the exact position of the Jurassic-Cretaceous boundary in the Khorat strata has not been pinpointed until now. The base of the Maha Sarakham Formation has been dated by means of palynological analysis as Mid-Albian-Cenomanian in age. Usually the Nam Phuong Formation is considered to be Late Triassic and the Phu Kradung, Phra Wihan and Sao Khua Formations are related to the Jurassic. However, the Phu Phan and Khok Kruat Formations are assigned to the

Early Cretaceous, for example by [15] and [18] . All these age determinations are primarily based on plant macrofossils, pollen and vertebrate remains, e.g. in [19] and [15] .

As mentioned above the Khorat Plateau almost consists of Cretaceous rocks (Figure 3). These are slightly deformed, fossiliferous and well-exposed. Thus, the Khorat Plateau received more attention than the Cretaceous rocks elsewhere in Thailand [12] . Most authors assume six formations within the Cretaceous deposits, which are named in ascending order according to their age and are as follows: the Phra Wihan, Sao Khua, Phu Phan, Khok Kruat, Maha Sarakham and Phu Thok Formations.

3.1. Phra Wihan Formation

The Phra Wihan Formation crops out in the Phu Phan Range and is also widely distributed around the western and especially the southern rims of the Khorat Plateau. At the southern rim it forms the Dongrek Range which marks the border to Cambodia. This formation is named after the temple mountain of Khao Phra Wihan located at the south-eastern border of the Khorat Plateau. The rocks usually consist of light buff to grey, fine- to coarse- grained quartzitic sandstones and to a lesser extent of siltstones and mudstones with occasional occurrences of conglomerates [12] . They were deposited in a fluvial environment dominated by a high-energy, shallow braided river system and associated floodplains [15] . This formation conformably overlies the Phu Kradung as well as the underlying Sao Khua Formation and has a gradational contact to both of them. Its thickness varies between 100 and 250 m. In the central and northern parts of the Phu Phan Anticline the formation overlies the Phu Kradung Formation with a sharp contact [12] [15] . The Phra Wihan Formation dates to around Middle Jurassic/Early Cretaceous in age by [20] on the basis of plants and arthropod fossils (crustaceans and insects). A recent study by [15] proposes a Lower Cretaceous age (Berriasian-Barremian), due to the analyses of palynomorphs which were collected near the base of the formation.

3.2. Sao Khua Formation

The Sao Khua Formation crops out throughout the whole Phu Phan Range. The rocks usually consist of an alternation of reddish-brown conglomeratic sandstone, siltstone and mudstone together with common calcretes. The thickness of this formation varies between 100 and 600 m. It was deposited in an environment with low- energy meandering channels and extensive floodplains. The contact of this formation with the underlying Phra

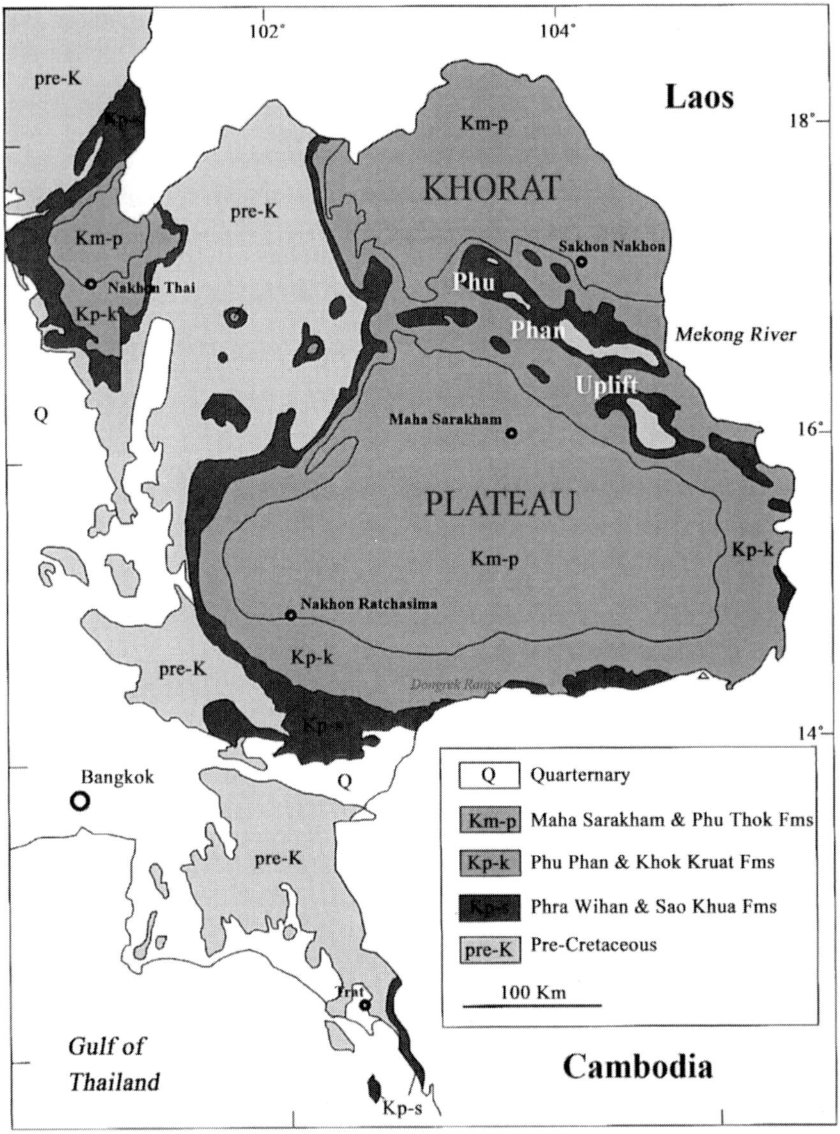

Figure 2. Distribution of Cretaceous strata in NE and SE Thailand. The pre-Cretaceous rocks may be Jurassic or Jurassic- Cretaceous in age [12] .

Wihan and overlying Phu Phan Formation is gradational and conformable [15] . At the beginning of the Sao Khua Formation it was considered to be Late Jurassic, for example by [21] on the basis of vertebrate remains. This area has the richest and most diverse vertebrate fauna found in Thailand

[12] . In contrast, [22] [23] as well as [15]) determined an Early Cretaceous age (Berriasian-Barremian) based on palynological investigations, similar to the Phra Wihan Formation.

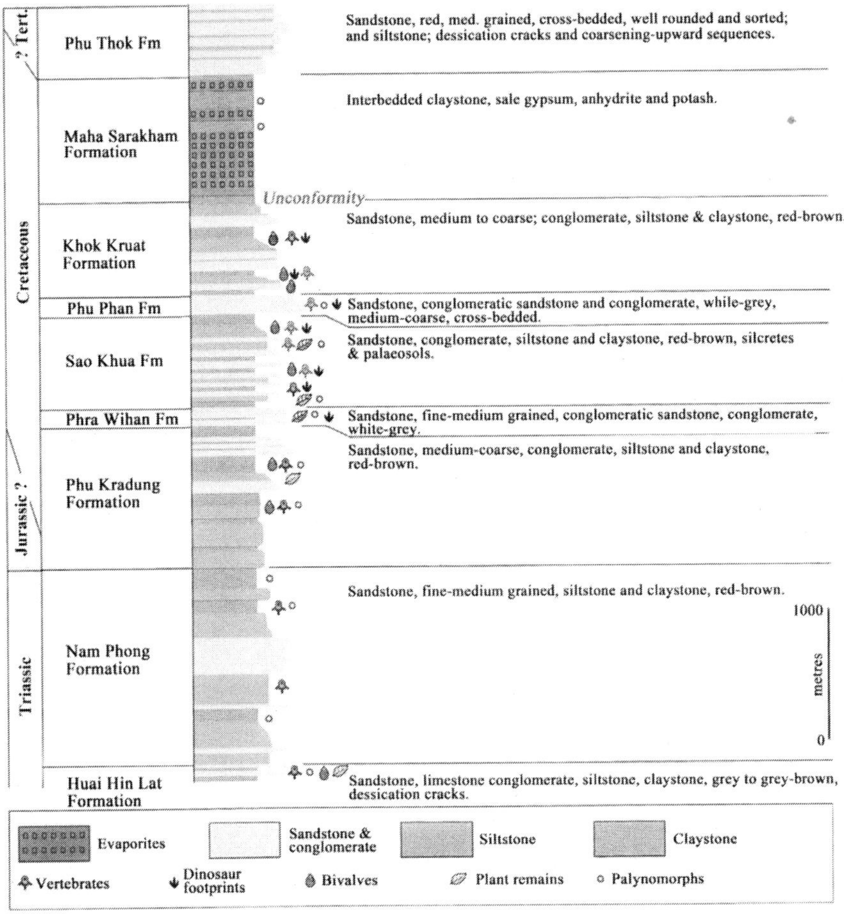

Figure 3. Stratigraphic column of the Mesozoic rocks of the Khorat Plateau, NE Thailand, [12] .

3.4. Phu Phan Formation

The rocks of the Phu Phan Formation usually consist of greyish-white, medium- to coarse-grained cross-bedded sandstones and lacustrine grey siltstones and mudstones with subordinated conglomerates [12] [15] . Generally, the formation is 75 - 150 m thick and the contact with the underlying Sao Khua Formation is locally erosive or distinctively sharp but mainly conformable and with the overlying Khok Kruat Formation. The supposed depositional environment was a high-energy, low-sinuosity braided river system with subordinate floodplains [12] [15] . According to [23] and [15] , the age of the formation is Barremian to Aptian, based on the ages of the over- and underlying formations and also on palynological studies.

3.5. Khok Kruat Formation

The Khok Kruat Formation is widely distributed in the outer parts of the Phu Phan Anticline. It usually comprises reddish-brown, fine- to medium-grained sandstone, siltstone, shale and mudstone with several conglomerate beds [12] [15] . The reddish colour gradually turns into greenish-grey at a few metres below the overlying Basal Anhydrite of the Maha Sarakham Formation. This is a result of subsequent diagenesis, probably from reducing subsurface brines that had percolated through the rocks during the time of burial [24] . The contact with the overlying Maha Sarakham Formation is unconformable [12] [15] . This sharp contact to the Basal Anhydrite has been reported for example by [25] and within seismic profiles by [26] . The depositional environment is expected to have been a meandering river system, less mature than the rivers that deposited the Sao Khua Formation [12] . This assumption of a fluviatile environment with possible marine influence is based on the associated fauna [15] . An age of Late Aptian for the upper parts of the formation was reported by [26] due to palynological evidence. An Aptian to Albian age based on dinosaur findings in the Khok Kruat Formation was also indicated by [27] .

3.6. Maha Sarakham Formation

The Maha Sarakham Formation comprises four different salt units which principally consist of evaporates separated by red-coloured siliciclastics: Basal Anhydrite, Lower Salt, Potash Zone, Several Colour Bands Salt Unit, Lower Clastic, Middle Salt, Middle Clastic, Upper Salt and Upper Clastic (see Figure 4 for stratigraphy). Its thickness ranges across the basin from 250 to 1100 m and its appearance varies upon the area controlled by rock salt structures [28] - [30] . The formation overlies the Khok Kruat Formation with a very sharp contact. All three units comprise beds which consist of halite-replaced pseudomorphs of bottom-growth gypsum. The formation was deposited in a hypersaline, landlocked lake within an arid continental desert, hence fossils are rare [15] [30] . In several places of the basin salt domes and salt anticlines are present. Reference [26] reported an Albian to Cenomanian age for them based on palynomorphs. The following is a description of all members including the siliciclastics based on descriptions by [8] [29] and modified after [16] . The rock salt on the Khorat Plateau in both the Khorat and the Sakhon Nakhon Basin has the same characteristics and correlations, which shows that these two basins once formed one large basin during the deposition until the uplift of the Phu Phan Range in the Early Paleocene, (refer to Chapter 2.3 in [29]).

3.7. Basal Anhydrite Unit

The Basal Anhydrite Unit represents the base of the Maha Sarakham Formation in both the Khorat and Sakhon Nakhon Basin. This formation overlies the Khok Kruat Formation disconformably and stylolitic surfaces are found [16] [28] . Most of the anhydrite is formed directly or from conversion of gypsum, which existed as beds and clusters with gradational contacts to enveloping sediments [28] . Its thickness varies from 0.7 - 6.2 m [24] .

3.8. Lower Salt Unit

The Lower Salt Unit overlies the Basal Anhydrite and the underlying Potash Zone Unit. This rock stratum consists mainly of coarse crystalline, transparent to semitranslucent halite and a minor amount of anhydrite in the form of thin partings [24] . The unit is separated by an anhydrite marker bed into two sequences. [16] named them the Halite L1 and Halite L2 Unit. This marker anhydrite and also the following are a primary one deposited directly from sea water [16] . The contact to the Basal Anhydrite Unit is very sharp and immediate. Reference [24] proposed that this can either be the result from the abrupt change in gypsum solubility as the brine was introduced into the halite saturation field or a result due to pressure solution under burial conditions. The thickness of this unit depends on the position within the basin as well as salt flowage and ranges between 30 - 500 m [29] [31] .

3.9. Potash Zone Unit

The Potash Zone Unit is located between the Lower Salt and the Several Colour Bands Salt Unit. This unit is composed of potassium and magnesium minerals which appear in single layers and is widely distributed throughout the whole Khorat Plateau [29] [31] . Its thickness and distribution is highly variable and may be due to patchy dissolution, deformation and subordinate depositional differences [16] . The potash minerals include sylvite (KCl), carnallite ($MgCl_2 \cdot KCl \cdot 6H_2O$) as well as the magnesium mineral tachyhydrite ($CaMgCl_6 \cdot 12H_2O$). [25] found that the sylvite is a secondary deposit derived from leaching of carnallite so that the sylvite follows the carnallite beds in intervals. In contrast, the potash minerals are of primary origin. After [16] , the Potash Zone Unit can be divided into three zones: 1) a lower zone consisting of massive halite with traces of red coloured carnallite filling dissolution cavities in halite; 2) a middle zone composed of massive and poorly bedded pale red or colourless carnallite and halite; and 3) an upper sylvite containing halite zones with traces of anhydrite and gypsum pseudomorphs. Each zone grades into the next one. Accessory minerals which are associated with sylvite and halite are borate minerals and these occur either in thin layers in the sylvite beds or as dispersed nodules and grains. The tachyhydrite usually appears as a single bed between the carnallite beds showing a gradational contact. The base of the Potash Zone Unit is defined by the first appearance of carnallite. In general, the thickness is highly variable from 10 to 80 m, in average about 50 m [16] [29] [31] .

3.10. Several Colour Bands Salt Unit

The Several Colour Bands Salt Unit is a very thin bed deposited on the Potash Zone Unit. It grades into the underlying Potash Zone. It's occurrence may be explained by ground water leaching. It is composed of multi-colour bands of grey, orange, red, honey, colourless and smoky dark halite. Some carnallite and tachyhydrite are also found (for details see [29] [31]).

3.11. Lower Clastic Unit

The Lower Clastic Unit is situated between the underlying Potash Zone and the overlying Middle Salt Unit. The rocks are very similar to the following Middle Clastic Unit. Both are semi-consolidated to damp mudstones showing a dissolution contact to the underlying rock unit, but gypsum is absent. The colour of the Lower Clastics is dark-red to reddish-brown but at the base there are locally greenish-grey portions. Carnallite veins and veinlets occur in this unit showing a mostly deep red-orange colour. Its thickness ranges from a few metres up to 70 m [24] [29] [31] .

3.12. Middle Salt Unit

The Middle Salt Unit overlies the Lower Clastic Unit and the underlying Middle Clastic Unit. The mean thickness is about 30 - 130 m [29] [31] . The whole Middle Salt Unit shows most of the depositional and diagenetic features like the Lower Salt Unit. Reference [16] divided this halite unit into two different sequences: the Halite M1 Unit and Halite M2 Unit. They are separated by an anhydrite marker bed. The Halite M1 Unit consists of well-bedded, dark honey coloured halite layers which are interbedded with thin anhydrite beds and layers of gypsum pseudomorphs. Some traces of a few sylvite and carnallite spots are found. The Halite M2 Unit contains dark honey coloured bedded halite, which is interbedded with dark smoky coloured halite beds. Anhydrite stringers and gypsum pseudomorphs are also found.

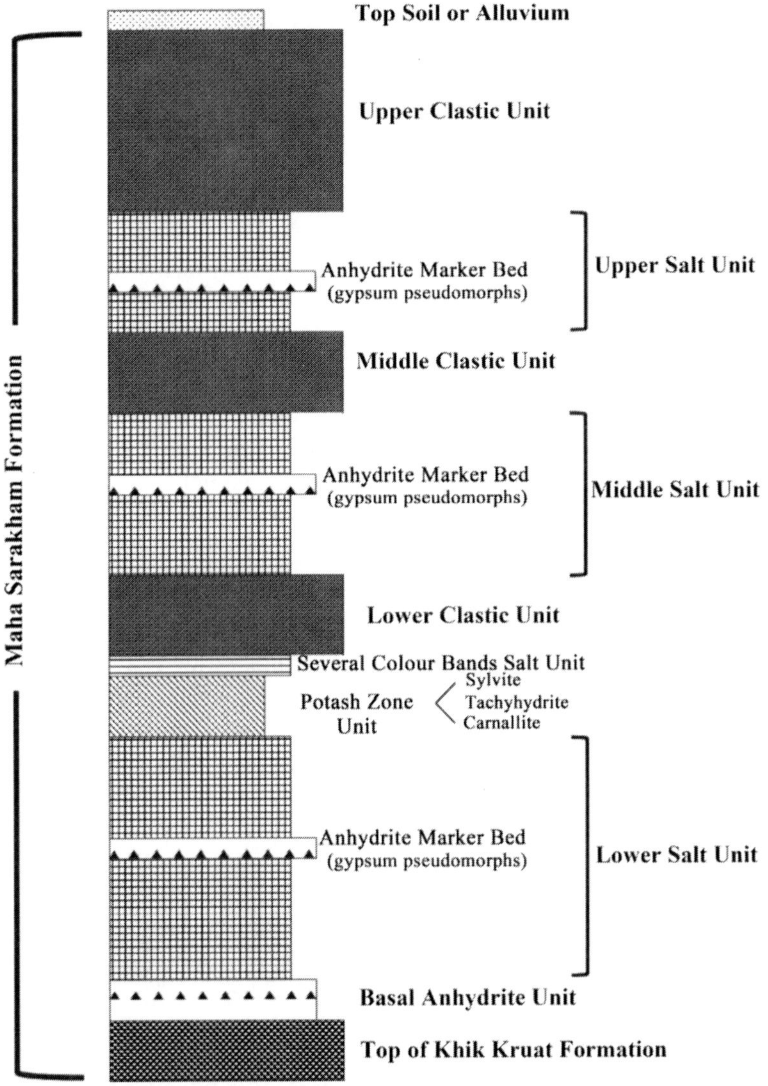

Figure 4. Lithostratigraphy of the complete sedimentary section of the Maha Sarakham Formation. After [16] [24] [29] .

3.13. Middle Clastic Unit

The Middle Clastic Unit overlies the Middle Salt Unit and the underlying Upper Salt Unit. The rocks are predominantly composed of clay, which is mostly dark reddish-brown in colour. In some cases, layers consisting of breccias are found. Furthermore, in some portions calcareous components and gypsum veins as well as veinlets near the bottom occur in this rock stratum [29] [31] . The thickness of this unit ranges between 20 - 70 m [16] . If no overlying Upper Salt Unit exists, due to dissolution and salt flowage, the Upper Clastic Unit overlies the Middle Clastic Unit directly. The contact of the Middle Clastic Unit with the underlying Middle Salt Unit shows either a sharp or a gradational contact [29] [31] .

3.14. Upper Salt Unit

The Upper Salt Unit, if not erased through salt structures or dissolution, generally overlies the Middle Clastic Unit and the underlying Upper Clastic Unit. Near the structural high, salt ridges and domes of the Upper Salt are dissolved and destroyed as a result of flowage. The thickness is highly variable and up to 20 m. It is usually composed of moderate to dark yellowish-brown or honey coloured halite, minor anhydrite layers, dark smoky coloured halite bands, milky-white coloured halite and a few orange coloured halite grains. Generally, the characteristics of the Upper Salt Unit are similar to the Middle Salt Unit, but with geochemical methods they can be separated by the KBr content [29] [31] . Several authors like [29] [31] and [16] divided this unit into two sequences: the Halite U1 Unit and Halite U2 Unit. These sequences are separated by an anhydrite marker bed, which is about 1.5 m in thickness. The Halite U1 Unit forms the contact to the Middle Clastic Unit and is composed of moderate to dark-honey or yellowish-brown coloured halite with minor anhydrite beds and a few layers of gypsum pseudomorphs, whereas the Halite U2 Unit mostly consists of dark-honey and interbedded dark- smoky and minor orange coloured halite. Local traces of potash are also found.

3.15. Upper Clastic Unit

The Upper Clastic Unit represents the uppermost rock stratum of the Maha Sarakham Formation. The Unit is mostly built up of pale to moderate reddish-brown or brick-red coloured sandstone, siltstone and claystone [29] [31] . Cross-beds and well-defined bedding are observable. The thickness is highly variable up to 680 m [16] . At middle to lower sections white clear gypsum veins and veinlets are found [31] . The contact with the overlying Phu Thok Formation, which is also composed of clastics, is not clearly defined. In comparison, the clastics of the Phu Thok Formation are coarser than the ones of the Upper Clastic Unit.

3.16. Phu Thok Formation

The Phu Thok Formation occurs in the northern and central parts of the Khorat Plateau. Its thickness varies between 200 - 350 m. The lower 100 - 200 m generally consists of a sequence of brick-red to red-brown coloured claystone with minor siltstone and rare fine-grained, thin sandstone beds, whereas the following meters consist of thick beds of brick-red coloured sandstone with cross-bedding. For the lower part of the formation a very low- energy fluvial system is assumed, while the upper part certainly represents an aeolian environment. The contact with the underlying Maha Sarakham Formation is unconformable. Based on its stratigraphic position, the Phu Thok Formation is considered to be Late Cretaceous in age although some authors have attributed it to the Early Cenozoic, e.g. [12] [30] . For a general stratigraphic correlation and detailed lithological description see also [32] .

4. SAMPLE LOCATIONS AND ANALYTICAL METHODS

The study area is located in the western part of the Khorat Basin on the Khorat Plateau, in northeastern Thailand. In general, the topography of northeastern Thailand consists of rolling plains bounded by mountains to the west and south. The exact working area is situated in a salt mine in Bamnet Narong (Chaiyaphum Province) where mining took place for decades. The samples were collected along a 1200 m profile in one shaft of the mine. Figure 5(a) shows the surface projection of this shaft with the respective sample points. The samples vary from halite and anhydrite to carnallite and one sample represents an intercalated clay layer from the Middle Clastic Unit. The vertical distribution of these different salt layers within the deposition is shown in the stratigraphic column in Figure 5(b).

Samples collected for rare earth elements were determined by ICP-MS on a PerkinElmer DRC II (Sciex, Canada) at the Department of Geochemistry. Bromide concentration was measured by anion chromatography DX320 (Dionex) with an auto-sampler AS11-HC at the Department of Applied Geology department. Sr concentrations and isotopic compositions were determined using a Thermo-Finnigan Triton© TIMS at the Department of Isotope Geology. Prior to digestion, samples were mixed with a tracer solution enriched in ^{87}Rb-^{84}Sr. Detailed descriptions of the analytical procedure for salt samples is given in [33] and [34] . Throughout this work, a value of 0.710267 ± 0.000041 (2σ) for the NBS 987 (n = 10) was observed. Instrumental mass fractionation was corrected with a $^{88}Sr/^{86}Sr$ ratio of 0.1194 using the exponential law. K-Ar analyses were also performed at the Department of Isotope Geology. The analytical procedure is given in [35] . All departments are part of the Geoscience Centre (GZG) at the University of Göttingen (Germany).

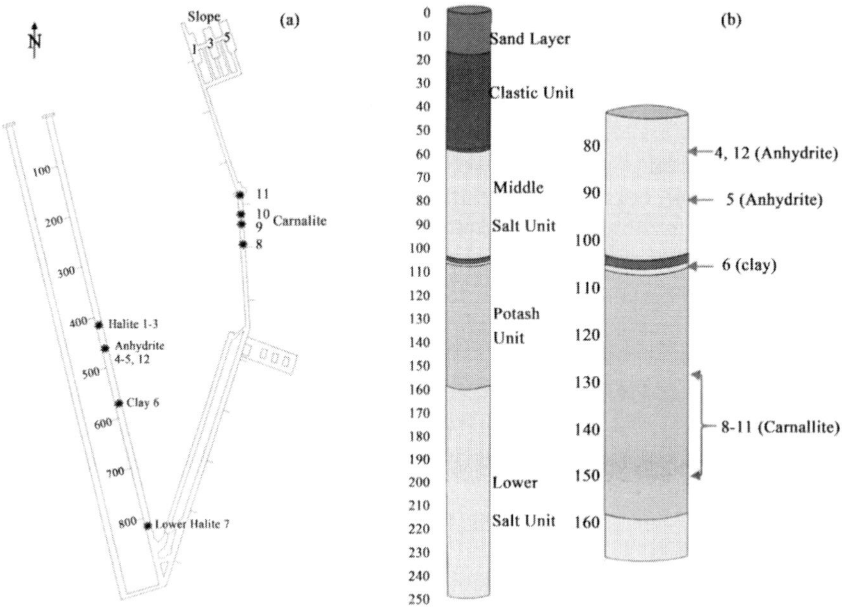

Figure 5. Surface projection of the 1200 m long shaft (a). The vertical distribution of the different salt layers (core sample) at the right in the diagram. Modified after URL: http://www.apmcnet.com/en/project_bamnet.php (query 20.09.2013).

5. ANALYTICAL RESULTS

5.1. Rb-Sr and Sr-Sr

Carnallites and anhydrites were analysed for their Rb-Sr values. Table 1 shows these data including the $^{87}Sr/^{86}Sr$ ratios corrected to an NBS987 value of 0.71024 used for the calculation of ages by means of the LOWESS Isotopic Database [7] in order to acquire the corresponding age of this stratigraphic unit.

The anhydrite samples are almost free of rubidium, except sample ST12, which in comparison to its strontium content includes more rubidium. Most probably it is not pure anhydrite, which is also recognizable in the plot in Figure 6(a). The $^{87}Sr/^{86}Sr$ ratios vary between 0.707461 and 0.707630. In contrast, the carnallites show a high range of strontium and rubidium content without an obvious trend. The lowest ratio by far is measured in sample ST11 with 0.707881, and in relation to the other samples, it contains much more strontium in comparison to the amount of rubidium. Generally, the carnallites scatter in the Rb-Sr diagram (Figure 6(b)).

The $^{87}Sr/^{86}Sr$ record can be used to date and correlate marine sediments [36] . The change in the marine $^{87}Sr/^{86}Sr$ ratio throughout the Phanerozoic is shown Figure 7. For the samples discussed, the Late Cretaceous time is important and needs to be looked at in more detail because of the stratigraphic relation of the formations. Details on how the LOWESS fit is derived are given in [36] . To determine the age by using the LOWESS fit, it is necessary to calculate the initial $^{87}Sr/^{86}Sr$ ratio for different ages by correcting for the amount of ^{87}Sr enrichment by ^{87}Rb decay. Is the initial value equal to the seewater curve, which is used for the calculation, then this is the age of the sample. This was done for all the anhydrites as well as for all the carnallites.

Table 1. Rb-Sr Data.

Sample	^{86}Sr (µmol)	Sr (µg/g)	^{87}Rb (µmol)	Rb (µg/g)	$^{87}Rb/^{86}Sr$	$^{87}Sr/^{86}Sr$ [fc]	2 se	$^{87}Sr/^{86}Sr$ [fc] corrected
Anhydrite ST4	0.024124	410.7	0.00019	1.12	0.007825	0.707595	0.000016	0.707595
Anhydrite ST5	0.030156	475.0	0.00017	1.01	0.005497	0.707484	0.000019	0.707461
AnhydritST12	0.001302	23.1	0.00007	0.463	0.057561	0.707653	0.000044	0.707630
Carnallite ST11	0.015876	284.0	69.8	69.8	0.706835	0.707904	0.000017	0.707881
Carnallite ST10	0.002626	46.8	53.0	53.0	3.230726	0.709293	0.000013	0.709270
Carnallite ST9	0.002016	35.7	60.6	60.6	4.886292	0.709171	0.000047	0.709148
Carnallite ST8	0.002483	43.9	45.6	45.6	2.986739	0.709636	0.000130	0.709613

fc. = ratio corrected for mass fractionation, corr. = ratio downscaled to an NBS987 value of 0.71024.

This method was not applicable for the carnallites due to their high contents of rubidium. The back-calculated initials for 95 Ma are extremely low (0.70256 - 0.70694), which would normally reflect a strong input of volcanic material into the basin. This cannot be true as the basin was filled from the west and the only volcanic material present crops out in the south and is Cenozoic in age. By plotting the $^{87}Sr/^{86}Sr$ ratios of the anhydrites against the so-called LOWESS composite strontium isotope curve from [7] , the authors fit the curve with limits of error corresponding to ages of the Upper Cretaceous. This was a period of worldwide sea level change (transgression) [37] .

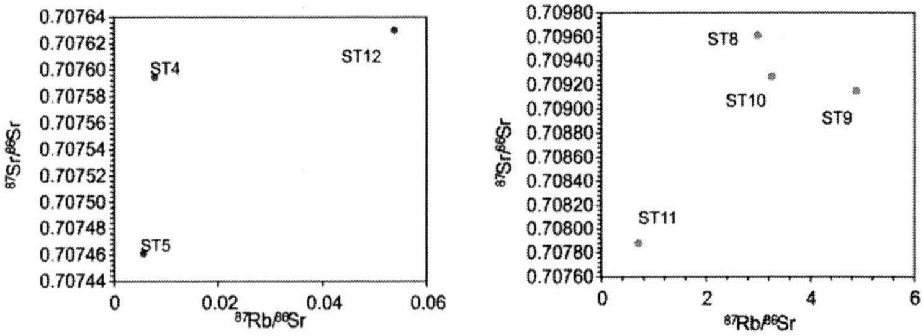

Figure 6. Rb-Sr-Plot of the anhydrites (a) and the carnallites (b).

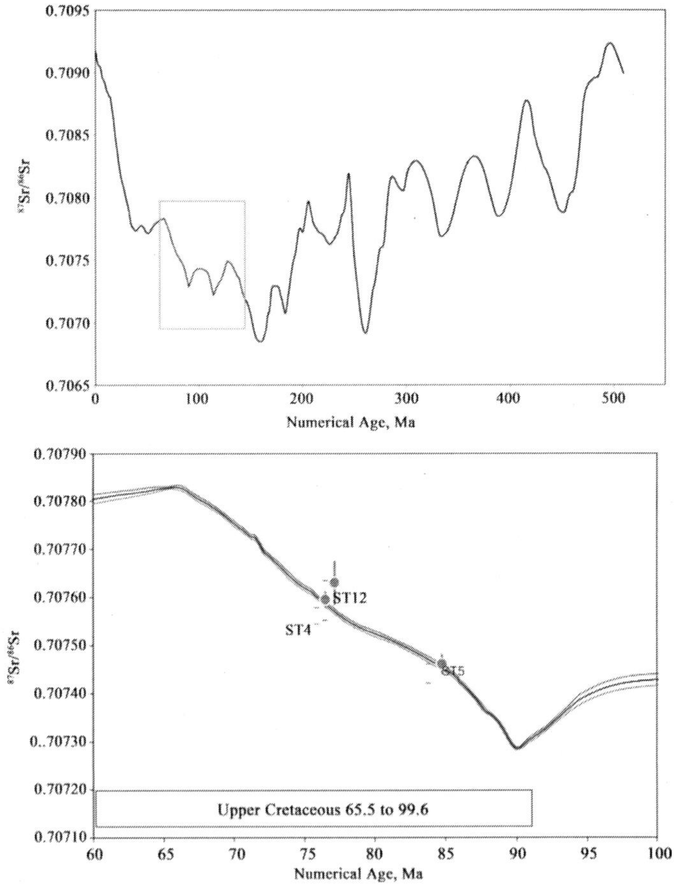

Figure 7. Sea water evolution curve, LOWESS, during the Phanerozoic with a more detailed section of the Cretaceous after [7] . Red coloured line means the upper limit of the $^{87}Sr/^{86}Sr$, the blue one analogues the lower limit and the black curve is the mean value.

5.2. K-Ca age Determination

Carnallite is an evaporitic double chloride mineral, a hydrated potassium magnesium chloride with the chemical formula: $KMgCl_3 \cdot 6(H_2O)$. It occurs within a sequence of potassium and magnesium evaporate minerals such as sylvite and kainite, and other salts like gypsum and anhydrite. Since this mineral is theoretically free of calcium, it should be possible to date these evaporates by means of the K-Ca method.

Four carnallites samples (Figure 5) have been analysed (Table 2 & Table 3) concerning their calcium and potassium content for the K-Ca age determination. Because of the immobility of calcium and its characteristic as a retentive daughter, the K-Ca ages are more resistant to resetting. However, in this case it turned out that the analysed samples (after loss of ignition) were inhomogeneous even on a small scale, and thus analytical results of the Ca-concentration could not be reproduced. To find out whether the calcium determinations are influenced by a possible heterogeneity of the salt, the sample ST11 was used to investigate each clearly recognisable phase of carnallite separately (in the rock association, Figure 8). This figure shows the hand-specimen of sample ST11 (a) and the parts (b)-(d), which were differentiated on the basis of their colour difference and then crushed in a mortar. The optical inhomogeneity was confirmed by LA-ICPMS investigations. The carnallites were analysed for their Li, B, Na, Mg, Al, Si, S, Cl, K, Ca, Mn, Br, Rb, Sr, Cs and Pb content by profiles on different fragments of the salts. Iron was not measured because it behaves like manganese. The profiles vary in length from 1 - 2.5 mm, since the length is correlated to the time in seconds, this is indicated as a reference. For these measurements the total magnesium amount was set to 400,000 ppm, and therefore it calculates everything in relation to it. This was later relativized during the normalisation. One encountered problem was that chlorine was immensely mobile during the measurement and shows a strong drift, so that the published values in ppm are incorrect as an absolute value. They are still relative if the proportions or ratios are considered, and for the change over time on a logarithmic scale. We here just show the most meaningful elevated ones (Figures 9-11). Sample ST8 was measured four times. The samples are all similar, therefore only some spectra from ST11 are given below. Apparently there are abrupt changes concerning the content of magnesium and potassium over four magnitudes (Figure 9) or the content of both elements can be constant along the profile. Again it can be seen that both magnesium and potassium can behave in the same way, but also contrary to each other (Figure 9(c) & Figure 9(d)). These changes in the concentration occur very suddenly and unexpectedly, which indicate different phases in the carnallite.

Table 2. Results of the LOWESS fit after [7] for all samples.

A	$^{87}Rb/^{86}Sr$	$^{87}Sr/^{86}Sr_{corr}$	Ini.75 Ma.	Mean	Ini.76 Ma.	Mean	Ini.77 Ma.	Mean	Ini.78 Ma.	Mean
ST4	0.007825	0.707595	0.707586	76.5	0.707586	76.5	0.707586	76.5	0.707586	76.5
ST12	0.053763	0.707630	0.707573	77.05	0.707572	77.10	0.707571	77.15	0.707571	77.15
ST5	0.005637	0.707661	0.707461	84.70	0.707455	84.70	0.707455	84.70	0.707455	84.7
ST11	0.706835	0.707881	0.707128	147,85	0.707118	148.20	0.707108	148.50	0.707098	148.85
ST10	3.230726	0.709270	0.705798	-	0.705752	-	0.705706	-	0.705660	-
ST9	4.886292	0.709148	0.703945	-	0.703875	-	0.703806	-	0.703736	-
ST8	2.986739	0.709613	0.706430	-	0.706388	-	0.706346	-	0.706303	-
B	$^{87}Rb/^{86}Sr$	$^{87}Sr/^{86}Sr_{corr}$	Ini.84 Ma	Mean	Ini.85 Ma.	Mean	Ini.86 Ma	Mean	Ini.100 Ma.	Mean
ST4	0.007825	0.707595	0.707585	76.55	0.707585	76.55	.707585	76.55	0.706878	76.60
ST12	0.053763	0.707630	0.707566	77.35	0.707566	77.35	0.707564	77.45	0.707554	77.95
ST5	0.005637	0.707661	0.707454	84.75	0.707454	84.75	0.707454	84.75	0.707453	84.8
ST11	0.706835	0.707881	0.707038	150.60	0.707038	150.60	0.707018	151.10	0.706878	155.20
T10	3.230726	0.709270	0.705382	-	0.705382	-	0.705289	-	0.704641	-
ST9	4.886292	0.709148	0.703320	-	0.703320	-	0.703181	-	0.702210	-
ST8	2.986739	0.709613	0.706049	-	0.706049	-	0.705964	-	0.705370	-

Table 3.ICP-OES data in ppm.

	Sample	Zn	Pb	Co	Ni	Mn	Fe	Cr	Mg	V	Ga	Al	Be	Ca	Cu	Ti	La	Sr	Ba	Na	K
	Mica Fe_23	1452	<DL	<DL	42	2806	196,741	79	32,922	222	218	103,760	4	2779	25	16,628	273	2	140	2301	103,476
	Mica Fe_24	1416	<DL	<DL	21	812	195,338	85	33,209	220	149	102,602	4	2759	17	16,569	266	2	139	2258	101,893
anhydrites	ST4_1	<DL	<DL	<DL	<DL	<DL	1094	<DL	18,451	<DL	<DL	1384	0	176,424	92	83	<DL	465	26	180,561	1076
	ST4_2	<DL	<DL	<DL	<DL	<DL	1014	<DL	23,579	<DL	<DL	1243	0	208,263	17	76	<DL	534	25	141,496	1065
	ST5_1	<DL	<DL	<DL	<DL	<DL	790	<DL	21,700	<DL	158	656	<DL	299,642	19	30	<DL	898	12	35,617	789
	ST5_2	<DL	<DL	<DL	<DL	<DL	865	<DL	21,565	<DL	163	619	0	304,822	33	36	<DL	903	12	35,145	846
	ST12_1	<DL	<DL	<DL	<DL	<DL	169	<DL	5263	<DL	149	<DL	0	17,040	10	9	<DL	46	14	434,236	2446
	ST12_2	<DL	<DL	<DL	<DL	<DL	113	<DL	3873	<DL	<DL	<DL	0	7860	<DL	8	<DL	22	8	395,595	798
carnallites	ST8_1	<DL	<DL	<DL	<DL	3	107	<DL	93,744	29	<DL	<DL	0	1854	17	14	62	38	<DL	150,088	203,516
	ST8_2	<DL	<DL	<DL	<DL	3	<DL	<DL	90,437	20	144	<DL	0	2125	<DL	7	<DL	60	<DL	166,053	178,016
	ST9_1	<DL	<DL	<DL	<DL	5	206	<DL	86,762	17	<DL	<DL	0	1187	<DL	16	<DL	26	10	63,461	158,077
	ST9_2	<DL	<DL	<DL	<DL	5	81	<DL	100,128	<DL	<DL	<DL	0	1936	<DL	<DL	<DL	50	<DL	131,568	179,790
	ST10_1	<DL	<DL	<DL	<DL	<DL	<DL	<DL	97,813	<DL	<DL	<DL	0	1580	<DL	<DL	<DL	60	<DL	119,066	193,269
	ST10_2	<DL	<DL	<DL	<DL	<DL	<DL	<DL	119,060	<DL	<DL	<DL	0	1021	<DL	<DL	<DL	38	<DL	56,251	223,500
	ST11_1	<DL	<DL	<DL	<DL	<DL	<DL	<DL	92,369	18	270	342	0	5483	<DL	14	<DL	486	<DL	142,841	171,521
	ST11_2	<DL	<DL	<DL	<DL	<DL	<DL	<DL	108,719	18	326	<DL	0	3605	<DL	9	<DL	264	<DL	77,525	209,677
	DRN_25	148	<DL	16	<DL	1703	67,556	41	31,567	253	162	92,617	2	53,259	35	7049	<DL	417	393	26,965	20,247
	DRN_26	138	<DL	38	<DL	1749	67,544	31	32,624	254	<DL	94,789	0	54,496	37	7176	<DL	428	388	27,558	20,772
	Mica Fe_23	1395	<DL	<DL	11	2893	192,839	97	35,257	201	272	102,248	4	2809	<DL	17,123	167	2	133	2319	103,287
	Mica Fe_24	1402	<DL	<DL	<DL	2900	197,862	89	35,097	223	185	102,661	4	2829	23	17,141	220	2	140	2354	104,622
	Detection Limit (DL)	10 - 400	155 - 400	28 - 200	29 - 400	6 - 2000	79 - 80,000	19 - 400	229 - 40,000	17 - 400	144 - 400	318 - 100,000	0 20	42 - 40000	10 200	3 - 5000	27 - 200	0 - 1000	7 - 2000	54 - 20,000	93 - 40,000

Figure 8. Pictures of sample ST11 and different portions of it. (a) The base material of sample ST11 with different phases highlighted; (b) Light orange-clear phase, crushed; (c) Intensive red phase, crushed; (d) Dark phase, crushed (in the rock association this phase was much darker than it appears in the picture).

Overall, sample ST11 is mostly subject to fluctuations in terms of the various measured elements. The changes of the elements manganese, sodium and calcium can be seen in the graphs of Figure 10. Sodium generally shows the strongest variations with respect to the changing high of magnitude. Compared with Figure 9, the sodium content always increases abruptly when the potassium and magnesium content decreases but with one exception: In part C of both figures the content of sodium increases immensely after about 235 seconds, while the potassium and magnesium content only shows a little change or fluctuation. Calcium itself also varies greatly, especially at points of increasing sodium content. In contrast to the change of the magnesium and potassium content the calcium content remains unchanged. The boron content in the carnallite varies strongly over four orders of magnitude, as shown in Figure 11. Furthermore, it is evident that each portion of the carnallite sample contains boron (B). The recognisable peaks of the boron content are in favour of a boron-containing phase, as already mentioned above. In a Rb/Br diagram [38] , the obtained ratios plot outside the field of primary carnallites except sample ST9 (Figure 12). Based on the facts presented, a dating of the investigated carnallites is not possible due to the inhomogeneity of the samples,. An attempt to have a more representative sample by using higher sample amounts failed due to the high concentration of sodium. The fact that the analysed samples do not originate from primary carnallite is confirmed by separate Br analyses (Table 4). If these data are plotted against the concentration of Rb in a Rb/Br diagram [38] , therefore also K was redistributed during later hydrothermal activity.

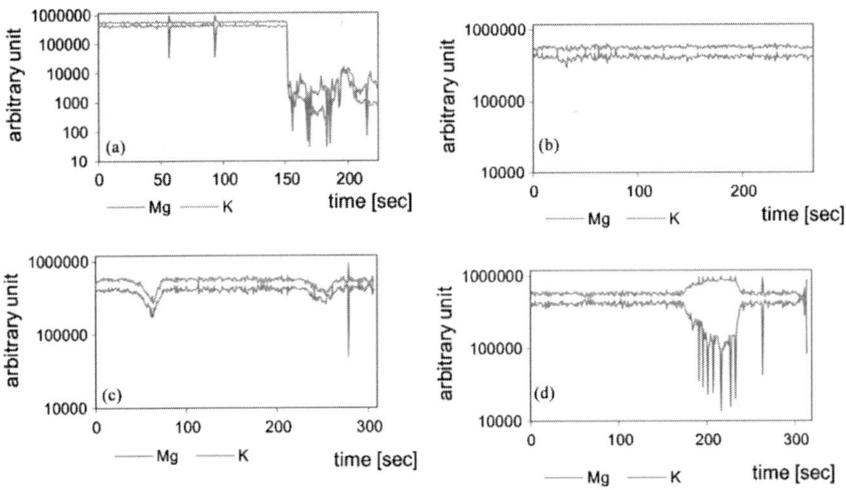

Figure 9. Mg and K results (using LA-ICPMS) of sample ST11with different measurements for different portions of the sample.

5.3. K-Ar

In 2002 the authors in [5] presented a K-Ar age for the clay layer in between the Middle Salt Unit and the Potash Unit (Figure 5) of 97.7 Ma. The analytical data are given in Table 4. The age was obtained in the fraction < 0.2 µm. The illite crystallinity of the sample corresponds to a Kübler Index

(KI) of 0.69 Δ°2Θ. This value is normally interpreted as reflecting diagenetic conditions. However, we cannot exclude that the sample contains tiny parts of detrital material, so therefore we interpret this age as a maximum age for the deposition of the clay Table 5.

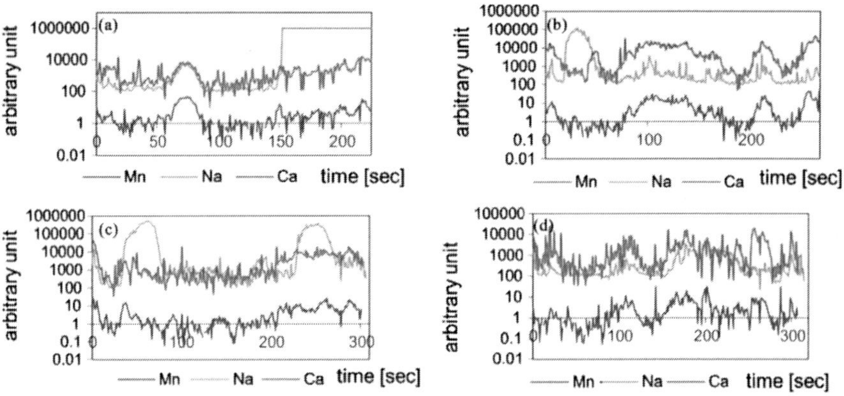

Figure 10. Mn, Na and Ca results (using LA-ICPMS) of sample ST11 with different measurements for different portions of the sample.

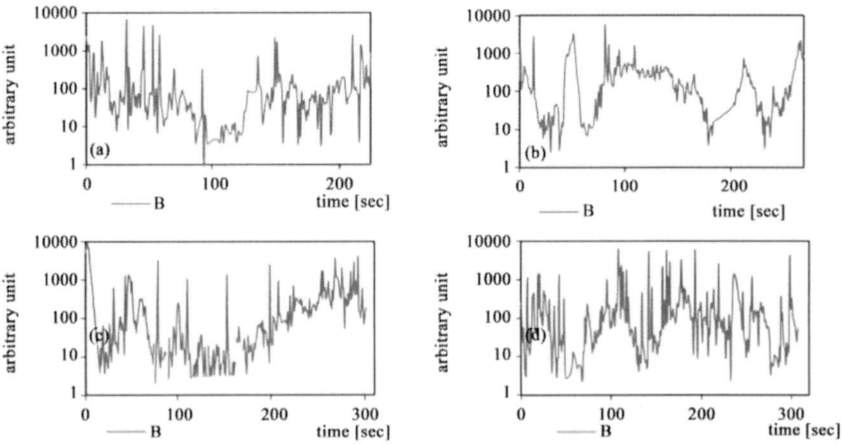

Figure 11. Boron results (using LA-ICPMS) of sample ST11 with different measurements for different portions of the sample.

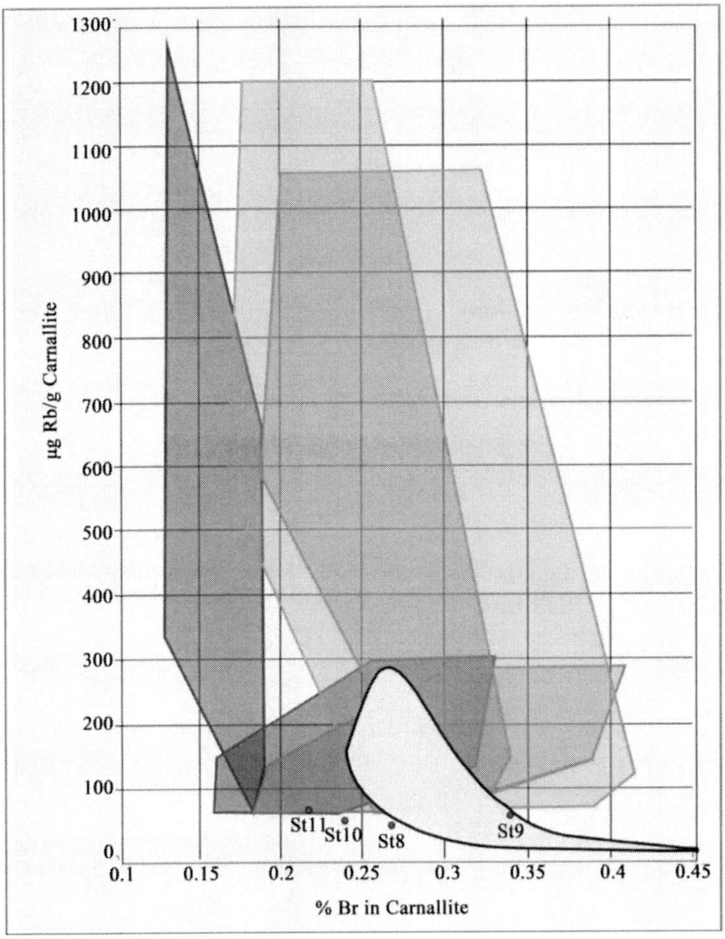

Figure 12. Rb distribution in primary carnallite (yellow area) after [38] .

Table 4. Bromide content.

Sample No.	Bromide in %	Bromide mean%
ST8 ST8	0.268 0.268	0.268
ST9 ST9	0.342 0.339	0.340
ST10 ST10	0.238 0.241	0.239
ST11 ST11	0.217 0.217	0.217

Table 5. K-Ar analytical data from clay sample ST6.

Sample No ST 6	K_2O (Wt. %)	$^{40}Ar^*$ (nl/g) STP	$^{40}Ar^*$ (%)	Age (Ma)	2σ-Error (Ma)	Illite crystallinity $\Delta°2\Theta$
<0.2 µm	6.12	18.98	78.99	93.7	2.7	0.69

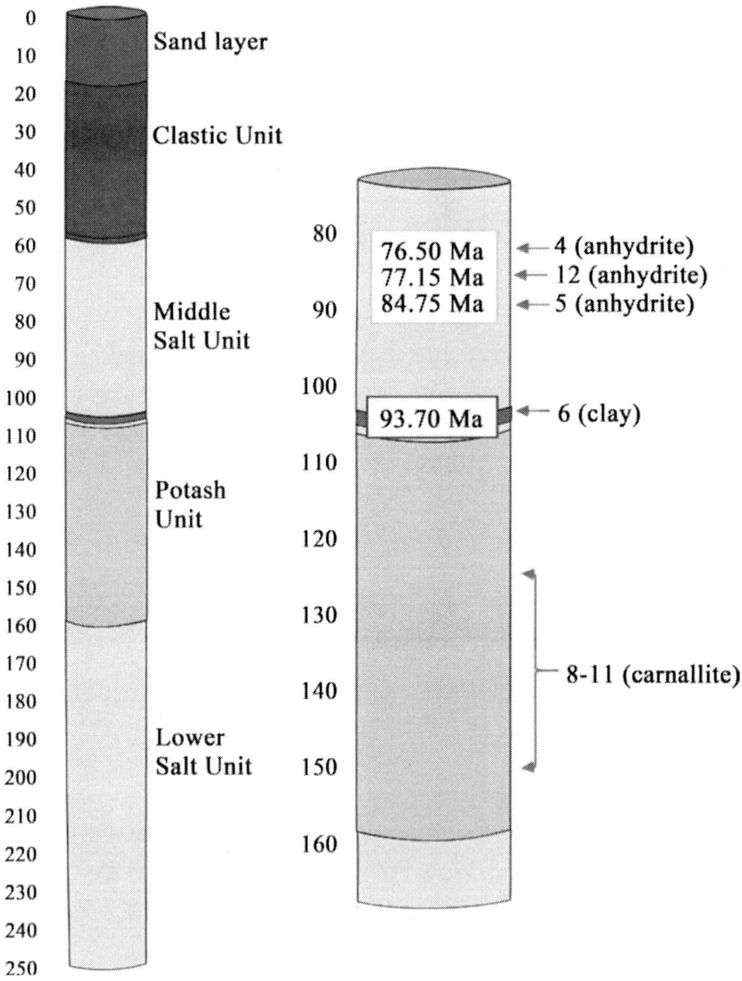

Figure 13. Vertical distribution of the different salt layers (core sample) with the data points and the obtained depositional ages.

6. DISCUSSION AND CONCLUSIONS

The age of the salt in Bamnet Narong has been debated since the late 1970's by using multiple stratigraphic and isotopic approaches. The goal of the present study was to obtain age data, particularly for the carnallites by using the K-Ca as well as the Rb-Sr decay. However, as discussed above the content of calcium is very inhomogeneous, and could therefore not be used for the age determinations. In the Rb-Sr isochrones diagram the carnallite samples scatter, therefore their isotope systems obviously are disturbed by later redistribution of rubidium and correspondingly potassium. The K-Br analysis of [39] corresponds with this assumption of redistribution. A possible dissolution of the carnallites effected by groundwater would increase the rubidium content in the solution, the result would be a phase with high $^{87}Sr/^{86}Sr$.

The obtained $^{87}Sr/^{86}Sr$ data indicates a deposition age for the anhydrites (using the LOWESS fit after [36]) to be between 76.50 Ma and 84.75 Ma. Figure 13 illustrates that these ages match the stratigraphy of the Asian Potash Mine as well as the K-Ar age of the clay layer, which represents the Lower Clastic Unit [5] . The age of the Clastic Unit has to be treated with caution because these clastics are erosional continental deposits, so they normally have to be older than the deposition in the basin. The age of the anhydrites of the Middle Salt Unit from the Maha Sarakham Formation is considered to be Santonian to Campanian (Upper Cretaceous). If the age of the clay layer is correct, it indicates that the underlying Potash Unit (carnallites) is at least Cenomanian in age. This would be in good agreement with the obtained ages from [6] who postulated, that the Khok Kruat Formation (which underlies the Maha Sarakham Formation) is of Aptian to Albian age.

ACKNOWLEDGEMENTS

We would like to thank the Open Access Publication Fund of the University of Göttingen for their financial support. We are grateful to C. Gross who greatly helped to improve the English text. We also appreciate the help and advice we received in the laboratory by S. Pawlig, N. Nolte, S. Metje, K. Simon, M. Rittmeier and Brigitte Dietrich (all GZG, Göttingen).

CITE THIS PAPER

Bent T. Hansen, Klaus Wemmer, Monique Eckhardt, Prinya Putthapiban, San Assavapatchara (2016) Isotope Dating of the Potash and Rock Salt Deposit at Bamnet Narong, NE-Thailand. *Open Journal of Geology*,**06**,875-894.

REFERENCES

1. Gardner, L.S. and Hayworth, P.N. (1967) Salt Resources of Thailand, Report of Investigation No.11. Department of Mineral Resources, Bangkok, Thailand, 100 p.
2. Harris Jr., R.W. (1977) Palynological Analyses of Outcrop and Well Samples from the Khorat Basin, Central Thailand. Mineral Fuels Division.
3. Sattayarak, N. and Polachan, S. (1990) Rock Salt Beneath the Khorat Plateau. Proceedings on Mineral Management, Department of Mineral Resources, Bangkok, 2-6 July 1990, 1-14.
4. Suwanich, P. (2010) Potassium Bromide (KBr) Contents in the Maha Sarakham Formation. Journal of Science and Technology Mahasarakham University, 29, 249-258.
5. Mattenklot, M. (1995) Modelberechnungen zur Brund Rb-Verteilung in Carnallitgesteinen. Kali und Steinsalz Band 11, Heft 10.
6. Mearon, S., Paytan, A. and Bralower, T.J. (2003) Cretaceous Strontium Isotope Stratigraphy Using Marine Barite. Geology, 31, 15-18.
7. Howarth, R.J. and McArthur, J.M. (1997) Statistics for Strontium Isotope Stratigraphy: A Robust LOWESS Fit to the Marine Strontium Isotope Curve for the Period 0 to 206 Ma, with Look-Up Table for the Derivation of Numerical Age. The Journal of Geology, 105, 441-456.
8. Wemmer, K. and Ahrendt, H. (1997) Comparative K-Ar and Rb-Sr Age Determinations of Retrograde Processes on Rocks from the KTB Deep Drilling Project. Geologische Rundschau, 86, 272-285.
9. Ridd, M.F., Barber, A.J. and Crow, M.J. (2011) Introduction to the Geology of Thailand. In: Ridd, M.F., Barber, A.J. and Crow, M.J., Eds., The Geology of Thailand, Geological Society, London, 1-18.
10. Satarugsa, P., Youngmee, W. and Meesawat, S. (2005) New Regional Boundary of Maha Sarakham Formation in the Northeastern Thailand: Results from 2D Seismic Mapping. International Conference on Geology, Geotechnology and Mineral Resources of Indochina (GEOINDO 2005), Khon Kaen, 28-30 November 2005, 212-220.
11. Tabakh, M.E., Utha-Aroon, C. and Schreiber, B.C. (1999) Sedimentology of the Cretaceous Maha Sarakham Evaporates in the Khorat Plateau of Northeastern Thailand. Sedimentary Geology, 123, 31-62.
12. Racey, A. and Goodall, J.G.S. (2009) Palynology and Stratigraphy of the Mesozoic Khorat Group Red Bed Sequences from Thailand. Geological Society, London, Special Publications, 315, 69-83.
13. Chonglakmani, C. (2011) Trias. In: Ridd, M.F., Barber, A.J. and Crow, M.J., Eds., The Geology of Thailand, Geological Society, London, 137-150.
14. Cooper, M.A., Herbert, R. and Hill, G.S. (1989) The Structural Evolution of Triassic Intermontane Basins in Northeastern Thailand. Proceedings of the International Symposium on Intermontane Basins: Geology and Resources, Chiang Mai, 30 January-2 February 1989, 231-242.
15. Meesook, A. (2011) Cretaceous. In: Ridd, M.F., Barber, A.J. and Crow, M.J, Eds., The Geology of Thailand, Geological Society, London, 169-184. [Citation Time(s):16]

16. Piyasin, S. (1995) The Hydrocarbon Potential of Khorat Plateau. Proceedings of the International Conference on Geology, Geotechnology and Mineral Resources of Indochina, Khon Kaen University, Khon Kaen, 22-25 November 1995, 551-562. [Citation Time(s):12]
17. Zhang, X., Ma, H., Ma, Y., Tnag, Q. and Yuan, X. (2013) Origin of the Late Cretaceous Potash-Bearing Evaporites in the Vientiane Basin of Laos: δ11B Evidence from Borates. Journal of Asian Earth Sciences, 62, 812-818.
18. Wannakomol, A. (2005) Soil and Groundwater Salinization Problems in the Khorat Plateau, NE Thailand—Integrated Study of Remote Sensing, Geophysical and Field Data. Dissertation an der Freien Universitat Berlin, 205 p.
19. Suwanich, P. (1986) Potash and Rock Salt in Thailand. Nonmetallic Minerals Bulletin NO. 2, Economic Geology Division, Department of Mineral Resources Bangkok, Thailand, 339 p.
20. McArthur, J.M., Howarth, R.J. and Shields, G.A. (2012) Strontium Isotope Stratigraphy. In: Gredstein, F.M., Ogg, J.G., Schmotz, M.D. and Ogg, G.M., Eds., A Geologic Time Scale, Elsevier, Amsterdam, 127-144.
21. Buffetaut, E. and Ingavat, R. (1986) The Succession of Vertebrate Faunas in the Continental Mesozoic of Thailand. Bulletin of the Geological Society of Malaysia, 19, 167-172.
22. Hansen, B.T., Wemmer, K., Pawlig, S., Klaus, J., Assavapatchara, S. and Nontaso, M. (2002) Isotopic Evidence for a Late Cretaceous Age of the Potash and Rock Salt Deposit at Bamnet Narong, NE-Thailand. Proceedings of the Symposium on the Geology of Thailand, Department of Mineral Resources, Bangkok, 26-31 August 2002, 26-31.
23. Pisuth-Arnond, V., Chiba, H. and Yamuang, S. (1986) A Preliminary Sulphur and Oxygen Isotope Study of the Maha Sarakham Evaporitic Anhydrite from the Bamnet Narong Area of Northeastern Thailand. Bulletin of the Geological Society of Malaysia, 19, 209-233.
24. Klaus, J.S., Hansen, B.T. and Buapeng, S. (2007) 87Sr/86Sr-Ratios: A Natural Tracer to Characterize the Groundwater of Bangkok Area, Thailand, with Respect to Public Water Supply and Freshwater Reinjection. Hydrogeology Journal, 15, 745-758.
25. Wiegand, B., Dietzel, M., Bielert, U., Groth, P. and Hansen, B.T. (2001) 87Sr/86Sr-Verhaltnisse als Tracer für Hydrochemische Prozesse in Einem Lockergesteinsaquifer (Libenau, NW Deutschland). Acta Hydrochimica et Hydrobiologica, 29, 139-152. Department of Mineral Resources (2013) Geological Standard Division. Bureau of Geological Survey. Lexicon of Stratigraphic Names of Thailand, 271 p.
26. Suwanich, P. (1983) Potash and Rock Salt in Thailand. Conference on Geology and Mineral Resources of Thailand,. Department of Mineral Resources, Bangkok, 19-28 November 1983, 244-252.
27. Booth, J. and Sattayarak, N. (2011) Subsurface Carboniferous-Cretaceous Geology of NE Thailand. In: Ridd, M.F., Barber, A.J. and Crow, M.J., Eds., The Geology of Thailand, Geological Society, London, 185-222.
28. Suwanich, P. (2007) Potash-Evaporite Deposits in Thailand. GEOTHAI 07 International Conference on Geology of Thailand: Towards Sustainable Development and Sufficiency Economy, Bankok, 252-262.
29. Tabakh, M.E., Schreiber, B.C., Utha-Aroon, C., Coshell, L. and Warren, J.K. (1998) Diagenetic Origin of Basal Anhydrite in the Cretaceous Maha Sarakham Salt: Khorat Plateau, NE Thailand. Sedimentology, 45, 579-594.
30. , E. and Suteethorn, V. (1992) A New Species of Ornithischian Dinosaur Psittacosaurus from the Early Cretaceous of Thailand. Palaeontology, 35, 801-812. [Citation Time(s):12]
31. Sattayarak, N., Srikulwong, S. and Patarametha, M. (1991) Subsurface Stratigraphy of the Non-Marine Mesozoic Khorat Group, Northeastern Thailand. Proceedings of GEOSEA 7th Conference, Bangkok, 5-8 November 1991.
32. Hite, R.J. and Japakasetr, T. (1979) Potash Deposits of the Khorat Plateau, Thailand and Laos. Economic Geology, 74, 448-458. Utha-Aroon, C. (1993) Continental Origin of the Maha Sarakham Evaporates, Northeastern Thailand. Journal of Southeast Asian Earth Sciences, 8, 193-203.

33. Racey, A., Love, M.A., Canham, A.C., Goodall, J.G.S. and Polachan, S. (1996) Stratigraphy and Reservoir Potential of the Mesozoic Khorat Group, North Eastern Thailand: Part 1, Stratigraphy and Sedimentary Evolution. Journal of Petroleum Geology, 18, 5-39.
34. Hahn, L. (1982) Stratigraphy and Marine Ingressions of the Mesozoic Khorat Group in Northeastern Thailand. Geologisches Jahrbuch, Reihe B, 43, 7-35.
35. Buffetaut, E., Cuny, G., Le Loeuff, J. and Suteethorn, V. (2009) Late Palaeozoic and Mesozoic Continental Ecosystems of SE Asia: An Introduction. Geological Society, London, Special Publications, 315, 1-5.
36. Heggemann, H., Kohring, R. and Schlüter, T. (1990) Fossil Plants and Arthropods from the Phra Wihan Formation, Presumably Middle Jurassic of Northern Thailand. Alcheringa, 14, 311-316.
37. Philippe, M., Suteethorn, V., Lutat, P., Buffetaut, E., Cavin, L., Cuny, G. and Barale, G. (2004) Stratigraphical and Palaeobiogeographical Significance of Fossil Wood from the Mesozoic Khorat Group of Thailand. Geological Magazine, 141, 319-328.

CHAPTER 17

Paleotsunami Inundation of a Beach Ridge Plain: Cobble Ridge Overtopping and Interridge Valley Flooding in Seaside, Oregon, USA

Curt D. Peterson,[1] Harry M. Jol,[2] Tom Horning,[3] and Kenneth M. Cruikshank[1]

[1]*Geology Department, Portland State University, 1721 SW Broadway, Portland, OR 77207, USA*

[2]*Geography Department, University of Wisconsin at Eau Claire, P.O. Box 4004, Eau Claire, WI 54702, USA*

[3]*Horning Geosciences, 808 26th Avenue, Seaside, OR 97138, USA*

ABSTRACT

The Seaside beach ridge plain was inundated by six paleotsunamis during the last ~2500 years. Large runups (adjusted >10 m in height) overtopped seawardmost cobble beach ridges (7 m elevation) at ~1.3 and ~2.6 ka before present. Smaller paleotsunami (6–8 m in height) likely entered the beach plain interior (4-5 m elevation) through the paleo-Necanicum bay mouth. The AD 1700 Cascadia paleotsunami had a modest runup (6-7 m height), yet it locally inundated to 1.5 km landward distance. Bed shear stresses (100–3,300 dyne) are estimated for paleotsunami surges (0.5–2 m depths) that flowed down slopes (0.002–0.017 gradient) on the landward side of the cobble beach ridges. Critical entrainment shear stresses of 1,130–1,260 dyne were needed to dislodge the largest clasts (26–32 cm diameter) in paleotsunami coulees that were cut (100–200 m width) into the landward side of the cobble ridges.

1. INTRODUCTION

In this paper, we present the results of investigating paleotsunami inundations that crossed multiple beach ridges and flooded intervening valleys in a low-elevation beach ridge plain. The resort town of Seaside, Oregon, with a summer population of up to 50,000, is developed on a prograded beach ridge plain (Figure 1).This low-elevation plain has experienced both historical (1964) farfield tsunami [1] and nearfield paleotsunami (4-5 Cascadia events during the last 2,500 yr) [2–5].

Inundation from the historical farfield tsunami (1964 Gulf of Alaska rupture) in Seaside was limited to flooding of the small Necanicum Bay mouth and two interridge valley channels [1]. Several of the nearfield Cascadia paleotsunami crossed the beach plain (1.5 km in width) [4]. However, previous work has not established whether the beach plain inundation occurred by (1) direct overwash of the seawardmost cobble beach ridges, or (2) bay mouth flooding and shore-parallel flow within the interridge valleys [5]. Discrimination of these two mechanisms is important to documenting paleotsunami runup heights and to proposing specific evacuation strategies for coastal populations in low elevation settings of beach ridge plains.

In this paper, we present records of paleotsunami deposits, which are distributed landward of stable cobble ridges in the Seaside beach plain [6]. These data establish the minimum overtopping heights and bed shear stresses of the paleotsunami flows. In this coble ridge plain the thin paleotsunami sand layers are locally preserved in intertidal wetlands and supratidal spruce-tree bogs that occur in shore-parallel interridge valleys. The deposits do not demonstrate simple trends of landward thinning or continuity along the interridge valleys [5].

Tsunami deposition is highly variable in such beach ridge settings [13], leading to uncertainty of paleoflow dynamics. Nevertheless, tsunami deposit mapping constrains predicted minimum flow hazards and is needed to supplement the numerically modeled inundations of beach plains, such as at Seaside [14]. Beach ridge plains proved to be amongst the most dangerous of tsunami inundation settings during the recent (2004) Indian Ocean tsunami [15].

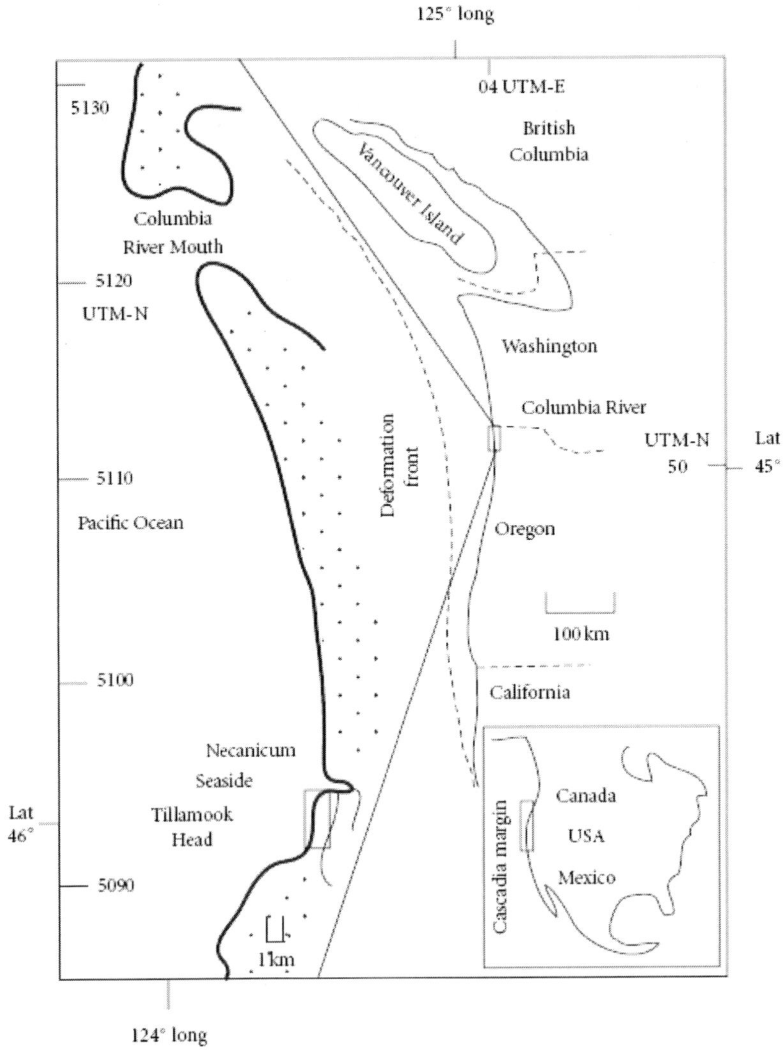

Figure 1: Location map of Seaside (boxed study area) at the southern end of the Clatsop beach-ridge plains (bold shoreline), located south of the Columbia River mouth. The study area is located near the center of the Cascadia subduction zone (inset) that extends ~1,000 km from Vancouver Island in British Columbia to northernmost California. The offshore deformation front of the subduction zone is shown by a bold dashed line. The Seaside population, up to 50,000 people, is expected to have ~25 minutes to evacuate following seismic shaking, before the onset of tsunami inundation [6]. Map coordinates include reference longitude and latitude and UTM Northings at 10 km intervals for the Clatsop Plains.

2. PREVIOUS WORK

Three factors contribute to a high tsunami risk in the Seaside beach ridge plain, including low elevation, large population, and interridge channels that restrict rapid evacuation at bridge crossings (Figure 2). These risk factors focused research attention on Seaside, leading to the first field evidence of paleotsunami overland inundation beyond tidal marshes in the Cascadia margin [2, 4]. However, the specific mechanisms of beach plain inundation by beach ridge overtopping, and/or by interridge valley flooding were left unresolved [5]. The uncertainty resulted from conflicting evidence of paleoflow indicators. Paleotsunami sand-sheets in Seaside are discontinuous across the cobble beach ridges, and are generally thin, averaging only several centimeters in thickness. Furthermore, the paleotsunami sand layers vary greatly in thickness (0.25–30 cm thick) along short distances (~100 m) of the interridge valley wetlands [4]. Cobbles are locally embedded in paleotsunami sand layers, such as between core sites NW2 and NW7, near the back-edge of the beach plain.

Sand fractions of the paleotsunami deposits (125–350 m in grain size) reflect both beach and river sand sources. The heavy-mineral fractions in selected paleotsunami deposits range from 25 to 87 percent beach sand component by weight (Figure 3) [3]. The river sand fractions were presumably entrained from the Necanicum and upper Neawanna channel bottoms by paleotsunami surges. Weathered gravel and oxidized soil peds occur in paleotsunami sand layers (#1, #2, and #3) in sites NW2 and NW7. The soil peds could represent backwash from hill slopes at the landward edge of the beach plain (Figure 2). The dated paleotsunami layers in the Neawanna wetlands extend back to 2 ka (Table 1). All core data presented in this paper are referenced to original core logs as shown in the caption of Figure 3.

Adding to the difficulties of paleotsunami deposit mapping in the Seaside beach plain are (1) bioturbation of paleotsunami deposits in supratidal soils, and (2) coseismic liquefaction of paleotsunami sand layers and underlying sand units [5]. Liquefaction disturbance including intruded contacts can occur in thin tsunami sand layers (1.0 cm thick), where they are buried by significant overburden (0.5–1.0 m depth) prior to seismic shaking. Dense sampling of target paleotsunami layers from multiple core sites at ~50 m intervals has helped to resolve these record preservation problems in nearby beach plains [7].

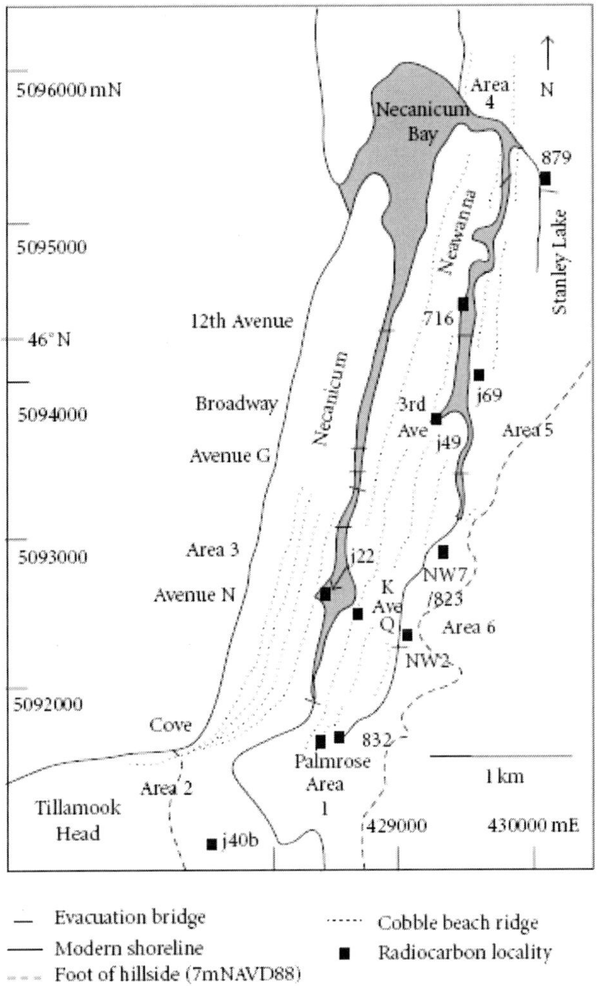

Figure 2: Map of Seaside beach ridge plain with modern shoreline (bold line), cobble beach ridges (dotted lines), and dated core sites (solid squares), and selected overtopping Areas 1–6. Evacuation routes, over bridges (short bold lines), cross the intertidal Necanicum and Neawanna channels. Map coordinates are in NAD83 in UTM (meters), with latitude and longitude reference points (degrees). Palmrose and Avenue Q are Native American midden sites that are preserved on cobble ridges (4-5 m elevation NAVD88). The seawardmost, cobble beach ridges are 6-7 m elevation. The fluvial-tidal wetlands in the interridge valleys (Necanicum and Neawanna) are of 2-3 m elevation. See Figure 4 for satellite image of study locality. See Figure 3 for core logs and cited sources of the core log data. See Table 1 for radiocarbon data.

Table 1: Radiocarbon dated samples from Seaside, Oregon.

Site name	Depth (cm)	Material type	C14 ± yr PB	Calibrated* cal yr BP 2σ	Lab no. beta-	Ref.
Covej40b	143	Cone	1230 ± 40	1060–1260	198700	4
Covej40b	228	Cone	2770 ± 40	2780–2950	198701	4
Palmrose	Base	Midden	~3,650			1
Palmrose	Top	Midden	~1,600			1
832	90	Peat	590 ± 70	515–668	96840	3
NW2	48	Peat	480 ± 60	326–644	42112	2
NW2	70	Peat	800 ± 60	657–904	42113	2
NW2	111	Peat	1100 ± 70	804–1230	42088	2
NW2	158	Peat	1370 ± 70	1094–1409	44595	2
NW2	268	Peat	2000 ± 70	1749–2148	42114	2
AveQ	Base	Midden	~3,250			1
AveQ	Top	Midden	~1,160			1
j22	72	Leaf	150 ± 30	60–290	194483	4
j4	230	Twig	210 ± 40	136–315^	194481	4
823	74	Peat	450 ± 60	312–625	96835	3
j49	410	Org.Mud	680 ± 40	540–660	198699	4
j69	70	twig	1530 ± 40	1320–1520	198698	4
716	81	Peat	360 ± 70	292–518	96837	3
879	54	Peat	480 ± 90	316–653	99234	3

Core sites and core logs are shown in Figure 2. Sample position is presented in depth (cm) below surface.
Calibrated years BP starts at 1950 AD.
Some event ages are approximated to 1000 yr BP (ka) in the text.
References (Ref): (1) [9, 11]; (2) [2]; (3) [4], (4) [6].
* Radiocarbon Calibration from CALIB 5.0.2 [12].
^ Radiocarbon calibrated age (j4) for older peak (81% probability distribution) of bimodal curve.

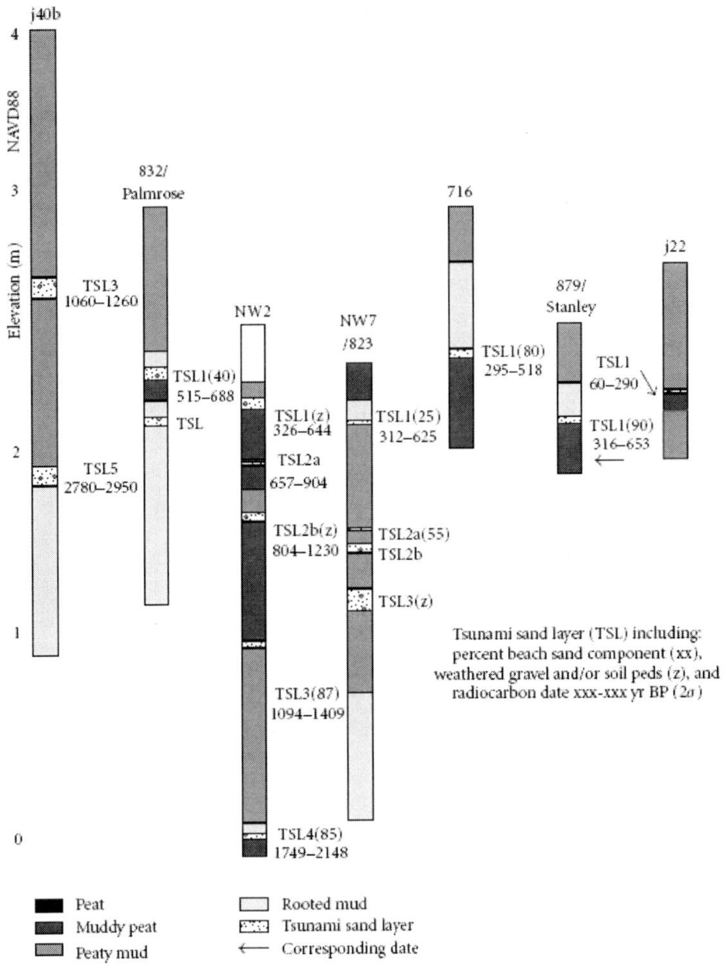

Figure 3: Core logs including dated paleosubsidence and paleotsunami sand layers in reported core sections from representative sites in the Seaside beach plain. Four of the five tsunami sand layers (TSL) are associated with regionally correlated coseismic subsidence horizons dated from ~0.3 to 2 ka. One weakly developed paleotsunami sand layer (TSL2a) is not associated with a paleosubsidence horizon [7, 8]. See Figure 2 for core site locations and see Table 1 for radiocarbon data. Core logs, radiocarbon dates, and paleotsunami sand source mineralogy data are from sites NW2, NW7, 832, 879 [2, 3, 9], sites 832, 716, and 879 [4], and from sites j40b and j22 [6]. All additional figures use the same core log citations: NWx for [2, 3] xxx for [4] jxx for [6].

In 2004 a pilot study of potential tsunami inundation in the Seaside study area was initiated by several US federal agencies [6]. As part of the multiagency study, the known AD 1700 paleotsunami sand layers were field checked and additional paleotsunami sites were mapped for evidence of beach ridge overtopping. In this paper, we integrate results form the several studies above to establish the range of flooding mechanisms and flow conditions that occurred in the Seaside beach plain locality.

The criteria used to identify paleotsunami deposits in beach plain settings of the study area are provided in previous papers [7, 8]. Data on paleotsunami deposit tracers in Seaside, including marine diatoms, sand mineralogy, and subsidence event stratigraphy are documented in earlier studies [2–4]. Details on paleotsunami core site coordinates, core logs, cut bank sections, and radiocarbon dating in Seaside are provided in the Tsunami Pilot Study [6]. Numerical-modeling has recently been used to estimate potential tsunami inundation produced by hypothetical probabilistic ruptures of the Cascadia megathrust fault [14]. The results from this presentation of the geologic inundation records that are preserved in the beach ridge plain can be compared to the numerically modeled tsunami hazards in Seaside.

3. BEACH PLAIN GEOMORPHOLOGY AND PALEOTSUNAMIS

Beach ridges in the Seaside study area were formed by cobbles and boulders that were derived from landslides shed off from the basaltic Tillamook Headland to the south of the beach ridge plain (Figure 2). The wave worked basaltic debris, typically ranging from 5 to 50 cm in diameter, was carried to the northeast alongshore (1–5 km distance) by large surf to form the beach cobble ridges [4]. Wave refraction focuses wave energy towards the Tillamook Headland, yielding cobble ridges of 7 m elevation NAVD88 at the Cove. (The NAVD88 datum (0 m) used in this study is equivalent to one meter below mean sea level, or −1 m relative to the earlier NAVD29 datum. The NAVD88 datum is also equivalent to mean low tide water (MLW) in the study area.)

Sustained storm-surge or set-up in northern Oregon reaches only 1-2 m above predicted tides [8]. Individual wave-swash runup extends above the reach of storm surge, helping to build the cobble ridges to 6-7 m elevation. Tidal inlet excursions (~0.5 km distance) to the north and south of its present position have occurred during the last several decades. Maximum tidal range at Seaside is 3 m. River floods impact the Necanicum River floodplain south of Seaside, but have not exceeded a height of 4.5 m

NAVD88 in the Necanicum or Neawanna fluvial-tidal channels in historical time [16].

Net shoreline progradation of the southern Clatsop Plains produced at least three sets of cobble beach ridges in the Seaside area (Figure 2). Abandoned foredunes do not extend south of the Necanicum River mouth [17, 18]. Shell middens at two cultural sites, Palmrose and Avenue Q, are preserved on top of the middle set of cobble ridges located between the Neawanna and Necanicum channels (Figure 2). The radiocarbon dated middens yield occupation dates of 3,650–1,650 yr BP (Palmrose) and 3,280–1,160 yr BP (Avenue Q) (Table 1) [9, 11]. The cobble beach ridges located landward of the present Necanicum River channel were abandoned by shoreline progradation prior to 3 ka.

Dated events of abrupt regional subsidence (0.5–1.5 m), as recorded in tidal marshes, and associated sand capping layers, identified as paleotsunami deposits (Figure 3), are correlated to regional Cascadia ruptures (Tables 1 and 2) [2, 4, 16]. Shorter records of paleosubsidence, including 2-3 buried marshes, extend the paleotsunami record back to 1,100 or 1,300 yr BP in the northern and southern Neawanna wetlands. Only the last Cascadia subsidence event at AD 1700 [19] has been identified in cores from the seawardmost Necanicum tidal wetlands.

The Necanicum River flowed westward along the north flank of the Tillamook Headland prior to bog development by ~2.8 ka in core 40 b (Figures 2 and 3). Following an avulsion and/or rivermouth blockage by beach ridge accretion the Necanicum River angled back to its present course through Seaside (Figure 2). During winter floods the overflow from the Necanicum River discharges into the interridge valley containing the Neawanna channel [16]. Both the Necanicum and Neawanna channels converge at the small Necanicum Bay. Narrow uppertidal to supratidal marshes occur in the Necanicum and Neawanna interridge valleys [4]. Spruce bogs occur in the abandoned, south Necanicum valley, the Palmrose Bogs, and at the back-edge of the beach plain. Peaty muds record a net rise of sea level (~1 m/1000 yr) in the tidal marshes of the intertidal Neawanna channel (Figure 3) [2].

Cobble ridge crests increase from elevations of 3-4 m elevation in the landwardmost beach ridges to 6-7 m in the seawardmost beach ridges (Figure 4). The seaward increase in cobble ridge crest height is due to the rise of relative sea level (3-4 m) since the onset of beach plain progradation at 4 ka [20]. The lower elevations of the beach plain interior greatly increase the flooding hazard from tsunami that overtop the seawardmost cobble ridges, and/or flow into the beach plain interior through the Necanicum tidal inlet (Figure 2). This study focuses on paleotsunami currents that overtopped shore-parallel beach ridges and flowed down the landward sides of the overtopped beach ridges.

Table 2: Central Cascadia subsidence and paleotsunami events for the last 2,500 years.

Event number	Regional subsidence	Seaside subsidence core site record	Regional tsunami	Seaside tsunami core site record	Event age
Historic	None	na	Farfield	j14	1964 AD
#1	Large	NW2	Nearfield	NW2, NW7	AD 1700
#2a	None	na	Source?	NW2	~0.8 ka
#2b	Medium	NW2	Nearfield	NW2, NW7	~1.1 ka
#3	Small	—	Nearfield	NM2, j40b	~1.3 ka
#4	Large	NW2	Nearfield	NW2	~1.7 ka
#5	Medium	—	Nearfield	j40b	~2.6 ka

Coseismic subsidence from event #1 at AD 1700 is estimated to be 1.0 ± 0.5 m in Seaside [16].
Regional central Cascadia data from [8].
Regional subsidence event dates [26].
Overland paleotsunami events [7, 27].
Seaside core site references: NW2, NW7 [2]; j14, j40b [6].
Seaside core sites are shown in Figure 2, except for site j14, which is shown later in Figure 11.
Seaside core logs and radiocarbon ages are shown in Figure 3, except for j14, which is shown later in Figure 10.

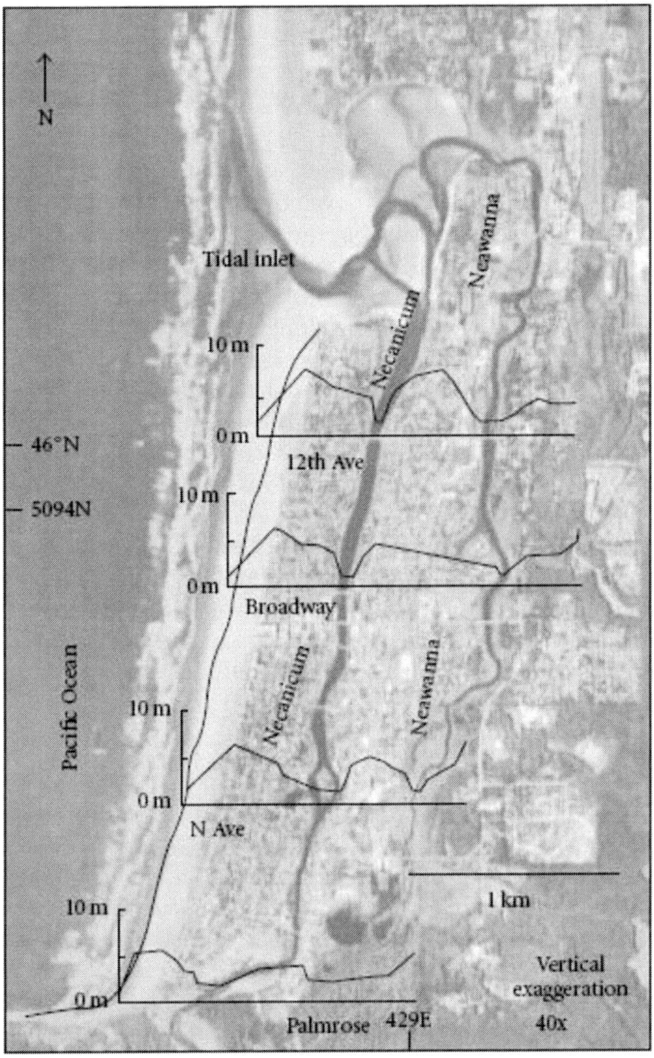

Figure 4: West-east topographic profiles across the Seaside beach plain at selected locations: 12th Avenue, Broadway Avenue, Avenue N, and the Palmrose midden site. See Figure 2 for location positions in the Seaside beach plain. Vertical elevation 1–10 m is based on the NAVD88 datum. Vertical to horizontal exaggeration of profile scale is 40. The interridge valleys of the Necanicum and Neawanna are particularly susceptible to tsunami flooding from (1) overtopping of the seawardmost beach ridges, and/or (2) sustained flow through the Necanicum Bay tidal inlet, north of 12th Avenue (Figure 2). Base image is from Google Earth.

4. PALEOTSUNAMI FLOW RELATIONS

4.1. Topographic Control

High-resolution topographic control is required for analysis of paleotsunami runup height and flow stress. High-resolution topography (0.6 m contour interval) is available for the Seaside beach plain (1 : 1,200 scale) from orthorectified photogrammetric methods [21]. The photogrammetric analysis reduces effects from artificial topography, such as buildings, in the densely developed beach ridges. Paleotsunami overtopping sites are compared to corresponding elevations along abandoned beach ridge crests. The restricted overtopping provides both maximum and minimum estimates of paleotsunami runup heights across the beach ridge plain (Figure 4). The differences between runup heights and ridge elevations yield estimated flow depths. Specifically, a ridge segment of known height that is, apparently overtopped provides a minimum runup height. An adjacent ridge segment of greater height, that is, not apparently overtopped provides a maximum limiting runup height. The topographic maps also provide slope angles on the landward sides of the abandoned beach ridges. The slope angles are needed to evaluate flow forces for different water column depths as outlined below.

4.2. Tractive Force Estimation

Paleotsunami shear stresses are estimated using the DuBoys tractive force relation for flow down the cobble ridge slopes (Table 3). Water slope is assumed to approximately follow ground surface gradients on the landward side of the beach ridges (Figure 4). Hydraulic head from tsunami surge bores could increase the water slope. Such potential bore effects are not included in this study, leading to more conservative estimates of the bottom shear stresses. Paleotsunami return flows to the ocean via the interridge valleys and the Necanicum Bay mouth are not addressed in this paper. We are primarily concerned with hazards resulting from the initial inundation(s) of the beach plain.

4.3. Critical Entrainment Relations

Bimodal transport and deposition of cobbles and/or boulders in paleotsunami sand-sheets are reported from many localities worldwide [22]. Such bimodal transport and deposition likely occur under sustained sheet flow conditions, where some gravel and abundant sand are in supply. We use published relations for the entrainment of large clasts to establish critical shear stresses in the Seaside beach ridge plain. A form of the Shields (1936) entrainment equation as modified by Miller et al. [23] and amended for bed material size by Komar [24, 25] is used to estimate the critical entrainment stresses for deposits containing a range of grain sizes (Table 3).

Table 3: Reported equations for calculating flow tractive forces from bed shear stress and critical entrainment shear stress.

Sources	Bottom shear stress
Modified DuBoys	$T_c = \rho g d S$
Shields* Miller et al. [23]	$T_c = 0.045(\rho_s - \rho)gD_m$
Komar [24, 25]	$T_c = 0.045(\rho_s - \rho)gD_B{}^{0.6}D_m{}^{0.4}$

The modified DuBoys equation for flowing water in broad shallow channels relates its bed shear stress T_c ($\mathrm{N\,m^{-2}}$) to the fluid density ρ ($1000\ \mathrm{kg\,m^{-3}}$), gravitational acceleration g ($9.8\ \mathrm{m\,s^{-2}}$), flow depth $d(m)$, and flow slope S (rise:run). The Shields entrainment relations, as modified by Miller [23], and Komar [24, 25], relate bottom shear stress to clast transport, where T_c is critical shear stress ($\mathrm{dyne\,cm^{-2}}$), ρ_s is grain density ($\mathrm{g\,cm^{-3}}$), ρ_s is fluid density ($\mathrm{g\,cm^{-3}}$), D_m is grain intermediate diameter of the largest clast transported (cm), and D_{50} is bed mean grain size (cm).

The calculations for critical entrainment stress used in Seaside are based on clast intermediate diameters (), as measured from exposed beach ridge crests, shallow trenches, or channel cutbanks.

Rounded, equant-to-ellipsoidal basalt clasts (2.85 g cm^{-3} density) are used for the clast size analysis. At least 25 clasts were measured from representative sample sites (0.25 m^{-2} surface area plots) to establish mean () and large () sizes of clasts. The maximum size clasts, observed in several sample plots from selected beach ridge entrenchment features, were also measured to the nearest centimeter.

5. RESULTS

5.1. Historical 1964 Farfield Tsunami

The inundations of the 1964 historical tsunami in the Necanicum and Neawanna channels provide comparative measures of flow inundation and resulting sand sheet deposition in the Seaside locality. Both the 1964 tsunami highwater flood lines and the limits of sand deposition in Seaside are documented in historic photographs and by eye-witness accounts (Figure 5) [5, 6]. The historical tsunami sand layers have been verified for presence, preservation potential, and thickness in the intertidal wetlands of the Necanicum and Neawanna valleys (Figure 5). The 1964 tsunami (4-5 m runup) produced substantial sand layers (1–10 cm thick) in the lower valley reaches that were characterized by high-velocity tsunami flow and abundant fine-sand supply in shallow channel bottoms.

The 1964 sand layers diminish in continuity and thickness with distance upstream in the narrow interridge valleys. Preserved tsunami sand layers reach only 50–60 percent of the inundation distance in the Necanicum valley and only 20–30 percent of the inundation distance in Neawanna valley (Figure 5) [6]. The lack of tsunami sand deposition in the upper valley reaches corresponds to decreased tsunami flow velocity and decreased sand supply in the gravelly channel bottoms. Based on the historical 1964 tsunami sand sheet distributions we expect a strong correspondence between paleotsunami flow and sand sheet deposition on the landward slopes of the overtopped beach ridges. We expect diminished sand sheet representation of paleotsunami inundation in the back-edge wetlands that (1) ramp-up against the beach plain hill slope, and (2) are distant from sand sources in the abandoned beach ridges and tidal channels.

5.2. Ridge Overtopping Areas

In this paper, the stratigraphic evidence of paleotsunami inundation by cobble ridge overtopping is presented for six areas (Areas 1–6) within the Seaside beach ridge plain (Figure 2). The representative areas are selected on the bases of (1) multiple paleotsunami layers located landward of stable ridges, (2) signs of scour into some the cobble ridges, and (3) wide ranges in ridge elevations and landward slopes. The overtopping areas are discussed below in a clock-wise order, starting with the Palmrose Bog (Area 1) at the southeast corner of the beach plain. Following the presentation of overtopping evidence, we address minimum overtopping heights and flow stresses in Areas 1, 3, 5, and 6.

5.3. Overtopping Area 1

Four representative cores from the selected freshwater bog sites near the Palmrose midden (Figure 2) contain 1 to 2 thin sand layers each (Figure 6). The sand layers range from 0.1 to 3 cm in thickness, and grade upwards into silt and organic debris. The uppermost paleotsunami sand layer is associated with an apparent subsidence contact, corresponding to the AD

1700 Cascadia rupture, as dated by the buried peat (515–688 yr BP) at site 832. The underlying peat predates the coseismic subsidence and tsunami event. A deeper paleotsunami layer at 1–1.5 m depth below the AD 1700 subsidence horizon in sites D2 and 730 could correlate to paleotsunami event #3 at about 1.3 ka (Table 2). This age is consistent with a dated paleotsunami layer (1094–1409 yr BP) at a similar depth (1.5 m) in nearby site NW2 (Figure 3).

The sand source mineralogy of the paleotsunami sand layers, ranging from 10 to 40 percent beach sand, reflects both fluvial and beach sand supply [3, 9]. The mixing of beach and fluvial sand sources in the Palmrose paleotsunami deposits could occur from surge inundation and sediment recharging in the Necanicum River channel, as recorded at site j70 (Figure 6). Paleotsunami surges transporting river sand could reach site j70 by propagating up the Necanicum channel from the Necanicum Bay mouth (Figures 2 and 4). A direct test of seawardmost beach ridge overtopping required additional investigations of wetland localities that are situated south and/or seaward of any Necanicum channel influence (see Sections 5.5 and 5.6).

5.4. Historical Bay Mouth Migrations

Intact wetlands occur directly landward of the seawardmost beach ridges in two localities of the southern Seaside beach plain, including the Cove and Avenue N, corresponding to overtopping Areas 2 and 3, respectively, (Figure 2). The earliest detailed coastal survey of the Seaside beach plain (Figure 7) shows both the beach ridge morphology at the Cove and the course of the nearby Necanicum River to be largely unchanged since AD 1874. Net beach accretion has preserved the earliest historical beach ridge morphology in the south Seaside beach plain. By comparison, substantial changes in shoreline position have occurred at the mouth of the Necanicum Bay since early historical time (1874), and probably since the last Cascadia rupture at AD 1700 [16].

5.5. Overtopping Area 2

Six cores are shown in a west-to-east transect to establish paleotsunami overtopping of the 7 m NAVD88 Cove beach ridge (Figure 8). Attempted coring in wetland sites immediately adjacent to the Cove beach ridges met with penetration refusal due to the presence of large clasts at shallow depths (1.0–1.5 m deep). Cores, taken at sites located more than 100 m distance from the Cove beach ridges, yielded deeper stratigraphic sections, including 2-3 anomalous sand layers. The quartz-rich beach-source sand from the anomalous sand layers generally ranges from 180 to 350 m in size.

The shallowest sand layer (~1 cm thick) in the most proximal site (c1) occurs below the youngest preserved subsidence horizon, thus predating the AD 1700 Cascadia rupture (Table 1). This shallowest sand layer, tentatively attributed to paleotsunami event #2, pinches out within a short distance (200 m) landward from the Cove beach ridges.

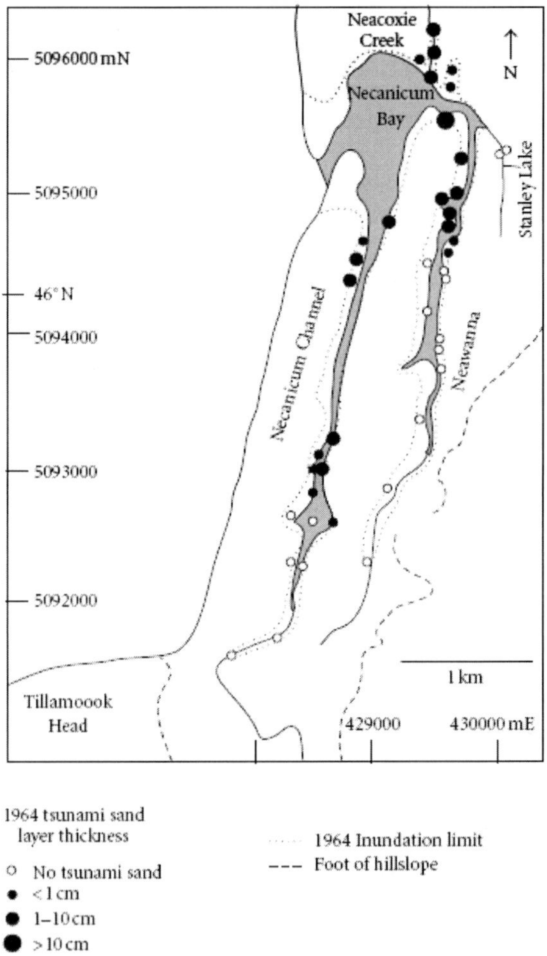

Figure 5: Map of 1964 maximum highwater line and tsunami sand layer (TSL) deposits in the Seaside beach ridge plain. Inundation was limited to the Necanicum Bay, and the Necanicum and Neawanna interridge valleys [5, 6]. Preserved sand layers from the 1964 event do not reach the full extent of the mapped tsunami flooding in either the Necanicum or Neawanna interridge valleys.

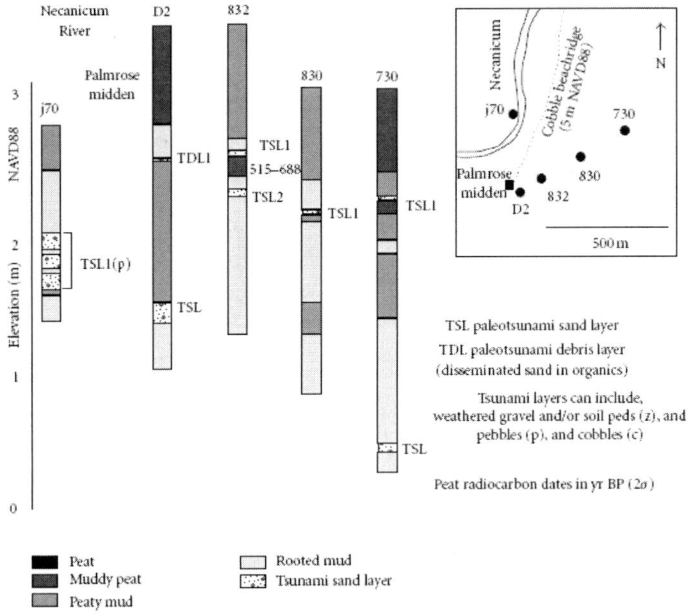

Figure 6: Core logs and inset site map from the Palmrose bogs (overtopping Area 1). Sites include j70 [6], D2 [2, 9], sites 832, 830, 730 [4]. See Figure 2 for area location and Table 1 for radiocarbon data.

More distal core sites in the Cove wetlands track the two deepest paleotsunami sand-sheets, between 1.5 and 3 m depth, to 750 m distance landward of the Cove beach ridges (Figure 8) [6]. Penetration refusal occurred at about 3 m depth in the most proximal core sites (c1, j37) due to large clasts (cobble or boulder size). Pebbles (1-2 cm diameter) were recovered in the 2.5 cm diameter gouge core from both of the lower sand-sheets at sites j38, j39, and j41. Marine shell fragments were recovered in the paleotsunami sand-sheet at 199–201 cm depth in core site j38.

The two deepest sand-sheets from the Cove wetlands (Figure 8) are dated by radiocarbon at 1060–1260 and 2780–2950 cal yr BP (Table 1). These two sand-sheets, produced by the #3 paleotsunami event (~1.3 ka) and the #5 paleotsunami event (~2.6 ka) (Table 2), represent the highest paleotsunami runups recorded in the Seaside beach plain. The furthest landward extents of these two sand-sheets in the Cove wetlands were not pursued during the Seaside Tsunami Pilot Study [6]. They likely extend beyond 1.0 km in distance from the Cove Beach ridges, thus merging with potential shore-parallel inundation up the Necanicum channel (Figure 4). Pebble transport and deposition in paleotsunami sand-sheets at sites c1, j38, j39, and j41 argue for sustained high-velocity flow over the back-ridge plain at the Cove.

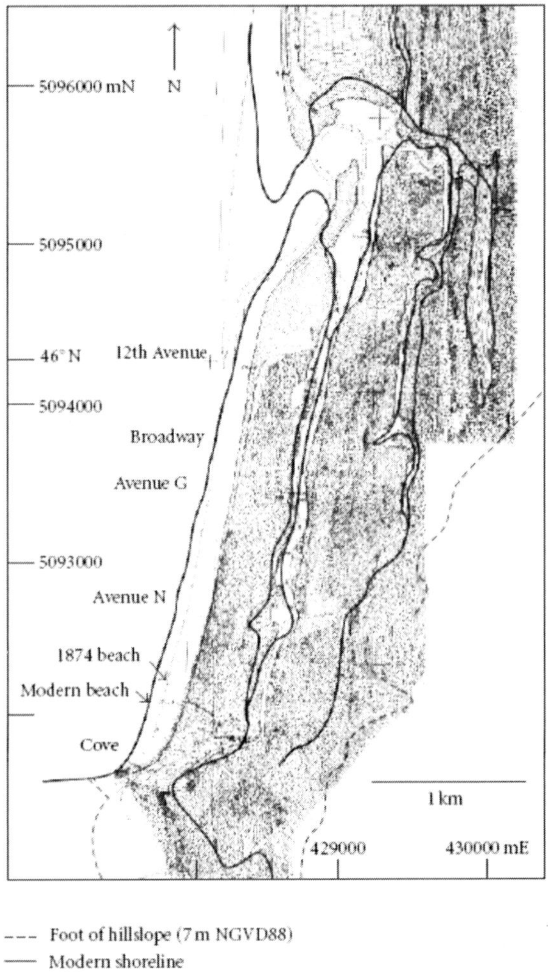

Figure 7: Early historic map of the Seaside area from Coast of Oregon Survey (1874) with overlays of modern beach and bay shorelines (bold lines). Map overlays are based on visual best-fit scaling of Necanicum River shorelines. The beach shorelines show relatively little change since 1874. However, forest cover patterns on the Necanicum Bay spit suggest that substantial shoreline changes occurred in the bay mouth prior to the 1874 survey. The seawardmost beach ridges show apparent decreases in forest cover north of Avenue N and north of 12th Avenue, suggesting recent northward extensions of the spit prior to 1874. The historic 1874 map base is from U.S. Coastal Survey [10].

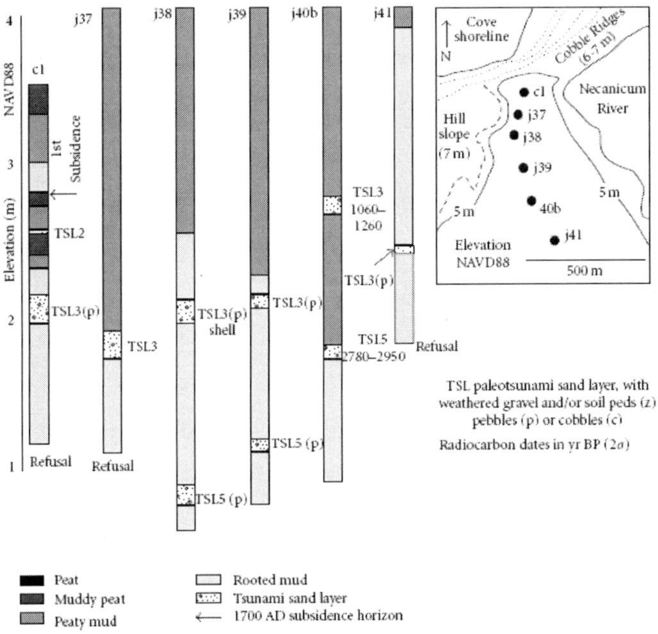

Figure 8: Core logs and inset site map from the Cove wetlands (overtopping Area 2). See Figure 2 for area location and Table 1 for radiocarbon data.

5.6. Overtopping Area 3

To further test for the event #1 paleotsunami runup height, some additional cores were taken in proximal wetlands at Avenue N (Figure 2). The Avenue N wetlands occur in a shallow entrenchment feature or gully, that is, oriented somewhat oblique to the beach ridges. The gully feature truncates two minor ridge crests at its seaward margin (Figure 9), therefore it post-dates the shoreline progradation. The paleotsunami gully or coulee feature is interpreted to have originated from localized paleotsunami erosion on the landward slope of the grouped cobble ridges, as discussed in greater detail in following sections. The intact seawardmost cobble ridges reach about 6 m elevation at the Avenue N in Overtopping Area 3.

Peaty mud sections cored in the Avenue N wetlands (Figure 9) contain a single subsidence event and an associated paleotsunami sand layer (grain size 180–350 m). The subsided peat surface is radiocarbon dated at 60–290 cal yr BP at core site j22 (Table 1). The shallow subsidence horizon and paleotsunami layer are interpreted to record the last Cascadia rupture and paleotsunami event #1 at AD 1700 (Table 2). The development of the pre-AD 1700 peaty horizon within the Avenue N paleotsunami gully post-dates the original downcutting of the cobble ridges, by previous paleotsunamis, likely including event #3 (see Section 5.5).

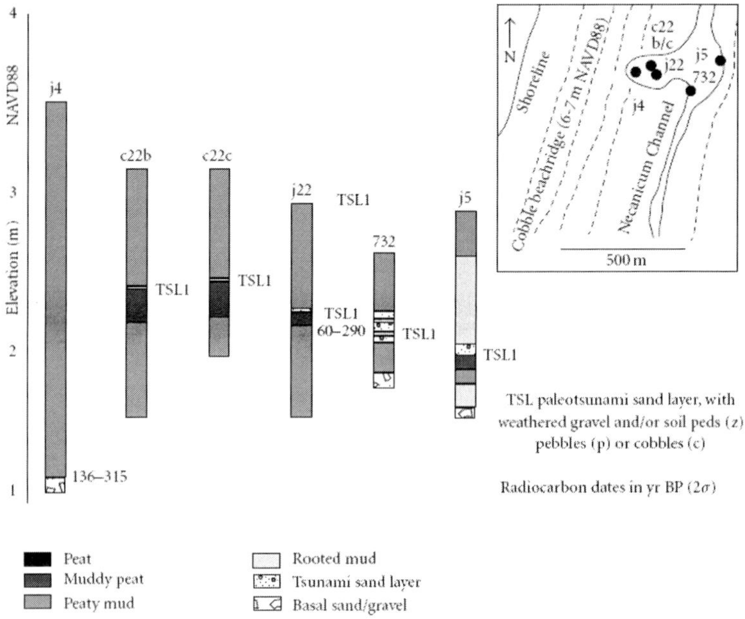

Figure 9: Core logs and inset site map from the Avenue N wetlands (overtopping Area 3). See Figure 2 for area location and see Table 1 for radiocarbon data.

The AD 1700 paleotsunami apparently propagated up the Necanicum interridge valley from the Necanicum Bay mouth as recorded by a substantial paleotsunami layer (~20 cm thick) of mixed beach and river sand mineralogy at site 732 (Figure 9) [4]. Three sand layers, separated by river/bay mud laminae in the Necanicum bank (site 732), could indicate multiple surges from the AD 1700 tsunami up the Necanicum channel from the bay mouth.

A young date for a twig in the basal mud at core site j4 (136–315 yr BP) could indicate scour by the AD 1700 paleotsunami at the head of the Avenue N wetlands (Figure 9). If the AD 1700 paleotsunami did overtop the seawardmost beach ridges at ~6 m elevation near Avenue N, then the resulting flow left only thin sand deposits at sites j22, c22b, and c22c. The interpreted predominant flow from the AD 1700 paleotsunami was directed shore-parallel up the Necanicum channel from the Necanicum Bay mouth (Figure 4).

5.7. Overtopping Area 4

Wetlands near the confluence of the Neawanna and Necanicum channels at Area 4 (Figure 2), record the historical (1964) tsunami sand-sheet (Figure 5) as well as paleotsunami from the Cascadia AD 1700 rupture (Figure 10). The geometry of the Necanicum Bay mouth at the time of the AD 1700 event is not well constrained due to apparent northward migration of the bay mouth in earliest historical time (Figure 7). Tsunami inundation from the Necanicum Bay mouth could have propagated northwards to the northeast bay wetlands through a Necanicum-Neawanna channel, as occurred during the historical 1964 tsunami in Seaside [1, 5].

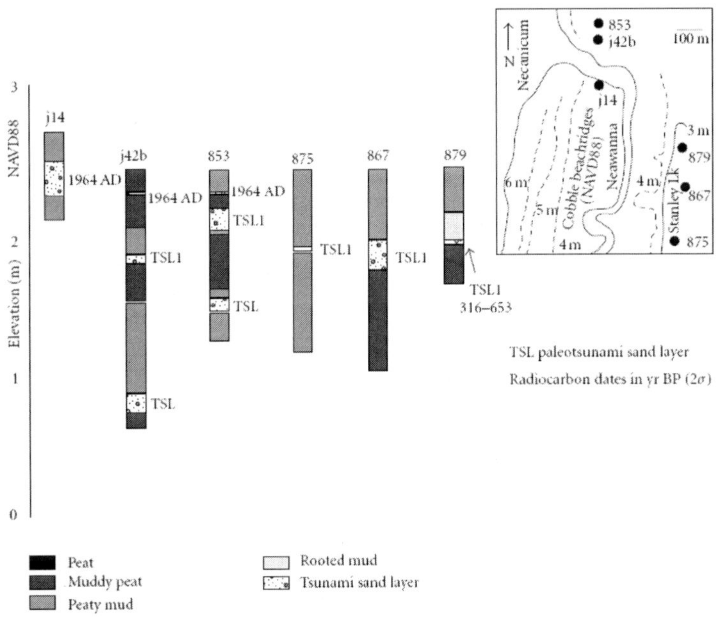

Figure 10: Core logs and inset site map from the Northeast Necanicum and Stanley Lake wetlands (overtopping Area 4). See Figure 2 for area location and see Table 1 for radiocarbon data. Cobble ridge crests (dotted lines) are shown with elevation contours.

Alternatively, the AD 1700 inundation could have crossed from southwest to northeast over the low gravel ridges (~5 m elevation) that separate the Necanicum and Neawanna channels (Figure 4). An overtopping of the intervening cobble ridges is supported by the occurrence of a prominent AD 1700 sand-sheet (23 cm thick) at site 867 in Stanley Lake on the landward side of the Neawanna channel (Figure 10). The AD 1700 sand-sheet thins abruptly to the north (site 879) and south (site 875) of a slight saddle, or breach, in the low cobble ridge (4 m elevation) that separates the Neawanna and Stanley Lake wetlands near site 867.

5.8. AD 1700 Tsunami Sand-Sheet Distribution

Localized distributions of AD 1700 sand layers at the back-edge of the Seaside beach plain have confounded earlier attempts at establishing flooding mechanisms [3–5]. Additional coring work confirmed the localized sand-sheet deposition [6], and raised the possibility of shallow channelized flow through passes in the cobble ridges. We summarize the localized sand-sheet deposition from the AD 1700 paleotsunami in the Neawanna wetlands to target potential areas of ridge overtopping.

The preserved AD 1700 subsidence sites, as recorded in the Seaside beach plain wetlands (Figure 11), are derived from the first subsidence horizons (buried peats) observed downcore [4, 6]. The absence, presence, and thickness of paleotsunami sand layers associated with the AD 1700 subsidence horizon (buried peat) are shown for each site. Other possible AD 1700 paleotsunami deposits [6] that are not confirmed by a corresponding subsidence horizon or by radiocarbon dating are not included in Figure 12.

Confirmed AD 1700 paleotsunami sand layers extend up the Necanicum channel wetlands to site j70 (Figure 11). No preserved AD 1700 subsidence records are found in the north Necanicum channel wetlands. The recent northward migration of the tidal inlet (Figure 7) appears to have initiated at a terminated seawardmost beach ridge mapped in AD 1874 at a distance of about 2.0 km south of the present tidal inlet. That position is consistent with the termination of older cobble beach ridges at Avenue G (Figure 2) and with a lack of preserved AD 1700 paleotsunami layers in the central Necanicum interridge valley.

Relatively thick AD 1700 paleotsunami sand layers (20–30 cm in thickness) occur at sites j43 and 712 in the Neawanna channel wetlands, and at site 867 on the seaward side of Stanley Lake (Figure 11). Interspersed between the Neawanna sites of thick AD 1700 paleotsunami deposits are sites of thin AD 1700 sand layers or apparent absences of the AD 1700 paleotsunami sand layer. Such variable presence and thickness of the AD 1700 sand layers suggest localized substantial overtopping at Areas 4, 5, and 6 (Figure

2). Overtopping Areas 4, 5, and 6 might have been located landward of the interpreted position of the Necanicum bay mouth at the time of the AD 1700 paleotsunami inundation(s). A broad bay mouth could have focused the AD 1700 paleotsunami inundation directly over the cobble ridges that bound the seaward side of the Neawanna wetlands.

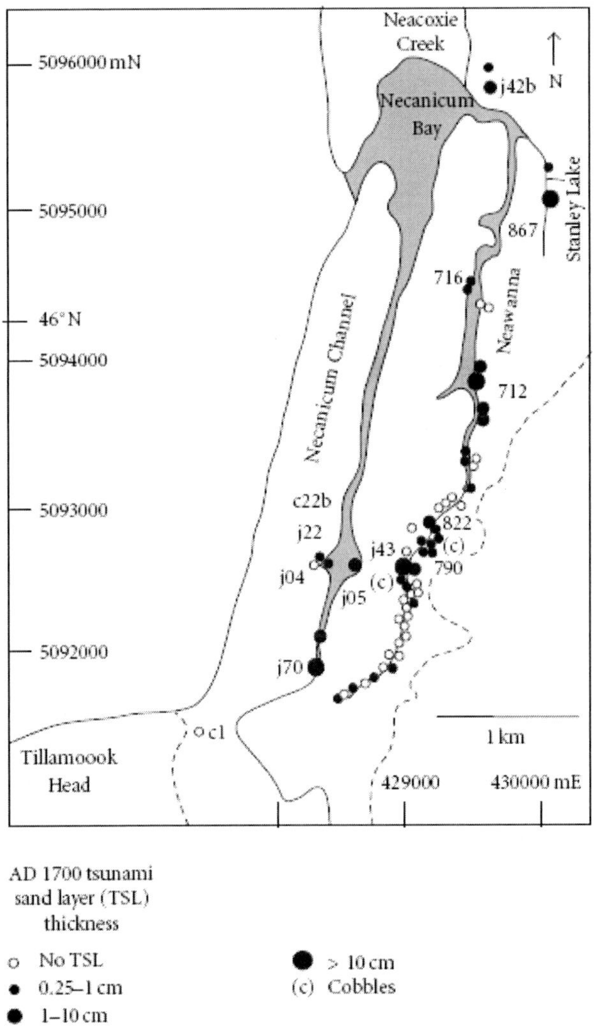

Figure 11: Map of AD 1700 paleotsunami sand layers, as confirmed by association with a coseismic subsidence contact and/or radiocarbon dating. Core site logs and dates are compiled elsewhere [4, 6].

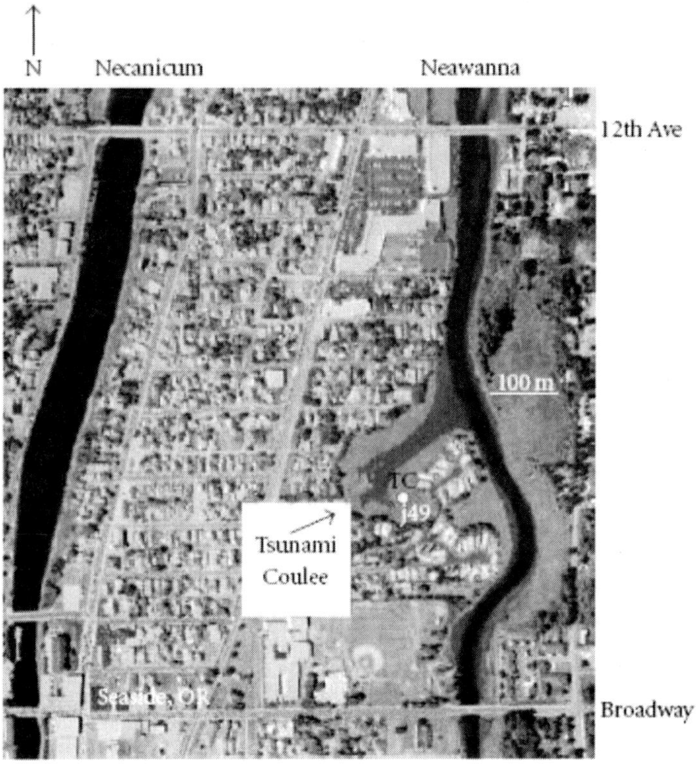

Figure 12: Google Earth image of a Seaside Neighborhood showing an anomalous entrenchment feature, termed a tsunami coulee (bold), located on the west side of the Neawanna tidal channel (see Figures 2 and 14 for location in Seaside). The tsunami coulee (~150 m in length) cuts two cobble ridges that are aligned north-to-south. The tsunami coulee is interpreted to represent headward erosion from paleotsunami flow(s) trending northeast over the cobble ridges (~5 m elevation).

We address the evidence for localized ridge overtopping in two contrasting sites from Areas 5 and 6 below.

5.9. Overtopping Area 5

An anomalous entrenchment feature (~100 m width) on the west bank of the Neawanna channel truncates a cobble beach ridge at 3rd Avenue in Area 5 (Figure 12). The deep gully feature, at least 5 m depth, is located due seaward of the 30 cm thick AD 1700 paleotsunami sand layer at site 712

(Figures 11, 12, and 13). This feature is representative of other cobble ridge gullies or headward eroded coulees, such as at Avenue N (Figure 9) in the Seaside beach ridge plain.

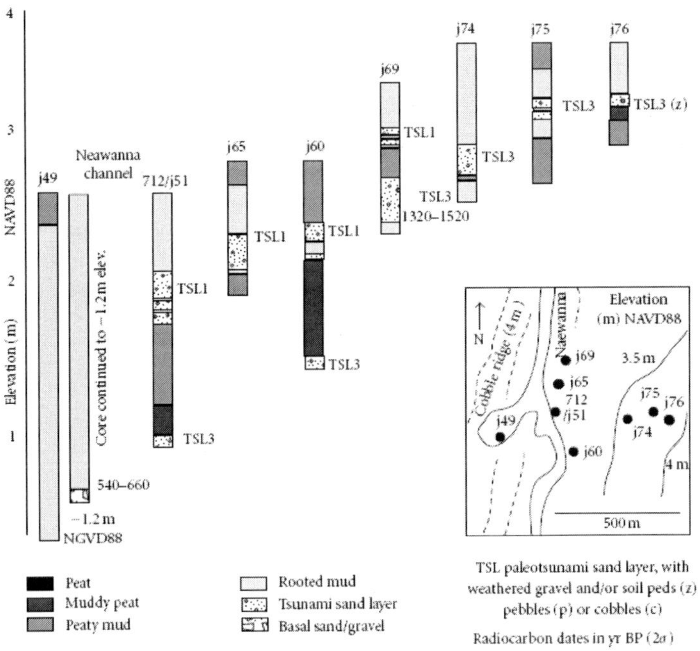

Figure 13: Core logs and inset site map from the 3rd Avenue wetlands, overtopping Area 5. See Figure 2 for area location and see Table 1 for radiocarbon data.

The 3rd Avenue paleotsunami gully has been cored to penetration refusal at 415 cm depth in site j49 (Figure 13). Mud fill in the channel feature overlies a sharp contact with sand/gravel at a depth of 410 cm depth. A radiocarbon date from a wood fragment in the basal mud (410 cm depth) in site j49 yields an anomalous "young" age (540–660 yr BP) for the 4 m deep gully (Table 1). Normal sediment accumulation by net sea level rise (~100 cm/1000 yr) should put a 4.0 m depth horizon at several thousand years in age in the Neawanna wetlands. We attribute the origin of the recent mud backfill to localized paleotsunami scour from the AD 1700 paleotsunami event.

Core transects originating at site 712 establish broad sand-sheets that extend ~500 m landward to the back-edge of the beach plain near site j76 (Figure 13). The age of the upper sand-sheet (site 712/j51) is assumed to correlate to the most recent scour of the 3rd Avenue feature at AD 1700. A lower sand-sheet, which extends to the back-edge of the beach plain in Area 5 (Figure 13) correlates to the Cascadia event #3 paleotsunami, based on a twig date of 1,320–1,520 yr BP at 70 cm depth in site j69 (Table 1) [6]. This lower sand-sheet is widely recorded along the full length of the back-edge wetlands from site j42b in the north (Figure 10) to sites NW2 and 730 in the south (Figure 3). Weathered gravel and soil concretions, or peds, in the paleotsunami deposit at site j76 indicate that the paleoflow(s) reached and scoured the base of the hillslope colluvium at greater than 5 m elevation. The hill slope soil debris that is, mixed with tsunami sand represents a tsunami backwash deposit. The headward erosion of the paleotsunami coulee at 3rd Avenue probably correlates to the extensive sand-sheet produced by the very large paleotsunami runup from rupture event #3 (Table 2).

5.10. Overtopping Area 6

Cutbank exposures of the AD 1700 paleotsunami deposits in the central reaches of the Neawanna wetlands reveal localized presence of pebbles and cobbles (Figure 2). Cobbles are rare in the Neawanna paleotsunami sand layers, but their presence demonstrates substantial shear stress for flows at the back-edge of the beach plain. Cobbles embedded within the AD 1700 sand-sheet (TSL1) at sites j44 and j45 on the west bank of the Newanna channel are associated with localized thickening of the AD 1700 paleotsunami sand-sheet there (Figure 14).

The AD 1700 paleotsunami sand-sheet thickens (20–45 cm thick) in the immediate vicinity of paleotsunami cobble emplacements at core sites j43, j44, and j45 (Figure 14). The AD 1700 paleotsunami sand-sheet pinches out to less than 0.25 cm thickness to the north and south of the embedded cobbles, respectively, at sites 799 and 728. The breached cobble ridges located landward of the Avenue K gully could be the source of the paleotsunami cobbles found in the AD 1700 paleotsunami sand layers. The narrow paleotsunami gullies directed high-velocity flow to the marsh sites j43, j44, and j45. The Avenue K overtopping site is representative of the smaller ridge breaches and associated cobble deposits that formed by paleotsunami inundation in the central and southern Neawanna wetlands.

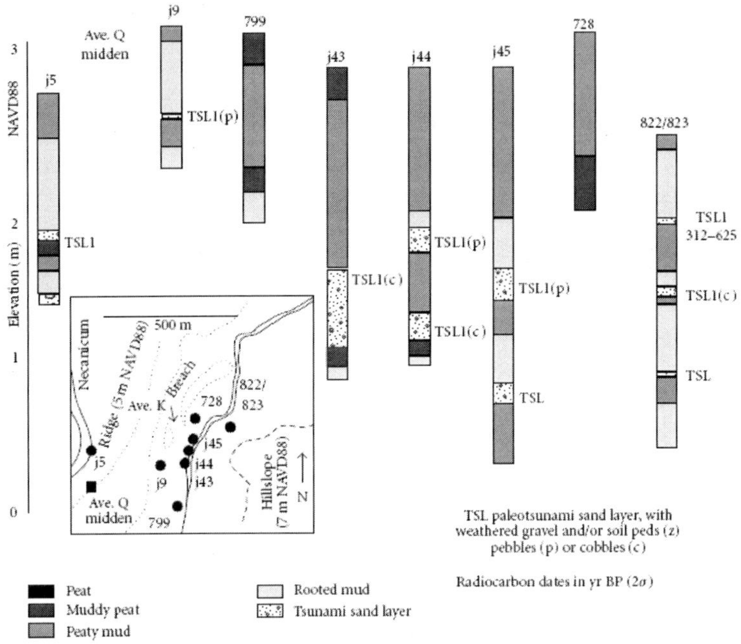

Figure 14: Core logs and inset site map from the Avenue K wetlands. See Figure 2 for area location and see Table 1 for radiocarbon data.

5.11. Clast Entrainment

In an effort to establish paleotsunami flow stresses over the cobble ridges in the Seaside beach plain the paleotsunami deposits and breached cobble ridges are analyzed for clast size. Tsunami flow over mixed sand and cobble beach ridges produce remobilized deposits having large clasts embedded in tsunami sand layers [28]. Measurements of pebbles, cobbles, and small boulders are used in later sections of this paper to constrain local conditions of critical entrainment shear stress. The large clasts are analyzed for mean and maximum sizes from paleotsunami sand-sheets and at gullied or breached beach ridges, as shown in Figure 15.

Large clast sizes deposited in paleotsunami sand-sheets range from pebbles (2 cm diameter) at sites j70, j38, and j41 to cobbles (5–12 cm diameter) at sites 822 and j44 (Table 4). One large cobble having a 20 cm diameter (Figure 15) was deposited in the AD 1700 tsunami sand-sheet, lining a broad trough at 160 cm depth at site j44. The cobble was likely eroded from the Avenue K ridge breach, and then transported to the Newanna wetland site by shallow high-velocity flow that was directed from the ridge breach (Figure 14).

Figure 15: Scaled photograph of cobbles from exposed crest of Avenue K beach ridge (a), (b) Neawanna cutbank section with AD 1700 paleotsunami sand layer and basalt cobbles (adjacent to site j44), and (c) largest tsunami cobble (20 cm intermediate axis diameter) in paleomarsh depression in Neawanna cutbank section (~160 cm depth) adjacent to site j44 (see Figure 14 for core site positions).

6. DISCUSSION

6.1. Strategy for Constraining Tsunami Flow Hazards in the Beach Plain

Locally varying presence and absence of paleotsunami deposits on the landward side of stable beach ridges in the Seaside study area are used to estimate minimum runup heights and water column depths. Representative water column depths and measured ridge slopes are then used to calculate flow tractive forces. Resulting predictions of shear stress are independently tested using depositional records of clast entrainment. The topographic controls on water depth and flow force are then used to target the sites of greatest flooding hazard for evacuees in the study area.

Table 4: Paleotsunami clast sizes from the Seaside beach plain.

Site	Event	Transport condition	Depth (m)	Bed clast size $D_B \pm 1\sigma$ cm	Largest clast size D_m (cm)	Maximum clast sizeD_{max} (cm)
867	#1	Deposited	0.5	0.02	0.02	
3rd Ave	#3,5?	Eroded	Ridge	9.1 ± 3.8	13.6	26
712	#1	Deposited	0.3	0.02	0.02	
822	#2	Deposited	0.9	0.02	5.2	6.0
AveK	#3,5?	Eroded	Ridge	8.1 ± 5.3	19	32
j44	#1	Deposited	1.2	0.02	12.4	20
j70	#1	Deposited	0.7	0.02	2.0*	
AveN	#3,5?	Eroded	Ridge	10 ± 5	20	30
832	#1	Deposited	1.1	0.02	0.02	
j38	#3	Deposited	2.0	0.02	2.0*	—
j38	#5	Deposited	3.2	0.02	2.0*	—
j41	#3	Deposited	1.5	0.02	2.0*	—

Paleotsunami events (#) are presented in Table 2.
Bed grain size includes mean (D_B) with ±1 std Dev (cm) from sample plot.
Largest clast sizes (D_m) are from the sample plots (0.25 m^2) analyzed for each site.
Maximum clast sizes (D_{max}) are from largest clast size observed in several sample plots from each site.
* Deposited pebble sizes (sites j70, j38, j41) are taken from cores (diameter 2.5 cm) thereby limiting large clast size to less than 2.5 cm diameter.
Core sites 867, 712, 822, 832 are reported elsewhere [4], as are core sites j44, j70, j38, j41 [6]. Entrenchment sites 3rd Ave, AveK, and AveN, are first identified in this paper.
Eroded cobble sizes are taken from breached cobble ridge crests at 3rd Ave, AveK, and southwest of AveN. Deposited cobble sizes are taken from cutbank exposures at sites 822 and j44. See core site locations in Figures 6–14.

6.2. Overtopping

Cobble ridge elevations and slope gradients are shown for several overtopping localities in Seaside. The seawardmost ridge crest at the Cove is 7 m elevation (Figure 16). GPR profiling [29] and topographic surveys of converging ridge crests at the northern edge of the Seaside Cove demonstrate that the Cove beach ridges have prograded (~100 m seaward distance) since closing off of the Cove wetlands before 2,780–2,950 yr BP (see core site j40b in Table 1). There is no significant elevation change across the ridge crests at the Cove, but the seawardmost ridge crests do decreases to 6 m elevation with 300–500 m distance northeast of the Cove. The substantial overtopping of the Cove beach ridges by paleotsunami events #3 and #5 (Figure 8) required flow elevations in excess of 8 m elevation at the Cove.

The anomalous entrenchment feature or gully at the Avenue N wetlands (Figure 9) is hypothesized to have resulted from paleotsunami overtopping and scour on the seaward slope of the Necanicum interridge valley (Figure 16). To test for such a gully erosion mechanism, the slope gradient (0.017) is established from the elevation change (five meters) between the seawardmost ridge crest (6 m elevation) and the Necanicum channel (1 m elevation) along a southwest-to-northeast profile (300 m distance). See the following for tractive force calculations from overtopping flows.

Beach ridge overtopping by the smaller AD 1700 paleotsunami is established at 3rd Avenue (Area 5), Avenue K (Area 6), and the Palmrose midden (Area 1) (Figures 2, 6, 13, and 15). Overtopping at these locations requires flow(s) over the middle set of cobble ridges (5 m elevation) as shown in Figure 17. The very localized overtopping by the AD 1700 paleotsunami suggests flow depths of ~1 m. The overtopping flows of the older paleotsunami events, particularly those that breached the cobble ridges at 3rd Avenue, and Avenue K, could have reached 2-3 m in water column depth. These flow depths are based on the seawardmost ridge overtopping height of at least 8 m elevation, as shown for paleotsunami events #3 and #5 at the Cove (Figures 8 and 16).

The gradient at the 3rd Avenue overtopping area (0.013) is taken from a southwest-northeast elevation change of 5 m to 1 m over a distance of 300 m (Figure 17). A lower gradient of 0.012 is measured from the Avenue K locality where the west-east slope of the beach ridge falls from 4 m to 1 m over a distance of 250 m. The gradient at the Palmrose overtopping locality is 0.002, based on an elevation drop from 4 m to 3.5 m at site 832 over a west-east distance of 250 m.

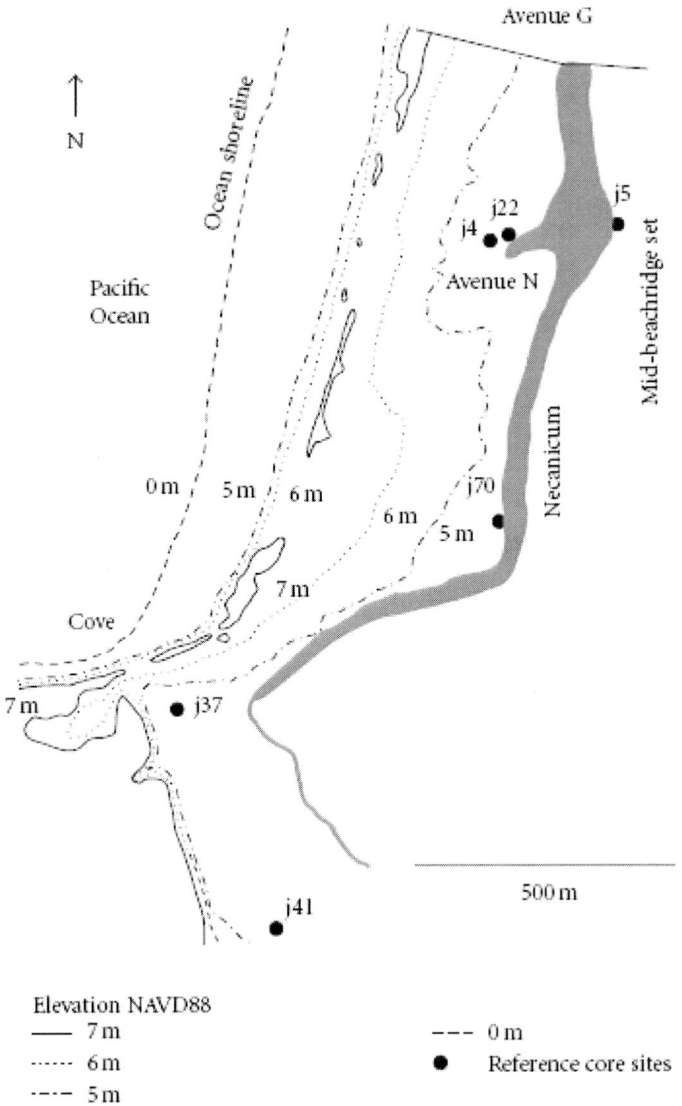

Figure 16: Seawardmost beach ridge topography in the south Seaside beach plain. Cobble ridge elevations are in meters NAVD88. The seawardmost beach ridge elevation at the Cove is 7.1 m, corresponding to the 20 ft NAVD29 contour. The seawardmost beach ridge elevation at Avenue N (6 m elevation) falls to the Necanicum channel (1 m elevation) in the Avenue N entrenchment feature, over a distance of 300 m.

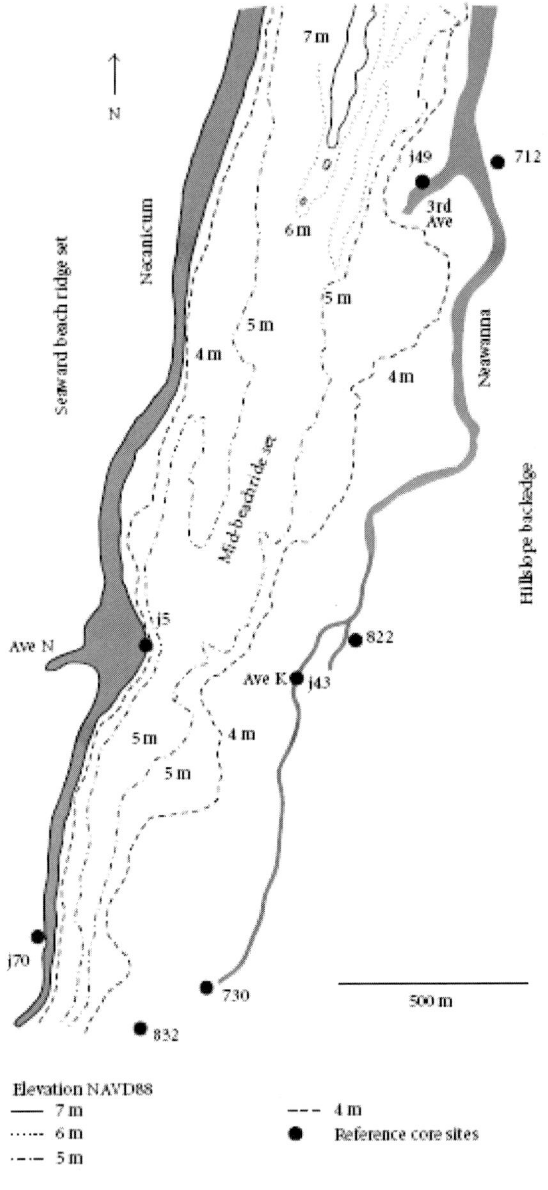

Figure 17: Topography of the middle beach ridge(s) separating the Necanicum and Neawanna interridge valleys in the Seaside beach plain. Cobble ridge elevations are in meters. The highest beach ridge divide (5 m elevation) falls to the Neawanna channel (1 m elevation) at the 3rd Avenue and K Avenue localities. A very broad ridge (4 m elevation) drops to the supratidal wetlands (3.5 m elevation) in the Palmrose bogs at site 832.

Table 5: Bed shear stress estimated for paleotsunami flow down beach ridge slopes at selected overtopping localities.

Locality	Flow depth (m)	Slope gradient	Shear stress ($N\ m^{-2}$)	Shear stress ($dyne\ cm^{-2}$)
Area 3				
Ave N	0.5	0.017	83	830
Ave N	1.0	0.017	167	1670
Ave N	1.5	0.017	250	2500
Ave N	2.0	0.017	333	3330
Area 5				
3rd Ave	0.5	0.013	64	640
3rd Ave	1.0	0.013	127	1270
3rd Ave	1.5	0.013	191	1910
3rd Ave	2.0	0.013	255	2550
Area 6				
Ave K	0.5	0.012	49	490
Ave K	1.0	0.012	98	980
Ave K	1.5	0.012	147	1470
Ave K	2.0	0.012	196	1960
Area 1				
Palmrose	0.5	0.002	10	100
Palmrose	1.0	0.002	20	200
Palmrose	1.5	0.002	29	290
Palmrose	2.0	0.002	39	390

Tractive forces or bed shear stresses (Newtons m^{-3} and dyne cm^{-3}) are calculated for four paleotsunami overtopping localities using a range of flow depths (1-2 m depth) and measured slopes (m:m gradients) in the modified DuBoys equation (Table 3). Ground surface gradients at the paleotsunami overtopping localities are shown in Figures 16 and 17. Larger flow depths (2.0 m) could represent paleotsunami events #3 and #5. Small flow depths (0.5–1.0 m) could represent paleotsunami event #1 at AD 1700.

6.3. Bed Shear Stress

Paleotsunami tractive forces or bed shear stresses are estimated for the landward slopes of the overtopped beach ridges at Avenue N (Area 3), 3rd Avenue (Area 5), Avenue K (Area 6), and Palmrose (Area 1) (Table 5) using the modified DuBoys equation (Table 3). For this analysis, the water surface gradients on the landward side of the ridges are assumed to follow the local ground surface slopes. The calculations shown here are for a set of flow depths (0.5–2.0 m depth), spanning the lower range of expected values based on flow overtopping evidence, and on measured slope angles (Figures 16 and 17). A constant fluid density (1000 kg m^{-3}) is assumed for the dilute flows, although bottom layers might be charged with sand-sheet flow. Estimated shear stress values increase with increasing flow depth (0.5–2.0 m depth) and decrease with decreasing slope (0.017–0.002 gradient) at Avenue N (830–3,330 dyne cm^{-2}), 3rd Avenue (640–2,550 dyne cm^{-2}), Avenue K (490–1,960 dyne cm^{-2}), and Palmrose (100–390 dyne cm^{-2}).

6.4. Critical Entrainment Shear Stress

Bottom shear stresses of paleotsunami flows are estimated from clasts deposited in paleotsunami sand layers or dislodged from breached ridges (Table 3). Plane bedded sand layers (5–10 cm thick) lack large clasts at sites 867, 712, and 832. These deposits are assigned bottom shear stress values of least 10 dyne cm^{-2} (Table 6), this minimum being based on deposition by sand-sheet flow. Pebble transport at sites j70, j38, and j41 also requires bottom shear stress values of at least 10 dyne cm^{-2} as calculated from the entrainment of measured pebbles using the critical entrainment equations [24, 25]. Cobbles transported over thin sand beds (~1 cm thick) at sites 822 and j44 yield estimated bottom shear stress values of 15–20 dyne cm^{-2}. The largest clast observed in a thin sand-sheet deposit (20 cm diameter cobble at site j44) indicates a critical shear stress of 26 dyne cm^{-2}. Relatively small shear stresses are required to roll rounded clasts over flat sand surfaces [30].

The erosion of cobbles in ridge entrenchments at Avenue N, 3rd Avenue, and Avenue K require very large bed shear stresses of 660–1080 dyne cm^{-2}, as calculated from mean cobble sizes in the eroded cobble ridges (Table 6). This is due to the large pivot angles that need to be overcome by the bed sheer stresses to roll cobbles over cobbles [23, 24]. Dislodgements of the largest boulder sizes (26–32 cm) observed in the cobble ridges indicate shear stresses of 1130–1260 dyne cm^{-2}. For this analysis we do not address the flow durations needed to unravel the flanks of the cobble ridges in the production of the tsunami breaches or coulees. However, rapid flooding of the interridge valleys probably limits the greatest flow gradients to the early phase(s) of inundation. Downslope flow velocity should approach low values when channels become filled with surge water. The timing of channel filling could depend on tsunami wave form(s), multiple wave trains, and backwash discharge between tsunami wave trains.

6.5. Paleotsunami Flooding Mechanisms

Paleotsunami inundation in the Seaside beach plain occurred by wide-scale overtopping of the seawardmost beach ridges (event #3 at ~1.3 ka) and by localized overtopping of the middle beach ridges (event #1 at AD 1700) (Figures 18and 19). Paleotsunami events #3 and #5 had shoreline runups of greater than ~10 m elevation NAVD88 at the Cove when adjusted to corresponding lower sea levels of −1.3 to −2.6 m at 1.3 to 2.6 ka (assuming a net rate of relative sea-level rise of 1 m/1000 yr). The AD 1700 paleotsunami did not substantially overtop the Cove beach ridge at 7 m elevation. Instead, it entered the interior beach plain through the paleo-Necanicum Bay mouth. The AD 1700 paleotsunami surges did overtop the middle beach ridges at 5 m NAVD88. In all cases the Necanicum and Neawanna interridge valleys (1–3 m NAVD88) were flooded to dangerous depths (3–5 m flow depth).

Paleotsunami tractive forces are established by bed shear stress (Table 5) and critical entrainment (Table 6) for several overtopping localities in Seaside. Flow depths of ~1.0 m are sufficient to dislodge the largest clasts from the high gradient Avenue N and 3rd Avenue beach ridge localities where tsunami coulees formed (Figure 13). Greater depths are likely required to entrench the lower gradient Avenue K beach ridge, which shows only minor breaching. The very low-gradient slope at Palmrose is consistent with a lack of any entrenchment features in that broad flat beach ridge.

6.6. Implications for Tsunami Hazards and Tsunami Evacuation

The high shear stresses demonstrated by large clast unraveling (660–1,260 dyne cm^{-2}) at tsunami coulee sites, Avenue N, 3rd Avenue and Avenue K (Figures 9, 13, and 15), indicate particularly dangerous flooding hazards in the steeper beach ridge slopes. Even the modest AD 1700 paleotsunami (6-7 m runup height) produced sand-sheet flow (at least 10 dyne cm^{-2}) at four back-edge localities of the 1.5 km wide beach plain. Large floating structural debris in currents reaching 1,000 dyne cm^{-2} bed shear stresses would likely destroy unreinforced structures and prove fatal to evacuees trapped at congested bridge crossings. In the Seaside beach ridge plain, the interridge valleys are associated with the greatest water depths and the highest flow shear stresses produced during initial tsunami inundation(s).

Table 6: Critical shear stress calculated from transported clast sizes.

Site	Event	Setting	Mean clast shear stress dyne cm^{-2}	Large clast shear stress dyne cm^{-2}	Maximum clast shear stress dyne cm^{-2}
867	#1	Deposit	10 (s)	—	—
3rd A	#3,5?	Eroded	740 (c)	870 (c)	1130 (b)
712	#1	Deposit	10 (s)	—	—
822	#2	Deposit	10 (s)	15 (c)	16 (c)
AveK	#3,5?	Eroded	660 (c)	930 (c)	1140 (b)
j44	#1	Deposit	10 (s)	21 (c)	26 (c)
j70	#1	Deposit	10 (s)	10 (p)	—
AveN	#3,5?	Eroded	820 (c)	1080 (c)	1260 (b)
832	#1	Deposit	10 (s)	—	—
j38	#3	Deposit	—	10 (p)	—
j38	#5	Deposit	—	10 (p)	—
j41	#3	Deposit	—	10 (p)	—
j41	#3	Deposit	—	10 (p)	—

Estimated ages of paleotsunami events (#) are shown in Table 2. Critical entrainment is evaluated for clasts in paleotsunami deposits (Deposit) and for representative clasts dislodged from cobble beach ridges (Eroded) as shown in Table 4. Critical entrainment shear stresses (dyne cm^{-2}) are estimated for sand-sheet-flow (s) and pebbles (p) in sand-sheet-flow (~10 dyne cm^{-2}), cobbles (c) in sand-sheet-flow (~15–26 dyne cm^{-2}), and cobbles and boulders dislodged from cobble beach ridges (660–1260 dyne cm^{-2}). Flow competence relations [24, 25] used for this analysis are shown in Table 3. Clast sizes are from Table 4.

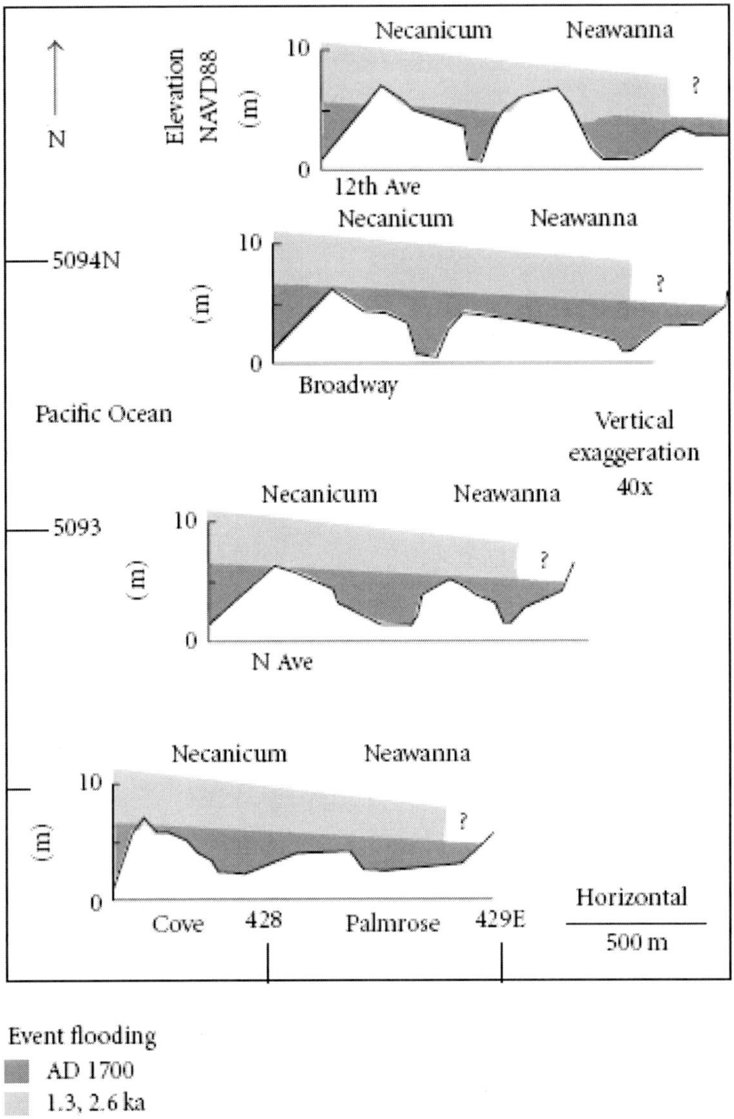

Figure 18: Paleotsunami flooding of selected topographic profiles in the Seaside beach ridge plain using the modest runup height of event #1 at AD 1700, and the larger runups of events #3 and #5 at 1.3 and 2.6 ka, respectively.

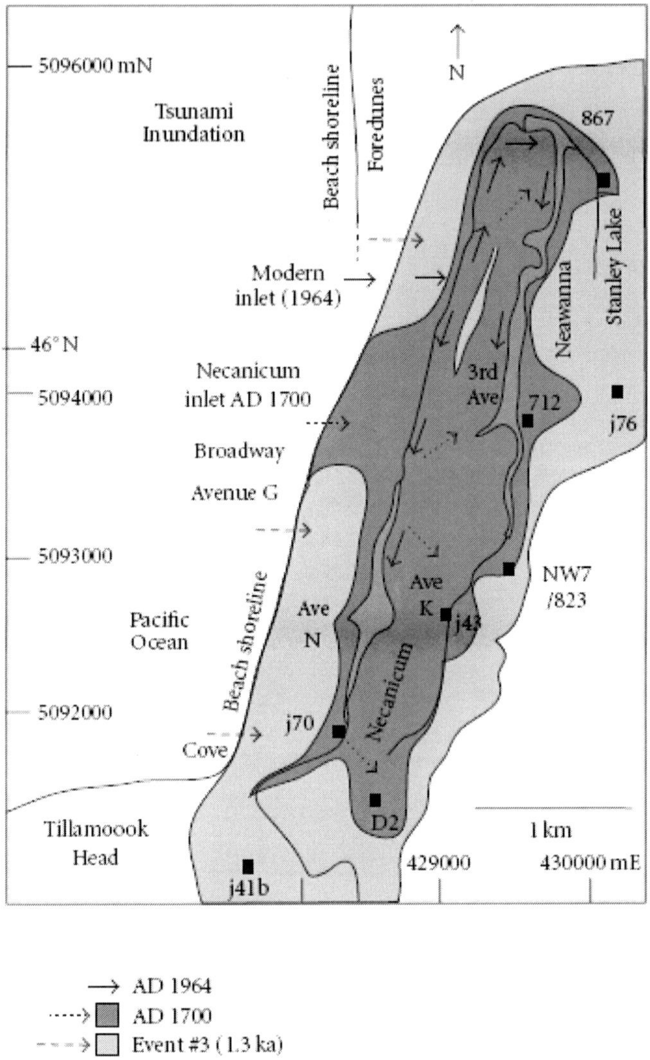

Figure 19: Map of tsunami inundation in Seaside, Oregon, for three events including historical farfield tsunami in 1964 (small arrows), the last Cascadia nearfield tsunami at AD 1700 (dark gray), and a larger Cascadia tsunami from rupture event #3 at 1.3 ka (light gray). The historical 1964 inundation was localized to modern tidal inlet (small arrows) and intertidal channels. The AD 1700 inundation entered a paleo-Necanicum tidal inlet to flood the interior beach ridge plain (medium arrows). The larger Cascadia paleotsunami at 1.3 ka flooded the entire beach ridge plain, including the highest cobble beach ridges on either side of the tidal inlet (large arrows).

7. CONCLUSIONS

Different paleotsunami inundations of the Seaside beach ridge plain occurred by (1) wide-scale overtopping of the seawardmost beach ridges, (2) flooding through the paleo-Necanicum bay mouth, and (3) localized overtopping of interior beach ridges. At least two paleotsunami reached runup heights of at least 10 m, as adjusted for paleo-sea level. High-velocity inundation distances exceed the width of the beach plain at 1.5 km. High bed shear stress is verified by transport of large clasts in the landward dipping slopes of several beach-ridge overtopping localities. Downslope currents combined with greater flooding depths in the interridge valleys could pose challenging conditions for evacuation over narrow bridges that span the beach plain channels. Vertical evacuation options in seismically- and tsunami-resistant structures that are built on the higher beach ridge crests might offer safer alternatives than attempting to cross interridge valley channels over crowded bridges.

ACKNOWLEDGMENTS

Bob Peters performed core logging and radiocarbon sample submittal for paleotsunami core localities revisited in 2004 under the Tsunami Pilot Study Working Group study. Neal Maine coordinated the permissions for coring in Nature Conservancy wetland areas along the Necanicum and Neawanna channels in Seaside. Paul Komar provided valuable reviews of critical entrainment analyses performed in this study. Kennett Peterson provided paper-editing assistance. Radiocarbon dating support was provided by NOAA, through grants to Oregon Graduate Institute and through the Tsunami Pilot Study Working Group, 2006.

REFERENCES

1. J. F. Lander, P. A. Lockridge, and M. J. Kozuch, Tsunamis affecting the west coast of the United States 1806–1992, NGDC Key to Geophysical Records Documentation No. 29, National Geophysical Data Center, Boulder, Colo, USA, 1993.
2. M. E. Darienzo, Late Holocene paleoseismicity along the northern Oregon coast, Ph.D. thesis, Portland State University, Portland, Ore, USA, 1991.
3. M. E. Darienzo, S. Craig, C. D. Peterson et al., "Extent of tsunami sand deposits landwarwd of the Seaside spit," Clatsop County, Oregon. Final Report to Clatsop County Sheriffs, 1993.
4. B. K. Fiedorowicz, Geologic evidence of historic and preshistoric tsunami inundation at Seaside, Oregon, M.S. thesis, Portland State University, Portland, Ore, USA, 1997.
5. B. K. Fiedorowicz and C. D. Peterson, "Tsunami deposit mapping at Seaside, Oregon, USA," Geoenvironmental Mapping, vol. 1, pp. 630–648, 2002.

6. Tsunami Pilot Study Working Group, "Seaside, Oregon Tsunami Pilot Study-Modernization of FEMA flood hazard maps," NOAA OAR Special Report, NOAA/OAR/PMEL, Seattle, Wash, USA, 2006.
7. R. B. Schlichting and C. D. Peterson, "Mapped overland distance of paleotsunami high-velocity inundation in back-barrier wetlands of the central Cascadia margin, U.S.A," Journal of Geology, vol. 114, no. 5, pp. 577–592, 2006.
8. C. D. Peterson, K. M. Cruikshank, H. M. Jol, and R. B. Schlichting, "Minimum runup heights of paleotsunami from evidence of sand ridge overtopping at Cannon Beach, Oregon, Central Cascadia Margin, U.S.A," Journal of Sedimentary Research, vol. 78, no. 5-6, pp. 390–409, 2008.
9. M. E. Darienzo, "The relationship between Holocene changes in sea level, geomorphology, and prehistoric Indian site abandonment and shell midden changes in the Seaside, Oregon area," Final Report to Research Heritage Associates, Eugene, Ore, USA, 1992.
10. U.S. Coastal Survey, "U.S. Coastal Survey Section X1 Coast of Oregon between Latitudes 45°58'and 47°03' including Nekanakum Creek & Sea Side House," Sheet No. 2 of 1874, scale 1:10,000, Register No 1381, 1875.
11. G. E. Phebus and R. M. Drucker, Archaeological investigations at Seaside, Oregon: An intermediate report on the excavations of two major archaeological sites at Seaside, Oregon, through September 1977, Seaside Museum and Historical Society, Seaside, Ore, USA, 1979.
12. M. Stuiver and P. J. Reimer, "Extended 14C data base and revised CALIB 3.0 14C age calibration program," Radiocarbon, vol. 35, no. 1, pp. 215–230, 1993.
13. M. Umitsu, C. Tanavud, and B. Patanakanog, "Effects of landforms on tsunami flow in the plains of Banda Aceh, Indonesia, and Nam Khem, Thailand," Marine Geology, vol. 242, no. 1–3, pp. 141–153, 2007.
14. F. I. González, E. L. Geist, B. Jaffe et al., "Probabilistic tsunami hazard assessment at Seaside, Oregon, for near-and far-field seismic sources," Journal of Geophysical Research C, vol. 114, no. 11, Article ID C11023, 2009.
15. J. C. Borrero, C. E. Synolakis, and F. Hermann, "Northern Sumatra field survey after the December 2004 great Sumatra earthquake and Indian Ocean tsunami," Spectra, vol. 22, pp. S93–S104, 2006.
16. E. T. Barnett, Potential for coastal flooding due to coseismic subsidence in the central Cascadia margin, M.S. thesis, Portland State University, Portland, Ore, USA, 1997.
17. D. K. Rankin, Holocene geologic history of the Clatsop plains foredune ridge complex, M.S. thesis, Portland State University, Portland, Ore, USA, 1983.
18. F. Reckendorf, C. Peterson, and D. Percy, "Dune Ridges of Clatsop County," Department of Oregon Geology and Mineral Industries, Open-File Report O-01-07, Text and Interactive Map on CDROM, 2001.
19. B. F. Atwater, S. Musumi-Rokkaku, K. Satake, Y. Tsuji, K. Ueda, and D. Yamaguchi, "The orphan tsunami of 1700: Japanese clues to a parent earthquake in North America," US Geological Survey Professional Paper, paper 1707, 2005.
20. L. K. Woxell, Prehistoric beach accretion rates and long-term response to sediment depletion in the Columbia River littoral cell, M.S. thesis, Portland State University, Portland, Ore, USA, 1998.
21. K. R. Wengler, "Photo Contour Map, Clatsop County, Oregon, North Coast Area, NW1/4 section. Scale 1:1,200; contour interval 2 feet," Prepared for Federal Emergency Management Administration, Flood Map Series, CH2M Hill, Corvallis, Ore, USA, 1973.
A. Scheffers and D. Kelletat, "Bimodal tsunami deposits–a neglected feature in paleo-tsunami research," in Geographie der Meere und Kusten, G. Schernewski and T. Dolch, Eds., pp. 67–75, Coastline Reports 1, 2004.
22. M. C. Miller, I. N. McCave, and P. D. Komar, "Threshold of sediment motion under unidirectional currents," Sedimentology, vol. 24, pp. 507–527, 1977.
23. P. D. Komar, "Selective gravel entrainment and the empirical evaluation of flow competence," Sedimentology, vol. 34, no. 6, pp. 1165–1176, 1987.
24. P. D. Komar, "Flow-competence evaluations of the hydraulic parameters of floods: an assessment of the technique," in Floods: Hydrological, Sedimentological and

Geomorphological Implications, E. Beven and P. Carling, Eds., pp. 107–134, John Wiley & Sons, New York, NY, USA, 1989.

25. B. F. Atwater, M. P. Tuttle, E. S. Schweig, C. M. Rubin, D. K. Yamaguchi, and E. Hemphill-Haley, "Earthquake recurrence, inferred from paleoseismology," in The Quaternary Period in the United States, A. R. Gillespie, S. C. Porter, and B. F. Atwater, Eds., pp. 331–350, Elvevier, Amsterdam, The Netherlands, 2004.
26. R. B. Schlichting, Establishing the inundation distance and overtopping height of paleotsunami from late-Holocene geologic records at open-coastal wetland sites, central Cascadia margin, M.S. thesis, Portland State University, Portland, Ore, USA, 2000.

A. G. Dawson and S. Shi, "Tsunami deposits," Pure and Applied Geophysics, vol. 157, no. 6–8, pp. 875–897, 2000.

27. H. M. Jol, D. G. Smith, and R. A. Meyers, "Digital ground penetrating radar (GPR): a new geophysical tool for coastal barrier research (examples from the Atlantic, Gulf and Pacific coasts, U.S.A.)," Journal of Coastal Research, vol. 12, no. 4, pp. 960–968, 1996.

A. Shields, "Application of similarity principles and turbulence research to bedload movement," 1936, Tranlated from Anwendung der Aehnlichkeitsmechanik und der Turbulenzforschung auf die Geschiebebewegung, Mitt. Preuss. Versuchsanst. Wasserbau Schiffbau, by W.D. Ott & J.C. von Vcheten. Publications of California Institute of Technology. Hydrodynamics Lab. No. 167, pp. 37. 2006.

Index